Nutrition

Pre- and Postnatal Development

Human Nutrition
A COMPREHENSIVE TREATISE

General Editors:
Roslyn B. Alfin-Slater, *University of California, Los Angeles*
David Kritchevsky, *The Wistar Institute, Philadelphia*

Volume 1	*Nutrition: Pre- and Postnatal Development* *Edited by Myron Winick*
Volume 2	*Nutrition and Growth* *Edited by Derrick B. Jelliffe and E. F. Patrice Jelliffe*
Volume 3A	*Nutrition and the Adult: Macronutrients* *Edited by Roslyn B. Alfin-Slater and David Kritchevsky*
Volume 3B	*Nutrition and the Adult: Micronutrients* *Edited by Roslyn B. Alfin-Slater and David Kritchevsky*
Volume 4	*Nutrition: Metabolic and Clinical Applications* *Edited by Robert E. Hodges*

Nutrition

Pre- and Postnatal Development

Edited by

Myron Winick

Columbia University
New York, New York

PLENUM PRESS · NEW YORK AND LONDON

Library of Congress Cataloging in Publication Data

Main entry under title:

Nutrition, pre- and postnatal development.

(Human nutrition; v. 1)
Includes bibliographical references and index.
1. Children–Nutrition. 2. Infants–Nutrition. 3. Malnutrition in children–Complications and sequelae. 4. Pregnancy–Nutritional aspects. I. Winick, Myron. II. Series. [DNLM: 1. Child nutrition. 2. Infant nutrition. 3. Maternal-fetal exchange. 3. Fetus–Growth and development. 4. Growth. QU145.3 H9183 v. 1]
QP141.H78 vol. 1 [RJ206] 612′.3′08s
[618.9′23′9] 78-26941

ISBN 978-1-4615-7212-1 ISBN 978-1-4615-7210-7 (eBook)
DOI 10.1007/978-1-4615-7210-7

Softcover reprint of the hardcover 1st edition 1979
A Division of Plenum Publishing Corporation
227 West 17th Street, New York, N.Y. 10011

This volume is dedicated to all the scientists whose work has advanced our understanding of how to apply knowledge to benefit human nutrition.

Contributors

Lewis A. Barness • Department of Pediatrics, University of South Florida, Tampa, Florida

Edith Liwszyc Cohen • Laboratory of Neuroendocrine Regulation, Department of Nutrition and Food Science, Massachusetts Institute of Technology, Cambridge, Massachusetts

Carolyn Cramoy • Department of Pediatrics and Institute of Human Nutrition, Columbia University College of Physicians and Surgeons, New York, New York

Patricia L. Engle • Division of Human Development, Institute of Nutrition of Central America and Panama (INCAP), Guatemala, Central America

Charles J. Glueck • General Clinical Research Center and Lipid Research Center, University of Cincinnati College of Medicine, Cincinnati, Ohio

Larry Goldberger • Department of Psychology, Cornell University, Ithaca, New York

Daryl B. Greenfield • Boston College, Boston, Massachusetts

Peter Hahn • Centre for Developmental Medicine, Departments of Pediatrics and Obstetrics and Gynaecology, University of British Columbia, Vancouver, B.C., Canada

Marc Irwin • Division of Human Development, Institute of Nutrition of Central America and Panama (INCAP), Guatemala, Central America

Derrick B. Jelliffe • Division of Population, Family and International Health, School of Public Health, University of California at Los Angeles, Los Angeles, California

E. F. Patrice Jelliffe • Division of Population, Family and International Health, School of Public Health, University of California at Los Angeles, Los Angeles, California

Michael Katz • Departments of Pediatrics and Public Health, Columbia University College of Physicians and Surgeons, New York, New York

Gerald T. Keusch • Department of Medicine, Tufts University, School of Medicine, Boston, Massachusetts

Robert E. Klein • Division of Human Development, Institute of Nutrition of Central America and Panama (INCAP), Guatemala, Central America

Sally A. Lederman • Institute of Human Nutrition, Columbia University College of Physicians and Surgeons, New York, New York

Rudolph L. Leibel • Rockefeller University, New York, New York.

David A. Levitsky • Division of Nutritional Sciences and Department of Psychology, Cornell University, Ithaca, New York

Thomas F. Massaro • Division of Nutritional Sciences, Cornell University, Ithaca, New York

Juan M. Navia • Institute of Dental Research, School of Dentistry, University of Alabama in Birmingham, Birmingham, Alabama

Donough O'Brien • Department of Pediatrics, University of Colorado Medical Center, Denver, Colorado

Ernesto Pollitt • School of Public Health, University of Texas, Houston, Texas

Pedro Rosso • Department of Pediatrics and Institute of Human Nutrition, Columbia University College of Physicians and Surgeons, New York, New York

S. Jaime Rozovski • Division of Diabetes and Metabolism, Department of Medicine, Boston University School of Medicine, Boston, Massachusetts

Giorgio Solimano • Institute of Human Nutrition and Center for Population and Family Health, Columbia University College of Physicians and Surgeons, New York, New York

John W. Townsend • Division of Human Development, Institute of Nutrition of Central America and Panama (INCAP), Guatemala, Central America

Reginald C. Tsang • Fels Division of Pediatric Research, Children's Hospital Research Foundation, Cincinnati, Ohio

Myron Winick • Institute of Human Nutrition, Columbia University College of Physicians and Surgeons, New York, New York

Richard Jay Wurtman • Laboratory of Neuroendocrine Regulation, Department of Nutrition and Food Science, Massachusetts Institute of Technology, Cambridge, Massachusetts

Charles Yarbrough • Computers for Marketing Corporation, Kenwood, California

Foreword

The science of nutrition has advanced beyond expectation since Antoine Lavoisier as early as the 18th century showed that oxygen was necessary to change nutrients in foods to compounds which would become a part of the human body. He was also the first to measure metabolism and to show that oxidation within the body produces heat and energy. In the two hundred years that have elapsed, the essentiality of nitrogen-containing nutrients and of proteins for growth and maintenance of tissue has been established; the necessity for carbohydrates and certain types of fat for health has been documented; vitamins necessary to prevent deficiency diseases have been identified and isolated; and the requirement of many mineral elements for health has been demonstrated.

Further investigations have defined the role of these nutrients in metabolic processes and quantitated their requirements at various stages of development. Additional studies have involved their use in the possible prevention of, and therapy for, disease conditions.

This series of books was designed for the researcher or advanced student of nutritional science. The first volume is concerned with prenatal and postnatal nutrient requirements; the second volume with nutrient requirements for growth and development; the third with nutritional requirements of the adult; and the fourth with the role of nutrition in disease states. Our objectives were to review and evaluate that which is known and to point out those areas in which uncertainties and/or a lack of knowledge still exists with the hope of encouraging further research into the intricacies of human nutrition.

Roslyn B. Alfin-Slater
David Kritchevsky

Preface

This is the first volume of a series about nutrition as it relates to all aspects of health and disease. It is appropriate, therefore, that this volume be devoted to nutrition and early development, for we have learned during the past three decades that during this critical early period of life nutrition plays a very important role.

During early life, while cells are actively dividing and enlarging, when organs are being formed and functions being developed, nutritional insults not only may alter the developmental pattern but also may permanently affect the organism. In the first chapter, the effects of early nutrition on cellular growth of liver, muscle, and brain are demonstrated. The mechanisms by which early malnutrition retards the rate of cell division and impairs nucleic acid and protein synthesis are discussed and a theory of metabolic adaptation to undernutrition at the cellular level is put forth. The next chapter deals specifically with the brain and discusses the effect of altered nutritional status on the synthesis and secretion of certain neural hormones. This is followed by a chapter on the overall physiologic adaptation undergone by the young mammal when exposed to a period of early nutritional stress.

The period of pregnancy is then extensively discussed from the standpoint of how the mother is affected by poor nutrition and how this effect is transferred to the fetus. Because of the rapid rate at which information is being collected in this field, this chapter represents an extensive review of the literature to date accompanied by interpretations by the author, himself a leader in the field.

The next two chapters address the breast-versus-bottle argument. Both authors agree that breast feeding is best. Dr. Jelliffe points out why in great detail. Dr. Barness discusses the theoretical and practical aspects of bottle feeding and the types of infant formulas that can be employed.

The next three chapters discuss what has become one of the most important issues in the area of developmental nutrition: early nutrition and subsequent behavior. The first deals with the animal data demonstrating that nutrition and environment both play a role in determining a number of behaviors.

The second describes a specific study in Guatemala showing how improved nutrition early in life can improve mental development. The third is a comprehensive interpretative review by the editor, who has long been interested in this field, of the evidence in humans implicating early nutrition and other early environmental factors in the genesis of retarded mental and behavioral development.

The next six chapters deal with nutrition in specific diseases. Dental development and disease are covered from both an experimental and a clinical standpoint. Nutrition and infection and the immunologic changes induced by early malnutrition are presented in depth. This is again an area in which research is moving rapidly, and the authors are extremely qualified to discuss the problem. The problem of dietary lipids and atherosclerosis is becoming more and more important in childhood and is covered in the next chapter. Finally, the last three chapters deal with iron deficiency, a serious problem in the United States; certain inborn errors of metabolism, a problem gaining more prominence as new forms of diagnosis and therapy become available; and infant diarrhea, one of the most serious problems in developing countries.

The book is not an attempt to cover all aspects of developmental nutrition but is rather a discussion of what the editor views as the most important aspects at present. The wealth of information presented is due to the excellence of the individual chapter authors, whereas any omissions can be blamed on the editor.

Myron Winick, M.D.

New York

Contents

Chapter 5

Nutrition and Pregnancy

Pedro Rosso and Carolyn Cramoy

Chapter 6
Early Infant Nutrition: Breast Feeding
Derrick B. Jelliffe and E. F. Patrice Jelliffe

Chapter 7
Early Infant Nutrition: Bottle Feeding
Lewis A. Barness

Chapter 8

Malnutrition, Learning, and Animal Models of Cognition

David A. Levitsky, Larry Goldberger, and Thomas F. Massaro

Chapter 9

Nutrition and Mental Development in Children

Patricia L. Engle, Marc Irwin, Robert E. Klein, Charles Yarbrough, and John W. Townsend

Chapter 10

Malnutrition and Infection

Gerald T. Keusch and Michael Katz

Chapter 11

Nutrition in Dental Development and Disease

Juan M. Navia

Chapter 12

Pediatric Nutrition: Potential Relationship to the Development of Atherosclerosis

Charles J. Glueck and Reginald C. Tsang

1

Nutrition and Metabolic Development in Mammals

Peter Hahn

1. Introduction

1.1. Development

As discussed here, development means the growth and differentiation of an individual. It is important to stress that all the information necessary for the finished product (the adult) is contained in a single fertilized cell. Why this cell starts dividing and differentiating is still a mystery. Basically, development consists of repression and derepression of certain genes. On the whole, development is a "one-way street." Under ordinary circumstances, there is no return to younger stages, although this is not an absolute impossibility. Thus, it is possible to take the nucleus from an epithelial gut cell of a tadpole, implant it into a fertilized frog cell from which the nucleus has been removed, and obtain from this new cell a whole new tadpole (Gurdon, 1962). On the other hand, it is not possible to alter fully differentiated organs (e.g., the completed brain of a tadpole) once they are made.

1.2. Environment

No development is possible without an appropriate environment. For mammals (and most other vertebrates) such environmental requirements include:

1. A certain temperature.
2. A certain oxygen tension.
3. A certain quality of food.

Peter Hahn • Centre for Developmental Medicine, Departments of Pediatrics and Obstetrics and Gynaecology, University of British Columbia, Vancouver, B.C., Canada.

4. A certain quantity of food.
5. Water (the supply of oxygen and water can be considered part of nutrition).

Several mammalian developmental periods can be distinguished according to the pathways along which nutrients are supplied (Fig. 1):

1. The fertilized egg, until blood supply is established, is fed directly from the surrounding tissues.
2. Once the placenta and the circulation are developed, fetal nutrition, as we understand it, sets in. Essentially, all nutrients are supplied from the mother via the blood and most fetal waste products are removed in the same way.
3. Early postnatal nutrition, which in mammals is a period when breast milk is consumed.
4. The weaning period, a transitional stage when milk is still consumed but other food is also eaten more and more frequently.
5. The postweaning stage.

During periods 1 to 3, the individual cannot choose its food. It must consume what is offered. It cannot, for instance, separately quench its thirst and satisfy its hunger. This is absolutely true for periods 1 and 2, and variably valid for period 3, depending on the species. Thus, most rodents (rats, mice, hamsters, rabbits, but not guinea pigs) are born so immature that they are

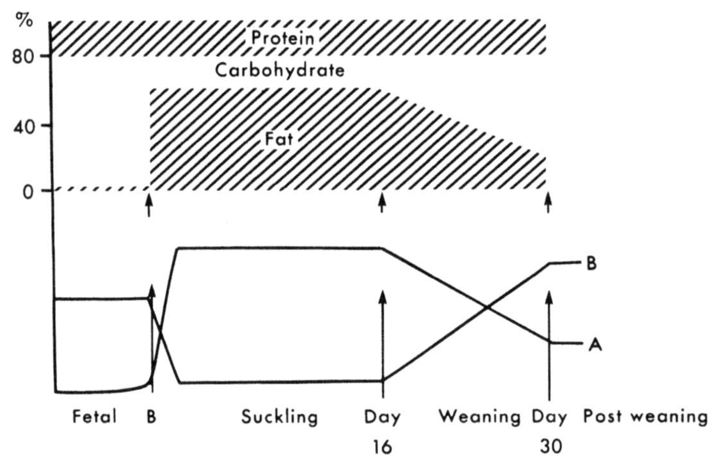

Fig. 1. Changes in nutrient composition of food consumed by fetal, suckling, weanling, and weaned rats, showing the relationship to changes in metabolic pathways. *Top part:* Calories (%) contributed by protein, carbohydrate, and fat during development. Note the high intake of carbohydrate by the fetus and weaned rat and the high fat intake in the suckling period. *Ordinate:* Percent of total calories. *Lower part:* Schematic curves representing lipogenesis, glycogen formation, and liver glycogen content (curve B) and gluconeogenesis and fatty acid oxidation (curve A). *Abscissa:* Developmental periods. B, Birth; day 16, start of weaning period; day 30, end of weaning period.

completely dependent on mother's milk. Other species (calves, kids, lambs, foals, etc.) are capable of, at least partially, fending for themselves from the moment they are born.

In the following pages, we shall consider metabolic development in relation to nutrient supply. We shall concentrate on the metabolism of the main nutrients, their interrelationship, and the possible way by which the quality and quantity of the nutrients supplied may affect the immediate state and further development of the individual. Inherent in our discussion is the assumption that the full genetic potential of an individual can only be realized in an optimum environment and hence that what is possible developmentally is not necessarily realized.

1.3. Early Postconceptional Period

Very little is known of the early metabolic development of the fertilized egg, even though in the thirties Needham devoted an entire book to reviewing the subject. Particularly, the nutritional requirements of the early embryo, its needs and the exact pathways by which it is fed, are still unknown in all details. Thus, it is possible to fertilize a mammalian egg *in vitro*. However, it will not develop to the adult stage unless transferred to a receptive uterus. It may be assumed that in analogy to isolated cells, an incubation medium containing the necessary amino acids, vitamins and glucose (e.g., Weymouth's medium) would essentially be adequate to support the growth of the fertilized egg, yet it is not. How these substances are supplied to the fertilized egg *in vivo* has not been examined in any detail, nor is it known whether the early embryo requires some or all of the known hormones for adequate growth. Again, in analogy with isolated cells, one would expect a requirement for growth factors and insulin. Recently, the nutrient requirements of the preimplantation embryos, particularly of mice, have been reviewed (Brinster, 1971).

1.4. The Embryo and the Fetus

Until recently, it was assumed that the nutrient supplies to the embryo were the same as those to the fetus. Particularly in man, fetal nutritional requirements have been extrapolated from umbilical blood analyses (i.e., perinatal data). It is becoming clear, however, that such extrapolation is probably unjustified since there are indications that the composition of the nutrients that pass via the placenta from maternal to fetal blood varies with the age of the fetus. Another factor that must always be taken into account is the species that is being studied. One that is born very mature will be capable already, *in utero,* of utilizing nutrients in a manner that is not yet available to a less mature species. Third, the metabolism of individual tissues (e.g., brain) must be considered, and fourth, the metabolic peculiarities of the adults of a species must be known (i.e., the low availability of glucose in ruminants).

In the following pages we shall discuss the metabolism of the main nutrients in the fetus, the perinatal, the suckling, and the weaning period. It will be clear to the reader that at times it was difficult to keep these periods separate and I have unhesitatingly ignored this separation when I felt that overlap would make things more comprehensible.

2. Fetal Period

2.1. Carbohydrate Metabolism

2.1.1. Glucose

It is generally accepted that glucose is the substance that principally meets the energy requirements of the developing fetus. This seems to be true even for ruminants which, as adults, use short-chain fatty acids instead (Ballard *et al.*, 1969). Variations in the maternal level of blood glucose are small throughout gestation. It is supplied via the placenta to the umbilical vein of the fetus. When considering substrate utilization it is usual to consider the liver as being representative of the body as a whole, which is of course erroneous at all ages, particularly so in the fetus, because much of the umbilical blood bypasses the liver and goes directly to the heart. Yet, the metabolic development of no other tissue is as well known as that of the liver. From the nutritionist's point of view, this emphasis on hepatic metabolism is justified in postnatal studies. The gastrointestinal tract and the liver are the two tissues that are the first to encounter food and, presumably also, the first to respond to specific nutritional stimuli. Prenatally, however, conditions are different. Nutrients that cross the placenta enter the umbilical venous blood and, either directly or indirectly via the liver, reach the right heart. They never reach the liver via the portal vein, nor is the gut the first to encounter them. It is also worth pointing out that nutrients reach the brain more directly pre- than postnatally (see, e.g., Dawes, 1968).

Let us consider the first step necessary for glucose utilization by any cell: the formation of glucose-6-phosphate. This step is catalyzed by several hexokinases. In addition, adult liver possesses a specific glucokinase that is functional only at very high concentrations of glucose. This enzyme is not present in fetal mammalian liver, nor is it found in the liver of ruminants. Teleologically, the absence of glucokinase is relatively easy to explain: both mammalian fetuses and adult ruminants of any age never have high levels of glucose in their portal blood—the former because their gut is not functional and they do not consume carbohydrates orally, the latter because their main source of energy is short-chain fatty acids. Walker and Eaton (1967) have examined the nutritional control of hepatic glucokinase in suckling rats in relation to glucose intake. It is apparent that both inborn (genetic) and environmental factors regulate the enzyme. In actual fact, however, the first messenger for glucokinase synthesis appears to be insulin. High glucose concentrations do not act only on the liver but on the β-cells of the pancreas as well, and the released insulin together with the glucose cause new glucokinase synthesis. The complex relationship between nutritional status, endocrine regulation, and development can be well illustrated for this particular enzyme. Glucokinase activity will not increase in the liver at the time of weaning in the rat if the animal is weaned to a high-fat diet. Glucokinase activity can be induced prematurely to some extent by giving a high-glucose diet and insulin injections to suckling rats. However, a much more potent stimulus for the premature induction of glucokinase appears to be thyroid hormone. This recent discovery (Partridge *et al.*, 1975) shows that one must be very careful before coming to definite conclusions about the control of any enzyme.

2.1.2. Glycogen

In most mammals, glycogen content—not only in liver but also in skeletal muscle, heart, adipose tissue, and brain—is high just before delivery (Shelley, 1969; Hahn and Koldovsky, 1966). Depending on the species, glycogen stores build up either before the middle of gestation (i.e., man, monkey) or just before term (rat, rabbit, dog). Again, much more is known about hepatic than about muscular development. It is assumed that hepatic glycogen synthesis is due to the steady supply of glucose from the mother and the activity of glycogen synthetase. Hormonal control has been demonstrated (Jost and Picon, 1970). Both the fetal adrenal gland and fetal insulin are required for glycogen synthesis (Jost and Picon, 1970; Picon, 1971; Manns and Brockman, 1969; Eisen *et al.*, 1973b). It has also been suggested that growth hormone plays a role (Jacquot, 1975). All these controls act both on glycogen synthetase (Gilbert and Vaillant, 1975—rat; Schwartz *et al.*, 1975—man) and on the key enzyme of glycogen breakdown—phosphorylase (Schwartz and Rall, 1975—man).

The interrelationship between glycogen breakdown and synthesis is shown in Fig. 2. At first sight, such control seems far removed from any nutritional effect. However, the level of blood glucose is regulated by insulin, glucagon, and other factors, and the blood level of these two hormones is in part controlled by the blood level of glucose acting on the pancreas. Thus, starvation, for example, will lead to a fall in blood glucose, a consequent fall in blood insulin and, perhaps, a rise in the blood glucagon level. An elevation in the latter is not absolutely necessary since the ratio of insulin to glucagon in the blood apparently is more important than the actual blood level of either hormone (Unger, 1974). The consequence of starvation is activation of phosphorylase and a greater rate of glucose production from glycogen. Feeding a high-carbohydrate diet has the opposite effects. From the developmental point of view, many questions arise. Let us first inquire why liver glycogen content is low in early fetal life and increases toward them. Many possibilities must be considered for the rat fetus.

1. *Insufficient glucose is supplied from the mother.* This possibility can probably be ruled out since, in the rat at least, the size of the fetus is determined to some extent by the number of fetuses in the uterus, yet glycogen content does not rise in inverse proportion to the number of fetuses. However, glycogen content after the time it is normally found in the liver does depend

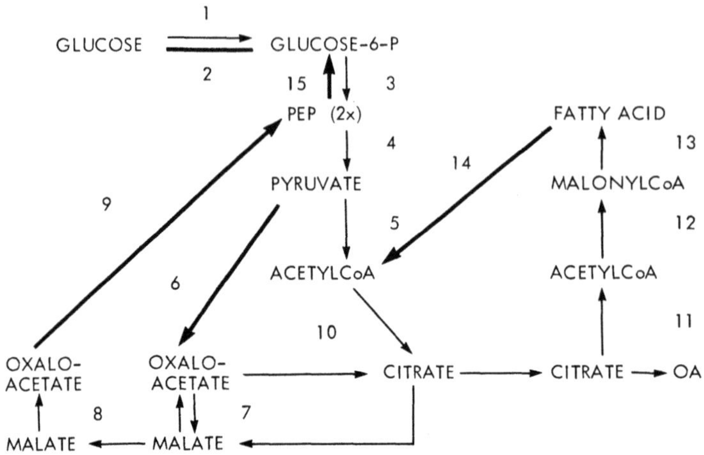

Fig. 2. Relationships among glycolysis, gluconeogenesis, fatty acid synthesis, and fatty acid oxidation in rat liver. 1, Hexokinase; 2, glucose-6-phosphatase; 3, glycolysis; 4, pyruvate kinase; 5, pyruvate dehydrogenase; 6, pyruvate carboxylase; 7 and 8, malic dehydrogenase; 9, phosphoenolpyruvate carboxykinase; 10, citrate synthase; 11, citrate enzyme (lyase); 12, acetyl-CoA carboxylase; 13, fatty acid synthetase complex; 14, fatty acid oxidation, including carnitine transferases; 15, gluconeogenic pathways. Strong arrows show reactions that are prominent in suckling rats. Reactions 5, 6, 7, and 10 occur in the mitochondria.

to some extent on the nutritional status of the fetus (Girard and Marliss, 1975; Girard *et al.*, 1973, 1974).

2. *Insulin is not made by the fetal pancreas.* This is probably not true since insulin is present in the blood of relatively young fetuses.

3. *Too much glucagon is produced by the fetal pancreas, causing glycogen breakdown.* This is not true (Girard and Marliss, 1975). Fetal blood levels of glucagon are low. Since insulin and glucagon cannot cross the placenta the mother cannot contribute to fetal blood levels (Chez *et al.*, 1974b) of these hormones.

4. *Liver cells do not possess the receptor for insulin.* This may be true but has really never been adequately illustrated, although a recent paper has shown low binding not only in fetal but also postnatal liver (Blazquez *et al.*, 1976).

5. *Liver cells do not possess glycogen synthetase.* This is partly true. Enzyme activity and content are low before day 16 in the rat, and then rise. In particular, glycogen synthetase (a) (the active form of the enzyme) increases considerably (Pines *et al.*, 1976) after day 16 postconceptionally.

6. *Liver cells only appear after day 16, together with glycogen synthetase.* This is not true.

7. *Other factors play a role.* This is very likely.

As already mentioned, glucocorticoids are required for the laying down of liver glycogen. In the fetal rat and probably other species, they "induce" glycogen synthetase [i.e., they cause glycogen synthetase to be synthetized (Jacquot and Kretchmer, 1964)]. How they do this is not clear (for a discussion, see Wicks, 1974). However, since glucocorticoids are required, it must be asked (a) whether they are produced by the fetal adrenal before day 16 (or by the maternal gland since in rats they do cross the placenta), or (b) whether fetal liver cells possess receptors for glucocorticoids (they do) (Feldman, 1974).

If we now summarize our knowledge of fetal glycogen synthesis, we find that after more than 20 years of intensive research, we still do not understand it completely. We know that, before a certain day of fetal development, it is not possible to initiate such synthesis by injecting hormones, but that just one or two days later, we can speed up synthesis by injection of, e.g., cortisonc, and prevent it by fetal (or fetal + maternal) adrenalectomy. Why liver cells do not respond before that day is not clear. Most probably, the lack of receptors for steroid, insulin, or growth hormone is involved. One can envisage, for instance, that normally glucocorticoids "prepare" the cell (permissive effect) for the action of insulin.

Whatever the mechanisms, it is clear that many steps may be interposed between the nutrients supplied and the response in a cell. It is also evident that the fetus may lack regulatory mechanisms, normally present in the adult. It seems worthwhile to underline that, to date, no apparent "nutritional" factors in the control of fetal glycogen deposition have been properly defined, except perhaps severe undernutrition of the mother. It has been suggested (Pines *et al.,* 1976) that glucose concentrations directly control glycogen synthetase (a) activation in fetal rats. Glucose was injected into the peritoneal cavity of pregnant rats or into the uterine vein. In both instances, synthetase (a) activity of fetal liver rose rapidly (within 6 minutes). This is considered to be a direct effect of glucose on the enzyme. However, it is equally possible, and in fact more probable, that glucose injection led to insulin release in the fetus and that the insulin was responsible for the synthetase activation, as suggested by Rabain and Picon (1974). [For a review of control of liver glycogen in general, see Hers (1976).] Support for this contention comes from a recent paper by Sparks *et al.* (1976). Fetal monkey liver was perfused with different substances, and glycogen synthetase activity was monitored. Varying the glucose concentration of the perfusion fluid between 50 and 500 mg/100 ml had no effect on the percentage of active synthetase in the liver. If insulin was added, activity was increased (see also Glinsman *et al.,* 1975). However, surprisingly enough, galactose added to glucose caused some activation and, if insulin was added to this sugar mixture, the percentage of active enzyme increased by a factor of up to 10, depending on the glucose concentration. It is speculated by the authors that galactose is uniquely necessary in the regulation of liver glycogen metabolism in the newborn primate. These data explain the previous finding of the same group (Glinsman *et al.,* 1975) that hyperglycemia per se did not lead to glycogen accumulation in fetal monkey liver, just as high glucose concentrations did not raise glycogen levels in isolated fetal rat liver explants (Eisen *et al.,* 1973b). In the same paper, Glinsman *et al.* (1975) speculate that "the inability to store glucose as glycogen makes the neonate vulnerable to the development of glycogen depletion . . . . " It is worthwhile to analyze this statement.

First we must stress that the author studied fetal and not neonatal liver. He could stimulate preterm glycogen synthesis if galactose was also present. Thus, before birth, conditions evidently favor glycogen formation. Immediately after birth conditions are quite different but not because the neonate is vulnerable to glycogen depletion (see perinatal period). The newborn has to use its own reserves immediately after delivery and it would be fair to say that it is eminently *capable* of glycogen depletion. Why that should be a sign of vulnerability is not at all clear.

2.1.3. Glycogenolysis

All the data indicate that the level of hepatic fetal glycogen is regulated by glycogen synthetase control, and very little by control of phosphorylase. Nevertheless, it is the sequence adenyl cyclase, protein kinase, and eventual activation of phosphorylase (Novak *et al.*, 1972) and inactivation of synthetase that is decisive. Again, it is very difficult to decide how far the presence of a certain mechanism indicates its usage. Under very special circumstances (explants, perfusion, injection of large doses of hormone into the fetus), liver glycogen can no doubt be broken down. However, under normal circumstances, even prolonged starvation of a mother rat will not lead to a large fall in fetal liver glycogen (Girard and Marliss, 1975), even though further accumulation of glycogen may be arrested. It has been shown by Girard (1975) that starvation of the preterm pregnant rat from day 17 to day 21.5 leads to a decrease in the level of glucose in both mother and fetus. At the same time, there is an increase in the fetal blood level of glucagon and a decrease in that of insulin (Table I).

In human liver explants, glucagon is effective in causing glycogen breakdown before the hormone can be detected in the fetal pancreas (Schwartz *et al.*, 1975).

At the moment, we would be hard pressed to consider other than genetic factors as being responsible for the temporal sequences of enzyme appearances and responsiveness during fetal life. However, this author is quite ready to be

Table I. Effect of 72 Hours of Starvation of the Mother on Plasma Levels of Some Metabolites and Hormones in the Mother and Her Term Fetuses

	Mother		Fetus	
	Fed	Starved	Fed	Starved
Insulin (μU/ml)	21	10	199	128
Glucagon (pg/ml)	153	237	318	746
Glucose (mM)	5	3.3	3.7	2.1
Lactate (mM)	2	1.3	7.8	3.5
Fatty acids[a] (μM)	230	740	160	175
β-OH-Butyrate[a] (μM)	70	7370	81	8250

[a] It is evident that fatty acids do not cross the placenta but that β-OH-butyrate does (Girard, 1975).

thus pressed and to conceive of subtle changes in nutrient and perhaps mineral supplies (e.g., calcium ions; see Schwartz *et al.*, 1975) that, together with the genetically determined development, cause individual differences in the actual time of appearance of a certain mechanism during development. Too little is known of the details of transplacental nutrition at various stages of fetal life to have a definite answer either way. Most data for sheep, monkey, and rat come from the latter part of pregnancy, and only in man has early fetal development been looked at in slightly more detail.

It should be stressed that the mammalian nonruminant fetus can be considerably altered metabolically by the opposite of starvation (i.e., a plethora of glucose). This excess glucose supply to the fetus occurs in maternal diabetes and is a relatively good example of the effect of nutrition on the fetus. Insulin (and glucagon) do not cross the placenta (Chez *et al.*, 1974b) and the fetus can and does react to a prolonged excess of glucose by insulin release and synthesis (R. Schwartz, 1975). This, however, does not occur immediately (Nakai *et al.*, 1976) but is apparently an adaptive process by which the fetus tries to handle the extra glucose, and it is only the fetus and newborn that has been repeatedly exposed to excess glucose that reacts rapidly. Essentially, fetuses from diabetic mothers (IDM) are overnourished fetuses that are born, not only overweight, but also larger than normal. The initial stimulus that causes these changes appears to be the high blood glucose level, which, it is claimed, leads to premature maturation of β-cell response to glucose, and thus to premature and excess insulin secretion. Since such fetuses also grow more rapidly, and since fetuses without insulin grow more slowly, it has been suggested that insulin is the real "growth" hormone of development. This label, however, serves only to confuse. Tradition has ascribed the name "growth hormone" to a polypeptide made in the anterior pituitary that acts on cartilage growth via somatomedin and has other more-or-less defined metabolic effects. In fact, it is no more a "growth" hormone than insulin, L-thyroxine, or vitamin D.

2.1.4. Glucose Formation

Activated phosphorylase breaks down glycogen to glucose-1-phosphate, which is then changed to glucose-6-phosphate. This substance is at the crossroads of glucose metabolism:

$$\begin{array}{ccccc} & & \text{glycogen} & & \\ & & \uparrow{\scriptstyle b} & & \\ \text{glucose} & \xleftarrow{\ a\ } & \text{glucose-6-p} & \xleftrightarrow{\ c\ } & \text{pentose shunt} \\ & & \downarrow{\scriptstyle d} & & \\ & & \text{glycolysis} & & \end{array}$$

Pathway b has been discussed. Pathway a is not present in all tissues. The reaction is catalyzed by the enzyme glucose-6-phosphatase (microsomal enzyme) and serves to make glucose available to peripheral tissues in times of glucose dearth. Such times are very rare during fetal life, and in fact occur only under extreme conditions [e.g., during prolonged starvation of the mother rat (Girard and Marliss, 1975)]. Hence, it is not surprising that in most species, glucose-6-phosphatase activity rises from nearly zero to very high levels immediately after birth. For man, relatively large amounts of the enzyme are found in the liver already at midterm. But whether this means that the human fetus controls its blood glucose rather more actively than other mammalian fetuses is not known. There is no doubt, however, that in explants from human or rat fetal livers, glucose-6-phosphatase activity can be raised by glucagon if a glucocorticoid is also present (Boxer *et al.*, 1974).

2.1.5. Glycolysis

All mammalian fetuses go through a stage when mitochondria are immature and few in number and when, as a consequence, most energy is obtained anaerobically. It has been tacitly assumed for nearly 40 years that, even in the newborn mammal, again depending on the degree of immaturity of the species at birth, anaerobic pathways predominate. Their predominance has been taken to explain the ability of the newborn to survive oxygen lack for longer periods than the adult. This prolonged survival can be dramatically shortened by monoiodoacetate, an inhibitor of anaerobic glycolysis. Again, it must be stressed that the above is a general rule only insofar as time, tissue, and species are specified. In man, for instance, mitochondria develop rapidly in several tissues after the eighth week of gestation, while in the rat most of this development occurs postnatally (for a review, see Hahn and Skala, 1971a). In addition, heart and, to some extent, brown fat mitochondria are developed prenatally while, in the brain and muscle, their postnatal evolution is even slower than in the liver. The best data in this field are probably those of Girard (1975), who showed that the blood level of lactate, the end product of anaerobic glycolysis, is nearly four times higher in the blood of 17-day-old fetuses than in that of their mothers (7.8 against 2.0 mM).

These questions were discussed again recently by Berger and Hommes (1975). They showed that, toward the end of gestation in fetal rat liver, only 10–15% of pyruvate is oxidized to CO_2, the rest being converted to lactic acid. Since the level or pyruvate dehydrogenase is low in fetal rat livers (Bailey *et al.*, 1976; Knowles and Ballard, 1974), it is concluded that, in term rat fetuses, energy is produced mainly by anaerobic glycolysis, particularly since mitochondrial enzymes in liver also develop only perinatally (Jakovcic *et al.*, 1971). However, the ratio of lactate to CO_2 produced from added pyruvate by liver cells increases with gestational age in the rat, suggesting that earlier in fetal life all energy-producing processes occur at a slow rate and that, as gestation proceeds, more and more energy is produced anaerobically. Unfortunately, there are no data available to substantiate this more thoroughly. We must again stress that this was studied in liver cells only, and that very little is known about other tissues. In addition, pyruvate can also give rise to ketone bodies in liver cells, a fact that has not been considered. Nevertheless, it is known, for instance, that pyruvate kinase activity in liver is higher pre- than postnatally (Walker, 1971), while, in brain and muscle of both rat (Hahn,

1970a) and man (Hahn and Skala, 1970), the activity of this enzyme is low in the fetus and rises only after birth to much higher values. We shall return to this later.

It must also be mentioned that a number of enzymes exist in several forms as isozymes. This is true, for example, for lactic dehydrogenase and pyruvate kinase. In most cases, one can distinguish one form that predominates in fetal tissues and another which appears only later in life.

2.1.6. *Gluconeogenesis*

From the nutritionist's standpoint, the rate of glycolysis in fetal tissues would be expected to be high, since glucose is being supplied at a relatively constant rate, is the main nutrient, and, as stated, is also the main energy substrate. From that point of view, the rate of *Gluconeogenesis* would also be expected to be low. It is well known that, essentially, only the liver is capable of producing glucose from nonglucose precursors (lactate, amino acids) and that it does so only if exogenous glucose supply is low. Hence, it is not surprising that the enzymes involved in gluconeogenesis [i.e., pyruvate carboxylase, phosphoenolpyruvate carboxykinase (PEPK), and fructose diphosphatase (Table IV)] are not found (or show very low activity) in fetal mammalian liver (Walker, 1971) excepting fetal ruminants (Ballard *et al.,* 1969). Under normal circumstances, the fetus is not called upon to make glucose. However, it must be inquired how far the fetus is capable of responding to a glucose lack. This has been tested by Girard (1975), who showed that, if a pregnant rat is fasted for 4 days up to day 21.5 of gestation, several changes occur in her fetuses (Table I).

It is worth considering this table in detail since it illustrates well how helpless the fetus is in comparison to the neonate. Even though blood glucose has decreased by nearly 40% as a result of the fall in the maternal blood glucose level, the fetus is incapable of mobilizing its liver glycogen. Nor is it capable of increasing its rate of gluconeogenesis to values found in the newborn. With fetal starvation PEPKinase activity is increased 5-fold, whereas the rise in the neonate is nearly 100-fold. Similar differences are found for glucose-6-phosphatase. This very severe and unnatural degree of starvation causes some decrease in the fetal insulin blood level and a rise in the level of glucagon. Again, however, these changes are small when compared to those that occur after birth.

The most striking change that occurs in the fetuses of fasting mothers is the tremendous rise in the blood level of ketones. This is due to a similar rise in maternal blood and may be of considerable significance (see below). It is also worth noting that the lactate level is decreased, suggesting that some gluconeogenesis does occur.

No data are available for human fetuses under such conditions. It is not known whether severe intrauterine malnutrition, resulting in small-for-date infants, causes similar changes as those shown in the table.

2.1.7. Control of PEPK in the Fetus

We shall pause here to consider this rate-limiting enzyme of gluconeogenesis in fetal liver as an example of the complexity of the interrelationships between nutrition and metabolic processes. The example serves to illustrate that changes in food intake (in this case, glucose transfer from maternal to fetal blood) do not directly cause (or very rarely cause) changes in metabolic pathways. Usually, a complex series of events occurs between the metabolite that causes the metabolic reaction and the enzymes eventually responsible for the reaction. In the case of hepatic cytoplasmic PEPK, it has been clearly demonstrated in fetal and adult rats that changes in enzyme activity reflect changes in the rate of synthesis of the enzyme itself and that these are brought about not by metabolites but by hormones (Hanson *et al.*, 1973).

As mentioned for glycogen synthetase and phosphorylase, PEPK activity is also regulated by the level of cyclic AMP. However, the preformed enzymes of glycogen metabolism are activated or inactivated, while in the case of PEPK the synthesis of new enzyme is regulated.

In the adult, it is thought that starvation or a high-fat diet raises liver PEPK activity as follows: hypoglycemia, decreased insulin, and increased glucagon levels → decreased I/G molar ratio → activation of hepatic adenyl cyclase → active protein kinase → ??? → increased PEPK synthesis.

When we now inquire why PEPK levels are low in fetal rat liver, we can, first, point to the greater insulin and lower glucagon levels in the blood. This high I/G ratio itself suffices to prevent any induction of PEPK synthesis. Second, we might consider that the hepatic cell receptors for glucagon are immature. This may be true to some extent prenatally (Bär and Hahn, 1971; Blazquez *et al.*, 1976) but it is also true postnatally (see below) and hence does not seem tenable as an explanation. Additional support for this negative conclusion comes from the fact that hepatic adenyl cyclase responds to glucagon at least 3 days before term (Bär and Hahn, 1971). Further, it is possible that any of the further links in the chain between cyclic AMP and the final enzyme protein are not fully developed.

Against this stands the fact that in fetal liver explants PEPK synthesis can be induced with cyclic AMP or glucagon, but not if the liver is taken from rat fetuses younger than 17 days. During fetal development, cyclic AMP is effective 1 or 2 days before glucagon, both *in vitro* (Wicks *et al.*, 1972) and *in vivo* (Yeung and Oliver, 1968a), again suggesting development of the receptor part of the adenyl cyclase. On the other hand, before day 17, not even cyclic AMP has an effect, which implies immaturity of some other mechanism and or repression of such a mechanism.

In contrast to the rat, fetal development in man proceeds more rapidly. Cyclic AMP can induce PEPK synthesis in liver explants as early as the ninth week of pregnancy, and a good response is obtained by the twenty-first week [i.e., halfway through pregnancy (Kirby and Hahn, 1974)]. However, in both rat and human fetal liver explants, even the highest activity obtained after maximum stimulation is only a fraction of that found postnatally. The very high PEPK activity found postnatally always *decreases* in liver explants from newborn rats (Frohlich *et al.*, 1976) suggesting a different control pre- and postnatally. Even in the guinea pig, which is born in a very mature state, cytoplasmic PEPK activity is low in the fetus and only rises after birth (Robinson, 1976; Arinze, 1975).

It is tempting to suggest the following sequence of events in the rat fetus: (1) PEPK activity is completely repressed (how much of the low activity is due to the diluting effect of the large number of red cells found in fetal liver?) and the cyclic-AMP-sensitive protein kinase for PEPK induction (derepression) is not present; and (2) on day 17, cyclic AMP becomes a relatively effective inducer, and on day 18 cyclase receptors to glucagon appear. However, until birth, PEPK activity remains very low, possibly because of the high levels of insulin (which represses PEPK) and constant levels of glucose. Even during fasting of the mother (Table I), the I/G molar ratio never falls to early postnatal levels and that may explain why when cyclic AMP is given *in vivo* to fetal rats PEPK levels never attain postnatal values. However, this explanation does not hold for *in vitro* experiments. Neither in man nor in primates or sheep do fetal levels of insulin and glucagon seem to be very different from those in the mother, nor is the change in the I/G molar ratio very striking in these species after delivery (Fig. 3, Table II). Yet, in the one species (ruminant) PEPK activity is already high prenatally. In man and monkey, although more abundant in fetal liver than in the rat, PEPK activity is low and, up to now, it has not been possible to raise it to postnatal values by *in vitro* techniques using cyclic AMP and prednisolone. It must be mentioned that PEPK exists in two forms, mitochondrial and cytoplasmic, and that it seems that only the latter is under hormonal control (Hanson *et al.*, 1973). In addition to glucose-6-phos-

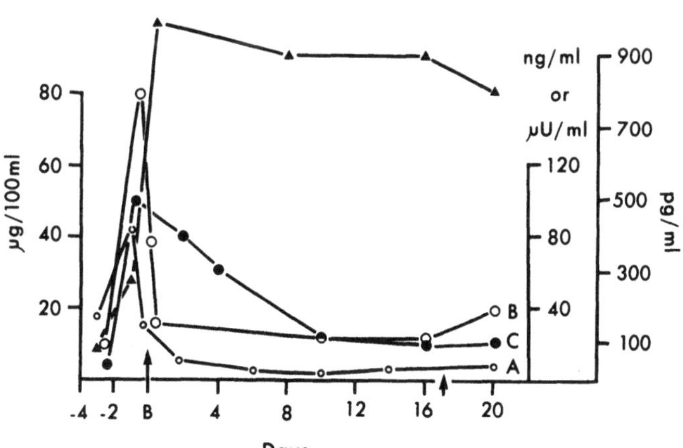

Fig. 3. Changes in the blood levels of some hormones in the perinatal period of the rat. *Abscissa:* Age in pre- and postnatal days; B, birth. *Ordinate:* Left: corticosterone in µg/100 ml—curve A (○); right: ng/ml—growth hormone—curve B (○), µU/ml—insulin—curve c (●), and pg/ml—glucagon—curve D (▲).

Table II. Selected Data on Levels of Insulin and Glucagon in Blood

Insulin				Glucagon					
Fetus	Newborn	Mother	Unit	Fetus	Newborn	Mother	Unit	Species	Reference
77–199[a]	20	21	μU/ml	105–270	1000	560	pg/ml	Rat	Girard *et al.* (1974)
35–150		35	μU/ml					Rat	Felix *et al.* (1969)
6	3	2	ng/ml					Rabbit	Milner (1969)
54	14	23	μU/ml					Rabbit	Hardman *et al.* (1971)
0.5	0.5	1.5	pg/ml (143–148 days of gestation)	80	350			Monkey	Chez *et al.* (1971) Chez *et al.* (1974b)
13	50	30	μU/ml					Sheep	Willes *et al.* (1969)
15.5	29		μU/ml					Sheep	Basset *et al.* (1973)
20	15		μU/ml	200 (400)[b]	300		pg/ml	Sheep	Fiser *et al.* (1974 a,b)
				138	189	124	pg/ml	Man, full term	
				143	239	131	pg/ml	Man, SGA	Bloom and Johnson (1972)
				125	148	125	pg/ml	Man, IDM	
15	7 (7 on day 3)		μU/ml	83	180 (240 on day 3)		pg/ml		
				260	5000		pg/ml (incl. gut glucagon)		Delamater *et al.* (1974)

[a] Days 18 and 21 after conception.
[b] Early and late gestation.

phatase, PEPK (and perhaps pyruvate carboxylase, a mitochondrial enzyme) fructosediphosphatase is essentially a "one-way" enzyme of gluconeogenesis. Its activity does not appear to be controlled by cyclic AMP, yet it also increases postnatally (Walker, 1971).

2.2. Fatty Acid Metabolism

2.2.1. Fatty Acid Synthesis

Before discussing fetal metabolism of fatty acids, it must again be pointed out that development is different in liver, adipose tissue, and brain, to take only three of the many tissues. Unless otherwise stated, we shall be referring to liver.

Glucose, which appears to be the main energy source of all mammalian fetuses examined, is transported across the placenta, probably by facilitated diffusion. This transport is probably independent of the type of placenta involved. In contrast, the ease of transport of lipids across the placenta, particularly of fatty acids, differs from species to species. Thus, in sheep, transport of palmitic acid from mother to fetus has been demonstrated (for a review, see Hahn, 1972). Nevertheless, even in this species, the fetal blood contains 0.1 meq/liter of fatty acid, while that of the mother is about 1.0 meq/liter (Alexander *et al.,* 1969). In man, the rate of placental transport of fatty acids decreases with increasing chain length (Dancis *et al.,* 1976). We have calculated elsewhere (Hahn, 1972) that even undetectable rates of fatty acid transport would suffice for the laying down of fat in the human preterm fetus. In rats, Hummel *et al.* (1975) calculate that half the fatty acids in the rat fetus are derived from the mother on the twenty-first day of gestation. Considerable transport has also been demonstrated in the rabbit (Elphick *et al.,* 1975) and is increased during starvation (Edson *et al.,* 1975).

Yet the actual amount of exogenous fat consumed by the fetus is small. Essentially, it is on a high-carbohydrate, high-protein, low-fat diet. In adults, such diets, if fed in sufficient amounts, raise the rate of lipogenesis, and this is also true for the fetus. Hence, the activities of the main enzymes of fatty acid synthesis—citrate cleavage enzyme, acetyl-CoA carboxylase, and fatty acid synthetase—are all higher in fetal than in neonatal rat liver and brown adipose tissue (for a review, see Hahn, 1972; Hahn, 1970b; Bailey and Lockwood, 1973). In contrast to gluconeogenesis and glycolysis, which are both controlled largely via activation of adenyl cyclase, the rate of lipogenesis seems to be substrate-controlled. Thus, a high-fat diet inhibits fatty acid synthesis and cyclic AMP has a similar effect. However, the mechanism of this suppression by cyclic AMP does not seem to be directly on the lipogenic enzymes but rather is indirect. Cyclic AMP, via a protein kinase, activates a lipase, which releases more fatty acids, and these are thought to suppress lipogenesis. This seems to be true both *in vivo* (Miguel and Abraham, 1976) and *in vitro* when isolated hepatocytes and liver explants (Goodridge, 1973) are used.

It must be stressed, however, that much of what has been described above is still speculative and that other mechanisms may well be involved. It is, for instance, well established that even a low-fat diet can suppress lipogenesis in mice, if essential fatty acids are included in the diet (Smith and Abraham, 1970). Whether these then give rise to prostaglandins, which act on enzyme activation or synthesis, is not known. It is also well established that the effect of a high-carbohydrate diet on lipogenesis is dependent on increased insulin release and levels. No increase in lipogenesis occurs in diabetic animals or man. Hence, in the fetal rat at least, we can perhaps explain the high rate of hepatic lipogenesis by the high blood levels of insulin, the high I/G ratio, and the constant supply of glucose. Recently, Miguel and Abraham (1976) showed that feeding pregnant rats corn oil, or fasting them, resulted in more linoleate being transferred to the fetuses. However, this had no effect on the high rates of fetal lipogensis or fetal liver glycogen levels.

2.2.2. Fatty Acid Oxidation

As expected from the high carbohydrate diet of the fetus and the low level of fatty acids in the fetal blood, the rate of fatty acid oxidation in fetal tissues is low or absent altogether. Again, however, this statement must be qualified, since although it applies to the rat, it is not absolutely applicable to human and monkey fetuses. Fatty acid oxidation occurs in mitochondria. In peripheral tissues it depends on the presence of carnitine and carnitine transferases (Fritz, 1967). The end product of fatty acid oxidation is CO_2 in peripheral tissues and ketone bodies in the liver. In the rat, several factors make it unlikely that the fetus can oxidize fatty acids to any large extent: (1) the number of mitochondria in fetal tissues is small (Hahn and Skala, 1971a); (2) the carnitine content of fetal tissues is low (Hahn and Skala, 1971a); (3) the activity of the carnitine transferases is low (Lee and Fritz, 1971; Augenfeld and Fritz, 1972; Hahn and Skala, 1972); and (4) the supply of fatty acids to the tissues is minimal.

In fact, since in the rat, fatty acid transfer across the placenta is only slight, any fatty acid in fetal blood must come from the fetuses' TG stores. These are dismally small, the only depot of fat being brown adipose tissue, which, at term, contains about 10% fat. In contrast, in the fetal guinea pig fatty acid transport across the placenta is considerable and the fetal liver then makes lipoproteins. This is the reason blood levels of triglycerides are very high in fetal guinea pigs (Bohmer and Havel, 1975). Yet the rate of fatty acid synthesis in the fetal liver is high, but occurs mostly from short-chain fatty acids. In adipose tissue glucose is the main precursor (Jones and Firmin, 1976).

In fetal rat liver, ketone production is very low (Drahota *et al.*, 1964), even from added fatty acids, suggesting a lack of carnitine transferases. However, other enzymes required for ketone formation also show very low activity in fetal rat liver. These are 3-hydroxyacyl-CoA dehydrogenase, palmityl-CoA synthase, hydroxymethylglutaryl-CoA synthase (mitochondria), hydroxymethylglutaryl-CoA lyase, acetoacetylthiolase, and 3-hydroxybutyrate dehydrogenase (Bailey and Lockwood, 1973). According to Lee and Fritz (1971) fetal rat liver homogenates produce 2.3 μmol/g/hr of ketone bodies from 0.5 mM of palmitate and 7.86 μmol from palmitoyl carnitine. The figures for adult liver are 22.6 and 29.6. Since ketone body formation from octanoylcarnitine is 26 μmol/g/hr in fetal and 40 in adult liver, it seems probable that the palmitoyl carnitine transferases, on the inside and on the outside of the mitochondrial membrane, are not fully developed but that the medium-chain transferase is considerably matured prenatally. In man only one paper has demonstrated that fetal liver from 12-week-old fetuses is capable of producing ketones (Hahn *et al.*, 1964), even though hepatic transferases are of relatively low activity (Hahn and Skala, 1973).

In contrast to fatty acids, the transport of which across the placenta has been shown to occur in most mammals but usually only to a very limited extent (Dancis *et al.*, 1976), ketone bodies seem to pass from mother to fetus much more easily, as demonstrated by Scow *et al.* (1964), Girard and Marliss (1975) for rats, and Sabata *et al.* (1967) for man. In the sheep, such transport does not seem to occur. However, in this species, acetate may take its place (Charlton-Char and Creasy, 1976).

During starvation of the mother and in pregnant diabetics the blood ketone levels are raised more than in nonpregnant individuals (Scow *et al.*, 1964). As a result, more ketones enter the fetal blood and apparently are well utilized. In fact, the fetal human brain is capable of using ketones very early in life (Adam *et al.*, 1975a), and it is even possible to induce prematurely enzymes of ketone utilization in rat fetuses by feeding a high-fat diet to the mother (for a review, see Bailey and Lockwood, 1973). The rat brain in the perinatal period also uses ketones preferentially, as first shown by Drahota *et al.*, (1965), and these are used not only for energy purposes (see Bailey and Lockwood, 1973; Hahn, 1972) but also for lipid synthesis (Edmond, 1974).

One point that should be mentioned here is that the rate of transport of any substance from mother to fetus also depends on the gestational age. For instance, the transport of linoleic acid from maternal to fetal blood in monkeys is greater in the term fetuses than in very young (37-day-old) fetuses. The difference is sevenfold. In a 37-day-old fetus, 10% of an injected dose appears in the fetus, whereas in 150-day-old animals 70% of the injected dose makes its way into the fetus (for a review, see Hahn, 1972). This in itself shows that, very probably, fetal nutrition changes in subtle ways during intrauterine development and makes it plausible that some of the changes that are observed during intrauterine development may be controlled not only genetically but also by changes in the composition of the nutrients supplied to the fetus.

2.3. Amino Acid Metabolism

The fetus has to assemble most of its proteins, since the majority of protein in the maternal blood cannot pass the placental barrier. Hence, amino acids from maternal blood are avidly taken up by the fetal circulation. This transfer is an active process, as follows from the fact that the level of most amino acids is considerably higher in fetal than in maternal blood (Dancis *et al.*, 1968; Snyderman, 1970). Thus, in a way, the fetus selects its food from the mixture offered it by the mother. Very probably the placenta itself also plays an important selective role.

Amino acids can be (1) used as building blocks for proteins, and (2) deaminated and transaminated.

1. The rate of proteosynthesis is of necessity high in the fetus. However, relatively little is known of the control mechanisms at work. It has been shown that amino acid transport into liver cells is accelerated by insulin in very young human and rat fetal livers *in vitro* (Schwartz *et al.*, 1975). This finding supports the contention that insulin has a direct effect on growth. It is also clear that limiting the supply of protein to the fetus will result in smaller fetuses and, depending on the tissue, a smaller number of cells or an equal number of smaller cells than normally (for a review, see Winick, 1975). It is obvious that amino acids can only be put to optimum use in the presence of a sufficient amount of energy. Hence, it is understandable that when the supply of proteins to the mother is marginal, improved fetal growth can be achieved by giving her extra calories without any additional source of amino acids (see, e.g., Lechtig *et al.*, 1975). It should be emphasized, however, that these effects of nutrition seem to be primarily on the mother, insofar as she can now make more amino acids available to the fetus, amino acids that she would otherwise have used for gluconeogenesis.

2. Amino acids are also de- and transaminated. Thus, they can be used for gluconeogenesis, particularly those that give rise to pyruvate. However, this does not occur to any extent in the rat fetus, where even if pyruvate were formed, the absence or very low activities of both pyruvate carboxylase and posphoenolpyruvate carboxykinase (PEPK) makes new glucose formation impossible. Similar arguments apply to other mammalian fetuses, including the human fetus.

Considering the above and the fact that the amino acid pattern of the fetus is to a large extent determined by the mother and accepting that growth is the most important homeostatic mechanism of the fetus, it is not surprising that only little ammonia is formed. It follows that urea formation is also low. This has been shown for both man and rats (for a review, see Snell, 1975). The activities of all enzymes of the urea cycle (carbamylphosphate synthetase, ornithine transcarbamylase, argininosuccinic acid synthetase, argininosuccinase, and arginase) are low in fetal liver. However, it must be stressed here that there are two carbamylphosphate synthetases, one to start the urea cycle, which is located in the mitochondria, and one that starts the synthesis of pyrimidines, which is located in the cytoplasm. The latter is very active in fetal liver (for a review, see Roux, 1973). Thus, some ammonia is undoubtedly produced from amino acids. However, I am not aware of any work on the transfer of purines and pyrimidines from mother to fetus nor on the blood levels of these substances. Quite possibly all have to be synthetized *de novo* by the fetus.

3. Perinatal Period

Delivery probably entails the most abrupt transition from one diet to another that a mammal ever experiences. Suddenly the continuous and steady transplacental supply of glucose, amino acids, and other substances via blood is interrupted forever, and after a shorter or longer time period of starvation, a high-fat diet is supplied via the gastrointestinal tract. Considerable efforts have been exerted to determine the sequence of events that occurs in the fetus during and after parturition. In fact, in man this is the period most accessible to examination, and hence more is known about it than about earlier stages of human fetal life.

The most obvious change that could be expected as soon as the supply of nutrients from the mother is curtailed is a decrease in the blood level of glucose of the term fetus or newborn. This, in turn, should lead to the activation of phosphorylase and the breakdown of glycogen, particularly in the liver. Consequently, as more glucose is required, hepatic gluconeogensis should be initiated. Starvation, followed by the feeding of a high-fat diet, should lead to a suppression of lipogenesis and an enhancement of fatty acid oxidation. In addition, trans- and deamination reactions should be more pronounced. In actual fact, all these changes do occur in the liver and adipose tissue, but not, for instance, in brain or muscle.

Detailed data are available for the rat and hence we shall first discuss this species. The rat is born in a very immature state, with no fur, sealed eyelids, and closed ears. Nevertheless, even this immature animal readies itself for delivery in numerous ways. The activities of hepatic enzymes destined to rise to very high levels very soon after birth commence to increase during the last three days of the 22-day gestation period.

It is not quite clear why these changes should be initiated already prenatally; in other words, whether they are just due to a genetic timetable or whether they are initiated by environmental (nutritional) changes. Some experiments have been designed to answer this question. It is possible to deliver rabbits or rats pre- or postmaturely and to look for changes that normally occur after delivery. Premature delivery of viable fetuses causes changes that are normally found after term delivery, while prolonged gestation, delayed some but not all of these changes until after birth (Yeung and Oliver, 1968b). Hence, it appears that it is the actual entry into the world that triggers the further development of the newborn. It would seem logical to assume that the prenatal changes observed toward term are due to environmental changes just as much as "genetic maturation"; in other words, they are "manipulated" by environmental factors.

Delivery for the fetus is a process involving considerable stress. It appears that stress (by pathways that are not clear but probably also involve cold exposure of the newborn) causes release of catecholamines (epinephrine and norepinephrine). These, in turn, activate hepatic adenyl cyclase and thus glycogen breakdown. They also increase pancreatic glucagon release and decrease insulin release from the pancreatic islets.

As mentioned above, an alternative stimulus could be the decrease in blood glucose levels, acting either directly on the liver or inducing glucagon release from the pancreas (but perhaps also from the gut). See also, however, Kervran and Girard (1974).

Here we must insert a few words on hormones and their receptors. It is generally accepted that hormones such as glucagon and insulin act on the cell surface, the former by activating adenyl cyclase. Blazquez *et al.* (1976) recently published data showing low glucagon binding to rat liver cell membranes during development. Binding was low prenatally, even though a rise from about 20 to 160 fmol/mg of protein was noted betweeen days 15 and 21 of conception. This figure increased only after the fifth postnatal day, attaining a value of 350 fmol/mg on day 20. At the same time Blazquez *et al.* (1976) again described increasing responsiveness of hepatic adenyl cyclase to glucagon with age (see Bär and Hahn, 1971). The authors conclude from their data that glucagon insensitivity is present in fetal rat liver. They suggest that "down regulation" of glucagon receptors by the relatively high circulating fetal levels of glucagon may play a role. Since perinatally glucagon levels suddenly increase (Girard, 1975) and stay elevated for about 18 days, one would expect binding to remain low (which it does only up to day 5) and liver glycogen to stay elevated after birth, which of course, it does not. What is much more likely is, as already mentioned, that the insulin/glucagon molar ratio is suddenly decreased perinatally by a rise in glucagon and a fall in insulin levels (see Girard, 1975; Blazquez *et al.*, 1975) and that the actual number of receptor sites is only of very minor, or perhaps no significance.

It is possible, to some extent at least, to quantitate the nutrient utilization in the newborn. For the rat this was done by Hahn and Koldovsky (1966). They concluded from their data and calculations that very soon after birth fat becomes the main supplier of energy. We can use the published data (Table III), particularly those of Girard (1975), to amplify, implement, and make more precise these calculations by Hahn and Koldovsky (1966) for the rat.

A 5.5-g newborn rat contains a total of 5.1 cal, of which (very generously calculated) 0.154 cal are derived from glycogen, 3.96 cal from protein, and 0.99 cal from fat (according to Girard, this figure is much smaller, since he

Table III. Effect of Starvation Immediately after Birth on Blood Levels of Some Metabolites in the Rat[a]

	Hours after birth			
Substrate	0	1	16 (starved)	16 (fed)
Glucose (mM)	3.0	1.0	1.0	6.5
Lactate (mM)	8.1	6.1	1.3	1.74
FFA (μM)	127	180	133	872
Glycerol (μM)	23	124	15	347
β-OH-butyrate (μM)	78	33	175	764

[a] Adapted from Girard (1975).

assumes all fat to be present in brown adipose tissue only). The caloric intake of a newborn rat is about 45 cal/100 g of body weight/day or approximately 0.1 cal/hr/5.5 g of body weight (Hahn and Koldovsky, 1966). It is evident that 1 hr after birth the newborn would use up to two thirds of its glycogen reserves if carbohydrates were the sole source of energy, and that within 2 hr after birth no carbohydrate would remain in the body. Very early after birth (within 1 hr) gluconeogenesis from lactate commences and at its peak (6 hr after birth) gives 20 μmol of glucose/hr/100 g of body weight or 0.047 cal/hr/newborn rat (i.e., about a third of the required amount of energy). The peak turnover rate of glucose (about 6 hr after birth) again is 20 μmol/hr/100 g, or 0.047 cal/hr/newborn rat.

During the first postnatal hour a maximum of 25 mg (0.1 cal) of glycogen is required. To supply this the liver would have to mobilize 78% of its glycogen stores. In fact, however, the level of hepatic glycogen remains unchanged during the first postnatal hour (Girard *et al.*, 1973). However, the blood glucose level decreases. Assuming that blood is 7% of body weight, the fall in blood glucose by 30 mg/100 ml is equal to 0.1 mg for a 5.5-g rat (i.e., a negligible amount). How can we explain these findings?

First, it is possible that the figure of 45 cal/100 g of body weight/day, representing the caloric requirements of the newborn rat, is exaggerated. This may especially be so under "normal" circumstances, when the mother rat is still delivering litter mates and the newborn is left in the cold. Its body temperature and thus its energy needs may decrease and may be as low as half of the amount suggested above. However, this is not valid for the numerical data presented here, since in Girard's experiments animals were delivered into 37°C ambient temperature.

Second, the rate of gluconeogenesis may rise much more rapidly after birth than is suggested by the figures in the table. However, during the first postnatal hour there is hardly any change in PEPK activity in the liver. It begins to rise only 1.5–2 hr after birth, and does not reach its maximum capacity until about 6–10 hr later. Even at that maximum this would provide 0.047 cal/hr (PEPK 100 μmol/min/mg of liver protein, liver being 4% of body weight), or again only about one-third of the total requirements. Admittedly, some authors report higher values for PEPK (and we ourselves have found values of up to 200 μmol/min/mg high-speed supernatant liver protein). However, these values certainly are not reached during the first 3 postnatal hours.

Gluconeogenesis from amino acids can also not be considered as an important factor in this very early postnatal period, simply because this again would have to proceed through PEPK, which at that time of life has still very low activity. Thus even though blood levels of most amino acids drop precipitously within 1 hr after birth, these are probably not used for gluconeogenesis to any large extent and immediately. This is also borne out by the fact that they accumulate in the liver during early postnatal life (Girard, 1975).

Hence, we are left with two further alternatives only, both depending on the breakdown of triglycerides. Lipolysis, initiated by activation of the hormone-sensitive lipase, gives rise to both glycerol and free fatty acids, and these may be the main energy donor immediately after birth. It is very difficult to quantitate the role played by free fatty acids and glycerol immediately after birth, since we have no data on the rate of triglyceride breakdown in adipose tissue at that time of life. However, the blood level of glycerol rises from 23 to 124 μmol/liter (i.e., more than fivefold), while that of free fatty acids only from 127 to 180 (i.e., only by about 50%). Since for every glycerol molecule released, three molecules of fatty acid must be formed, it is obvious that theoretically we should have 66 μmol/liter fatty acid to start with, and 372 μmol 1 hr later. Very roughly, this is an increase of 306 μmol/liter, from which we can subtract 53 to obtain 253 μmol/liter or about 85 nmol/5.5-g rat (0.00019 cal), which is not found in the blood after 1 hr.

Naturally, this kind of calculation tells us very little concerning actual fatty acid turnover. In order for fatty acids to cover all the energy needs of the newborn rat for the first postnatal hour (0.1 cal/5.5 g/hr) the newborn requires approximately 0.87 mg/min of fatty acids, or 0.73 μmol/min. Assuming that 2% of the body consists of fat (Hahn and Koldovsky, 1966), this works out as 0.73 μmol/min to be produced by 110 mg of fat or 6.6 nmol/min/mg of fat or about 0.66 nmol/min/mg of wet tissue of fat tissue = *0.66 μmol/g/min* of fatty acid or 0.22 μmol/g/min of glycerol. This is a rather high rate of lipolysis for adipose tissue. For the newborn rabbit the rate *in vitro* is 0.058 μmol/min/g of brown adipose tissue weight (Harding, 1971), for the newborn rat a figure of 0.03 μmol/min/g of tissue has been reported (Hemon, 1976). The figure of 0.73 μmol/rat/min is in rather close agreement with the activity of carnitine palmitoyl transferase in liver at that age (i.e., 20 nmol/min/mg of protein or 4 nmol/mg of wet liver weight). This equals 800 nmol (liver weight is 200 mg × 4 = 800 nmol), or 0.8 μmol/min/newborn rat.

Let us now consider the glycerol moiety released from triglycerides. If 0.73 μmol of fatty acids are released/rat/min this would correspond to about 0.24 μmol of glycerol, provided, of course, that the triglycerides are broken down completely and no di- and monoglycerides are formed. Thus, 0.24 μmol/min or 14.4 μmol/hr or 1.29 mg or 0.0048 cal could be supplied by the glycerol, and this is only a small part of the calories required by the newborn rat.

Finally, we must consider ketone bodies. Their level decreases after birth in the crucial first postnatal hour and then rises again, suggesting either decreased supply or increased utilization. As we shall see below, the latter is more likely. Interesting data have been presented by Haymond *et al.* (1974) for man. They compared term newborns with small-for-gestational-age (SGA) infants weighing about 2000 g. The latter were delivered with higher blood levels of both lactate and alanine than were term babies, and the levels fell more slowly. In both groups there was a fall in the level of ketones. In the control group the level started to rise after the sixth hour but no such rise was observed in the SGA group. It is important to note that controls were starved for 12 hr postnatally and then fed a 5% glucose solution, while the SGA group received this solution at the latest 6 hr after birth. Thus, starting from the sixth hour after delivery, the two groups are no longer comparable. It is, for instance, possible that the absence of a rise in ketones in the SGA group is due to the early feeding of glucose and later of formula. In the control group, the rise in ketones ceases in the twelfth hour, again at the moment when glucose is fed. It was demonstrated some time ago that feeding glucose to full-term newborn infants decreases their blood level of free fatty acids (Melichar and Novak, 1966) and it seems likely that the same mechanism is at work here.

In general, published data on newborn infants have to be carefully examined, since details of feeding procedures immediately after birth are frequently overlooked, both by the investigator and by the reader when results are compared and interpreted (for a discussion, see Hahn, 1975).

Having said the above, the conclusion of the authors is nevertheless justified (i.e., that the postnatal development of gluconeogenesis is delayed in the SGA group). In addition, it has also been shown that early feeding, particularly of prematures, prevents endogenous protein breakdown and enables the newborn to deal more effectively with its new environment (Melichar *et al.*, 1974).

These data are in good agreement with some recent work on newborn puppies (Adam *et al.*, 1975b; Chlebowski and Adam, 1975). Even in fed newborn puppies, 25% of the glucose used is recycled (i.e., gluconeogenesis is well in evidence). In the immediate postnatal period, glycogen is mobilized (by norepinephrine?) within 3 hr after birth, but only after mitochondrial CO_2 fixation had been induced (pyruvate carboxylase) together with gluconeogenesis (phosphoenolpyruvatecarboxykinase). Unfortunately, the authors determined PEPK in the whole homogenate. Hence, they could not distinguish between the mitochondrial and cytoplasmic enzymes, so the postnatal rise in activity appears only slight, although it is considerable for the cytoplasmic PEPK (Arinze, 1975).

It has also been shown for dogs that gluconeogenesis from glycerol is highest in fed newborns and decreases with age. However, fasting of the newborn causes a fall in the rate of glucose formation from glycerol (Hall *et al.*, 1976). What seems to emerge from recent studies is that in most mammalian species examined so far, there is a fall in the blood glucose level very soon after birth, which is accompanied by a rise in the level of glucagon and a fall in insulin (Table II). However, this does not hold true for the newborn lamb, in whom glucose and insulin blood levels rise postnatally.

Thus, three factors may be thought to play a decisive role immediately after and probably already during delivery: (1) deprivation of the constant food supply; (2) stress (which, in the lamb at least, also causes an increase in the blood level of triiodothyronin; Sack *et al.*, 1976); and (3) hypoxia.

Each of these factors separately would have similar effects: mobilization of liver (and other) glycogen and an increase in the rate of lipolysis, particularly in adipose tissue.

If we accept that the second messenger for these perinatal changes is cyclic AMP, formed from ATP by adenyl cyclase stimulated by glucagon (at least in the liver), then we must inquire what the stimulus for glucagon release is in the newborn. The least effective stimulus is the lower glucose level itself (Girard, 1975), while arginine (Sperling *et al.*, 1974) and alanine

(Chez *et al.*, 1974a) are more effective. Since amino acid levels are higher pre- than postnatally, and in fact decrease rapidly after birth, it seems unlikely that they are responsible for the rise in glucagon levels. It seems more probable that the catecholamines are the first messenger for glucagon release (via islet cell adenyl cyclase?), since they can cause elevated blood glucagon levels perinatally (Girard, 1975) and are known to inhibit insulin release. In addition, a very good stimulator of glucagon release is hypoxia (Girard, 1975—rat; Johnston *et al.*, 1972—man), which of course causes catecholamine release. Hypoxia, at least to some extent, does accompany most deliveries.

In summary, delivery causes metabolic changes that result in gluconeogenesis, lipolysis, and ketone formation, which in all mammalian newborns manifest themselves as a very rapid rise in the blood levels of fatty acids, ketones, and glycerol (Hahn and Koldovsky, 1966; Hahn, 1972) and a fall in the levels of glucose and amino acids (Girard *et al.*, 1975; Sperling *et al.*, 1974; Reisner *et al.*, 1973). These changes are accompanied by a rise in the blood level of glucagon and a fall in that of insulin and growth hormone (Table III).

4. Suckling Period

After the initial shock of encountering a new environment unfed, the newborn commences to adapt to a high-fat milk diet (Hahn and Koldovsky, 1966). Essentially, this diet is a continuation of the postnatal starvation period, since again fat is predominantly utilized (now, however, derived from the food instead of from the body), and glucose is synthesized since not enough is usually found in milk. However, in contrast to the neonatal period of fasting, proteins are now mainly used for growth.

An interesting difference may be noted between species that are born relatively mature with a large amount of fat in their body (man, 16% fat at birth), and species that are born very immature (rat, <2% fat at birth). The former can survive for a longer period of time, immediately after birth, on their own fat reserves than the latter. Hence, although both live off fat as soon as they are born, man can use his own fat for a long period of time, whereas the newborn rat very soon has to utilize the fat contained in its mother's milk. This is, of course, well borne out by the fact that newborn rats commence to suckle very soon after birth, much sooner than man usually does.

Hahn and Koldovsky (1966) calculated on the basis of data obtained on starving infant rats that most of the energy of these animals is obtained from fat and that milk protein is mainly used for growth. Even though this conclusion is more or less valid today, it has to be modified. It has been demonstrated that gluconeogenesis occurs at a very high rate soon after delivery (Girard *et al.*, 1973; Snell, 1975), and that even though lactate is the main source of the newly formed glucose, some is also derived from amino acids, particularly alanine. It has also been calculated by Kreutler-White and Miller (1976) that 25% of the daily caloric requirements in neonatal rats is met by dietary proteins. This seems an exaggerated figure and, in fact, recalculating their figures it appears that their artificially fed rats consumed 69 cal/100 g of body weight/day, which is considerably in excess of any published data (45–50 cal/100 g/day is the accepted figure). In addition, their experimental design makes it very difficult to compare neonates to older animals. Nevertheless, there is little doubt that gluconeogenesis from amino acids occurs in the neonatal period even in rats, as it certainly does in man [Melichar and Novak, 1966 (cited in Hahn and Koldovsky, 1966)]. However, there is a large difference between 3-day-old and 19-day-old animals. In the latter, glycogen formation from alanine is sixfold greater after a fast than in the former. It is important to note that urea formation is very low in neonatal rats, that the enzymes of the urea cycle only rise toward weaning (Snell, 1975), and that ammonia formation, although higher than in the adult, is equally high on day 20, when the rate of urea formation has increased considerably. In addition, cytoplasmic alanine transaminase activity is also low up to day 18 postnatally (Snell, 1975). Kreutler-White and Miller (1976) suggest that the unaccounted for nitrogen is used for nucleotide synthesis, the rate of which decreases with age (Roux, 1973). Even so, it seems well established that lactate is the main precursor for new glucose formation (Snell, 1975). Particularly in the newborn immediately after delivery, lactate levels are very high and decrease rapidly, faster than those of amino acids. Very possibly, in the rat at least, more lactate is available for gluconeogenesis, because mitochondrial decarboxylation of pyruvate (pyruvate dehydrogenase activity) occurs at a low rate (Bailey *et al.*, 1976; Knowles and Ballard, 1974), while pyruvate carboxylase activity rises rapidly after birth (Ballard and Hanson, 1967; Snell, 1975). If this is true, then fatty acids would supply the acetyl-CoA for the Krebs cycle and the pyruvate carboxylase would supply oxaloacetate for both the Krebs cycle and gluconeogenesis. This would explain the relatively high activity of pyruvate carboxylase very soon after birth.

Snell (1975) also showed that after perfusion of the liver with alanine, only 32% of the nitrogen could be recovered as urea plus ammonia in 10-day-old starved rats, while the figures were 65% and 74% for 20-day-old and adult rats, respectively. Thus, Snell concludes, as many did before him, that most of the alanine in suckling rats is used for protein synthesis.

Fat, as pointed out above, is the main energy source in the suckling period. It has several advantages over carbohydrate: (1) per unit volume, it contains more than twice the calories found in carbohydrate; (2) it is stored as such and not hydrated; and (3) it also is a good insulating material. In fact, there is a good correlation between the amount of fat in the milk of a species and the degree of maturity at birth; the more mature, the less fat. However, this is only valid if special demands are not made on the newborn as far as thermoregulation is concerned. Species that are born very mature and have to expend a large amount of their energy on keeping warm (seal, moose) have a high percentage of fat in their milk (see Hahn and Koldovsky, 1966). This, of course, again confirms the generalization that fat is the main energy substrate of the newborn mammal. The two species that consume relatively the largest amount of carbohydrate postnatally are man and horse. Very little is known about the horse's development. Yet the early postnatal period in man is metabolically very similar to that of other mammals (i.e., again fat utilization is high in the suckling period).

It must again be stressed, however, that the degree of maturity at birth, and consequently during the early postnatal period, is very important. Thus, a newborn infant can be fed a very-low-fat diet, rich in carbohydrate and protein, and grow normally (but see below). This does not seem to be possible in the rat, although it has really never been tried. Miller and Dymsza (1963) did feed suckling rats artificially, but not without fat. The reason for the need of fat in the rat may be bulk only. To obtain the same number of calories from carbohydrate that is obtained from fat, twice as much of the former is needed. Thus, the milk would be either hyperosmotic (i.e., more than 20% glucose), or nearly twice the volume of milk would have to be consumed, and this probably exceeds the capacity of the infant rat's gut. Things are much less critical in man, since being more mature at birth, the baby can probably handle such changes in food composition more adequately, and the ratio between energy consumed and the rate of growth is much more favorable than in the rat.

The high-fat diet (milk) of the suckling period results in high-fat utilization, not only in the form of fatty acids, but also in the form of ketones. In the rat this is reflected in a very high blood level of ketones (Hahn, 1972; Bailey and Lockwood, 1973) and a high utilization rate of these substances by the brain (e.g., Kraus *et al.*, 1974). Yeh and Zee (1976) suggest that the low insulin level found in suckling rats permits a high rate of ketogenesis, since injection of insulin with glucose decreased ketone blood levels. It is, of course, also possible that the high level of fat intake suppresses insulin secretion.

The high fat intake of suckling rats is reflected in high blood levels of triglycerides and a high level of lipoprotein lipase in adipose tissue. Both parameters decrease at weaning (Hahn, 1972; Hemon *et al.*, 1975).

Cholesterol blood levels are also high in the suckling period in all species examined (Carrol and Hamilton, 1973). In man, it is apparently possible to lower the blood cholesterol level of the newborn by feeding the mother diets rich in unsaturated fatty acids (Potter and Nestel, 1976).

4.1. Gastrointestinal Tract

Special mention should be made of the gastrointestinal tract, since it is the first to encounter any food consumed after birth and hence would be expected to react rapidly and extensively. Excellent reviews on its development have been published (Koldovsky, 1969, 1972). Surprisingly enough, however, very little work has been done on the metabolic development of the gut with regard to its utilization of the main nutrients (but see Hahn and Skala, 1971b). Much more is known about the digestive enzyme, which in the suckling period is geared to breakdown lactose (β-galactosidase). This shows very high activity postnatally but tends to disappear at weaning. Activity cannot be maintained, but the postweaning decrease can be somewhat delayed, by feeding a diet containing lactose (Koldovsky, 1969). Surprisingly, pancreatic and intestinal lipase activity is low in the suckling rat (Rokos *et al.*, 1963). Yet the rate of esterification of fatty acids is higher in suckling than adult rat gut and the same is true for the *in vitro* uptake of linoleic acid (Holtzapple, Smith, and Koldovsky, 1975). These authors suggest that perhaps there is uptake of triglycerides as such by the gut of the infant rat.

5. Weaning Period

In the rat, the weaning period has been defined as the time between the moment when the infant rat first consumes food other than breast milk (and can survive without the mother), and the moment when it no longer consumes any breast milk at all. It stretches between postnatal days 14 and 18 and day 30 (Hahn and Koldovsky, 1966; Babicky *et al.*, 1970; Krecek, 1963). It is much more difficult to define the weaning period in species that are born more mature, such as man. According to the definition above, weaning in man starts at birth or very soon after, and ends 1 week to 3 years or more later, depending on cultural and economic conditions.

The rat will again be taken as our example. An attempt will be made to compare it to the little that is known regarding the weaning period in man.

5.1. Weaning in the Laboratory Rat

The normal diet of the laboratory rat is Purina Chow or a diet close to it in composition. Essentially, this is a high-carbohydrate diet, so that as the rat is gradually weaned, it consumes more and more carbohydrate and less and less fat. This is reflected, as has been innumerably times described for adult animals (e.g., Tepperman, 1968), in adaptive metabolic changes that can be summarized as a decrease in the rate of gluconeogenesis (since sufficient glucose is consumed) and an increase in the rate of lipid synthesis. In other words, in some respects the metabolic pattern is now similar to what it was prenatally. There are, however, important differences, as is evident from Table IV. No malic enzyme is found in fetal livers and it starts to appear at the time of weaning. Apparently, the fetus can obtain sufficient NADPH from the pentose shunt and does not require malic enzyme, but this is no longer true for the weanling animal. The fetus also possesses hardly any alanine transaminase, which appears in the liver at the time of weaning.

Since in this chapter we are discussing developmental changes as related to nutrition, there is no space to go into changes in brain, muscle, and heart that are only indirectly related to nutrition. Suffice it to say that the rate of fatty acid synthesis in brain is highest in the suckling period, that it goes down between days 10 and 18, and that it does not seem to be affected by diet (for a review, see Hahn, 1972). Also, the rate of glycolysis, as measured by pyruvate kinase activity, increases steadily after birth in both muscle and brain, yet very little work has been done on the effect of nutrition on brain

Table IV. Rates of Enzyme Activities in the Liver of Fetal, Suckling, and Weaned Rat

	Period		
	Fetal	Suckling	Weaned
Phospheonelpyruvate carboxykinase Pyruvate carboxylase Fructose-diphosphatase Glucose-6-phosphatase Carnitine transferases Enzymes of ketone metabolism	Low or absent	High	Decrease
Pyruvate kinase Acetyl-CoA carboxylase Fatty acid synthetase Citrate cleavage enzyme Acetyl-CoA synthetase β-Methyl, β-hydroxyglutaryl-CoA reductase	High	Low	High
Alanine transaminase (cytoplasmic) Urea cycle enzymes	Absent	Low and rising	High
Malic enzyme	Absent	Absent, then rising	

and muscle enzyme development. There are indications (Hahn and Kirby, 1973) that diet can have an effect on brain in early postnatal life (see below). In fact, the effect of nutrition on muscle and brain metabolic development is a wide-open field. The only work in this area has been on very early undernutrition, which is a somewhat different approach.

5.2. Premature Weaning

In the rat it is possible to wean the young suddenly and prematurely between days 14 and 18. It should perhaps be pointed out parenthetically that the time of weaning in many laboratories is day 21, which of course is still premature weaning. Infant rats can be weaned to various diets with expected results.

Premature weaning to a high-carbohydrate diet will rapidly initiate metabolic changes adequate for the new situation (i.e., a decrease in gluconeogenesis and a rise in lipogenesis). It will accelerate considerably the normal course of events seen when rats are weaned naturally and gradually (Table IV).

Premature weaning to a high-fat diet, on the other hand, will maintain and prolong the state found in the suckling period (i.e., a high rate of gluconeogenesis and fatty acid oxidation and a low rate of fatty acid synthesis). A particularly striking demonstration of the effect of diet composition during weaning is the blood levels of cholesterol and glucagon and insulin. The blood level of cholesterol in all suckling mammals is high (see, e.g., Carrol and Hamilton, 1973). It rises after birth and decreases at weaning (for a review, see Hahn, 1970 a,b, 1972). If rats are weaned to a high-carbohydrate diet on day 18, their blood levels of cholesterol fall very rapidly, together with that of glucagon, while on a high-fat diet the blood cholesterol level stays high (although no cholesterol is contained in the diet) and the level of glucagon does not fall (Fig. 4) (Hahn and Koldovsky, 1976; Hahn *et al.*, 1977). Changes in the level of insulin are much less pronounced, but obviously there are profound changes in the insulin/glucagon molar ratio. Rats weaned to a high-fat diet on day 18 maintain a very high blood level of acetoacetate and a very high rate of hepatic ketone production, much higher than is found under similar circumstances in the adult (Hahn *et al.*, 1966). This difference in response between a newly weaned rat and an adult animal placed on a different diet is well demonstrated by an experiment reported by Hynie and Hahn (1972). They weaned rats prematurely on day 16 to a fat-free diet and determined acetyl-CoA carboxylase activity in the small intestine 5 days later. Table V shows that enzyme activity in the proximal and distal parts of the small gut increased about sevenfold in weanling rats. In adult animals only the proximal part of the gut reacted. Thus, this seems an example of the fact that when the young animal encounters the new diet for the first time, adaptation to the diet is different from that found in the adult.

5.3. Premature Weaning to a High-Protein Diet

Hahn and Koldovsky (1966) showed that weaning rats prematurely to a diet of protein only, but including vitamins and minerals, leads to their very rapid death. The same diet offered 12 days later will maintain these animals and will even permit them to grow. This was considered to be further evidence that protein utilization for energy purposes is not an important and possible function in the suckling and early weaning periods. Eighteen-day-old rats on

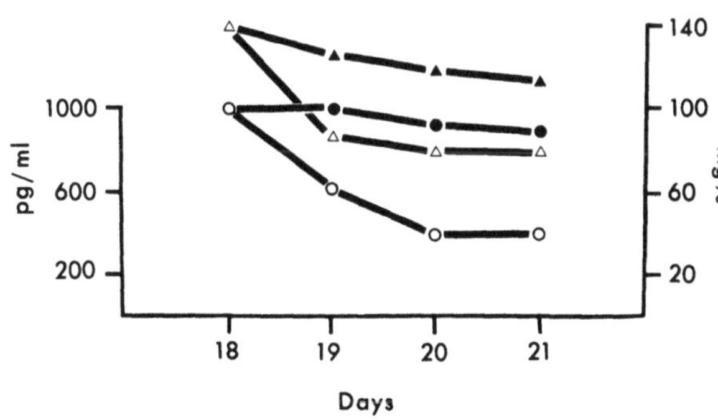

Fig. 4. Changes in the blood levels of glucagon and cholesterol at the time of premature weaning to a high-carbohydrate diet. *Abscissa:* Age in days. *Ordinate:* Left: glucagon (○), right: cholesterol (△). Black: rats remaining with their mother; white: rats weaned on day 18 to a high-carbohydrate diet. (Data from Hahn and Koldovsky, 1976, and Hahn *et al.*, 1977.)

Table V. Effect of Diet on AcetylCoA Carboxylase Activity[a] in the Proximal and Distal Parts of the Small Intestine of 21-Day-Old Rats and Adult Animals

		3-month-old		21-day-old	
	Diet	Purina Chow	5-day fat-free diet	With mother	5-day fat-free diet
Small intestine portion	Proximal	1.4 ± 2.6	34.1 ± 4.5	3.0 ± 0.4	22.3 ± 3.7
	Distal	1.3 ± 0.2	2.4 ± 0.4	2.4 ± 0.4	17.5 ± 3.8

[a] Activity in nmol/mg of protein/min. Note that activity in the distal portion is raised by the fat-free diet only in the weanling rats (Hynie and Hahn, 1972).

the high-protein diet probably died rapidly because (1) they could not produce carbohydrate in sufficient amounts from the protein consumed, and (2) their kidneys probably could not deal with the extra urea and ammonia.

6. Permanent Effects of Early Nutritional Changes

Permanent effects of early nutritional changes can be divided into those that are apparent immediately and those that are obvious only long after a dietary change has occurred.

1. The first group is well represented by two classical dietary diseases: undernutrition and overnutrition. The effects of undernutrition during pregnancy and in the perinatal period have been discussed and described repeatedly (for a review, see Brasel, 1974; Winick, 1975; Hahn, 1978). Much work has been done on experimental intrauterine growth retardation—with, from the metabolic point of view, rather disappointing results (Minkowski *et al.*, 1974; Hill, 1974). Most parameters examined in growth retarded newborn rats (e.g., enzyme activities were unaltered. The main finding was a delay in the return of blood glucose levels to normal. This is in sharp contrast to structural changes, particularly in the brain, caused admittedly by very severe pre- and postnatal undernutrition (Winick, 1975). Pre- and, particularly, postnatal overnutrition, on the other hand, does seem to lead to better defined metabolic alterations, again best defined in the rat. The method used to achieve overnutrition can be criticized, since it consists of comparing 13–15 newborns in one litter with 3 in another (Parkes, 1926; Kennedy, 1957). Hence, one might be justified in speaking of either overnutrition (3 rats) or undernutrition (14 rats), whatever your point of view when comparing one group with the other. What is "normal" remains for the investigator to decide. Nevertheless, it is quite clear that the small litter group grows faster, lays down more fat, ends up with a larger number of fat cells (Knittle and Hirsch, 1968), and shows a greater tendency toward diabetes. In fact, one can also cause accelerated growth and obesity in rats by pre- and postnatal injections of insulin (for a review, see Hahn and Novak, 1975).

2. The second group represents early effects that manifest themselves only later in life (Krecek, 1963). Many of these have been reviewed repeatedly (Hahn and Koldovsky, 1969; Hahn, 1978; Hahn and Novak, 1975). Here we want to discuss only one phenomena, the level of blood cholesterol in the rat. As already mentioned, premature weaning of rats to a high-carbohydrate diet results in a rapid fall in their level of blood cholesterol. If such prematurely weaned rats are compared with normally weaned rats at the age of 10 months and after 2 months of feeding them a high-cholesterol atherogenic diet, it is found that the prematurely weaned group has a significantly higher level of blood cholesterol than the normally weaned group (Kubat, 1966). Hahn and Koldovsky (1976) showed that there is an inverse relationship between the blood level of cholesterol in the weanling period and that in the adult rat fed a high-cholesterol diet for 2 months, or in male rats aged 6 months and fed a normal diet (Hahn and Kirby, 1973). This is similar to the finding of Reiser and Sidelman (1972), who showed that male rats whose blood level of cholesterol was decreased in the suckling period by feeding the mothers cholesterol-free diets showed a higher level of blood cholesterol 100 days later. In other words, both Reiser and Hahn and Koldovsky demonstrated that in the rat, at least, it is preferable to have a high level of blood cholesterol in the suckling period.

The mechanisms by which early metabolic changes cause permanent altered responses later in life are not known. It seems probable that control points are particularly affected (e.g., the maturation of the hypothalamus, together with the release of insulin and glucagon from the islets of Langerhans). On the whole, relationship between nutrition and hormonal control during development has not been explored extensively. For example on about day 18 postnatally, considerable changes occur in the levels of hormones in rat plasma: a fall in glucagon content (Girard, 1975) and a rise in corticosterone (Diez *et al.*, 1976) and insulin (Girard, 1975) contents. It is not at all clear how far these changes are related to the early weaning practiced in many laboratories and how they are affected by the subsequent diet. Hahn *et al.* (1977) have shown that, e.g., a high-fat diet will maintain glucagon content at high values usually found only in the suckling period. Whether such a diet would suppress the surge in corticosterone content remains to be determined.

In this connection it is of particular interest that the obesity induced by a high-fat diet in weanling hereditary obese mice develops without elevated insulin levels (Genuth, 1976). Perhaps the I/G molar ratio is increased instead.

7. Summary

In summary, the metabolic states of the various stages of development are closely related to the quality and quantity of nutrients consumed. This finding does not, in essence, differ from that found in the adult situation. What is different is the degree of maturity of some enzyme systems, tissues, and organs at any given moment. Also, it is becoming more and more apparent that the rates of hormone production and secretion and their control and the responsiveness of the target tissues (e.g., adenyl cyclase receptors) develop and may play a decisive role. Thus, in mammals at least, the simple idea that nutrients act directly on most cells must be discarded and should be replaced by the concept that, mostly, nutrients have immediate and long-term effects by acting on some selected endocrine cells, which then release their hormone, which then acts on other cells.

This concept might also explain the late effects of early adaptation to certain diets. A high-carbohydrate diet, for instance, fed to prematurely weaned animals may "fix" the response of glucagon and insulin and probably other hormones in a certain way which would not occur during natural weaning, and hence future responses may be conditioned by this early adaptation. This is not pure speculation, since it has been shown that a single injection of testosterone to 1- to 5-day-old female rats causes more or less permanent changes in the hypothalamus and the pituitary gonadotropin content later in life (for a review, see Hahn, 1978).

It seems to me that the elucidation of the mechanisms leading to such permanent changes is one of the main challenges of present-day nutrition.

8. References

Adam, P. A. J., Raiha, N., Rahiala, E.-L., and Kekomaki, M., 1975a, Oxidation of glucose and D-β-OH-butyrate by the early human fetal brain, *Acta Paediatr. Scand.* **64:**17.

Adam, P. A. J., Glazer, G., and Rogoff, F., 1975b, Glucose production in the newborn dog. I. Effects of glucagon *in vivo*, *Pediatr. Res.* **9:**816.

Alexander, D. P., Britton, H. G., Cohen, N. H., and Nixon, D. A., 1969, Foetal metabolism, in: *Foetal Autonomy* (G. E. W. Wolstenholme, ed.), pp. 95–112, Churchill Ltd., London.

Arinze, I. J., 1975, On the development of phosphoenolpyruvatecarboxykinase and gluconeogenesis in guinea pig liver, *Biochem. Biophys. Res. Commun.* **65:**184.

Asplund, K., 1972, Effects of postnatal feeding on the functional maturation of pancreatic islet B-cells of neonatal rats, *Diabetologia* **8:**153.

Augenfeld, J., and Fritz, I. B., 1970, Carnitine palmityl transferase activity and fatty acid oxidation by liver from foetal and neonatal rats, *Can. J. Biochem.* **48:**288.

Babicky, A., Ostadalova, I., Parizek, J., Kolar, J., and Bibr, B., 1970, Use of radioisotope techniques for determining the weaning period in experimental animals, *Physiol. Bohemoslov.* **19:**457.

Bailey, E., and Lockwood, E. A., 1973, Some aspects of fatty acid oxidation and ketone body formation and utilization during development of the rat, *Enzyme* **15:**239.

Bailey, K., Hahn, P., and Palaty, V., 1976, Pyruvate dehydrogenase activity in liver and brown fat of the development rat, *Can. J. Biochem.* **54:**534.

Ballard, F. J., and Hanson, R. W., 1967, Phosphoenolpyruvatecarboxykinase and pyruvate carboxylase in developing rat liver, *Biochem. J.* **105:**866.

Ballard, F. J., Hanson, R. W., and Kronfeld, D. S., 1969, Gluconeogenesis and lipogenesis in tissue from ruminant and nonruminant animals, *Fed. Proc.* **28:**218.

Bär, H. P. and Hahn, P., 1971, Development of rat liver adenylcyclase, *Can. J. Biochem.* **49:**85.

Basset, J. M., Thorburn, G. D. and Nicol, D. H., 1973, Regulation of insulin secretion in the ovine foetus *in utero*, *J. Endocrinol.* **56:**13.

Berger, R., and Hommes, F. A., 1975, Regulation of pyruvate metabolism in fetal rat liver, in: *Normal and Pathological Development of Energy Metabolism* (F. A. Hommes and C. J. Van den Berg, eds.), pp. 97–108, Academic Press, Inc., New York.

Blazquez, E., Lipshaw, L. A., Blazquez, M., and Foa, P. P., 1975, The synthesis and release of insulin in fetal, nursing and young adult rats: studies *in vivo* and *in vitro, Pediatr. Res.* **9**:17.

Blazquez, E., Rubalcava, B., Montesano, R., Orci, L., and Unger, R. H., 1976, Development of insulin and glucagon binding and the adenylate cyclase response in liver membranes of the prenatal, postnatal, and adult rat: Evidence of glucagon "resistance," *Endocrinology* **98**:1014.

Bloom, S. R., and Johnson, D. I., 1972, Failure of glucagon release in infants of diabetic mothers, *Br. Med. J.* **4**:453.

Bohmer, T., and Havel, R. J., 1975, Genesis of fatty liver and hyperlipemia in the fetal guinea pig, *J. Lipid Res.* **16**:454.

Boxer, J., Kirby, L. T., and Hahn, P., 1974, The response of glucose-6-phosphatase in human and rat fetal liver cultures to dibutyryl cyclic AMP, *Proc Soc. Exp. Biol. Med.* **145**:901.

Brasel, J. A., 1974, Cellular changes in intrauterine malnutrition, in *Nutrition and Fetal Development,* Vol. 2, *Current Concepts in Nutrition* (M. Winick, ed.), pp. 13–26, John Wiley & Sons, Inc., New York.

Brinster, R. L., 1971, Biochemistry of the early mammalian embryo, in *The Biochemistry of Development* (P. Benson and R. A. McCance, eds.), pp. 161–174, J. B. Lippincott Company, Philadelphia.

Carrol, K. K., and Hamilton, R. M. G., 1973, Plasma cholesterol levels in suckling and weaned calves, lambs, pigs and colts, *Lipids* **8**:635.

Charlton-Char, V., and Creasy, R. K., 1976, Acetate as a metabolic substrate in the fetal lamb, *Am. J. Physiol.* **230**:357.

Chez, R. A., Mintz, D. H., and Hutchinson, D. L., 1971, Effect of theophylline on glucagon and glucose mediated plasma insulin responses in subhuman primate fetuses and neonates, *Metabolism* **20**:805.

Chez, R. A., Mintz, D. H., and Epstein, M. F., 1974a, Fetal hormonal mechanisms for plasma glucose homeostasis in normal and glucose intolerant pregnancy, in: *Early Diabetes in Early Life* (R. A. Camarini-Davalos and H. S. Cole, eds.), pp. 141–163, Academic Press, Inc., New York.

Chez, R. A., Mintz, D. H., Epstein, M. F., Fleischman, A. R., Oakes, G. K., and Hutchinson, D. L., 1974b, Glucagon metabolism in nonhuman primate pregnancy, *Am. J. Obstet. Gynecol.* **120**:690.

Chlebowski, R. T., and Adam, P. A. J., 1975, Glucose production in the newborn dog. II, *Pediat Res.* **9**:821.

Clark, C. M., Beatty, B., and Allen, D. O., 1973, Evidence for delayed development of the glucagon receptor of adenylate cyclase in the fetal and neonatal heart, *J. Clin. invest.* **52**:1018.

Clinkenbeard, K. D., Reed, W. D., Mooney, R. A., and Lane, M. D., 1975, Intracellular localization of the 3-hydroxy-3-methyl-glutaryl coenzyme A cycle enzymes in liver, *J. Biol. Chem.* **250**:3108.

Dancis, J., Money, W. L., Springer, D., and Lecritz, M., 1968, Transport of aminoacids by placenta, *Am. J. Obstet. Gynecol.* **101**:820.

Dancis, D., Jansen, V., and Levitz, M., 1976, Transfer across perfused human placenta. IV. Effect of protein binding on free fatty acid, *Pediat. Res.* **10**:5.

Dawes, G. S., 1968, *Foetal and Neonatal Physiology,* Year Book Medical Publishers, Inc., Chicago.

Delamater, P. V., Sperling, M. A., Fiser, R. H., Phelps, D. L., Oh, W., and Fisher, D. H., 1974, Plasma alanine: Relation to plasma glucose, glucagon, and insulin in the neonate, *J. Pediatr.* **85**:702.

Diez, A. J., Sze, P. Y., and Ginsburg, B. E., 1976, Postnatal development of mouse plasma and brain corticosterone levels, *Endocrinology* **98**:1434.

Drahota, Z., Hahn, P., Kleinzeller, A., and Kostolanska, A., 1964, Acetoacetate formation by liver slices from adult and infant rats, *Biochem. J.* **93**:61.

Drahota, Z., Hahn, P., Mourek, J., and Trojanova, M., 1965, Effect of acetoacetate on oxygen consumption of brain slices from infant and adult rats, *Physiol. Bohemoslov.* **14**:134.

Dweck, H. S., and Cassody, G., 1974, Glucose intolerance in infants of very low birthweight, *Pediatrics* **53**:189.

Edmond, J., 1974, Ketone bodies as precursors of sterols and fatty acids in the developing rat, *J. Biol. Chem.* **249**:72.

Edson, J. L., Hudson, D. G., and Hull, D., 1975, Evidence for increased fatty acid transfer across the placenta during a maternal fast in rabbits, *Biol. Neonate* **27**:50.

Eisen, H. J., Glinsman, W. H., and Sherline, P., 1973a, Effect of insulin on glycogen synthesis in fetal rat liver organ culture, *Endocrinology* **92**:584.

Eisen, H. J., Goldfine, I. D., and Glinsman, W. H., 1973b, Regulation of hepatic glycogen synthesis during fetal development: roles of hydrocortisone, insulin and insulin receptors, *Proc. Natl. Acad. Sci. USA* **70**:3454.

Elphick, M. C., Hudson, D. G., and Hull, D., 1975, Transfer of fatty acids across the rabbit placenta, *J. Physiol.* **252**:29.

Feldman, D., 1974, Ontogeny of rat hepatic glucocorticoid receptors, *Endocrinology* **95**:1219.

Felix, J. M., Jacquot, R., and Sutter, B. C. J., 1969, Influence du jeûne les insulinémies maternelles et foetales chez le rat, *J. Physiol.* **61**(suppl. 1):129.

Fiser, R. H., Erenberg, A., Sperling, M. A., Oh, W., and Fisher, D. A., 1974a, Insulin-glucagon substrate interrelations in the fetal sheep, *Pediatr. Res.* **8**:951.

Fiser, R. H., Phelps, D. L., Williams, P. R., Sperling, M. A., Fisher, D. A., and Oh, W., 1974b, Insulin-glucagon substrate interrelationships in the neonatal sheep, *Am. J. Obstet. Gynecol.* **120**(7):944.

Fritz, I. B., 1967, Factors influencing the rate of long chain fatty acid oxidation and synthesis in mammalian systems, *Physiol. Rev.* **41**:52.

Frohlich, J., Hahn, P., Kirby, L., and Webber, W., 1976, Rat fetal brown adipose tissue *in vitro*: effects of hormones and ambient temperature, *Biol. Neonate* **30**:40.

Genuth, S. M., 1976, Effect of high fat vs. high carboyhdrate feeding on the development of obesity in weanling ob/ob mice, *Diabetologia* **12**:155.

Gilbert, M., and Vaillant, R., 1975, Contrôle de la synthèse glycogène dans le foie foetal de rat, *Biochimie* **57**:597.

Girard, J., 1975, Régulation du métabolisme énergétique pendant la période périnatale chez le rat, Ph.D. thesis, University of Paris.

Girard, J. R., and Marliss, E. G., 1975, Circulating fuels in late fetal and early neonatal life in the rat, in: *Early Diabetes in Early Life* (R. A. Camarini-Davalos and H. S. Cole, eds.), pp. 185–194, Academic Press, Inc., New York.

Girard, J. R., and Zeghal, N., 1975, Adrenal catecholamines content in fetal and newborn rats, *Biol. Neonate* **26**:205.

Girard, J. R., Cuendet, G. S., Marliss, E. B., Kervran, A., Rieutort, M., and Assan, R., 1973, Fuels, hormones and liver metabolism at term and during the early postnatal period in the rat, *J. Clin. Invest.* **52**:3190.

Girard, J. R., Kervran, A., Soufflet, E., and Assan, M., 1974, Factors affecting the secretion of insulin and glucagon by the rat fetus, *Diabetes* **23**:310.

Glinsman, W. H., Eisen, H. J., Lynch, A., and Chez, R. A., 1975, Glucose regulation by isolated near-term monkey liver, *Pediatr. Res.* **9**:600.

Goodrich, A. G., 1973, Regulation of fatty acid synthesis in isolated hepatocytes prepared from the livers of neonatal chicks, *J. Biol. Chem.* **248**:1924.

Gurdon, J. B., 1962, Adult frogs derived from the nuclei of single somatic cells, *Develop. Biol.* **4**:256.

Hahn, P., 1970a, Fetal and postnatal development of lipid and carbohydrate metabolism, in: *Fetal Growth and Development* (H. Waisman and J. Kerr, eds.), McGraw-Hill Book Company, New York.

Hahn, P., 1970b, Lipids, in: *Physiology of the Perinatal Period* (U. Stave, ed.), pp. 457–492, Appleton-Century-Crofts, New York.

Hahn, P., 1972, Lipid metabolism and nutrition in the prenatal and postnatal periods, in: *Nutrition and Development* (M. Winick, ed.), J. Wiley & Sons, Inc., New York.

Hahn, P., 1975, Nurture of the newborn, *New Engl. J. Med.* **292**:642–643.

Hahn, P., 1978, Nutrition of the newborn, in: *Perinatal Phsyiology* (U. Stave, ed.), pp. 397–423, Plenum Press, New York.

Hahn, P., and Kirby, L., 1973, Immediate and late effects of premature weaning and of feeding a high fat or high carbohydrate diet to weanling rats, *J. Nutr.* **103:**690.

Hahn, P., and Koldovsky, O., 1966, *Utilization of Nutrients during Postnatal Development,* Pergamon Press, Inc., Oxford.

Hahn, P., and Koldovsky, O., 1969, Development of metabolic processes and their adaptations during postnatal life, in: *Physiology and Pathology of Adaptation Mechanisms* (E. Bajusz, ed.), Pergamon Press, Inc., Oxford.

Hahn, P., and Koldovsky, O., 1976, Late effects of premature weaning on blood cholesterol levels in adult rats, *Nutr. Rep. Int.* **13:**87.

Hahn, P., and Novak, M., 1975, Development of brown and white adipose tissue, *J. Lipid Res.* **16:**79.

Hahn, P., and Skala, J., 1970, Some enzymes of glucose metabolism in the human fetus, *Biol. Neonate* **16:**362.

Hahn, P., and Skala, J., 1971a, Development of enzyme systems, *Clin. Obstet. Gynecol.* **14:**655.

Hahn, P., and Skala, J., 1971b, The development of some enzyme activities in the gut of the rat, *Biol. Neonate* **18:**433.

Hahn, P., and Skala, J., 1972, Carnitine and brown adipose tissue metabolism in the rat during development, *Biochem. J.* **127:**107.

Hahn, P., and Skala, J., 1973, Carnitine transferases in human fetal tissues, *Biol. Neonate* **22:**9.

Hahn, P., Vavrouskova, E., Jirasek, J., and Uher, J., 1964, Acetoacetate formation by livers from human fetuses aged 8–17 weeks, *Biol. Neonate* **7:**348.

Hahn, P., Drahota, Z., and Novak, M., 1966, Triglyceride and fatty acid metabolism in liver and adipose tissue of suckling rats, in: *Development of Metabolism as Related to Nutrition* (P. Hahn and O. Koldovsky, eds.) pp. 82–92, Karger, Basel.

Hahn, P., Girard, J., Assan, R., Frohlich, J., and Kervran, A., 1977, Control of blood cholesterol levels in suckling and weanling rats, *J. Nutr.* **107**(11):2062.

Hall, S. E., Hall, A. J., Layberry, R. A., Berman, M., and Hetenyi, Jr., G., 1976, Effect of age and fasting on gluconeogenesis from glycerol in dogs, *Am. J. Physiol.* **230:**362.

Hanson, R. W., Fisher, L., Ballard, F. J., and Reshef, L., 1973, The regulation of phosphoenolpyruvatecarboxykinase in fetal rat liver, *Enzyme* **15:**97.

Harding, P., 1971, The metabolism of brown and white adipose tissue in the fetus and newborn, *Clin. Obstet. Gynecol.* **14:**685.

Hardman, M. J., Hull, D., and Milner, A. D., 1971, Brown adipose tissue metabolism *in vivo* and serum insulin concentrations in rabbits soon after birth, *J. Physiol.* **213:**175.

Haymond, M. W., Karl, I. E., and Pagliari, A. S., 1974, Increased gluconeogenic substrates in the small-for-gestational-age infant, *New Engl. J. Med.* **291:**322.

Hemon, P., 1976, Some aspects of rat metabolism in the brown adipose tissue of normal and hypothyroid rats during early postnatal development, *Biol. Neonate* **28:**241.

Hemon, P., Ricquier, D., and Mory, G., 1975, The lipoprotein lipase activity of brown adipose tissue during early postnatal development of the normal and hypothyroid rat, *Horm. Metab. Res.* **7:**481.

Hers, H. G., 1976, The control of glycogen metabolism in the liver, *Biochem. Rev.* **56:**167.

Hill, D. E., 1974, Experimental growth retardation in rhesus monkeys, in: *Size at Birth* (E. Wolvestholme, ed.), CIBA Foundation Symposium 27, pp. 99–126, Elsevier, Excerpta Medica, Amsterdam.

Holtzapple, P. G., Smith, G., and Koldovsky, O., 1975, Uptake, activation, and esterification of fatty acids in the small intestine of the suckling rat, *Pediatr. Res.* **9:**786.

Hummel, L., Schirrmeister, W., and Zimmermann, T., 1975, Transfer of maternal plasma free fatty acids into the rat fetus, *Acta Biol. Med. Ger.* **34:**603.

Hynie, I., and Hahn, P., 1972, Changes in the activity of acetylCoA carboxylase in the intestinal mucosa of the rat during development, *J. Nutr.* **102:**1311.

Jacquot, R., and Kretchmer, N., 1964, Effect of fetal decapitation on enzymes of glycogen metabolism, *J. Biol. Chem.* **239:**1301.

Jakovcic, S., Haddock, J., Getz, G. S., Rabinowitz, M., and Swift, H., 1971, Mitochondrial development in liver of foetal and newborn rats, *Biochem. J.* **121:**341.

Johnston, D. I., Bloom, S. R., Greene, K. R., and Beard, R. W., 1972, Plasma pancreatic glucagon relationship between mother and foetus at term, *J. Endocrinol.* **55**:xxv.
Jones, C. T., and Firmin, W., 1976, Lipid synthesis *in vivo* by tissues of the maternal and foetal guinea pig, *Biochem. J.* **154**:159.
Jost, A., and Picon, L., 1970, Hormonal control of fetal development and metabolism, *Adv. Metab. Disord.* **4**:123.
Kennedy, G. C., 1957, The development with age of hypothalamic restraint upon the appetite of the rat, *J. Endocrinol.* **16**:9.
Kervran, A., and Girard, J. R., 1974, Glucose-induced increase of plasma insulin in the rat foetus *in utero, J. Endocrinol.* **62**:545.
Kirby, L., and Hahn, P., 1974, Enzyme response to prednisolone and dibutyryl adenosine 3′,5′-monophosphate in human fetal liver, *Pediat. Res.* **8**:37.
Knittle, J. L., and Hirsch, J., 1968, Effect of early nutrition on the development of rat epidydimal fat pads: Cellularity and metabolism, *J. Clin. Invest.* **47**:2091.
Knowles, S.J., and Ballard, F. J., 1974, Pyruvate dehydrogenase activity in rat liver during development, *Biol. Neonate* **24**:41.
Koldovsky, O., 1969, *Development of the Functions of the Small Intestine in Mammals and Man,* S. Karger, Basel.
Koldovsky, O., 1972, Hormonal and dietary factors in the development of digestion and absorption, in: *Nutrition and Development* (M. Winick, ed.), J. Wiley & Sons, Inc., New York.
Kraus, H., Schlenker, S., and Schwedesky, D., 1974, Developmental changes of cerebral ketone body utilization in human infants, *Hoppe-Seyler's Z. Physiol. Chem.* **355**:164.
Krecek, J., 1963, Premature weaning in the rat, *Cesk. Fysiol.* **12**:347.
Kreutler-White, P., and Miller, S. A., 1976, Utilization of dietary amino acids for energy production in neonatal rat liver, *Pediatr. Res.* **10**:158.
Kubat, K., cited in Hahn and Koldovsky (1966).
Lechtig, A., Uarbrough, C., Delgado, H., Habicht, J.-P., Martorell, R., and Klein, R. E., 1975, Influence of maternal nutrition on birth weight, *Am. J. Clin. Nutr.* **28**:1223.
Lee, L. P. K., and Fritz, I. B., 1971, Hepatic ketogenesis during development, *Can. J. Biochem.* **49**:599.
Lifrak, I. L., Lev, R., and Loud, A. V., 1976, Substrate induced acceleration of lactose synthesis in fetal rat intestine, *Pediatr. Res.* **10**:100.
Lindblad, R., 1974, Free amino acid levels in venous plasma, in: *Size at Birth* (E. Wolvestholme, ed.), CIBA Foundation Symposium 27, p. 111, Elsevier/Excerpta Medica, Amsterdam.
Manns, J. G., and Brockman, R. P., 1969, The role of insulin in the synthesis of fetal glycogen, *Can. J. Physiol. Pharmacol.* **47**:917.
Melichar, V., Razova, M., Janovsky, M., and Polacek, K., 1974, Nitrogen balance in low birth weight newborns during the adaptation period of life, *Physiol. Bohemoslov.* **23**:161.
Miguel, S., and Abraham, S., 1976, Effect of maternal diet on fetal hepatic lipogenesis, *Biochim. Biophys. Acta* **424**:213.
Miller, S. A., and Dymsza, H. A., 1963, Artificial feeding of neonatal rats, *Sciences* **141**:517.
Milner, R. D. G., 1969, Plasma and tissue insulin concentrations in foetal and postnatal rabbits, *J. Endocrinol.* **43**:119.
Minkowski, A. Roux, J., and Tordet-Caridroit, C., 1974, Pathophysiological changes in intrauterine malnutrition, in: *Nutrition and Fetal Development* (M. Winick, ed.) Vol. 2, pp. 45–78, John Wiley & Sons, Inc., New York.
Nakai, Hayashi, M., Kanazawa, Y., Kosaka, K., Kigawa, T., and Sakamoto, S., 1976, Alterations of insulin-secreting response to glucose in human infants during the early postnatal period, *Endocrinol. Jpn.* **23**:61.
Novak, E., Drummond, G. I., Skala, J., and Hahn, P., 1972, Developmental changes in cyclic AMP, proteinkinase, phosphorylase kinase and phosphorylase in liver, heart, and skeletal muscle of the rat, *Arch. Biochem. Biophys.* **150**:511.
Parkes, A. S., 1926, The growth of young mice according to the size of the litter, *Arm. Appl. Biol.* **13**:374.

Partridge, N. C., Hoh, C. H., Weaver, P. K., and Oliver, I. T., 1975, Premature induction of glucokinase in the neonatal rat by thyroid hormone, *Eur. J. Biochem.* **51**:49–54.

Picon, L., 1971, Insulin and fetal growth in the rat, in: *Hormones in Development* (M. Hamburgh, ed.), Appleton-Century-Crofts, New York.

Pines, M., Bashan, N., and Moses, S. W., 1976, Glucose effect on glycogen synthetase and phosphorylase in fetal rat liver, *FEBS Lett.* **62**:301.

Plas, C., and Nunez, J., 1976, Role of cortisone on the glycogenolytic effect of glucagon and on the glycogenic response to insulin in fetal hepatocyte culture, *J. Biol. Chem.* **251**:1431.

Pollak, J. K., and Duck-Chong, C. G., 1973, Changes in rat liver mitochondria and endoplasmic reticulum during development and differentiation, *Enzyme* **15**:139.

Potter, J. M., and Nestel, P. J., 1976, The effects of dietary fatty acids and cholesterol on the milk lipids of lactating women and the plasma cholesterol of breast-fed infants, *Am. J. Clin. Nutr.* **29**:54.

Rabain, F., and Picon, L., 1974, Effect of insulin on the maternofetal transfer of glucose in the rat, *Horm. Metab. Res.* **6**:5.

Reiser, R., and Sidelman, Z., 1972, Control of serum cholesterol homeostasis by cholesterol in the milk of the suckling rat, *J. Nutr.* **102**:1009.

Reisner, S. H., Aranda, J. V., Colle, E., Papageorgiou, A., Schiff, D., Scriver, C. R., and Stern, L., 1973, The effect of intravenous glucagon on plasma amino acids in the newborn, *Pediatr. Res.* **7**:184.

Robinson, B. H., 1976, Development of gluconeogenic enzymes in the newborn guinea pig, *Biol. Neonate* **29**:48.

Rokos, J., Hahn, P., Koldovsky, O., and Prochazha, P., 1963, The postnatal development of lipolytic activity in the pancreas and small intestine of the rat, *Physiol. Bohemoslov.* **12**:213.

Roux, J. F., 1973, Nucleotide supply of the developing animal, *Enzyme* **15**:361.

Roux, J. F., and Myers, R., 1974, *In vitro* metabolism of palmitic acid and glucose in the developing tissue of the rhesus monkey, *J. Obstet. Gynecol.* **118**:385.

Sabata, V., Hahn, P., and Drahota, Z., 1967, The role of glucose and of ketone-substances in the metabolism of foetuses of mothers suffering from diabetes, in: *Intrauterine Dangers to the Fetus* (J. Horsky and Z. Stembera, eds.), Elsevier/Excerpta Medica, Amsterdam.

Sack J., Beaudry, M., DeLamater, P. V., Oh, W., and Fisher, D. A., 1976. Umbilical cord cutting triggers hypertriiodothyroninemia and nonshivering thermogenesis in the newborn lamb, *Pediatr. Res.* **10**:169.

Schwartz, A. L., and Rall, T. W., 1975, Hormonal regulation of metabolism in human fetal liver. II. Regulation of glycogen synthase activity, *Diabetes* **24**:1113.

Schwartz, A. L., Raiha, N. C. R., and Rall, T. W., 1975, Hormonal regulation of glycogen metabolism in human fetal liver. I. Normal development and effects of dibutyryl cyclic AMP, glucagon and insulin in liver explants, *Diabetes* **24**:1101.

Schwartz, R., 1975, Islet responsiveness of the human fetus in utero, in: *Early Diabetes in Early Life* (R. A. Camarini-Davalos and H. S. Cole, eds.), pp. 127–134, Academic Press, Inc., New York.

Scow, R. D., Chernick, S. S., and Brinley, M. S., 1964, Hyperlipemia and ketosis in the pregnant rat, *Am. J. Physiol.* **206**:796.

Shelley, H. J., 1969, Carbohydrate metabolism in the foetus and the newly born, *Proc. Nutr. Soc.* **28**:42.

Smith, S., and Abraham, S., 1970, Fatty acid synthesis in developing mouse liver, *Arch. Biochem. Biophys.* **136**:112.

Snell, K., 1975, Gluconeogenesis in the neonatal rat: the metabolism and disposition of alanine during postnatal development, in: *Normal and Pathological Development of Energy Metabolism* (F. A. Hommes and C. J. Van den Berg, eds.), pp. 97–108, Academic Press, Inc., New York.

Snyderman, S. E., 1970, Protein and amino acid metabolism, in: *Physiology of the Perinatal Period* (U. Stave, ed.), pp. 441–456, Appleton-Century-Crofts, New York.

Sparks, J. W., Lynch, A., Chez, R. A., and Glinsman, W. H., 1976, Glycogen regulation in isolated perfused near-term monkey liver, *Pediatr. Res.* **10**:51.

Spellacy, W. N., and Buhi, W. C., 1976, Glucagon, insulin and glucose levels in maternal and umbilical cord plasma with studies of placental transfer, *Obstet. Gynecol.* **47**:291.
Spellacy, W. N., Buhi, W. C., Bradley, B., and Holsinger, K. K., 1973, Maternal, fetal and amniotic fluid levels of glucose, insulin and growth hormone, *Obstet. Gynecol.* **41**:323.
Sperling, M. A., DeLamater, P. V., Phelps, D., Fiser, R. H., Oh, W., and Fisher, D. A., 1974, Spontaneous and amino acid stimulated glucagon secretion in the immediate postnatal period, *J. Clin. Invest.* **53**:1159.
Stave, U., 1975, Perinatal changes of interorgan differences in cell metabolism, *Biol. Neonate* **26**:318.
Strosser, M. T., and Mialhe, P., 1975, Growth hormone secretion in the rat as a function of age, *Horm. Metab. Res.* **7**:275.
Sveger, T., Lindberg, T., Weibull, B., and Olsson, U. L., 1975, Nutrition overnutrition and obesity in the first year of life in Malmo, Sweden, *Acta Paediatr. Scand.* **64**:635.
Tepperman, J., 1968, *Metabolic and Endocrine Physiology,* Year Book Medical Publishers, Inc., Chicago.
Unger, R. H., 1974, Alpha and beta cell interrelationships in health and disease, *Metabolism* **23**:581.
Walker, D. G., 1971, Development of enzymes for carbohydrate metabolism, in: *The Biochemistry of Development* (P. F. Benson and R. A. McCance, eds.), J. B. Lippincott Company, Philadelphia.
Walker, D. G., and Eaton, S. W., 1967, Regulation of development of hepatic glucokinase in the neonatal rat by diet, *Biochem. J.* **105**:771.
Wicks, W. D., 1974, The mode of action of glucocorticoids, *Biochem. Horm.* **8**:212.
Wicks, W. D., Lewis, W., and McKibbin, J. B., 1972, Induction of phosphoenolpyruvatecarboxykinase by N^6, O^2 dibutyryl cyclic AMP in rat liver, *Biochim. Biophys. Acta* **264**:177.
Willes, R. F., Boda, J. M., and Manns, J., 1969, Insulin secretion by the ovine fetus *in utero*, *Endocrinology* **84**:520.
Winick, M., 1974, Maternal nutrition and intrauterine growth failure, in: *Modern Problems in Paediatrics, Proceedings,* Vol. 14 (F. Falkner *et al.*, eds.), p. 48, Phiebig.
Yeh, Y., and Zee, P., 1976, Insulin, a possible regulator of ketosis in newborn and suckling rats, *Pediatr. Res.* **10**:192.
Yeung, D., and Oliver, I. T., 1968a, Induction of phosphopyruvatecarboxylase in neonatal rat liver by adenosine 3′,5′-cyclic monophosphate, *Biochemistry* **7**:3231.
Yeung, D., and Oliver, I. T., 1968b, Factors affecting the premature induction of phosphoenolpyruvatecarboxylase in neonatal rat liver, *Biochem. J.* **108**:325.

9. Recommended Reading

Benson, P. F., and McCance, R. A. (eds.), 1971, *The Biochemistry of Development,* J. B. Lippincott Company, Philadelphia.
Camerini-Davalos, R. A., and Cole, H. S. (eds.), 1975, *Early Diabetes in Early Life,* Academic Press, Inc., New York.
Fomon, S. J., 1974, *Infant Nutrition,* W. B. Saunders Company, Philadelphia.
Hodari, A. A., and Mariona, F. G. (eds.), 1972, *Physiological Biochemistry of the Fetus,* Charles C. Thomas, Publisher, Springfield, Ill.
Hommes, F. A., and Van den Berg, C. J. (eds.), 1975, *Normal and Pathological Development of Energy Metabolism,* Academic Press, Inc., New York.
Shafrir, E (ed.), 1975, *Contemporary Topics in the Study of Diabetes and Metabolic Endocrinology,* Academic Press, Inc., New York.
Smith, C. A., and Nelson, M. N., 1976, *The Physiology of the Newborn Infant,* Charles C Thomas, Publisher, Springfield, Ill.

Stave, U. (ed.), 1972, *Physiology of the Perinatal Period,* Vols. I and II, Appleton-Century-Crofts, New York.
Winick, M. (ed.), 1972, *Nutrition and Development,* John Wiley & Sons, Inc., New York.
Winick, M. (ed.), 1974, *Nutrition and Fetal Development,* John Wiley & Sons, Inc., New York.

2

Malnutrition and Mental Development

Myron Winick

1. Introduction

Over the past 15 years, several studies involving different scientific disciplines have pointed to an association between malnutrition during a critical period of development and permanent changes in brain function. This association has been described in complex human situations that do not permit isolation of any single causal factor. Malnutrition usually occurs in a milieu where low socioeconomic status, limited education, poor sanitary conditions, and recurrent infections are common. In an attempt to isolate the nutrition factor and to explore the interrelations between malnutrition and other aspects of this environment, a number of epidemiological surveys of human populations have been made, and animal models have been sought. The human studies have the obvious advantage of immediate relevance. In addition, psychological testing of children is highly developed and well standardized and there is voluminous literature on the interpretation of many psychological tests. But human studies suffer from the same limitations in interpretation as do animal studies when the testing is done on very young infants. The younger the child, the more difficult it is to use the results of any test as an index of future performance. As we shall see, this has restricted studies of infant malnutrition; only recently have attempts been made to correlate the results of behavioral measurements made in infants below 2 years of age with tests performed at a later age.

In comparison with field studies of humans, animal models have several advantages. The diet can be closely controlled and frequently analyzed to ensure uniform composition. The environment can be regulated so that, except for the nutritional variable, conditions are nearly the same for experimental and control animals. Ethical considerations do not prevent the study of ex-

Myron Winick • Institute of Human Nutrition, Columbia University College of Physicians and Surgeons, New York, New York.

treme or long-term effects and nutritional deprivation of various degrees can be maintained for prolonged periods. Perhaps the greatest advantage of using animal models is the opportunity to correlate behavioral changes with neurophysiologic changes and with histological, histochemical, and biochemical changes in the central nervous system

One of the disadvantages of animal models is the difficulty in interpreting what a given behavior really means.

Another problem in interpreting animal behavior derives from what was thought to be the chief advantage of using animals, the ability to control the environment except for the nutritional variable. While this may appear possible theoretically, in practice it has never been achieved. For example, producing malnutrition in neonatal rats by allowing 18 pups to suckle from a single mother certainly changes the pups' environment in many ways other than simply restricting their food intake. Malnourishing the mother and having her nurse a normal-size litter does not solve this problem, since a malnourished mother is apt to care for her pups quite differently from a well-nourished mother.

Since alterations of the neonatal environment, especially the interaction of mother and pups, will cause behavioral changes quite independent of the state of nutrition, the "pure neonatal malnutrition" experiment has not been done. Even with prenatal malnutrition, where this is less of a problem, maternal behavior may influence fetal development independent of maternal nutritional status.

Thus, neither animal nor human studies have been able to answer the crucial questions in this field fully. Does malnutrition per se during critical periods of development permanently alter behavior? If so, in what way? Is the ability to learn properly lost and, if so, is it a permanent loss?

2. *Animal Studies*

2.1. *Malnutrition and Learning*

Early research on rats was directed toward the effects of malnutrition on learning ability either during the period of malnutrition or after refeeding (Cowley and Griesel, 1959, 1963, 1964, 1966). The standard methods employed were all based on performance in one or another type of maze as a test of the animals' ability to learn. With the Hebb-Williams maze, for instance, the animal being tested was permitted only one trial after an initial period to familiarize itself with the chamber. The malnourished animals negotiated the maze less well than did the well-fed animals. This inability to perform in the maze persisted to adulthood even if the animal was malnourished only for a brief period after birth. Because the use of only one trial was criticized as a poor test of learning, these experiments were repeated with multiple trials (Levitsky and Barnes, 1970). Again both the malnourished and the previously malnourished animals performed poorly.

A major criticism of both of these studies is that a food reward was used as an incentive for negotiating the maze (Barnes, 1967). Certainly malnourished animals will respond differently to such an incentive, and this altered response to food may persist for long periods after the animal has been rehabilitated. Thus, it is impossible to conclude from these experiments that malnutrition early in life impairs learning.

In order to remove food as an incentive, other types of mazes that do not require a food incentive, such as water mazes, were employed (Barnes *et al.*, 1966; Kerr and Waisman, 1970). These studies were all based on some sort of stressful situation to initiate the animal to negotiate the maze. Again the malnourished and the previously malnourished animals did poorly, but again learning ability per se could not be implicated. Malnourished animals react differently to almost all types of stress, and this altered response may in itself make the negotiation of the maze more difficult. Other types of studies to measure learning, such as the standard bar-pressing experiment with all its variations, have been employed. However, either a reward (usually food) or a punishment (usually electric shock) was used to "motivate learning," and the same criticism can be raised. The problem is that to measure the ability of an animal to learn, you must get him to do something; what you actually measure is his ability to perform. This may or may not be impaired because of a learning deficit.

The early animal experiments on malnutrition, then, highlighted two areas of difficulty. The first was that although several types of behavior could be measured in a variety of animal species, the meaning of given behavioral changes was difficult to understand. The second was the isolation of malnutrition as the only variable. These are exactly the same two problems which face researchers studying the effects of malnutrition on subsequent behavior in human populations.

2.2. Malnutrition and "Emotionality"

Classically, "emotionality" in rats has been tested by simply observing them in an open field. A rat is placed in a box with squares outlined on its floor. He is observed for a given period of time and the number of squares traversed (horizontal movement), as well as the number of times the animal rears up on his hind limbs (vertical movement), are recorded. The number of times the animal urinates and defecates is also recorded. Variations of this open-field technique have also been employed in which a "novel stimulus," such as a rubber ball on a pendulum or a loud noise, is introduced. The advantage of this experimental design is that the animal is neither rewarded nor punished—his natural behavior is simply observed.

When rats were malnourished during gestation and lactation, during lactation and for a period after weaning, or only during the postweaning period, and then tested as adults, they showed decreased locomotor activity (horizontal movement); decreased rearing, head raising, and pivoting (vertical activity); and increased excretion of urine and feces. In addition, the adult rats remained at the periphery of the field, seldom venturing into the central squares (Cowley and Griesel, 1959, 1964; Simonson *et al.*, 1971; Lat *et al.*, 1961; Altman *et al.*, 1971; Levitsky and Barnes, 1970; Frankova and Barnes, 1968; Guthrie, 1968; Barnett *et al.*, 1971). This behavior has been interpreted by some investigators as increased emotionality, and by others, perhaps more accurately, as decreased exploratory activity. All the changes described could be markedly intensified by introducing a loud noise during the period of observation (Cowley and Griesel, 1964; Levitsky and Barnes, 1970). Similar observations have been made in pigs tested in a modified open field after recovery from total food restriction or from a low-protein diet during the first 11 weeks of life (Barnes *et al.*, 1970). Nonhuman primates fed a low-protein diet played significantly less and showed less sexual behavior, less grooming, and more aggressive behavior than animals that had been fed a high-protein diet (Zimmermann *et al.*, 1972). Monkeys on a low-protein diet also showed less curiosity and puzzle-solving activity. It is interesting that as soon as a food reward was introduced, the monkeys became more curious and solved puzzles as well as the controls. When the reward was withdrawn, they returned to their apathetic state. In tests of social dominance, monkeys on a low-protein diet were submissive and always dominated by the control animals (Wise, unpublished master's thesis, University of Montana). Again, when food was introduced as a reward, their behavior changed and they became more aggressive and domineering but returned to their passive state when the reward was no longer offered.

These studies have led investigators to conclude that the primary behavioral abnormality induced by early malnutrition is a breakdown in attention or observation. Many workers have shown that an appropriate observing response must develop in the monkey before he can learn to solve a discrimination problem (Zimmermann *et al.*, 1972; Wise, unpublished master's thesis; Strobel and Zimmermann, 1971). It would appear that animals subjected to a poor diet in early life do not develop this response adequately.

Whether this deficiency persists throughout life is not yet known for the primate. In the rat and pig, similar deficiencies have been shown to persist. In human adults recovering from malnutrition, attention span improves as the preoccupation with food declines. On the other hand, similar behavioral abnormalities have been shown to persist in infants who have recovered from previous malnutrition.

2.3. *Malnutrition and Other Behaviors*

The development of certain reflexes (i.e., startle, grasp, visual placing) and certain physical characteristics (eye opening, incisor eruption) is delayed in malnourished animals. This retardation will naturally result in delayed development of behavior, which depends on the maturation of these reflexes and physical characteristics, and it is assuming more and more importance in our understanding of the behavioral consequences of early malnutrition. The developmental process is one of continuing interaction between an organism and his environment at critical times. If the timing of this interaction is disturbed, profound behavioral abnormalities may occur. Thus, if an animal cannot receive visual stimuli at the proper time because his eyes have not yet opened, his entire subsequent development may be altered. This concept extends not only to alterations in the appearance of the physical abilities necessary to receive environmental stimuli, but also to the institution of behavioral changes, such as those just described in motivation and attention, which, in turn, preclude proper integration of necessary environmental input. Thus, those holding this view on the mechanism by which malnutrition alters behavior would argue that there are three elements involved in determining behavior: the infant, the environment, and the time when the two interact.

2.4. *The Early Environment and Behavior*

A large body of literature (which we will only touch on) demonstrates that in a variety of animal species, profound behavioral changes will occur if the early environment is disturbed. Moreover, many of these changes persist into adult life. A "stimulatory environment" for rats has been created by frequent handling of the animals during early life, by electric shock, and by exposure to brief periods of cold from birth to 21 days of age (Levine and Denenberg, 1969). A severely deprived environment has been produced by rearing rats in the dark in a soundproof room and in single cages. Less severe deprivation has been applied to monkeys by isolating the young from the mother and feeding them by simply placing a bottle in the cage.

Stimulation of well-nourished rats decreases emotionality, increases exploratory behavior, and decreases reactions to adverse stimuli. Isolation of well-nourished rats causes the reverse. Thus, the well-nourished but isolated rat shows behavioral characteristics quite similar to the rat malnourished in early life. Isolation of monkeys produces a bizarre behavioral pattern which includes apathy, decreased exploratory behavior, withdrawal from the environment, and heightened emotional responses when confronted with stress. Again, we see behavior in the well-fed isolated animals that is similar to the behavior of animals that were poorly fed in early life.

The similarity between the effects of early malnutrition and early isolation becomes even more striking if certain physiological and biochemical data are examined. In rats, isolation will reduce the rate of cell division, the rate of myelination, and the number of dendritic arborizations and will elevate the activity of acetylcholinesterase. Stimulation produces the reverse. The biochemical changes associated with isolation are like those found in the brains of animals subjected to early malnutrition. This suggests not only that there is an obvious need for strict environmental control in experiments on malnutrition, but also that environmental deprivation and malnutrition may interact during the early postnatal period.

One approach to producing early malnutrition in the rat, as we have seen, has been to increase litter size during lactation. By increasing litter size, however, one not only reduces the amount of milk for each pup but also increases the amount of sibling stimulation and decreases the amount of attention given by the dam to each pup (Seitz, 1954). Exploring this problem, Frankova (1974) reported that there is considerable behavioral difference among pups from litters of different size. Exploratory behavior and spontaneous activity in the open field proved to be greatest in intermediate litter sizes (9–13 pups), less in the smallest (4 pups), and lowest in the largest litters (17 pups). There was also an inverse relationship between body weight and litter size (the smallest body weights were found in the largest litters). Frankova concluded that it is the interaction between sibling stimulation and neonatal nutrition that determines exploratory behavior and "emotionality" in the adult.

Levitsky and Barnes have given this hypothesis further support in experiments with rats that were either well nourished or protein-malnourished for 7 weeks and were then subjected to environmental stimulation, isolation, or normal social conditions (Levitsky and Barnes, 1972). Behavioral testing after refeeding revealed a highly significant interaction between isolation and early malnutrition with regard to locomotor activity and exploratory behavior. The isolated malnourished group showed much less of both than either the well-nourished or the other malnourished group. Environmental stimulation seemed to compensate for early malnutrition, as the stimulated malnourished group was rated very close to all the well-nourished animals in locomotor and exploratory behavior. Cines and her colleagues have reported similar results with stimulated and nonstimulated rats undernourished for the first 3 weeks of life and subsequently rehabilitated (Cines, 1972). Studying the same rats, Coombs and his colleagues reported that the biochemical abnormalities of the brain usually produced by neonatal undernutrition did not fully appear when stimulation was introduced (J. Coombs, personal communication). These studies demonstrate that some of the effects of malnutrition on development of the central nervous system of the rat can be reversed by enriching the animal's early environment.

More recently, Frankova has attempted a new means of increasing stimulation: introducing a trained virgin female as an "aunt" into the cage with the mother and pups from 8 A.M. to 4 P.M. (Frankova, 1974). The aunt assists the mother in caring for the pups, retrieves them, grooms them, and generally provides a heightened level of stimulation. The result was an increased interaction between individual pups and more interaction between the pups and their mother when the aunt was present. This form of stimulation changes the mother's as well as the pups' behavior, and the improvement was more marked in malnourished than in well-fed animals.

Levitsky and Barnes have suggested that malnutrition, by producing apathy and decreased curiosity may make the animal less susceptible to environmental programming (Levitsky and Barnes, 1970). Malnutrition may therefore indirectly prevent the proper stimulus from arriving at the proper time and, since experience is cumulative, the animal, unable to build on previous experience, would be retarded in development. Malnutrition reduces the range and amount of information available to the developing animal. Having been exposed to fewer earlier experiences, the adult animal is limited in his ability to cope with a normal variety of environmental stimuli. This limitation may, in turn, inhibit the processes of mental development and socialization.

It should be pointed out that although this explanation—which proposes a "final common pathway," isolation from the environment, as the actual cause of the behavioral abnormalities seen in malnutrition—is attractive, it is not necessarily correct. An alternative explanation is possible if one postulates that the brain has a limited number of responses which it makes to alterations in the environment. Thus, malnutrition and isolation may affect the brain quite independently and through different mechanisms. The response is similar, however, and the behavioral manifestations of that response are indistinguishable.

In general, then, it can be said that the animal experiments have not proved an association between early malnutrition and learning. They have shown that animals malnourished in a variety of ways, both prenatally and postnatally, show behavioral abnormalities best described as increased emotionality and decreased exploratory activity. This behavioral pattern persists even after rehabilitation. Similar behavioral abnormalities have been induced in young animals by isolating them from their environment, and partial reversal of the effects of malnutrition is possible if the environment is enriched. After more than two decades of experiments, the basic problems are still present. The animal experiments have certainly not solved the question of early malnutrition and later mental development, nor in my opinion can they ever solve this problem until the actual biochemical and neurophysiological mechanisms controlling specific behaviors are understood and the effects of malnutrition on these mechanisms studied. Thus, the importance of animal behavior work in the future will be directly proportional to its use in deriving a better understanding of the basic mechanisms controlling behavior. As part of this approach, the effects of early malnutrition may be better understood, but, perhaps more important, early malnutrition may provide a useful experimental model by which this mechanistic approach can be facilitated.

3. Human Studies

Investigators in several countries have tried to examine the effects of malnutrition early in life on subsequent human behavior. Most of these studies have focused on measuring intelligence, because testing procedures are readily available and because the demonstration of persistent intellectual deficits would have immediate social impact. Rather than attempting to review all such studies, I will group them into several categories and discuss representative studies in each category in detail.

3.1. Early Malnutrition and Intelligence

3.1.1. Malnutrition in Deprived Populations

The intellectual development of children severely malnourished as infants has been studied both retrospectively (Cabak and Najdanvic, 1965; Chase, 1969; Garrow and Pike, 1967; Graham, 1967) and prospectively (Cravioto and Robles, 1965; Cravioto *et al.*, 1966, 1967a; Kugelmass *et al.*, 1944). In both types of studies, however, it is difficult to isolate malnutrition as the cause of any mental deficiencies found, since the malnourished children invariably come from a lower socioeconomic class and a generally more deprived environment than even the most carefully matched control groups. Both types of studies also suffer from the lack of standardization of intelligence tests. It has been pointed out many times before that tests developed in industrialized nations may have little meaning in developing countries with very different cultures.

In addition, with retrospective studies one can never be sure of the criteria used to establish the diagnosis of malnutrition or of what other social factors may have affected the development of the malnourished children.

In a retrospective analysis, Cabak and Najdanvic demonstrated that Serbian children with a history of marasmus had significantly lower intelligence quotients than do Serbian children in general (Cabak and Najdanvic, 1965). They made no real attempt to control other environmental factors but selected individuals of the same racial or genetic stock for comparison. One important aspect of this study is that it deals with malnutrition during the first year of life. Precise time distinctions are often not made in such studies. Compared to other investigations, the Cabak and Najdanvic study showed one of the largest intelligence quotient deficits, and its subjects were the youngest when nutritionally deprived. The major weakness of the study was the lack of an adequate control group. The children selected for comparison were not only better nourished but came from higher socioeconomic strata.

Retrospective studies in developing countries throughout the world in which better control groups have been chosen do suggest that early malnutrition interferes with subsequent learning ability. In a study of 107 Indonesian children between 12 and 15 years of age from lower socioeconomic groups, including 46 who had been previously classified as malnourished, the Wechsler Intelligence Scale for Children and Goodenough tests were used (Liang *et al.*, 1967). The better nourished, taller children scored higher than the previously malnourished, shorter children. The lowest IQs were associated with the poorest prior nutritional status. At the Nutrition Research Laboratories in Hyderabad, India, a rather thorough study was conducted to determine whether school children treated for kwashiorkor some years before were retarded in comparison to other children attending the same school (Champakam *et al.*, 1968).

Differences in the test scores were greatest in the younger children (aged 8–9 years). The previously malnourished children were more retarded in their perceptual and abstracting ability than in their memory and verbal ability. Their performance was also poorer in the intersensory tests, particularly the visual–haptic test. The previously malnourished children were smaller and lighter than the controls in every age group, but their head circumferences were not significantly different.

The test score differences in this study are greater than those reported in other studies (about a 35-point IQ difference), and these differences exist 6 or more years after clinical recovery from kwashiorkor.

While it is tempting to conclude that kwashiorkor, and the protein deficient diet that preceded it, caused the poor mental performance of the Indian children who had the disease, other factors may have been involved. All the children had been treated in the hospital for a period of at least 6 weeks, and many were probably either bedridden or relatively inactive for long periods of time before and after the episode of the serious illness. This prolonged period of relative immobilization could have resulted in a loss of "learning time," while the stress of separation from home and family during hospitalization might possibly have had a long-term effect. Among the uncontrolled variables in the study were the motivation and responsiveness of the parents, educational levels of the parents, child spacing, and infectious diseases. The authors themselves conclude that "although the differences in mental performance between the two groups of children investigated in this study are clear-cut, it is not easy at this stage to determine to what extent this is a result of the episode of kwashiorkor and to what extent it is due to other factors."

More recently, Chase (1969) has reported that infants in the United States who were severely malnourished early in life performed consistently poorly when later tested. Though his study does not isolate malnutrition as the only important variable, it does make clear that the complex of social problems and nutritional deprivation operating in developing countries is prevalent among certain groups in our own country and is associated with the same type of retarded development.

A number of prospective studies of malnutrition and mental development have been done or are currently under way. Stoch and Smythe (1967), studying South African children, have shown that those malnourished early in life are smaller than a control population and have reduced head circumferences and intelligence quotients even after long-term follow-up. The intelligence quotient testing was adapted for South African children and would appear to be valid. Again, however, the control population leaves much to be desired. The malnourished children lived in inadequate housing with no sanitary facilities, came from poverty-stricken and often broken homes, and were generally neglected. Control families chosen from an industry-built project lived in neat brick houses which had sanitary facilites; all the fathers and mothers were employed and all the children had attended nursery school.

The problem of finding an adequate control population is not easily solved by matching socioeconomic backgrounds. In a Jamaican study, Garrow and Pike (1967) used siblings without a history of hospitalization for malnutrition as a control group for those who had such a history and found that the malnourished group reached the same height and weight and IQ score as the control group. Here we see the opposite problem. In this study the control children probably were also malnourished, subclinically if not clinically. Both groups of children had poor growth and development compared to generally accepted Jamaican norms and significantly retarded growth by U.S. standards. In a more extensive study carried out recently in Jamaica, 74 male children who had been hospitalized for severe malnutrition before they were 2 years of age were compared with their closest-aged brothers and classmates (Hertzig *et al.*, 1972). All the children were between 6 and 11 years of age when studied. Neurological status, intersensory competence, intellectual level, and a variety of language, perceptual, and motor abilities were evaluated. Intellectual level was significantly lower in the index cases than in their brothers or classmates. As might be expected, the classmate comparison group did best and the index cases worst, with the siblings in between. The difference in the brothers' and classmates' scores again points out a disadvantage of studies employing only siblings as controls: the presence of one child hospitalized for severe malnutrition might be expected to identify a high-risk family for chronic undernutrition.

One of the best series of nutrition studies to date is that of Cravioto and others in Mexico and Guatemala (Cravioto and Robles, 1965; Cravioto *et al.*, 1966, 1967a). In populations of uniform socioeconomic backgrounds, performance on psychological tests was found to be related to dietary practice and not to differences in personal hygiene, housing, cash income, crop income, proportion of income spent on food, parental education, or other social or economic indicators. Moreover, performance of both preschool and school children on the Terman, Merrill, Gesell, and Goodenough Draw-A-Man tests was positively correlated with body weights and heights. These tests had been adapted for the population studied. Further investigations in collaboration with the Institute of Nutrition of Central America and Panama (INCAP) in Guatemala again showed a positive correlation between size and performance. The tests included placing blocks in openings, tracing block shapes, and differentiating block shapes by touch alone. These tasks were considered to be measures of visual, haptic, and kinesthetic sensory integration, respectively. To confirm that the differences in height reflected differences in previous nutrition and not familial tendencies, the child's height was correlated with the height of his parents. This correlation proved to be extremely poor, a sharp contrast to the significant correlation in affluent populations between the height of children and that of their parents. (In populations where malnutrition is not prevalent, it is also true that short children perform as well as tall children on tests such as those used in the INCAP studies.) Since the shorter children they studied did not come from families significantly lower in socioeconomic status, housing, and parental education than those of the taller children, Cravioto *et al.* (1966) concluded that the most important variable reflected by the short stature was poor nutrition during early life, and that this also led to the lag in development of sensory integrative competence.

A number of other studies have expanded on these observations. Examining another aspect of neurointegrative competence and auditory and visual integration in Mexican children of school age from communities where malnutrition is common, Cravioto *et al.* (1967b) found that the taller children could integrate information received from both stimuli better than the shorter children of the same age. This observation is particularly important since integrative ability is essential in acquiring primary reading skills. A major consideration in interpreting the findings of this and other studies is the fact that antecedent malnutrition is being inferred from differences in height rather than by direct observation of dietary intake during the growing years. Much evidence suggests that this interpretation is valid. Observations by Boas on growth differences in successive generations of the American-born children of Jewish immigrants, of Boyd-Orr on secular trends in the height of British children, of Grulich on the height of Japanese immigrants, of Mitchell on the relation of nutrition to stature, of Buderkeun-Young on Italian children, as well as the recent study of heights of 12-year-old Puerto Rican boys in New York City by Abromovitz all support the inference. It is significant that in the study made by Cravioto and his associates in Mexico, the earlier the malnutrition, the more profound the psychological retardation. The most severe retardation occurred in children admitted to the hospital under 6 months of age and did not improve on serial testing even after 220 days of treatment. Children admitted later in life with the same socioeconomic background and the same severe malnutrition did recover after prolonged rehabilitation. This recovery of older children even when severely malnourished has been observed before. Kugelmass *et al.* (1944) found retardation in a group of malnourished children over 6 years of age; with prolonged rehabilitation, the children significantly increased their intelligence quotient scores.

Although it has been shown that early malnutrition is more likely to produce lasting effects than malnutrition occurring later in life, the exact time span when malnutrition has the most serious effect is not yet known. In the Jamaican study mentioned above, for example, Birch and his associates found that in all the children malnourished at any time during the first 2 years of life, significant behavioral abnormalities persisted at school age. Moreover, the severity of these functional difficulties did not differ with the time during the first 2 years of life that the malnutrition occurred. On the surface, the results of this study would seem to differ from the results of the studies by Cravioto *et al.* in which children under 6 months of age recovered less well than children malnourished after reaching 6 months of age. Closer examination of the two studies can probably explain the differences. In the Mexican study the children were followed only until 22 months of age, whereas in the Jamaican study they were examined 6–10 years later. The results of the two studies are compatible if one assumes that malnutrition during the first 6 months of life requires a longer period of rehabilitation for the children to achieve their maximum functional potential than malnutrition occurring later in the first 2 years of life. This presumes that both groups of children ultimately are able to recover to the same point, but that point would appear to be at a lower level of functioning than children who had never been malnourished.

3.1.2. Malnutrition in More Affluent Populations

In nearly all research on humans to date, the malnourished subjects (and sometimes the control children) have come from the lower socioeconomic stratum of a population. The need to study malnourished children who were not raised in a deprived environment has not been adequately recognized, although it has been suggested that the survivors of severe famines and of conditions such as celiac disease should be studied (Latham, 1968). Only very recently have the first reports of such studies become available.

The 1944–1945 famine in The Netherlands was sharply circumscribed in place and time, the nutritional deprivation was well documented, and extensive data were available for subsequent analysis. For 6 months 750 or fewer calories were available per person each day in the famine area of western Holland, whereas food rations provided at least (and often much more) than 1300 calories per person per day in the nonfamine areas in the rest of Holland. In the famine areas death rates from starvation were high, famine edema was prevalent, and many subjects lost 25% or more of their original body weight.

A retrospective cohort study of male inductees into the armed forces from both areas of Holland was recently undertaken (Stein *et al.*, 1972). Those born between early 1944 and the end of 1946 were divided into separate cohorts according to whether they were conceived or born before, during, or after the famine. Among these groups of survivors the frequency of severe or mild mental retardation (International Classification of Diseases 3250, 3251, 3252, and 3254) was not related to conception, pregnancy, or birth during the famine. Test scores of several thousand young men on the Dutch version of the Raven progressive matrices showed no differences between those coming from famine and nonfamine areas when they were medically examined at the time of induction into the armed services. The intelligence tests, which failed to show differences between famine and nonfamine subjects, were sufficiently sensitive to show highly significant differences in rates of mental retardation between inductees of two social classes, manual and nonmanual workers. Thus, although no effect on mental development could be detected, a very significant association between the social class of the father and both mental retardation and intelligence test scores was found. It was also found that birth weights in the famine areas were significantly lower than in the nonfamine areas of Holland and that a decline in fertility affected the manual workers more than the nonmanual workers.

The conclusions of this study differ markedly from research findings in most other countries. It has already been stressed, however, that the malnourished children in all previous investigations suffered many deprivations other than nutritional deficiencies. The Dutch were generally well nourished before the famine occurred. After the famine, dietary and other serious deprivations were relatively uncommon in their country. The Dutch research suggests that if there is an impairment in fetal development due to maternal starvation, it is not of a degree that cannot be overcome by standard child-rearing practices as they exist in Holland.

Children with cystic fibrosis represent another nutritionally deprived population that is more or less free of socioeconomic deprivation. Malnutrition occurs very early in life as a result of malabsorption of nutrients. A study was made of middle-class children with cystic fibrosis who suffered prolonged malnutrition during their first 6 months of life, with weights below the third percentile on the Boston (Stuart) growth charts for at least 4 of those first 6 months (Lloyd-Still *et al.*, 1972). All the children had other evidence of severe malnutrition, resembling the symptoms of kwashiorkor or marasmus. Twenty-nine siblings of these malnourished children served as controls, and IQ tests were given to 27 of the parents.

The malnourished children who were under 5 years of age showed lower scores on the Merrill-Palmer tests than the control children. There were no significant differences between the scores of study and control children over five on the Wechsler Intelligence Scale for Children (WISC), the Vineland Scale of Social Maturity, or the Wechsler Adult Intelligence Scale (which was given to those children over 14 years of age at the end of the study).

The results seem to show that malnutrition in the early months of life is associated with poorer scores on psychological tests while a child is ill and during the early years of life immediately following, or during recovery from malnutrition. Differences in IQ apparently disappear after 5 years of age.

The studies on nondeprived populations strongly suggest that early malnutrition may retard development temporarily but that recovery is possible given the proper subsequent environment. These conclusions underline the importance of two other types of investigations currently being carried out; serial studies of the effects of malnutrition on mental development and studies that attempt to enrich the environment of children who have been malnourished as infants.

3.2. Serial Studies on Intellectual Development of Malnourished Children

Serial studies of intellectual development will allow observations of the time when recovery from malnutrition occurs and the conditions necessary to facilitate recovery. The major obstruction to these studies is the difficulty in structuring "psychometric tests" for infants and young children which can predict later intelligence. Only a few such tests are useful with normal children, and none has had careful trials with malnourished children. Recently, however, Klein and his associates at INCAP have attempted to evaluate the usefulness of an "orienting response" in predicting later intellectual deficits in malnourished children (Klein, personal communication). In a preliminary study these investigators presented 40 trials of a pure-tone auditory stimulus to 8 marginally nourished and 8 malnourished 13½-month-old male infants. Initially, the marginally malnourished children had a greater orientation response (as measured by a slowing of their heart beat) than the malnourished children. When the stimulus tone was changed, the marginally nourished children again showed a greater response. The authors cite evidence that the lower response seen in the malnourished children is a sign of poor attention, which in other children tested by this method led to poor learning later. Whether this will prove to be true in malnourished children is still unknown, but follow-up studies of this population should answer the question.

3.3. Environmental Enrichment for Malnourished Children

Several studies have indicated that enriching the environment in deprived populations of children can improve their subsequent development (Campbell and Stanley, 1966; Schaeffer and Aaronson, 1972; Caldwell, 1967). Generally children from a low socioeconomic class have been randomly assigned to either a treatment or a control group. The treatment group receives a variety of special care and instruction through home visits and day-care-center experience. The results in general have shown that the "stimulated children" develop better and show a higher IQ at the end of the experience. Unfortunately, after most of these studies have ended, the child has been "returned" to his normal environment and his IQ usually reverts to the level of the control group.

One study in which long-term stimulation was applied is noteworthy (Skeels, 1966). Two groups of mentally retarded, institutionalized children whose mothers' average IQ was under 70 were studied. Thirteen of these children were transferred at age 2 from an orphanage to a state institution for the mentally retarded. The children were placed in the care of older female inmates in a one-to-one mother–child relationship. After 1½ years these children had gained 28 IQ points, whereas the children left at the orphanage had lost 26 IQ points. After 2½ years of legal adoption, the study children reached a mean IQ of 101. These children were followed up 30 years later. They were self-supporting and most had completed twelfth grade. Four of the children had one or more years of college. By contrast, most of the children left at the orphanage had completed only third grade and were at institutions for the mentally retarded. In addition, a number of them had died.

In order to examine the effects of "environmental enrichment" on the development of malnourished children, a retrospective analysis of a population of Korean children was undertaken. Some of these children were severely malnourished during the first year of life and were then adopted by families in the United States (Winick *et al.*, 1975). A second prospective study is presently under way in Colombia, where malnourished infants are identified in the hsopital and then placed in a special nursery school environment at 2 years of age (McKay *et al.*, 1974).

The 141 Korean girls were divided into three groups. Forty-two were severely malnourished, below the third percentile for both height and weight when compared to Korean standards. Fifty-two were marginally nourished, between the third and twenty-fifth percentile for height and weight, and 47 were well nourished, above the fiftieth percentile for height and weight. Only infants falling into these categories before their first birthdays were selected. All the infants were adopted before their second birthday by American families. The adoptions were entirely random, on a first come/first served basis. Parents had no idea of a child's previous nutritional history. All the families were carefully screened to ensure an adequate home environment for the adopted child. Questionnaires were sent to all the families and records were obtained from the schools the children were attending. Their ages at the time of follow-up ranged from 7 to 16 years. Intelligence tests were administered by the schools. Achievement based on a number of other test scores which were available was evaluated by two psychologists who had no prior knowledge of the children's history.

By the time these children reached 7 years of age there were no differences in average weight among the three groups. All exceeded normal by Korean standards but were significantly below normal by American standards. Changes in height were similar to those in weight except that the malnourished children remained slightly but significantly smaller. The mean IQ of the previously malnourished group was 102.05. The marginally nourished children achieved a mean IQ of 105.95. This is not a statistically significant difference. By contrast, the previously well-nourished children reached a mean IQ of 111.68, which does represent a significant difference from the malnourished children. When achievement in these three groups was compared, the results were similar. Both the severely malnourished and the marginally malnourished children were achieving exactly at expected norms for American children of the same age and the same grade. The previously well-nourished children were achieving slightly but significantly better.

These findings show that severely malnourished children, when reared in a middle-class environment, can catch up in height and weiht and reach an IQ and school achievement level which is perfectly normal for well-nourished children raised in an industrialized nation. They demonstrate in this unique population exactly what has been shown in animal studies—that environmental stimulation will reverse many, if not all, of the behavioral deficits elicited by early malnutrition. The data also suggest that when well-nourished children are placed in this more stimulating environment, they do even better. Their IQ scores and achievement scores are not only higher than those of the malnourished children but are also higher than the norms for American children in general. From a practical standpoint the importance of this study lies in its pointing out the reversibility of effects of early malnutrition. In all previous studies when the child was returned to his or her previous environment, the IQ was 70 or below at school age.

Recently, a similar cohort of children have been studied except that all were adopted after 3 years of age. IQ and achievement was lower in all three groups, with the previously malnourished now performing below U.S. norms and the well-nourished children just about at U.S. norms. The difference between the previously malnourished and the previously well nourished children was again significant. Thus, the time of adoption (and presumably of environmental enrichment) is extremely important in determining the final outcome. Examining the results of all the children in both studies by multiple regression analysis reveals that although both early nutrition and the time of adoption influences subsequent IQ and achievement, the latter is a stronger determinant than the former.

In the prospective study in Colombia, severely malnourished children after recovery have been placed in an "enriched" environment at about 2 years of age (McKay *et al.,* 1974). The children are exposed to all types of stimulating experiences in terms of both play and learning. Their nutrition has been kept adequate. These children are being compared with randomly picked similarly malnourished children who were not placed in this program and with previously well nourished children of higher socioeconomic class both in and out of the program. Preliminary results suggest that stimulation will improve the learning of these children. The test levels of the stimulated malnourished children were higher than those of the nonstimulated malnourished children and approached those of the children from the higher socioeconomic group who were not stimulated. The well-nourished, stimulated children had the highest learning capacity, but as the study progresses their lead shortens.

3.4. Dietary Supplementation

A final group of studies of human populations has focused on improvement in the nutrition of pregnant women and young infants. The basic design of these studies involves comparing a population undergoing severe malnutrition with a similar population that is given nutritional supplements. After a given period of time the results are evaluated in terms of whatever outcome measures the investigators choose. Not only are such studies technically difficult to carry out, but they involve ethical problems, especially with regard to the nonintervention groups. For these reasons only a few have been carried out and these under conditions and in populations that have been carefully selected. One study is being done by INCAP in three rural Guatemalan villages (Canosa, 1968). Great pains have been taken to ensure that these villages have comparable populations. One village receives a food supplement in the form of a high-protein supplement drink that is consumed by the young children and the pregnant mothers. A second village is supplied with a supplement of some caloric value but no protein content. Both villages are given medical care. The third village receives medical care only. Preliminary results show that growth rates are increased in the children receiving the high-protein supplement. Moreover, the women who get a protein supplement during pregnancy have babies whose birth weight is significantly higher than that of babies born in the nonsupplemented village. Finally, it would appear that the development of the children is better in the protein-supplemented village than in the other villages.

In a small study in rural Mexico, food supplements have been given to children of families carefully selected to represent the norms of their community (Chavez *et al.,* 1974). When the growth rate of the supplemented children was compared with that of their previously studied, unsupplemented siblings, a marked increase was observed. The supplemented children also demonstrated marked superiority in physical strength, independence, attentiveness, and ability to perform certain behavioral tests. They tended to explore their environment more thoroughly, play with toys more frequently, and interact with adults better than the nonsupplemented children. It must be emphasized that the children in this study who did not receive the supplement were picked because they were representative of the village and not because they were malnourished.

The sparse information that is available from intervention studies, then, would suggest that improvement in the diet during either pregnancy or early life will significantly alter birth weight, subsequent growth rate, and subsequent behavior.

In summary, the data in both animals and humans suggest that early malnutrition either before or shortly after birth will curtail the rate of cell division, resulting in a brain with fewer cells. In addition, myelination, arborization of dendrites, and secretion of certain neural hormones are reduced. The behavioral changes produced by early malnutrition are more difficult to assess. Pure malnutrition with adequate stimulation produces few, if any, long-term effects. Malnutrition in conjunction with other forms of environmental deprivation produces profound and long-term behavioral deficits, which in the human result in severe learning disabilities. Just what type of "environmental enrichment" is necessary to prevent the behavioral consequences of early malnutrition and exactly when it must be started are still being investigated.

4. References

Altman, J., Sudarsham, K., Das, G. D., McCormick, N., and Barnes, D., 1971, The influence of nutrition on neural and behavioral development. III. Development of some motor, particularly locomotor patterns during infancy, *Dev. Psychobiol.* **4:**97–114.

Barnes, R. H., 1967, Reported at Gordon Research Conference, New London, N. H.

Barnes, R. H., Cunnold, S. R., Zimmermann, R. R., Simmons, H., MacLeod, R. B., and Krook, L., 1966, Influence of nutritional deprivations in early life on learning behavior of rats as measured by performance in a water maze, *J. Nutr.* **89:**399–410.

Barnes, R. H ., Moore, A. V., and Pond, W. G., 1970, Behavioral abnormalities in young adult pigs caused by malnutrition in early life, *J. Nutr.* **100:**149–155.

Barnett, S. A., Smart, J. L., and Widdowson, E. M., 1971, Early nutrition and the activity and feeding of rats in an artificial environment, *Dev. Psychobiol.* **4:**1–15.

Cabak, V., and Najdanvic, R., 1965, Effect of undernutrition in early life on physical and mental development, *Arch. Dis. Child.* **40:**532–534.

Caldwell, B. M., 1967, Descriptive evaluations of child development and of developmental settings, *Pediatrics* **40:**46–54.

Campbell, D. T., and Stanley, J. C. (eds.), 1966, *Experimental and Quasi-Experimental Designs for Research,* 84 pp., Rand McNally & Company, Chicago.

Canosa, C., 1968, Nutrition, physical growth and mental development, Paper presented to the Pan American Health Organization Advisory Committee Meeting, Washington, D. C.

Champakam, S., Srikantia, S. G., and Gopalan, C., 1968, Kwashiorkor and mental development, *Am. J. Clin. Nutr.* **21:**844–852.

Chase, H. P., 1969, Paper presented at the 39th Ann. Meet. Soc. Pediat. Res., Atlantic City, N. J.

Chavez, A., Martinez, C., and Yaschine, T., 1974, The importance of nutrition and stimuli on child mental and social development, in: *Symposia of the Swedish Nutrition Foundation XII, Early Malnutrition and Mental Development* (J. Cravioto, L. Hambraeus, and B. Vahlquist, eds.), pp. 211–225, Almqvist & Wiksell, Uppsala, 244 pp.

Cines, B., 1972, Reported at the Symposium on Nutrition and Fetal Development, presented by the Institute of Human Nutrition and sponsored by the National Foundation—March of Dimes, New York.
Cowley, J. J., and Griesel, R. D., 1959, Some effects of a low-protein diet on a first filial generation of white rats, *J. Genet. Psychol.* **95:**187–201.
Cowley, J. J., and Griesel, R. D., 1963, The development of a second generation of low-protein rats, *J. Genet. Psychol.* **103:**233–242.
Cowley, J. J., and Griesel, R. D., 1964, Low-protein diet and emotionality in the albino rat, *J. Genet. Psychol.* **104:**89–98.
Cowley, J. J., and Griesel, R. D., 1966, The effect on growth and behavior of rehabilitating first and second generation low-protein rats, *Anim. Behav.* **14:**506–517.
Cravioto, J., and Robles, B., 1965, Evolution of adaptive and motor behavior during rehabilitation from kwashiorkor, *Am. J. Orthopsychiat.* **35:**449–464.
Cravioto, J., DeLicardie, E. R., and Birch, H. G., 1966, Nutrition, growth and neurointegrative development: An experimental ecologic study, *Pediatrics* **38:**(No. 2, Part II):319–372.
Cravioto, J., Birch, H. G., and DeLicardie, E. R., 1967a, Influéncia de la desnutrición en la capacidad de apprendizajo del niño escolar, *Bol. Med. Hosp. Infant.* (*Span. Ed.*)**24:**217.
Cravioto, J., Espinoza, C. G., and Birch, H. G., 1967b, Early malnutrition and auditory-visual integration in school age children, *J. Spec. Ed.* **2:**75–82.
Frankova, S., 1974, In: Interaction between early malnutrition and stimulation in animals, *Symposia of the Swedish Nutrition Foundation XII, Early Malnutrition and Mental Development* (J. Cravioto, L. Hambraeus, and B. Vahlquist, eds.), pp. 202–210, Almqvist & Wiksell, Uppsala, 244 pp.
Frankova, S., and Barnes, R. H., 1968, Influence of malnutrition in early life on exploratory behavior of rats, *J. Nutr.* **96:**477–484.
Garrow, J. S., and Pike, M. C., 1967, The long-term prognosis of severe malnutrition, *Lancet* **1:**1–4.
Graham, C. G., 1967, The effect of infantile malnutrition on growth, *Fed. Proc.* **26:**139–143.
Guthrie, H. A., 1968, Severe undernutrition in early infancy and behavior in rehabilitated ablino rats, *Physiol. Behav.* **3:**619–623.
Hertzig, M. E., Birch, H. G., Richardson, S. A., and Tizard, J., 1972, Intellectual levels of school children severely malnourished during the first two years of life, *Pediatrics* **49:**814–824.
Kerr, G. R., and Waisman, H. A., 1970, A primate model for the study of malnutrition during early life, in: *Feeding and Nutrition of Non-Human Primates* (R. S. Harris, ed.), pp. 65–85, Academic Press, Inc., New York, 310 pp.
Kugelmass, I. N., Poull, L. E., and Samuel, E. L., 1944, Nutritional improvement of child mentality, *Am. J. Med. Sci.* **208:**631–633.
Lat, J., Widdowson, E. M., and McCance, R. A., 1961, Some effects of accelerating growth. III. Behavior and nervous activity, *Proc. R. Soc. London Ser. B* **B153:**347–356.
Latham, M. C., 1968, Short comments, in: *Malnutrition, Learning and Behavior* (N. S. Scrimshaw, ed.), pp. 299–300, MIT Press, Cambridge, Mass, 566 pp.
Levine, S. , and Denenberg, V. (separate reviews), 1969, in: *Stimulation in Early Infancy* (A. Ambrose, ed.), pp. 21–63, Academic Press, Inc., London, 289 pp.
Levitsky, D. A., and Barnes, R. H., 1970, Effect of early malnutrition on the reaction of adult rats to aversive stimuli *Nature* **225:**468–469.
Levitsky, D. A., and Barnes, R. H., 1972, Nutritional and environmental interactions in the behavioral development of the rat: Long-term effects, *Science* **176:**68–71.
Liang, P. H., Hie, T. T., Jan, O. H., and Glok, L. T., 1967, Evaluation of mental development in relation to early malnutrition, *Am. J. Clin. Nutr.* **20:**1290–1294.
Lloyd-Still, J. D., Wolff, P. H., Horwitz, I., and Shwachman, H., 1972, in: *9th Proc. Int. Congr. Nutr., Mexico City,* **2:**357–364.
McKay, H., McKay, A., and Sinisterra, L., 1974, Intellectual development of malnourished preschool children in programs of stimulation and nutritional supplementation, in: *Symposia of the Swedish Nutrition Foundation XII, Early Malnutrition and Mental Development* (J. Cravioto, L. Hambraeus, and B. Vahlquist, eds.), pp. 226–233, Almqvist & Wiksell, Uppsala, 244 pp.

Schaeffer, E. S., and Aaronson, M., Infant education research project: Implementation and implication of a home tutoring program, in: *The Preschool in Action: Exploring Early Childhood Programs* (R. K. Parker, ed.), Allyn and Bacon, Boston, 450 pp.

Seitz, P. F. D., 1954, The effects of infantile experiences upon adult behavior in animal subjects: I. Effects of litter size during infancy upon adult behavior in the rat, *Am. J. Psychiat.* **110:**916-927.

Simonson, M., Stephen, J. K., Hanson, H. M., and Chow, B. F., 1971, Open field studies in offspring of underfed mother rats, *J. Nutr.* **101:**331–336.

Skeels, H. M., 1966, Adult status of children with contrasting early life experiences, *Monogr. Soc. Res. Child Dev.* **31**(3):1–65.

Stein, Z., Susser, M., Saenger, G., and Marolla, F., 1972, Nutrition and mental performance, *Science* **178:**708–713.

Stoch, M. B., and Smythe, P. M., 1967, The effect of undernutrition during infancy on subsequent brain growth and intellectual development, *S. Afr. Med. J.* **41:**1027–1035.

Strobel, D. A., and Zimmermann, R. R., 1971, Manipulatory responsiveness in protein malnourished monkeys, *Psychon. Sci.* **24:**19–20.

Winick, M., Meyer, K. K., and Harris, R. C., 1975, Malnutrition and environmental enrichment by early adoption, *Science* **190:**1173–1175.

Zimmermann, R. R., Steere, P. O., Strobel, D. A., and Hom, H. L., 1972, Abnormal social development of protein malnourished rhesus monkeys, *J. Abnorm. Psychol.* **80:**125–131.

3

Nutrition and Cellular Growth

S. Jaime Rozovski and Myron Winick

1. Introduction

Traditionally, growth of the whole body or of individual organs has been measured by weight gain alone (Donaldson, 1924; Dunn *et al.*, 1947). This measurement, however, does not take into account the contribution made by various cellular and noncellular components. More recently, the finding that deoxyribonucleic acid (DNA) is located almost entirely within the nucleus and that the amount of DNA is constant within the diploid nucleus of any species (Boivin *et al.*, 1948; Mirsky and Ris, 1949; Thomson *et al.*, 1953) has enabled Enesco and LeBlond (1962) to reexamine postnatal growth. They were able to calculate the number of diploid nuclei in the various organs of the rat using the following formula:

$$\text{number of nuclei (millions)} = \frac{\text{total organ DNA (mg)} \times 10^3}{6.2}$$

where 6.2 is the amount of DNA, expressed in pg, in a single diploid rat nucleus (Enesco, 1957). This figure was used to represent cell number. They then determined weight per nucleus by dividing total organ weight by the number of nuclei:

$$\text{weight/nucleus (ng)} = \frac{\text{total organ wt (g)} \times 10^3}{\text{number of nuclei (millions)}}$$

This figure was used to estimate cell size. Thus, once the number of cells is determined, the weight per cell, protein content per cell, ribonucleic acid (RNA) content per cell, or lipid content per cell can be determined by either

S. Jaime Rozovski • Division of Diabetes and Metabolism, Department of Medicine, Boston University School of Medicine, Boston, Massachusetts. *Myron Winick* • Institute of Human Nutrition, Columbia University College of Physicians and Surgeons, New York, New York.

weighing the organ or ascertaining the total protein, RNA, or lipid content of the organ and dividing by the number of cells. This can be expressed as weight/DNA, protein/DNA, RNA/DNA, or lipid/DNA ratio.

By these relatively simple biochemical techniques it has been possible to follow growth by monitoring the contribution made by the increase in the number of cells and the contribution attributable to increase in cell size. As Enesco and Leblond (1962) have pointed out, these calculations are valid only in organs containing one diploid nucleus per cell. However, most organs and tissues meet this criterion. It should also be emphasized that total DNA content, while accurately reflecting cell number, in no way differentiates one cell type from another. In addition, although the ratios as outlined above give an overall average for these materials per cell, no single cell may actually contain this quantity of material. Individual cells, especially when differing in type, may vary widely in their composition of either proteins or lipids.

Within these limitations, however, this "chemical" approach to cellular growth has allowed certain generalizations to be made which have given rise to an overall picture of growth at a cellular level.

Careful examination of all nonregenerating organs by these methods reveals three distinct phases of growth (Fig. 1). The first is characterized by a proportional increase in weight, protein, and DNA content; the number of cells is increasing whereas the ratios or the size of the individual cells is not changing. Simple hyperplasia is occurring. This phase ends as the rate of net DNA synthesis begins to slow while weight and protein content continue to increase at the same rate, resulting in a transitional phase of hyperplasia and concomitant hypertrophy which lasts until net DNA synthesis stops. After this, all further growth is by hypertrophy. Finally, when weight stabilizes and net protein synthesis stops, growth is finished (Winick and Noble, 1965).

These data allow us to view the overall growth of any organ as a continuous accretion of protoplasm made up of water, proteins, and in some cases lipids. The ultimate packaging of this protoplasm into individual cells depends on the rate of DNA synthesis. At present, the mechanisms controlling the period during which DNA may be synthesized by an organ and the mechanisms governing the rate of synthesis during that period are largely unknown.

In this chapter we will attempt to describe the changes in cellular growth in brain, liver, muscle, placenta, and fetus, and the effect of malnutrition on this growth.

2. Normal Cellular Growth

2.1. Brain

In whole rat brain, DNA synthesis and hence cell division stops at about 20 days of age. Total protein continues to increase until about 99 days of age, when the brain reaches its final size (Winick and Noble, 1965). However, more

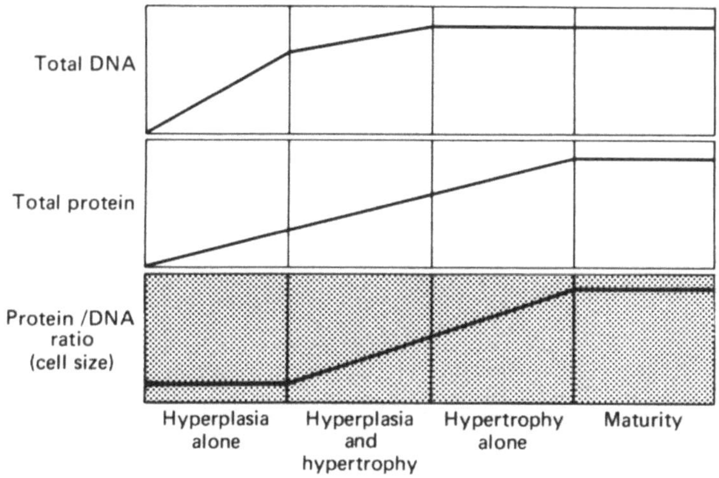

Fig. 1. Periods of cellular growth. Plotted are the relationships between DNA and protein during the three phases of organ growth. It will be observed that DNA content crests and levels off well before organ size, as determined by protein accretion and weight gain, reaches its maximum. (From Winick, 1970e.)

detailed examination reveals that different regions have their own pattern of cellular growth (Fish and Winick, 1969). In cerebrum, DNA synthesis continues until about 21 days postnatally. After this the cells continue to accumulate protein and lipid. Total cerebral lipid content is achieved somewhat later, and total protein content around 99 days of age. In cerebellum, DNA synthesis stops at 17 days postnatally. Net protein synthesis actually becomes negative for a short period after this, and the size of the individual cerebellar cells decreases. This decrease in cell size probably reflects the maturation of larger, more primitive cells into smaller, more mature cells. In brain stem, total cell number is increased to 14 days of age. Thereafter there is an enormous increase in the protein/DNA ratio. This increase probably not only reflects an increase in the size of the brain stem cells but also in growth, myelination, and enlargement of neuronal processes from other brain regions into the brain stem. Hippocampus is an area which demonstrates a type of cellular growth somewhat unique to central nervous system. There is a discrete rise in DNA content between days 14 and 17. The increase corresponds to a migration of neurones from under the lateral ventricle into the hippocampus, which occurs on day 15 in the rat (Altman and Das, 1966).

The ultimate cellular makeup of the various regions depends, then, on the rate of cell division within the particular region, the time that cell division stops, the type of cells dividing, and whether or not cells are migrating to or from the region.

In human brain the sequence of events is not as clearly defined as in rat brain. Studies initially indicated that DNA synthesis was linear prenatally,

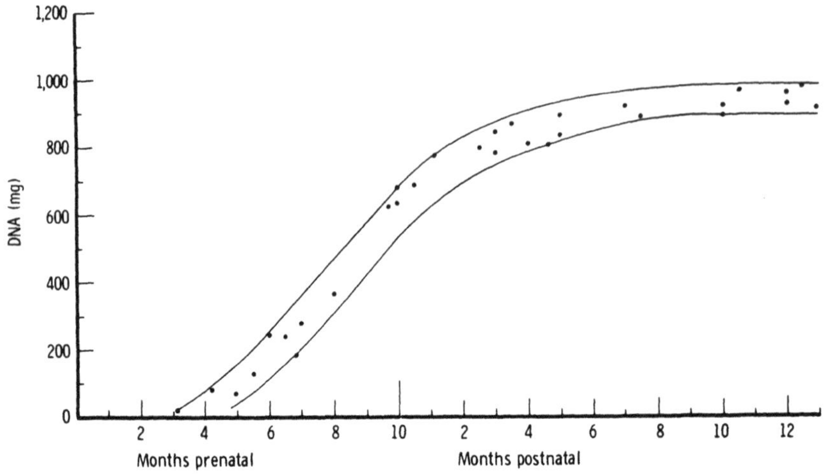

Fig. 2. Amount of DNA in the brain of normal human fetuses and children. Each dot represents one brain. (From Winick, 1969.)

began to slow down shortly after birth, and reached a maximum at about 8–12 months of age (Winick, 1968a) (Fig. 2). More recent studies have tended to modify these results somewhat and would extend the time beyond the first year of life (Dobbing and Sands, 1970). Moreover, Dobbing and Sands (1973) have shown that two peaks of DNA synthesis may occur normally in human brain. The first peak is reached at about 26 weeks of gestation and the second around birth (Fig. 3). They have interpreted these results as corresponding to

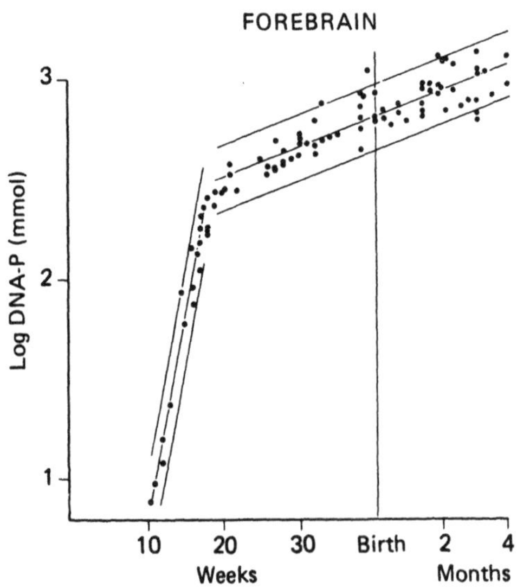

Fig. 3. DNA content of human fetal brain. (Adapted from Dobbing and Sands, 1973.)

the peak rate of neuronal division and the peak rate of glial division, respectively.

There are still very few data on the cellular growth of various regions of human brain. What data are available would indicate that the rate of cell division postnatally is about the same in cerebrum and cerebellum and stops at about the same time in both areas, that is, between 12 and 15 months of age (Winick *et al.*, 1970) (Fig. 4). The number of cases studied is, however, too small to attempt a precise statement.

In brain stem, DNA synthesis continues at a slow but steady rate until at least 1 year of age. The exact cell types involved and the migratory patterns of the cells in the developing human brain are not as clearly worked out as in the rat brain. For obvious reasons, radioautography cannot be done. What is known, then, is the result of careful histological and histochemical examination of brains of fetuses of various ages. In a series of elegant studies, Duckett and Pearse (1966) have shown that during fetal life, the brain not only increases linearly in weight but undergoes a series of biochemical changes. Glycolysis is present during the second month of fetal life; oxidative mechanisms appear during the third month; and activity and localization of a number of enzymes reach a mature pattern during the seventh month of fetal life.

In addition, there is evidence that the presence of acetylcholinesterase indicates tissue excitability (Nachmansohn, 1952). The activity of this enzyme is localized in neurons of the anterior horn of the spinal cord as early as the tenth week of embryonic life, according to Duckett and Pearse (1969). This

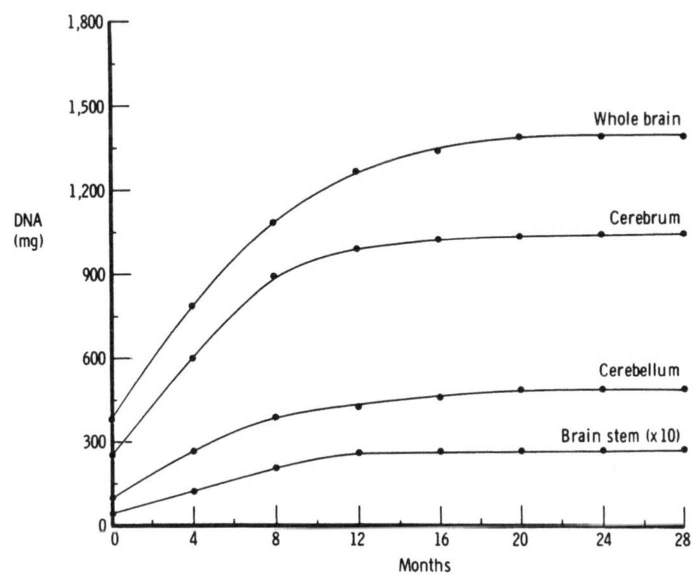

Fig. 4. DNA content in various regions of human brain during normal growth. (From Winick and Rosso, 1970.)

correlates well with the time that movement of the lower limb can be elicited by proper stimulation (Augustinsson, 1950).

Two specific cell types, the Cajal-Retzius cells (Duckett and Pearse, 1968) and the monamine oxidase cells (Duckett and Pearse, 1967), are present only in fetal life, disappearing before birth. Their function is unknown. Serial analysis of lipids in human brains would indicate that the lipid/DNA ratio rises shortly after birth until at least 2 years of age. This is reflected in a rise in both the cholesterol/DNA and phospholipid/DNA ratios (Rosso *et al.*, 1970). Thus, postnatal lipid synthesis is occurring at a more rapid rate than is DNA synthesis. This is undoubtedly related to the rapid myelination which is occurring during this period of life.

Although the descriptive work regarding the human brain would suggest that cellular growth is governed by the same principles as those governing cellular growth in rat brain, more data are needed to complete the picture. Indirect measurements have been used to follow the normal growth of human brain. The most common of these is cranial circumference. Some correlations have been made between increase in cranial circumference and cellular growth of the brain. Approximate formulas have been worked out relating head circumference to brain weight, protein, and DNA content during the first year of life (Winick and Rosso, 1969a).

Studies in normal rat brain have demonstrated that the activity of the enzyme DNA polymerase correlates with the rate of cell division. This is true not only in whole brain but in the various brain regions studied (Brasel *et al.*, 1970) (Fig. 5).

Thus, during the hyperplastic phase brain growth is accompanied by high activity of enzymes associated with cellular proliferation. In addition, there are data which suggest that the rate of RNA turnover increases in the growing brain. This is accompanied by an increase in the activity per cell of alkaline RNase, an enzyme involved in RNA degradation (Rosso and Winick, 1975). Finally, two species of cytoplasmic RNA have been described in developing rat brain which are not present in adult brain. These have been shown to disappear at about the time of cessation of cell division (Lewis and Winick, 1977). The exact nature and function of these RNAs has not yet been determined.

In summary, normal cellular growth of mammalian brain is made up of an early proliferative phase, in which cell division predominates and the quantity of protein and lipid per cell remains relatively constant. The rate of proliferation of cells appears to be separated into two peaks, one probably neuronal and the other glial. At the same time, cells migrate from certain regions of the brain to other regions. In human brain, there is evidence that certain cell types appear and disappear during this early phase of growth.

Later growth is characterized by slowing and finally by a cessation of cell division, in spite of a constant rate of net protein synthesis, and an increasing rate of myelin synthesis. Finally, myelination is completed and net protein synthesis stops. The mechanisms controlling the rate of cell division and the migratory patterns of cells are just beginning to be investigated.

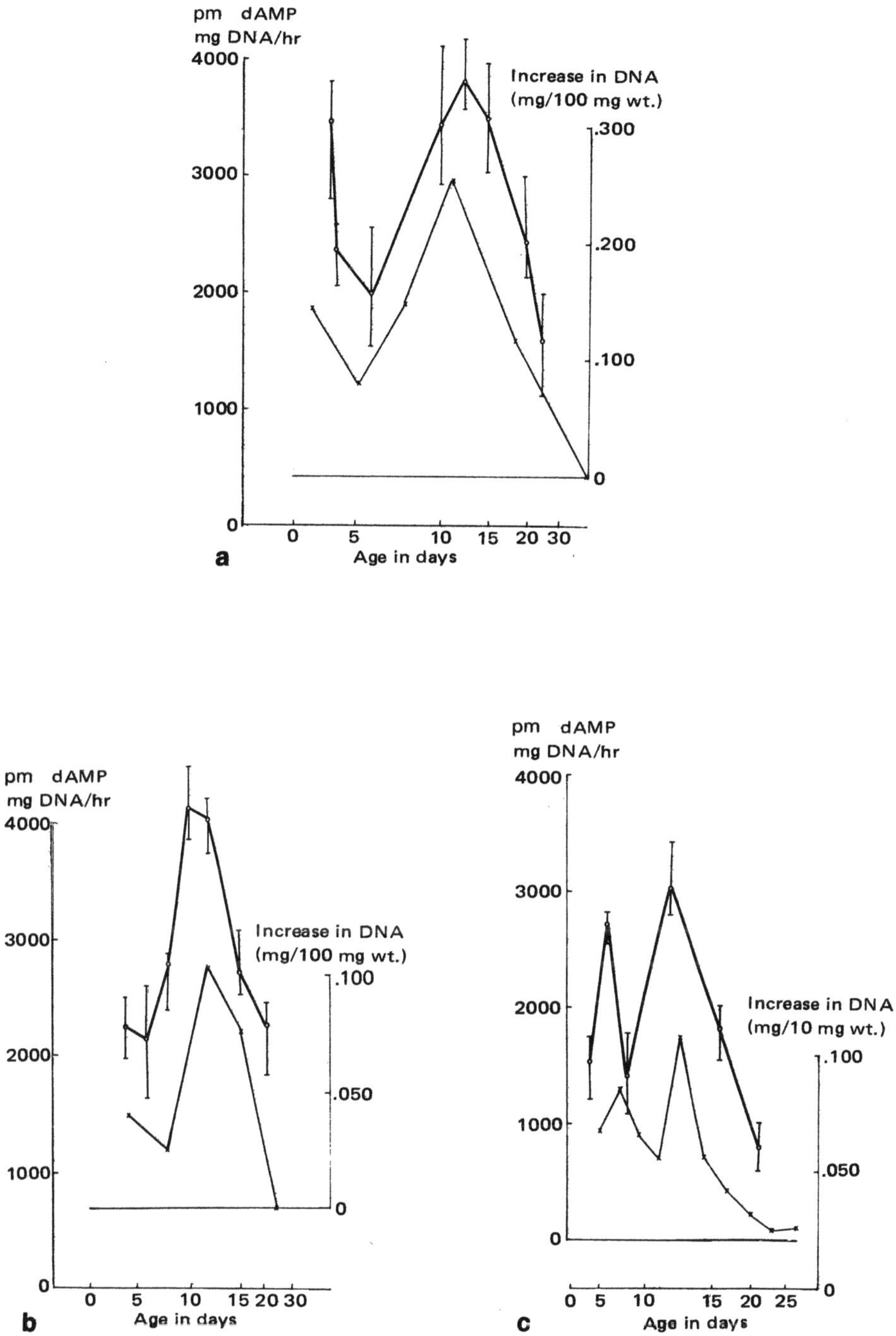

Fig. 5. DNA polymerase activity in normal rat brain during development. Polymerase activity is shown by open circles, with brackets representing the range of data and not SDs. Rate of increase in DNA is shown by closed circles. DNA rate curve parallels curve for enzyme activity. (a) Whole brain. (b) Forebrain. (c) Cerebellum. (From Brasel and Winick, 1972.)

2.2. Liver

Liver DNA in the rat increases approximately until 3 months of age (Enesco and Leblond, 1962; Winick and Noble, 1965) (Fig. 6). However, the liver is different from most of the other organs since the mean DNA content of the nuclei changes with age. Therefore, the number of nuclei could not be accurately derived from the curve of DNA vs. age. The change in DNA content per nucleus in liver is due to polyploidy (Jacobj, 1925, 1935; Swift, 1950, 1953). Actually, in the liver of rats and other mammals, five cytogenetic cell types develop after birth: mononuclear diploid, binuclear diploids, mononuclear tetraploids, binuclear tetraploids, and mononuclear octopoids (Alfert and Geschwindt, 1958; Naora, 1957). The changes in the proportion of these different types of cells with age can be seen in Fig. 7 (Najdal and Zajdela, 1966a). The data demonstrate that up to 3 weeks of age, liver cells are mainly mononuclear diploid.

To calculate the number of nuclei in liver of rats, Enesco and Leblond (1962) divided the total liver DNA at a particular time by the average DNA content per nucleus at that time. Even though the existence of polyploidy in liver makes measurements of DNA difficult to interpret, Epstein (1967) has shown that as ploidy increases, the cytoplasmic mass of the cell increases proportionally. Thus, a tetraploid liver cell nucleus "governs" twice as much cytoplasm as a diploid liver nucleus.

Liver growth can therefore be examined by using total DNA content to measure "cell number" as long as we remember that although a doubling of the chromosome number will double the nuclear DNA content, two discrete nuclei may not result. However, the cytoplasm will also double in size but will not divide into two separate cells.

Total liver DNA increases rapidly up to 40 days. After 90 days, total liver

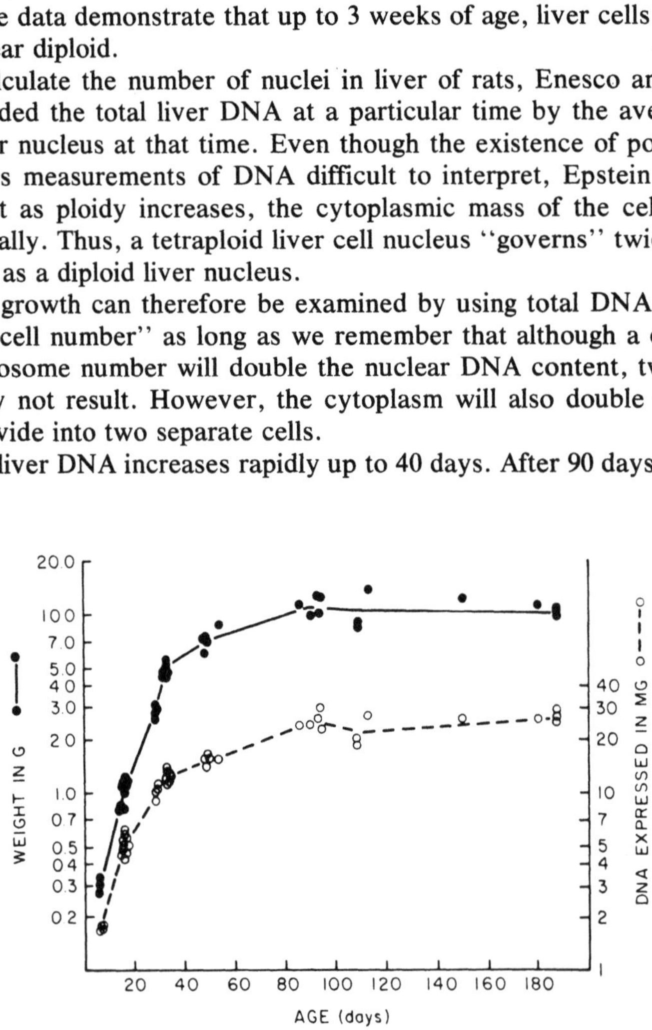

Fig. 6. Increase in rat liver weight and DNA with age. (Adapted from Enesco and Leblond, 1962.)

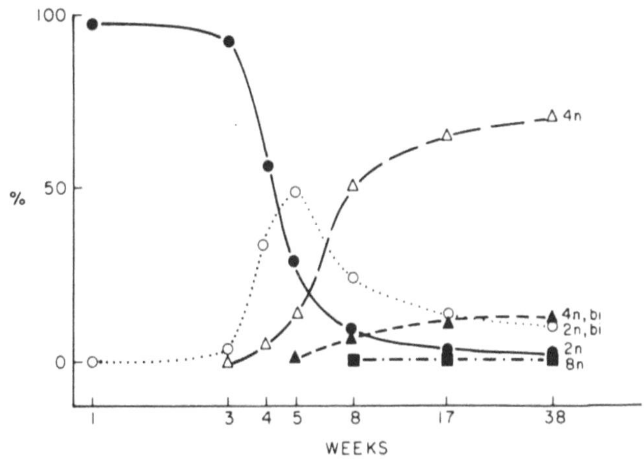

Fig. 7. Age-dependent changes in the proportion of different types of mononuclear and binuclear cells in rat liver (age in weeks, logarithmic scale). (Adapted from Najdal and Zajdela, 1966a.)

DNA content remained unchanged (Enesco and Leblond, 1962) (Fig. 6). The rapid increase in DNA content of liver coincides with an increased incorporation of [^{14}C]thymidine into DNA in liver (Winick and Noble, 1965). Klemperer and Haynes (1968) and Machovich and Greengard (1972) found the highest activities of thymidine kinase in rat liver at 17 days of gestation. Enzyme activity increased between birth and 2 days of age and then decreased, reaching adult values by 12 days of age (Fig. 8). A good correlation was found between enzyme activity and the number of nuclei in the liver during early postnatal life.

The increase in DNA content in liver during early life has also been

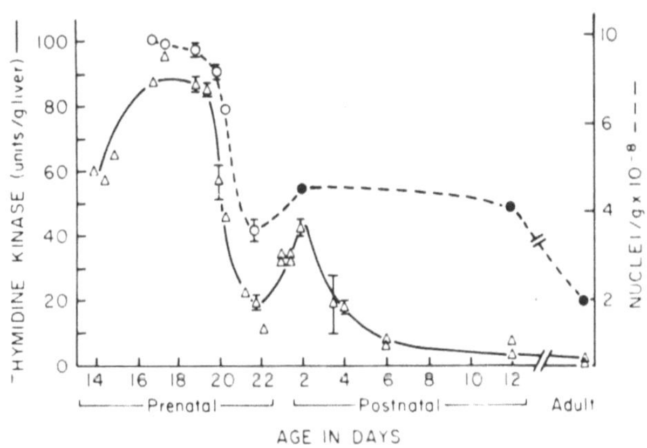

Fig. 8. Changes in liver thymidine kinase and cellularity during development. (Adapted from Machovich and Greengard, 1972.)

observed in sheep (Johns and Bergen, 1976) and in pigs (Widdowson and Crabb, 1976). In humans, the multiplication of diploid mononucleate cells is the major mode of liver growth until puberty (Swartz, 1956).

In studies done in fetuses obtained from therapeutic abortions, Widdowson *et al.*, 1972) observed a rapid increase in total liver DNA during the first 30 weeks of gestation. Cell size (expressed as protein/DNA ratio) did not change until 30 weeks of gestation, when it increased rapidly.

As previously pointed out, the weight per nucleus and protein per nucleus have been suggested as indices of cell size (Enesco and Leblond, 1962; Winick and Noble, 1965). The most rapid increase in total body protein in the rat occurs during lactation (Winick and Noble, 1965).

Figure 9 shows the increase in protein per nucleus in the whole body and in different organs of the rat. In heart, kidney, and lung, cell size increases rapidly during early life and then remains constant. Total protein in the liver increases steadily during gestation. The rate of increase is rapid at birth but slows down shortly thereafter. After weaning, the increase in total liver protein again accelerates (Winick and Noble, 1965). The increase and subsequent decrease in total liver protein which occurs around the time of birth is confirmed by changes in the rate of protein synthesis per unit DNA, which shows a peak at the time of birth (Miller, 1969). Thus, DNA synthesis occurs in early postnatal life while net protein synthesis remains relatively low for about 10 days postpartum.

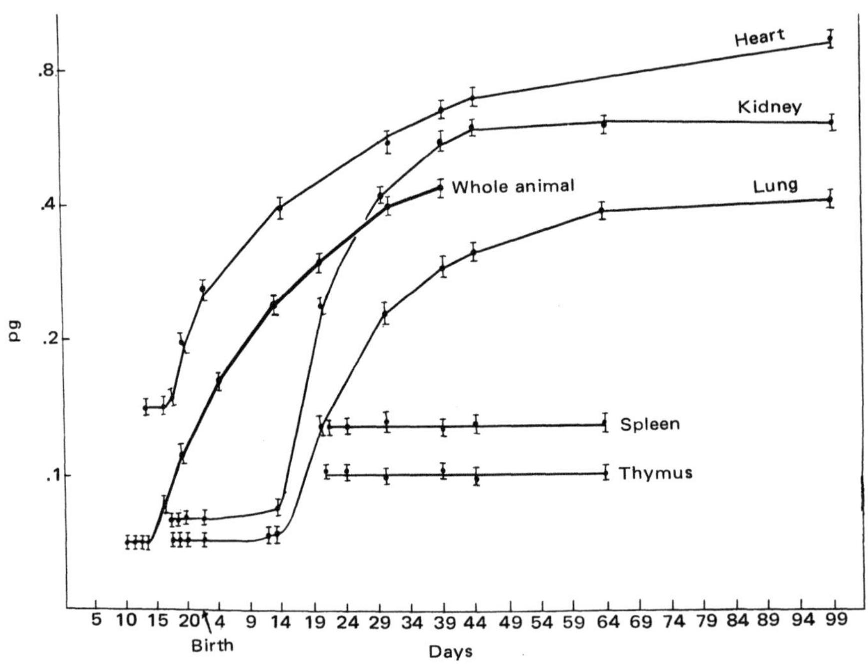

Fig. 9. Protein content per nucleus during normal growth in the rat. (From Winick and Noble, 1965.)

Geschwindt and Li (1949) observed low RNA/DNA ratios in rat liver during the prenatal period. The ratio increased between birth and day 15 and reached a maximum by day 40. An increased RNA/DNA ratio from day 10 to adult age was noted by Fukuda and Sibatani (1953), and Winick and Noble (1965) confirmed these results but saw no increase between 14 and 44 days of age. Thus, early liver growth in the rat is accompanied by a rapid accumulation of cellular RNA. Oliver and Blumer (1964) reported a small peak of incorporation of [^{14}C]orotic acid into nuclear and cytoplasmic RNA at 2 days of age and a larger one at 10 days of age. This coincides with the higher ratio of polyribosomes/monomers and dimers observed at 10 days (Otten, 1968) and suggests that the synthesis of RNA may be increased.

Recent findings in our laboratory indicate that cytoplasmic RNA synthesis is in general more rapid in immature liver than in adult rat liver (Lewis and Winick, 1977). Sereni *et al.* (1967) observed a rapid increase in incorporation of labeled precursors into nuclear RNA during gestation and around birth, reaching maximum values at 3 days of age and remaining elevated thereafter. We have shown a decrease in the incorporation of [^{14}C]orotic acid into nuclear RNA after 40 days of age (C. G. Lewis and S. J. Rozovski, personal communication). The activity of RNA polymerase has been shown to increase with age during the postnatal period, reaching a plateau around 30 days of age (Sereni *et al.*, 1967; Bagshaw and Bond, 1974; Novello and Stirpe, 1970).

Thus, it would appear that the rate of RNA synthesis increases during the prenatal period and reaches a maximum sometime after weaning. The quantity of cellular RNA may vary not only as a result of changes in synthesis but as a result of changes in the rate of degradation. The activity of alkaline RNase has been shown to be important in regulating the rate of degradation (Kraft and Shortman, 1970). RNase activity decreases in rat liver during development (Bardon and Paumůla, 1967; Bresnick *et al.*, 1966; Rahman *et al.*, 1969), and this decrease occurs both in liver nuclei and cytoplasm (Rosso *et al.*, 1973). RNase activity in cytoplasm remained constant after 2 weeks of age (Rosso *et al.*, 1973), supporting the findings of Fujii and Villee (1968), who observed no changes in RNA catabolism between 2-week-old and adult animals. Since at 30–60 days of age the rate of accumulation of cellular RNA (increased RNA/DNA) was maximal and since RNA polymerase activity is constant during this period of time (see above), the authors concluded that the increment in RNA/DNA ratio is due to increased RNA transport from nucleus to cytoplasm (Rosso *et al.*, 1973). This stresses the importance of nuclear hydrolysis of RNA in determining the amount of cellular RNA in rat liver. The observation by Faiferman *et al.* (1970), demonstrating that different types of nuclear RNA have different degrees of resistance to RNase, would suggest a quantitative and qualitative contol of RNA transported to cytoplasm during development. Thus, cellular RNA content is regulated during development not only by changes in the rate of RNA synthesis but also by changes in the rate of catabolism and by alterations in transport from nucleus to cytoplasm.

Recent results from our laboratory (Lewis and Winick, 1977) revealed two electrophoretic peaks in liver cytoplasm of 14-day-old rats which were not present in adult rats. Characterization of these peaks indicated that they were RNA with S values of about 40 and 34. These two peaks disappear from liver at about 28 days of age. At present, studies are being undertaken to further characterize these peaks. It is interesting to note that the time of their disappearance correlates with the fall in DNA polymerase activity in liver (Jasper and Brasel, 1974).

Thus, cellular growth in liver is characterized by an early hyperplastic phase during which RNA content of liver cells increase. This is due in part to increased rates of RNA synthesis and in part to decreased catabolic rates. In addition, changes in transport from nucleus to cytoplasm also occur during this period. Finally, two species of cytoplasmic RNA appear to be present in early life and to disappear with maturity. Later growth is characterized by cellular hypertrophy, with RNA content per cell remaining high as the rate of synthesis and degradation reach equilibrium.

2.3. *Placenta*

Since placenta is readily available for study, abnormalities in fetal growth that are paralleled in the placenta could more easily be investigated using this tissue. With this in mind, placental growth has been examined in the normal rat and human and under certain abnormal conditions known to affect the growth of the fetus.

Using radioautography, Jollie has demonstrated that labeled mitotic figures do not appear in the trophoblastic layer of rat placenta after day 18 of gestation (Jollie, 1964). Our own studies demonstrate that although weight, protein, and RNA rise linearly until day 20, DNA fails to increase after day 17, owing to a cessation of DNA synthesis (Winick and Noble, 1966a) (Fig. 10).

Thus, three phases of cellular growth may be described in rat placenta just as in the other organs of the rat. From 10 days until about 16 days of gestation, DNA synthesis and net protein synthesis are proportional and cell number increases, whereas cell size is unchanged. This is the period of pure hyperplasia. From 16 to 18 days, as a consequence of a slowing in the rate of DNA synthesis with protein synthesis continuing at the same rate, hyperplasia and hypertrophy occur together. Finally, at about 18 days, cell division stops altogether, while weight and protein continue their linear rise. The ratios rapidly increase. Hypertrophy occurs alone.

Maturational changes occur throughout gestation. Therefore, growth by cell division is not necessary for certain of these maturational changes to occur. During the final period of hypertrophy, certain electron microscopic changes take place in the rat placenta. There is a reduction of the "placental barrier," with the appearance of endothelial and trophoblastic fenestrations. Increased micropinocytotic activity, irregularities at the inner plasma membrane, and the appearance of large vacuoles can all be seen in element III. There is also approximation of inner and outer membranes at points of constriction and formation of pediclelike foot processes (Jollie, 1964).

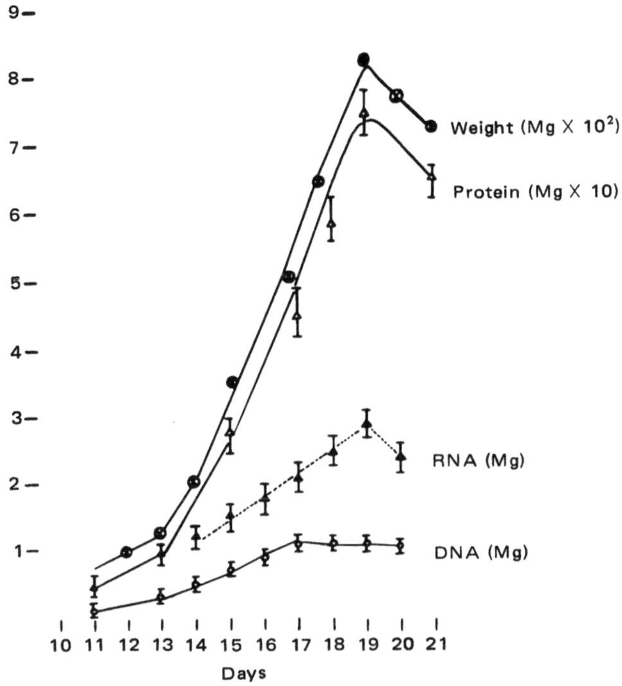

Fig. 10. Total weight, protein, DNA, and RNA during development of rat placenta. Each point represents the average of at least 15 separate determinations. The bars on the figure represent the range. (From Winick and Noble, 1966a.)

Concomitant with these morphologic changes, profound functional changes also take place. There is a change in the selectivity of transportable materials and an increase in the transport rate of certain materials. For example, the total amount of α-amino isobutyric acid or glucose transported to the fetus increases markedly throughout gestation. However, if this is related to fetal weight, which is also increasing during gestation, there is an increase, reaching a peak around 16 days and then a decrease (Rosso, 1975). The activities of ornithine decarboxylase (ODC) and S-adenosylmethionine decarboxylase (SAMD), two enzymes involved in polyamine synthesis and probably in RNA synthesis, are increased to day 17 and then decline. Polyamine concentrations follow the same course as does the accumulation of cellular RNA (Wasserman *et al.*, 1976). Also, glycogen, which had previously been deposited in copious amounts, rapidly becomes depleted (Correy, 1935). Thus, a series of biochemical and functional changes take place in developing placenta from which one may deduce that it is functioning most efficiently at the end of the second and beginning of the third week of pregnancy, and that the efficiency drops thereafter.

Although the exact timing of events is not as clear as with the rat, available data indicate that the human placenta grows in a qualitatively similar manner (Winick *et al.*, 1967). Placenta is the only human tissue in which cellular growth has been studied throughout its entire life-span. Therefore, it is not known whether the sequence to be described is characteristic for other human tissues. However, studies cited in the previous section would indicate that human brain grows in a qualitatively similar manner.

At least until the fetus reaches 3500 g, fetal weight gain is accompanied by a linear increase in the weight of the placenta. In addition, both total protein and RNA increase linearly to term. DNA, however, ceases to increase after the placenta reaches about 300 g. This corresponds to a fetal weight of about 2400 g, or a gestational age of 34–36 weeks (Winick *et al.*, 1967). Thus, as previously demonstrated in the rat, cell division ceases before term. In the human this appears to be about week 35 of gestation.

Although the cellular events are similar during the growth of human and rat placenta, there is one quantitative difference. The RNA/DNA ratio is twice as high in the rat. The reason for this difference is unknown, but it may be due to increased connective tissue within the human placenta. Fibroblasts contain relatively little RNA. Possibly the trophoblasts contain equal quantities of RNA in both species.

In summary, the normal cellular growth of placenta proceeds through an orderly sequence of changes as gestation progresses. Therefore, the time at which a stimulus is exerted may be as important as the nature of the stimulus itself. The same stimulus acting early might interfere with cell division, whereas later it cannot. Conversely, the nature of the cellular effects produced might give a clue to the time an unknown stimulus was most active. In any event, the DNA, RNA, and protein content of the placenta can be examined under conditions known to affect both fetal and placental growth. The similarity in the growth pattern between rat and human placenta also permits using the rat as an experimental model.

2.4. Skeletal Muscle

Enesco and Leblond (1962) observed an increase in DNA in rat gastrocnemius muscle with age. Since polyploidy is not a factor contributing to the increase in DNA with age in muscle (Lash *et al.*, 1957; Enesco and Puddy, 1964), this increase in DNA represents a true increase in the number of nuclei. Enesco and Puddy (1964) determined that 35% of the nuclei in samples of four different limb muscles lay outside the muscle fibers in male Sherman rats at 16, 36, and 86 days of age. Since the percentage of nuclei outside the fiber did not change over the age span studied, an increase in DNA content, even if uncorrected for nonfiber nuclei, will proportionally reflect growth in muscle cell nuclei. Munro and Gray (1969) observed that the relationship between muscle fresh weight and its total DNA content varies by twofold or less in mammals ranging in size from mice to horses. Since DNA is present mainly in nucleus, this observation indicates that the number of nuclei in muscle may determine muscle mass in different species. A study by Enesco and Puddy (1964), in which combined histometric and chemical techniques were utilized, shows that the number of individual muscle fibers in rats does not increase postnatally but that the number of nuclei within the fiber, as well as the fiber size, show significant increments with age. Note, however, that this study was carried out in normal animals and that the constant proportion with age may not hold true in animals growing abnormally. The DNA content of various striated muscles in rats has been reported to rise to fixed levels by 90–95 days of age, while increase in weight, myofibrillar proteins, and sarcoplasmic proteins continued until 140 days of age (Gordon *et al.*, 1969). Thus, it would appear that in the rat, hyperplasia of muscle fiber nuclei continues until some 90 days postnatally, and hypertrophy of muscle fiber continues to approximately 140 days. However, more recent data by Millward *et al.* (1975) indicates that there is an increase in weight and protein and in the weight/DNA and protein/DNA ratio in gastrocnemius and quadriceps muscle of male rats beyond that time. Analysis of the changes in protein and weight in the different types of muscle protein were not made in this study (Millward *et al.*, 1975). In the pig it has been argued that the number of muscle fibers in muscle remains constant shortly after birth (Staun, 1963, 1972; Stickland and Goldspink, 1973; Hegarty *et al.*, 1973; Ezekwe and Martin, 1975; Davies, 1972). However, recent studies by Swatland (1976), who estimated the number of fibers appearing in transverse sections of whole sartonius muscle in pigs by a histological technique, showed an increase in the number of muscle fibers between 56 and 168 days of postnatal age.

Cheek and his collaborators (Cheek *et al.*, 1965, 1968; Graystone and Cheek, 1969) have measured total muscle mass and total muscle nuclear number in the Sprague–Dawley rat. Using an ingenious technique, they measured DNA concentration in a sample of skeletal muscle and assessed the total muscle mass from determinations of total noncollagen protein or potassium in pulverized, defatted, dried carcass or from determination of total muscle intracellular water or calcium content of bone. Total muscle nuclear number can be calculated by the following formula:

$$\text{total muscle nuclear number} = \frac{\text{DNA concentration (mg/g)} \times \text{muscle mass (g)}}{\text{DNA content per nucleus}}$$

The validity of these methods depends on the muscle samples being representative of muscle throughout the body and on the accuracy of noncollagen protein or potassium or intracellular water as reflections of total muscle mass. These investigators point out that the four different methods give similar results for muscle in normal rats at different ages, and that these values are similar to values obtained in studies in which creatinine excretion was used to determine muscle mass and to values obtained by dissection techniques (Graystone, 1968).

Using the noncollagen method, Cheek and his collaborators (Graystone, 1968) have measured total muscle mass in normal male and female Sprague–Dawley rats from 3 to 14 weeks of age. In males muscle mass increases linearly during this period, from approximately 15 to 144 g. The female begins with the same amount of muscle mass at birth, but the rate of growth is less rapid, especially after 8 weeks of age. At 14 weeks of age, the adult female achieves a muscle mass of 90 g. When muscle mass for male and female are compared with total body weight rather than age, the difference in muscle mass disappears. When the development of the individual cell mass was studied in growing rats, an initial difference between male and female was also observed. From 3 to 14 weeks the individual cell mass in males increased tenfold, Gordon *et al.* (1966) and Millward *et al.* (1975) using total muscle analysis, but agrees with values reported by Hubbard *et al.* (1974). In females, individual cell mass increases at the same rate as in males up to weaning (3 weeks). Afterward, cells mass increases more rapidly in females, until 9 weeks. By 13 weeks the value for the male is again equal to that of the female (Cheek *et al.*, 1968, 1971; Elliott and Cheek, 1968a).

On the basis of a single muscle sample for determination of DNA content (Cheek *et al.*, 1968), the total number of muscle nuclei is achieved by 8 weeks of age in normal male Sprague–Dawley rats. A spurt in the rate of DNA accumulation occurs during puberty (6–8 weeks). In the female, after 3 weeks of age the number of nuclei increase much more slowly than in the male. In addition, there is little acceleration in the rate of cell division during puberty. This results in a final nuclear number of approximately two-thirds of the male value at 14 weeks, (Fig. 11).

The discrepancy between the figure of 90–95 days of age, cited earlier, and this value of 8 weeks, or 56 days, of age of cessation of DNA accumulation in rat skeletal muscle may relate to the problems of extrapolation from a single specimen to total muscle DNA content.

Although Cheek *et al.* (1971) report agreement for DNA concentration in several muscles of normal Sprague–Dawley rats and in five muscles from young *Macaca mulatta* monkeys, Enesco and Puddy (1964) find agreement in only

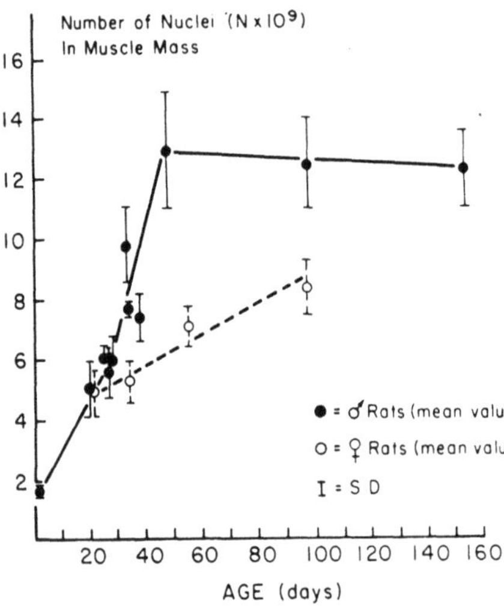

Fig. 11. Number of nuclei in muscle mass for male and female rat against age. (Adapted from Cheek *et al.*, 1971.)

certain muscles in male Sherman rats. In addition, there are differences noted with age. In hypophysectomized rats, DNA concentration of various muscles may differ (Beach and Kostyo, 1968). Hubbard *et al.* (1974) found marked differences in DNA concentration between "fast-twitch" (plantaris and gastrocnemius) and "slow-twitch" (soleus) muscles in untrained Sprague–Dawley rats. Widdowson and Crabb (1976) found similar concentrations of DNA in gastrocnemus and quadriceps in neonatal pigs. These data suggest that DNA concentration may vary between different muscle groups. Therefore, the calculation of total muscle nuclear population from the DNA content of a single sample may not provide entirely accurate values, although trends with growth, disease, or therapy might well be assessed in this way.

In studies done in fetal *Macaca mulatta* monkeys, Cheek (1971) observed a significant increase in muscle nuclear number during the last trimester of pregnancy. The number of nuclei doubled and the ratio of protein to DNA increased markedly. The authors suggest a correlation between protein/DNA or DNA content of muscle mass to fetal age or fetal weight, while in the postnatal period the number of nuclei correlate better with age, and the size of muscle cell (protein/DNA) is closely related to muscle mass (Cheek, 1971).

Methods of assessing muscle cell growth in the human must be adapted to biopsy sampling for biochemical measurements of DNA and protein. All the reservations we have noted about the reliability of a single sample for the assessment of DNA content of the entire skeletal musculature pertain to human studies as well as to rat data. However, such biopsy data are the only ones available for the human.

Creatinine excretion has been used to measure total muscle mass. Graystone (1968) ably reviews early studies that support the high correlation between the fat-free body mass and urinary creatinine levels. Under the conditions of a low creatinine/low hydroxyproline diet for 3 days before and including 3 consecutive days of urine collection, muscle mass can be calculated by the following formula:

$$\text{muscle mass (kg)} = \text{mean urinary creatinine excretion per day (g)} \times 20$$

The derivation of this factor of 20 is well documented in Graystone's paper. Linear relationships with high coefficients of correlation are obtained in normal children with no apparent sex difference when creatinine excretion in milligrams per day is plotted against body weight. When height is used as a baseline, normal males and females have similar amounts of muscle per unit height until a height of 137.5 cm is reached (Fig. 12). Thereafter, growth in muscle mass in boys accelerates rapidly, achieving values at early adolescence of 1½–2 times the values per unit height noted in normal females. Sex differences, especially in later childhood, occur when creatinine excretion is compared with chronological or bone age. Graystone (1968) has also determined the mathematical relationships between creatinine excretion and total body water, extracellular volume, total body chloride, total body potassium, and intracellular water. This remarkable investigation documents normal growth in muscle mass in childhood and describes its relation to the other major body

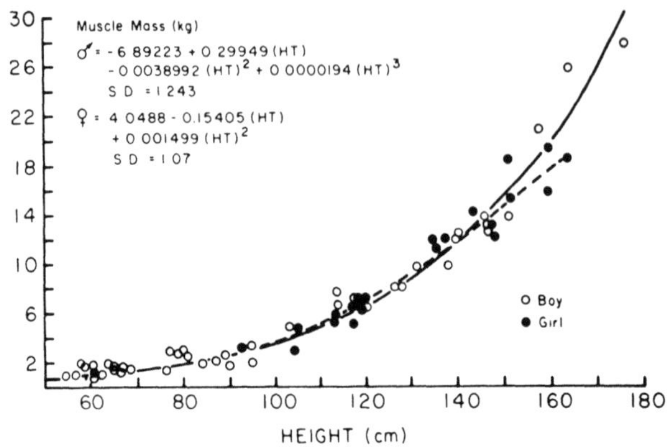

Fig. 12. Relationship between muscle mass (obtained from urinary creatinine excretion) and length for males 0.1–17 years and females 3–17 years. (Adapted from Cheek and Hill, 1970.)

compartments. It provides valuable baseline information for the study of abnormal growth.

Using a single muscle biopsy and 24-hr creatinine excretion, Cheek (1968) calculated total muscle nuclear population with the equation on page 75. He determined that muscle nuclear number in male infants increases linearly with age, length, total body water, and basal oxygen consumption. The equations are given in Table I. No data are available for males between 1½ and 5 years of age. From 5 to 10½ years the mathematical relationship between muscle nuclear number and age is again linear. The slope of the line is less, however; that is, the rate of DNA replication in muscle tissue is less rapid in the older boys. At the age of 10½ years the rate of DNA replication in muscle tissue again accelerates (Fig. 13).

From 5 to 16 years, growth in muscle nuclear number is linear with total body water and basal oxygen consumption. Estimates of the number of muscle nuclei in males at various ages indicates a 14-fold increase between 2 months and 16 years of age. In normal females from 6 to 17 years, growth in muscle nuclear number is linear with age, but in contrast to males there is no accel-

Table I. Muscle Nuclear Number in Male Infants of 0.18 to 1.5 Years of Age[a]

Nuclear number × 10^{12}	*N*	Correlation coefficient	SD
Number = 0.0895 (CA)[b] + 0.206	17	0.66	0.043
= 0.00503 (HT) − 0.076	17	0.73	0.0396
= 0.02915 (TW) + 0.116	14	0.70	0.043
= 0.0076 (cal/hr) + 0.118	13	0.93	0.025

[a] Adapted from Cheek (1968).
[b] *Abbreviations.* CA: chronologic age in years; HT: height in centimeters; TW: total body water in liters.

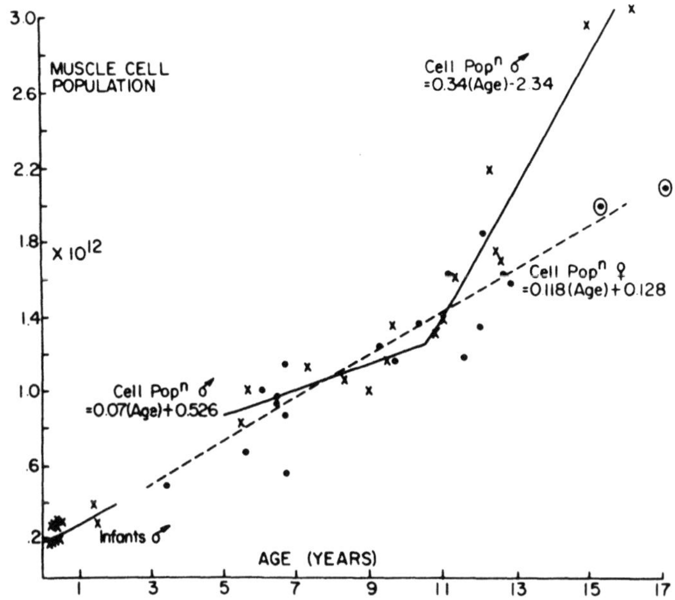

Fig. 13. Muscle nuclear number in children versus age for normal children. The ×'s represent individual points of data for males. The dots represent individual points of data for females. The equations for the separate lines are recorded. (From Cheek, 1968.)

eration in the rate of DNA replication during adolescence (Fig. 13). Muscle nuclear number is also linear with height, total body water, and basal oxygen consumption. There are sex differences with age and height but not with total body water or basal oxygen consumption.

Using the protein/DNA ratio, Cheek (1968) has followed growth in "cell mass" in normal children. Muscle cell size in girls increases faster than in boys, and by 10 years of age the girls have reached the maximum cell size. Boys do not catch up until 15 years of age (Fig. 14). It is possible that further increments in cell size occur later in age for males (Cheek and Hill, 1970).

These data indicate that the male adolescent growth spurt is associated with an increase in muscle nuclear number.

These extensive studies of Cheek and his co-workers assume a constant DNA content and a constant ratio of fiber to nonfiber nuclei in all skeletal muscle and at all ages from birth to adolescence. The validity of these assumptions remains to be substantiated, and the numerical results may change with further studies. Nevertheless, the trends described in cell growth during development cannot be expected to change significantly.

3. *Effect of Diet on Cellular Growth*

The commonest method employed in altering the nutritional status of neonatal rats is to vary the number of pups nursing from a single mother. The

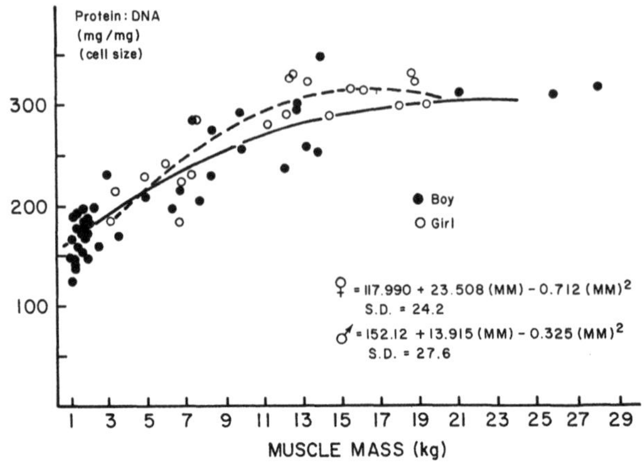

Fig. 14. Muscle cell size (protein/DNA related to muscle mass) for boys and girls during growth. (Adapted from Cheek and Hill, 1970.)

normal rat litter consists of 8–12 pups, and therefore a nursing group of 10 animals has arbitrarily been considered normal. Malnutrition is imposed by increasing the size of the nursing group to 18 animals, and overnutrition by decreasing the size to 3.

More recently, other methods of undernutrition have been employed. Protein restriction in the lactating mother reduces the quantity of milk produced without altering its composition. Allowing the animals to nurse for only a single 8-hr period per day also reduces the quantity of milk consumed. All these methods produce a total caloric restriction as well as a restriction in individual nutrients, the most notable of which is protein. So far all three methods have produced comparable results on organ growth; therefore, we shall examine them together.

To produce qualitative changes in the milk without changing the quantity produced, the nursing animal must be artificially fed. The two procedures that have been employed are repeated tube feedings and gastrostomy with continuous infusion of liquid. Both are time consuming, extremely tedious, and technically difficult. Miller (1969) has used repeated tube feeding extensively to study protein synthesis in the developing liver. Another technique that is being used currently is the clamping of the uterine artery during gestation (Wigglesworth, 1964). This method reduced the blood flow to the fetuses, thereby curtailing the nutrient supply.

The "large and small" litter technique (Kennedy, 1957) was used extensively by Widdowson and McCance (1960), who demonstrated that the growth rate of nursing pups was inversely proportional to the number of animals in the nursing group. In addition, those pups malnourished during lactation by being reared in large litters remained stunted for the rest of their lives, regardless of how they were fed after weaning (McCance and Widdowson, 1962).

Previous studies had indicated that undernutrition later in the growing period of the rat would retard growth but that nutritional rehabilitation could restore normal body weight (Jackson and Steward, 1920). Later, Winick and Noble (1966b) expanded these studies by examining the changes in cell number and cell size in various organs of rats subjected to malnutrition at various stages of postnatal development. Their experiments clearly demonstrated that if malnutrition were imposed during the proliferative phase of growth, the rate of cell division was slowed and the ultimate number of cells was reduced. Moreover, this change was permanent and could not be reversed once the normal time of cell division has passed. In contrast, undernutrition imposed during the period when cells are normally enlarging will curtail the enlargment, but on subsequent rehabilitation the cells will resume their normal size.

3.1. Brain

The experiments described above (Winick and Noble, 1966b) demonstrated that total brain cell number could be permanently reduced by undernourishing the rat during the first 21 days of his life and that no matter what is attempted thereafter, this reduction in cell number would persist.

If the reduction in brain size in the animals reared in litters of 18 was due to a reduced cell number, how about those reared in litters of 3? When these experiments were performed (Winick and Noble, 1967), it became clear that these overnourished animals had an increased number of brain cells when compared to brains of animals nursed in normal size litters. Thus, the number of cells attained by the developing rat brain depends, in part, on the nutrition of the animals during the period of time when brain cells are actively undergoing proliferation. Subsequent experiments have demonstrated that the rate of cell division can actually be manipulated in either direction by changing the state of nutrition during the proliferative phase (Winick *et al.*, 1968). Thus, nutrition for the first 9 days of life produced a deficit in brain cell number which can be entirely overcome by overnourishing the animal for the next 12 days. It should be noted here that, as pointed out earlier, we cannot differentiate one cell type from another with these methods. It is therefore possible that the deficit is made up by proliferation of a different cell type than that which was inhibited during the earlier restriction.

Malnutrition during the first 21 days of life also inhibits lipid synthesis in whole rat brain. The rate of cholesterol and phospholipid synthesis is reduced and the total brain quantity of these materials is lowered (Davison and Dobbing, 1966). This reduction is proportional to the reduction in DNA or cell number, and hence the ratios of these lipids to DNA or the amount of these lipids per cell is unchanged. If the malnutrition continues beyond the proliferative phase of growth, the continued inhibition of lipid synthesis will result in a reduced lipid content per cell. Enzymes involved in lipid synthesis, such as galactocerebroside sulfokinase, are also reduced in activity by malnutrition during the first 10 days of life (Chase *et al.*, 1967). In addition, specific lipids, such as cerebrosides and gangliosides, are reduced both in content and concentration, suggesting that the dedritic arborizations (where these lipids are localized) are reduced in number.

Malnutrition early in life has also been shown to increase the rate of RNA turnover in brain. This is accompanied by an elevation in the activity of alkaline RNase and a reduction in the concentration of polyamines.

Regional patterns of cellular growth are also modified by malnutrition during the nursing period (Fish and Winick, 1969). Cerebellum, where the rate of cell division is most rapid, is affected earliest (by 8 days of life) and most markedly. Cerebrum, where cell division is occurring at a slower rate, is affected later (at 14 days of life) and less markedly. The effects produced include a reduced rate of cell division in both areas, as well as a reduction in overall protein synthesis and in the synthesis of various lipids. In addition to these effects on areas of rapid cell division, the increase in DNA content which normally appears in hippocampus between days 14 and 17 is delayed and perhaps even partially prevented (Fig. 15).

It would appear from these data that those regions in which the rate of cell division is highest are affected earliest and most markedly and that cell migration is also curtailed. Whether this is actually an interference with migratory patterns or an inhibition of cell division at the source below the lateral ventricle is not fully known. But data to be discussed shortly strongly suggest that the latter accounts for at least some of the reduced cell number in hip-

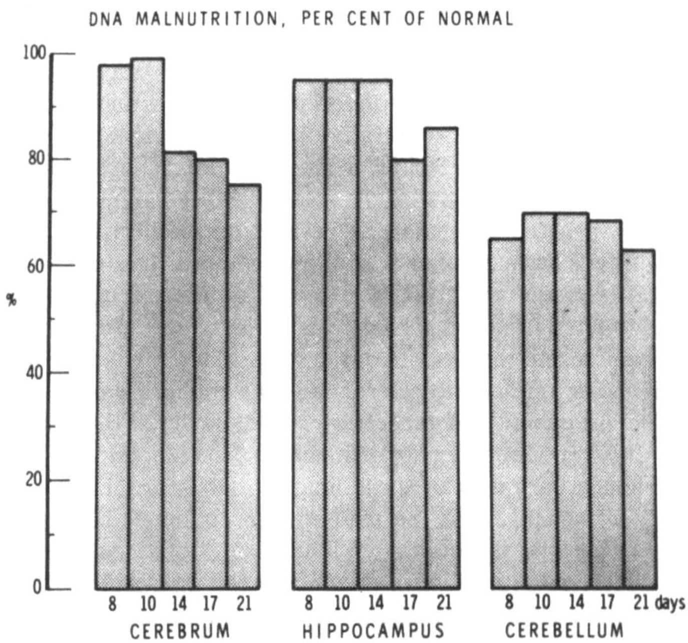

Fig. 15. Effect of malnutrition on regional growth of the rat brain. (From Winick, 1970c.)

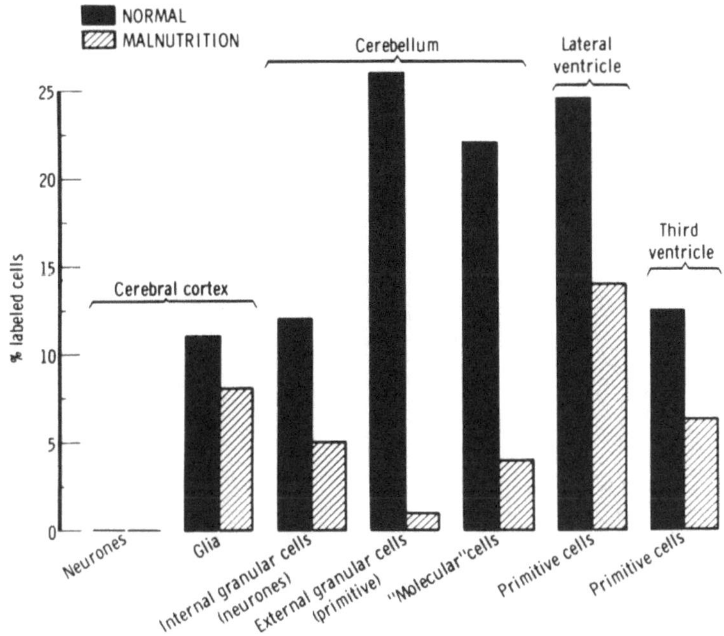

Fig. 16. Effect of malnutrition on individual cell types in rat brain. (From Winick, 1970.)

pocampus. Regional patterns of lipid synthesis and the effects of malnutrition on these patterns have not been clearly established. What data are available, however, suggest that areas where myelination is most rapid are most vulnerable to the effects of early malnutrition (Culley and Lineberger, 1968).

In all the discussion to this point, individual cell types have not been considered. At present, three types of studies have been conducted on malnourished animal brains during rapid growth. The first is careful histologic examination, employing a variety of special stains; the second is histochemical examination in an attempt to differentiate effects on patterns of specific enzyme development; and the third is radioautographic study, to determine the effect of undernutrition on the division of particular cell types. Unfortunately, the same species have not been employed in all these studies, which makes cross-comparison difficult.

Histologic changes have been observed in the central nervous systems of rats (Platt, 1962), pigs, and dogs (Platt *et al.*, 1964) reared after weaning on protein-deficient diets. Both neurones and glia in spinal cord and medulla degenerate. These changes persist even after intensive rehabilitation with a protein-rich diet lasting for as long as 3 months. The changes could be made more severe either by beginning the restriction at an earlier age or by extending the duration of the deficient diet. In pigs it has also been demonstrated that severe undernutrition early in life produces histologic changes in the cortex itself. Neurons in the gray matter are reduced in number and appear swollen. More recently, histochemical changes have been described in the brains of

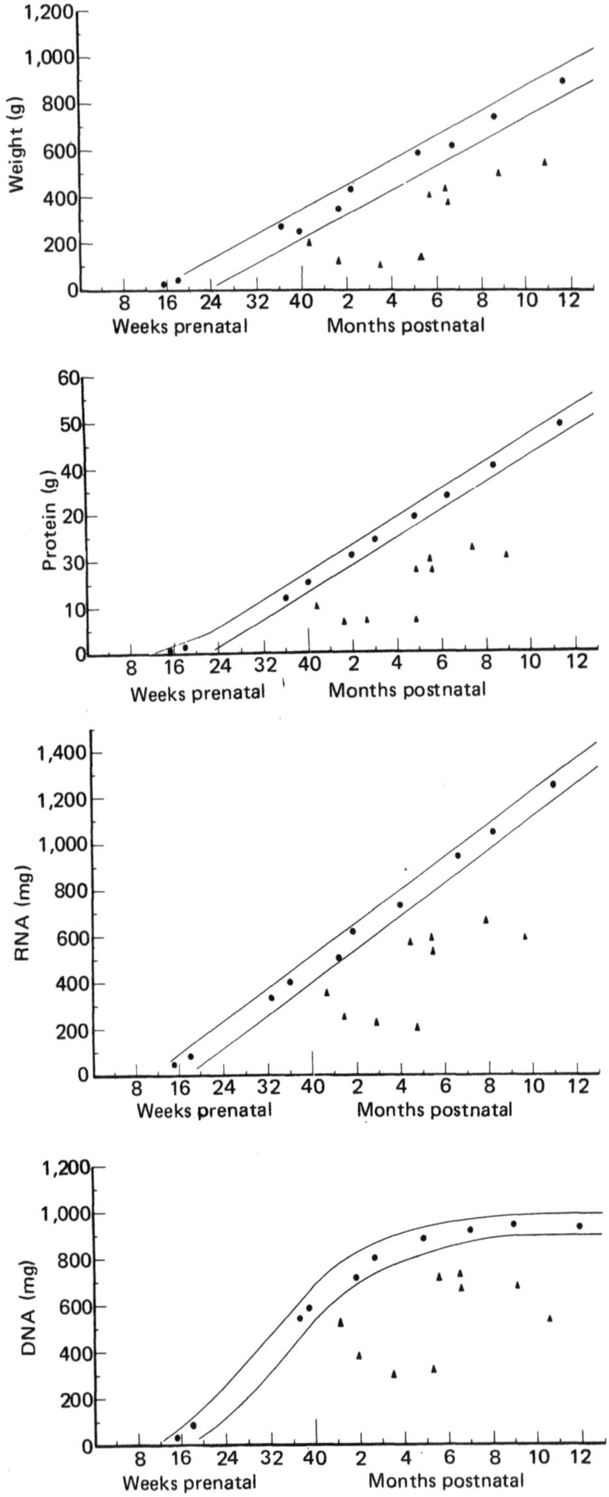
1,200
1,000
800
600
400
200
0
Weight (g)
8 16 24 32 40 2 4 6 8 10 12
Weeks prenatal
Months postnatal
60
50
40
20
30
10
0
Protein (g)
8 16 24 32 40 2 4 6 8 10 12
Weeks prenatal
Months postnatal
1,400
1,200
1,000
800
600
400
200
0
RNA (mg)
8 16 24 32 40 2 4 6 8 10 12
Weeks prenatal
Months postnatal
1,200
1,000
800
600
400
200
0
DNA (mg)
8 16 24 32 40 2 4 6 8 10 12
Weeks prenatal
Months postnatal

rats submitted to early malnutrition (Zeman and Stanbrough, 1969). The appearance of a variety of enzymes, demonstrable by special staining techniques, is delayed and the ultimate quantity obtained is reduced. Thus, early malnutrition produces specific histologic and histochemical changes within the cells of the central nervous system. Again, the earlier the malnutrition, the more severe the damage and the more likely it is to persist.

Radioautographic studies indicate that in neonatal rats malnourished for the first 10 days of life, only glial cell division is inhibited in cerebrum, since neuronal cell division ceases prior to birth. In cerebellum, the rate of cell division of external granular cells, internal granular cells, and molecular cells is reduced.

In addition, the rate of cell division in neurons under both the third and lateral ventricle is decreased (Winick, 1970a) (Fig. 16). This reduction in neurons under the lateral ventricle explains, at least in part, the reduced DNA content in hippocampus 5 days later, since these are the cells that are destined to migrate into the hippocampus.

The effect of malnutrition on the human brain has only been studied to a limited extent (Winick *et al.*, 1970; Rosso *et al.*, 1970). The data indicate that in marasmic infants who died of malnutrition during the first year of life, wet weight, dry weight, total RNA, total cholesterol, total phospholipid, and total DNA content are proportionally reduced. Thus, the rate of DNA synthesis is slowed and cell division curtailed, resulting in a reduced number of cells (Fig. 17). Since the reduction in the other elements was proportional to the reduction in DNA content, the ratios are unchanged, hence, the size of cells or the lipid content per cell is not altered. It is again emphasized that these are "average" cells we are describing; it is quite possible that certain cells (i.e., those with lipid being actively deposited) are being affected differently from those in which this is not occurring. If the malnutrition persists beyond about 8 months of age, not only are the number of cells reduced, but the size of individual cells is reduced. In addition, the lipid per cell is reduced. Recent data suggest that the ganglioside content and concentration of human brain may be reduced by early malnutrition. Since gangliosides are localized within dendritic arborizations, these data suggest that there may be a reduction in such arborizations in the brains of children who suffer severe early malnutrition (J. Dickerson, personal communication). Thus, in human brain there is a similar type of response to malnutrition. During proliferative growth, cell division is curtailed; during hypertrophic growth, the normal enlargement in cells is prevented.

It is obviously not possible to collect recovery data in infants, but indirect data suggest that the situation may be similar to the situation in animals. Since head circumference was correlated with these cellular parameters and was reduced in proportion to the reduction in the number of cells in these infants, their head circumferences were appropriate for their brain sizes and cell num-

←

Fig. 17. Cellular growth in the brains of normal and malnourished children. Lines indicate normal range for U.S. population; circles indicate normal Chilean children; triangles indicate Chilean children who died of severe malnutrition during the first year of life. (From Winick and Rosso, 1969b.)

ber and reduced for their ages. In similar children recovering from this type of severe marasmus, this reduced head circumference persisted even after maximum rehabilitation until they were at least 5 years old. Regional effects of malnutrition in the human have been studied and the data indicate that reduction in cell number occurred in cerebrum, cerebellum, and brain stem in children who died of marasmus during the first year of life (Winick *et al.*, 1970) (Fig. 18). Thus, the available data indicate that the effects of malnutrition on human brain are qualitatively quite similar to the effects on rat brain. However, the quantitative events have still not been worked out. Recently, polyamine content in human brain during development and the effects of early malnutrition on polyamine content have been studied. The data indicate that putrescine is present in highest concentrations and increases in a manner parallel to the increase in nucleic acid content. In addition, areas in which myelination is rapid contain high concentrations of putrescine. Finally, early malnutrition results in a lowering of putrescine and spermidine concentration, with no change in spermine concentration.

3.2. Liver

Some effects of early malnutrition on cell division in liver were noted in the last century. Morpurgo (1889) reported that during starvation there was a reduction in the number of mitotic figures in the livers of young rabbits. In some cases they were completely lacking but reappeared upon subsequent alimentation. Leduc (1949) reported similar findings in young mice during starvation and observed an increase in mitotic activity during rehabilitation. Similar results were obtained when mice were reared on a low-protein diet and then transferred to one containing a higher level of protein (Leduc, 1949; Argyris, 1971). Later experiments have shown that both the appearance of polyploidy and of binuclear cells, which normally occurs during liver development (Fig. 7), is retarded in animals reared on a protein-free diet (Najdal and Zajdela, 1966b; Wheatley, 1972; Mariani *et al.*, 1966).

Thus, slowdown or cessation of cell division during malnutrition has been confirmed by indirect studies. Dallman (Dallman, 1971; Dallman and Manies, 1973a), showed a decrease in incorporation of radioactive thymidine into nuclear DNA in the liver of young rats placed on a low-protein diet. By contrast, incorporation into mitochondrial DNA was increased after 14 days of malnutrition (Dallman and Manies, 1973a). The authors suggest that this represents an adaptation in the liver cell, preserving function at the expense of growth.

Although Dallman and Manies (1973a) found no change in DNA polymerase in liver nuclei of rats malnourished postweaning, others have found levels of this enzyme and of thymidine kinase reduced in livers of animals malnourished before weaning (Tagliamante *et al.*, 1972; Jasper and Brasel, 1974). Using a double-isotope technique, Dallman and Manies (1973b) also found increased reutilization of amino acids in the livers of growing rats. Activity of amino

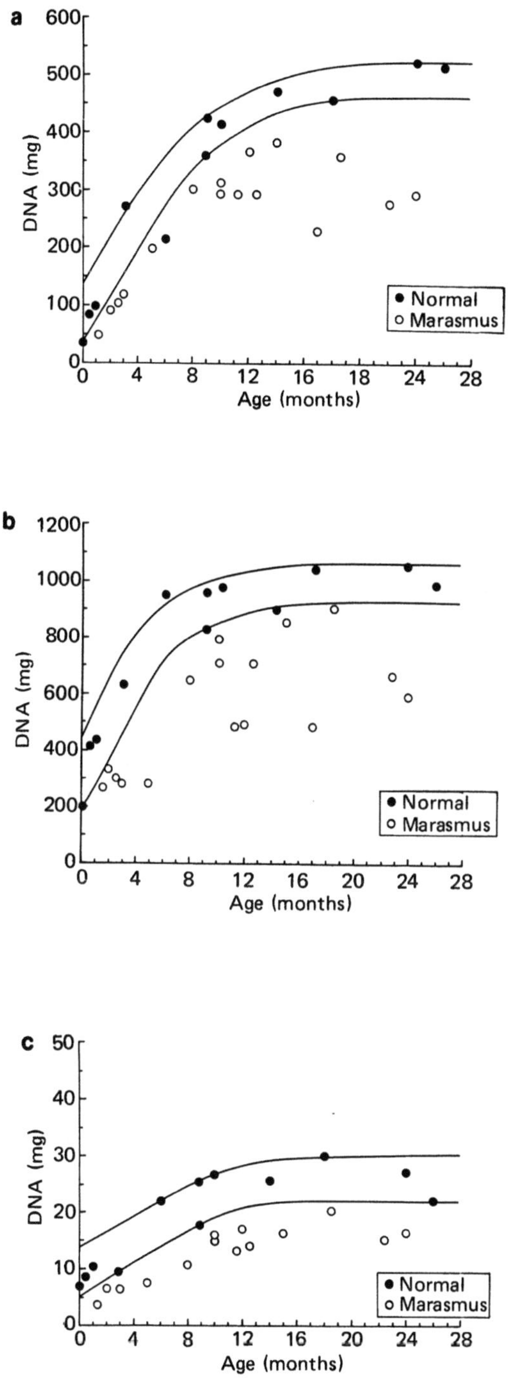

Fig. 18. DNA content in various regions of the human brain. (a) Cerebellum. (b) Cerebrum. (c) Brain stem. (From Winick *et al.*, 1970.)

acids activating enzymes in young rat liver increases proportionately with the increase in DNA content during early growth, so that the activity per milligram of DNA remains constant during the growing period (Mariani *et al.*, 1966). In malnourished animals, the activity of amino acid-activating enzymes was increased, suggesting preferential utilization of amino acids for protein synthesis.

RNA metabolism is profoundly altered during malnutrition. Total RNA content is reduced as much as 50% in the liver of protein-deficient rats (Kosterlitz, 1947; Munro *et al.*, 1953; Winick and Noble, 1966b). By contrast, incorporation of radioactive precursors into RNA is increased (Munro *et al.*, 1953; Wannemacher *et al.*, 1968; Shaw and Fillios, 1968; Lewis and Winick, 1978; Rozovski *et al.*, 1978), as is the activity of DNA-dependent RNA polymerase (Shaw and Fillios, 1968; Quirin-Stricker and Mandel, 1968). Lewis and Winick (1978) observed that short-term feeding of a diet low in proteins leads to cellular events which stimulate the nucleus to synthesize and process nucleolar and nucleoplasmic RNA more rapidly.

The discrepancy between a decrease in RNA content and an increase in RNA synthesis during malnutrition can be explained by an increased rate of degradation of RNA, reflected by an increased activity of ribonuclease (Girija *et al.*, 1965; Enwonwu and Sreebny, 1971; Rosso and Winick, 1975).

Studies in our laboratory have examined the effect of prolonged malnutrition on the metabolism of RNA and polyamines in the liver of young adult rats placed on a 6% casein diet for 5 weeks (Rozovski *et al.*, 1976, 1978). Polyamines have been shown to be involved in RNA metabolism (Russell, 1973; Cohen, 1971; Bachrach, 1973), and recently a close relationship has been found between ornithine decarboxylase (ODC), the first enzyme in polyamine synthesis, and RNA polymerase (Mannen and Russell, 1977). Our studies have demonstrated that the incorporation of orotic acid into nuclear RNA increased after 1 week of malnutrition, but by 5 weeks it had returned to control values (Fig. 19). The activity of ornithine decarboxylase, *S*-adenosylmethionine decarboxylase, and the cellular content of putrescine and spermidine followed a similar pattern (Fig. 20). Total liver DNA content increased in the malnourished animals up to the third week of malnutrition and then this increase stopped (Rozovski *et al.*, 1978). We have interpreted these results as indicating a process of adaptation to malnutrition where certain indispensable functions are preserved (i.e., RNA and polyamine synthesis) at the expense of other dispensable functions (growth or DNA synthesis). Once this is accomplished and the requirements are reduced, RNA metabolism returns to normal. This interpretation is strengthened by the observation of Shaw and Fillios (1968), who noted an increase in leucine incorporation into protein and an increase in the activity of RNA polymerase in liver of rats during the first days on a 5% protein diet. However, when malnutrition was continued for a prolonged period, these parameters returned to normal or near-normal values. Thus, malnutrition affects the growing liver by initially increasing the rate of protein synthesis and the turnover of RNA. At the same time, DNA synthesis and hence cell division is slowed and finally ceases. This cessation of cell division

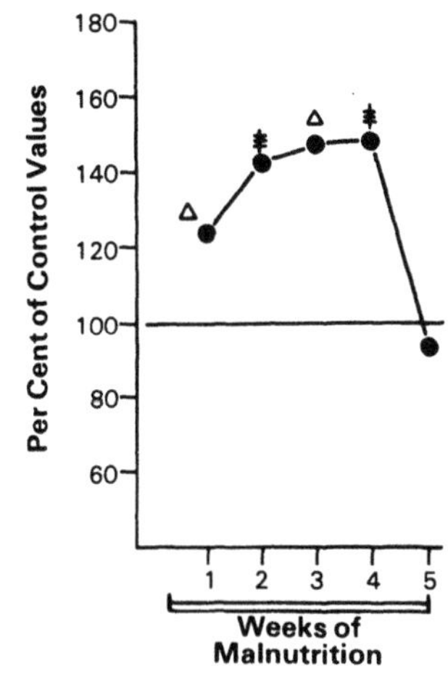

Fig. 19. Incorporation of [^{14}C]orotic acid into nuclear RNA (dpm/μg of RNA) in rat liver during malnutrition. Values expressed as percent of control. Statistically significant from control: ‡, $p < 0.01$; △, $p < 0.005$. (Adapted from Rozovski *et al.*, 1978.)

is accompanied by a return to normal in the rate of RNA and protein synthesis. This sequence of events may represent a biochemical adaptation to early malnutrition.

3.3. *Placenta and Fetus*

During intrauterine life, all organs of the fetus are in the hyperplastic phase of growth. At no other time should the organism be more susceptible to nutritional stresses. And yet only recently has any information about fetal malnutrition been forthcoming. This is true probably for two reasons, one operational and the other philosophical. The first was the relative inaccessi-

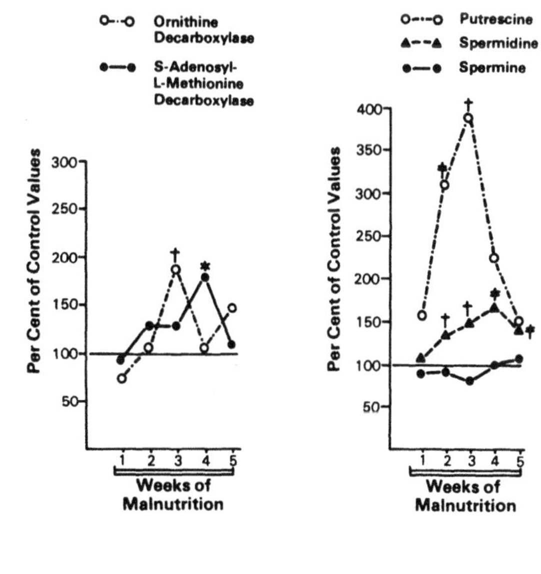

Fig. 20. Activity of polyamine-synthesizing enzymes (pmol of CO_2/30 min/mg of protein) and cellular content of polyamines (nmol/mg of DNA) in rat liver during malnutrition. Values are expressed as percent of control. Statistically significant from control: †, $p < 0.05$, ‡, $p < 0.02$; *, $p < 0.001$. (Adapted from Rozovski *et al.*, 1978.)

bility of the fetus for experimental manipulation. The second has been the generally accepted view of the fetus as the perfect parasite, extracting its needs from its mother. Recently, as researchers have ventured to study the uterus, this widely accepted viewpoint is being challenged. Fetal malnutrition may result from reduced maternal circulation, inadequate nutrients within the maternal circulation, or faulty placental transport of specific nutrients. The first two situations are now being extensively investigated in experimental animals.

The supply of blood to a single fetus in an animal delivering a litter of fetuses may be reduced spontaneously. It is not uncommon to see a "runt" in a litter of dogs or cats, and it is common knowledge that these animals will survive only with special care and that they will never reach the same final size as their littermates even if this special care is given. Occasionally, the same situation occurs in a litter of pigs. Widdowson has studied the cellular changes which take place in the organs of these spontaneously occurring "runt" pigs. Her findings indicate that cell division has been curtailed in heart, kidney, brain, and skeletal muscle, the only organs studied so far. Cell size was also reduced in all organs studied when compared to littermate controls (Widdowson, 1970).

In the rat, blood supply can be artificially reduced by clamping the uterine artery supplying one uterine horn. Using this technique, Wigglesworth (1964) has compared the growth of the fetuses in the ligated horn to that of the fetuses in the unligated horn. Growth rate was reduced in proportion to the distance of the particular fetus from the ligated artery. Those at the uterine end closest to the ligation generally died. As one progressed farther away from the ligated uterine artery and closer to the intact ovarian artery, growth rate increased. More recently, the cellular growth of various fetal organs, including placenta, has been studied in surviving animals within the ligated horn. Ligation on the thirteenth day of gestation will affect the rate of cell division in placenta and all fetal organs except the brain. Ligation on the seventeenth day will curtail cell division in the fetal organs, again sparing brain, but in the placenta, cell size will be reduced with cell number remaining normal (Winick, 1968b). Thus, in currently available animal studies in which blood supply has been either artificially or spontaneously curtailed, the rate of cell division in fetal organs, excluding brain, has been retarded. Placenta, moreover, responds in a manner that might have been predicted from the earlier studies involving early postnatal malnutrition. Ligation during the period of hyperplasia results in reduced cell number, whereas ligation during hypertrophy results in reduced cell size. Therefore, by determining the final effect on placental growth at delivery, it may be possible to pinpoint the time at which a stimulus producing such a result must have been active. As we shall see, this possibility may have relevance in the human, where placenta is the only tissue readily available for study. Another abnormality was defined in placentas from the ligated horns: elevation of total organ RNA content and hence an elevation of the RNA/DNA ratio or RNA per cell (Winick, 1968b). Such elevations in tissue RNA/DNA have been described in several tissues under a variety of circumstances. Clamping the aorta results in an increased RNA/DNA ratio in the left ventricle (Gluck *et al.*, 1964). Repeated nerve stimulation results in an elevation of the RNA/DNA ratio in the innervated muscle (Logan *et al.*, 1952), injection of estrogen results in an increased RNA/DNA ratio in the uterus, and removal of one kidney will result in an increased RNA/DNA ratio in the contralateral kidney (Karp *et al.*, 1971). The exact significance of this change is unknown, but it has been described under conditions requiring increased protein synthesis. This increase in placental RNA/DNA ratio may therefore represent an abortive attempt by placental cells to increase their rate of protein synthesis secondary to the stress of vascular insufficiency. Further evidence that this may be occurring comes from the fact that the activity of alkaline RNase, an enzyme involved in RNA catabolism, is elevated in placentas within the ligated horn. This suggests that the overall turnover of RNA is increased under these conditions (Velasco *et al.*, 1973).

Maternal protein restriction in rats will also retard both placental and fetal growth. In placenta, cell number (DNA content) was reduced by 13 days after conception, cell size (protein/DNA ratio) remained normal, and the RNA/DNA ratio was markedly elevated. Retardation in fetal growth first became apparent at 15 days, followed by a progressive decrease in cell number in all the organs studied. By term there were only about 85% the number of brain cells in control animals (Winick, 1968b). These data agree with previous data of Zamenhof and co-workers (Zamenhof *et al.*, 1968), which showed a similar reduction in total brain cell number in term fetuses whose mothers were exposed to a slightly different type of nutritional deprivation. Maternal protein restriction will also result in a lowering of placenta DNA polymerase activity, reflecting the reduced rate of cell division. In addition, alkaline RNase activity is elevated. The activities of ornithine decarboxylase (ODC) and *S*-adenosyl methionine decarboxylase (SAMD), two enzymes involved in the synthesis of polyamines, are reduced, as are the concentrations of putrescine and spermidine.

The transport of certain nutrients across the placenta, such as amino acids and certain sugars, is also reduced by maternal malnutrition. Thus, the cellular changes produced by severe prenatal food restriction are reflected in the placenta even earlier than in the fetus, but retardation of cell division in all fetal organs, including brain, can be clearly demonstrated.

By employing radioautography after injecting the mother with tritiated thymidine, cell division can be assessed in various discrete brain regions. Differential regional sensitivity can be demonstrated in this way by the sixteenth day of gestation in the brains of fetuses of protein-restricted mothers. The cerebral white and gray matter are mildly affected. The area adjacent to the third ventricle and the subiculum are moderately affected, whereas the cerebellum and the area directly adjacent to the lateral ventricle are markedly affected (Winick *et al.*, 1969; Winick, 1971). These data again demonstrate that the magnitude of the effect produced on cell division is directly related to the actual rate of cell division at the time the stimulus is applied. Moreover, they demonstrate that the maternal–placental barrier in the rat is not effective in protecting the fetal brain from discrete cellular effects caused by maternal food restriction.

The subsequent course of these animals born of protein-restricted mothers can be examined. Chow and Lee have reported that even if these animals are raised normally on foster mothers, they demonstrate a permanent impairment in their ability to utilize nitrogen (Chow and Lee, 1964). Data from our own laboratory demonstrate that if these animals are nursed on normal foster mothers in normal-sized litters, they will remain with a deficit in total brain cell number at weaning. Thus, we can again see early programming of the ultimate number of brain cells. This program, moreover, is written *in utero* in response to maternal nutrition.

These same newborn pups of protein-restricted mothers may be subjected to postnatal nutritional manipulation. If they are raised in litters of three on normal foster mothers until weaning, the deficit in total number of brain cells may be almost entirely reversed (Winick *et al.*, 1969). Although quantitatively the number of cells approaches normal, qualitatively the deficit at birth might very well be made up by an increase in cell number in different areas from those most affected *in utero*. Thus, although it may appear that optimally nourishing pups after exposing them to prenatal undernutrition will reverse the cellular effects, this may not actually be so in specific brain areas.

Perhaps the most analogous comparison to the situation in humans is to expose these pups, malnourished *in utero*, to subsequent postnatal deprivation. One can raise these animals on foster mothers in groups of 18. Animals so reared show a marked reduction in brain cell number by weaning. This effect is much more pronounced than the effect of prenatal or postnatal undernutrition alone (Winick *et al.*, 1968). Animals subjected to prenatal malnutrition alone, as previously described, show a 15% reduction in total brain cell number at birth. Animals subjected only to postnatal malnutrition show a similar 15–20% reduction in brain cell number at weaning. In contrast, the "doubly deprived" animals demonstrate a 60% reduction in total brain cell number by weaning (Fig. 21). These data demonstrate that malnutrition applied constantly throughout the entire period of brain cell proliferation will result in a profound reduction in brain cell number, greater than the sum of effects produced during various parts of the proliferative phase. It would appear that the duration of malnutrition, as well as severity during this early critical period, is extremely important in determining the ultimate cellular makeup of the brain.

Recent experiments by Widdowson (1970) in the guinea pig demonstrate that caloric restriction during gestation markedly reduces the birth weight of the offspring and curtails the rate of cell division in the brain. In the skeletal muscle, not only is there a reduction in cell number, but the actual number of muscle fibers is reduced, and each muscle fiber has an increased number of nuclei. These animals, when fed normally after birth, fail to recover normal height or weight (Widdowson, 1970).

The animal data, then, clearly demonstrate that undernutrition due to either reduced blood supply or reduced availability of nutrients will curtail placental and fetal growth, retard the rate of cell division in various fetal

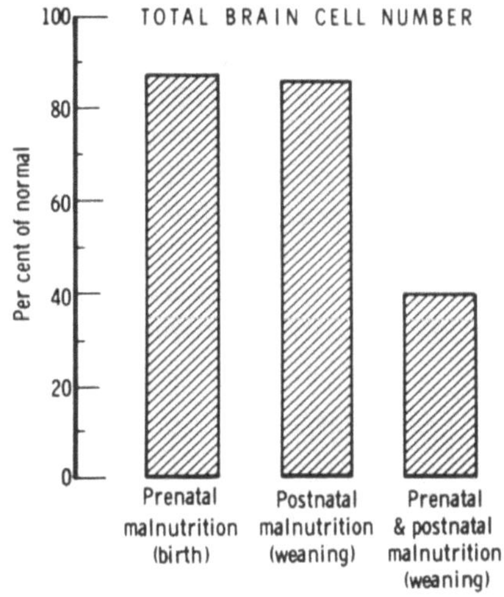

Fig. 21. Comparison of food restriction after birth, protein restriction during gestation, and "combined" prenatal and postnatal restriction. (From Winick, 1970c.)

organs, and result in an animal whose organs contain fewer cells. Evidence also indicates that animals born after developing in this type of intrauterine environment will carry these cellular deficits for the rest of their lives. However, the two types of "malnutrition" produce very different effects on brain. In the vascular insufficiency model, brain is spared, whereas in the maternal malnutrition model, brain is markedly affected. Human placenta goes through the same three phases of growth as those described for the organs of the rat. Cell division ceases at about 34–36 weeks of gestation, whereas weight and protein increase until nearly term (Winick *et al.*, 1967).

Placentas from infants with "intrauterine growth failure" show fewer cells and an increased RNA/DNA ratio when compared to controls (Winick, 1967). Fifty percent of placentas from an indigent population in Chile showed similar findings (Winick, 1970b). Placentas from a malnourished population in Guatemala had fewer cells than normal (Dayton *et al.*, 1969). In a single case of anorexia nervosa in which a severely emaciated mother carried to term and gave birth to a 2500-g infant, the placenta contained less than 50% of the expected number of cells (Winick, 1970c). Thus, both vascular insufficiency and maternal malnutrition will curtail cell division in human placenta. The cellular makeup of the placenta in both these situations strongly suggests that both stimuli have been active for some time prior to weeks 34–36 of gestation. In addition, more recent data demonstrate that placentas from malnourished women show an increased activity of alkaline RNase, very similar to what has been described in the rat models (Velasco *et al.*, 1975).

The effects of these stimuli on the cellular growth of the fetus are more difficult to assess. Indirect evidence suggests that cell division in the human fetus may be retarded by maternal undernutrition. Fetal growth is retarded and birth weight reduced (Smith, 1947). If one examines available data on infants who died after exposure to severe postnatal malnutrtion, three separate patterns emerge. Breast-fed infants malnourished during the second year have a reduced protein/DNA ratio but a normal brain DNA content. Full-term infants who subsequently died of severe food deprivation during the first year of life had a 15–20% reduction in total brain cell number. Infants weighing 2000 g or less at birth who subsequently died of severe undernutrition during the first year of life showed a 60% reduction in total brain cell number (Winick, 1970d). It is possible that these children were deprived *in utero* and represent a clinical counterpart of the "doubly deprived" animal. It is also possible that these were true premature infants and that the premature is much more susceptible to postnatal malnutrition than the full-term infant.

3.4. Skeletal Muscle

Several studies have described the effect of malnutrition on rat skeletal muscle growth (reviewed by Trenkle, 1974, and Young, 1974). When malnutrition is imposed during the period of hyperplastic growth, it will result in a retardation in the rate of DNA synthesis in skeletal muscle (Winick and Noble, 1965; Elliott and Cheek, 1968a,b; Graystone and Cheek, 1969; Cheek, 1968; Millward *et al.*, 1975; Srivasta *et al.*, 1974; Howarth, 1972; Howarth and Baldwin, 1971). In muscle of young rats fed a diet low in both calories and proteins, Hill *et al.* (1970) found a marked reduction in both DNA content and protein/DNA ratio. By contrast, when the animals were restricted in calories but were fed adequate proteins, less reduction in muscle growth or muscle nuclear number was noted. In these animals the protein/DNA ratios were either normal or slightly increased. These authors (Cheek and Hill, 1970; Hill *et al.*, 1970) postulate that reducing calorie intake will affect DNA replication primarily, but the reduction of calories and proteins has more serious effects on intracellular protein synthesis as well.

In studying the influence of dietary protein on rat skeletal muscle, Howarth (1972) placed weanling rats on 0, 6, 12, 18, and 24% protein diet for 14 days. Total DNA accumulation in gastrocnemius muscle was directly related to the protein content of the diet, and no increase was observed in muscle DNA over the 14-day period in rats placed in the protein-free diet. Since proteins were affected in the same way as DNA in the different diets, no change was observed in cell size (protein/DNA ratio) except in the protein-free diet, where cell size seems to be reduced. This finding was later confirmed (Nnanyelugo, 1976). Moss (1968) has shown that during normal growth of the pectoral muscle of chickens, the mean cross-sectional area of the fibers increased in proportion to the total number of nuclei and in proportion to the two-thirds power of the weight of muscle and that a mathematical relationship was maintained. A continuous restriction of food did not affect this relationship. However, starvation or severe undernutrition for a few days, which reduced muscle weight and fiber cross-sectional area, caused no loss of nuclei, hereby disturbing the relationship (Moss, 1968). The low-protein/low-calorie diet in Hill's experiment (6% casein, 60% of intake of normal rats) may have been severe enough to decrease the protein/DNA ratio.

In laboratory studies of malnutrition, we can control the diet the animals receive and can usually produce a specific type of malnutrition. In clinical studies, however, patients rarely exhibit pure marasmus or pure kwashiorkor but present a picture of mixed protein–calorie malnutrition. In the literature on muscle growth in human malnutrition, the age of onset, duration of malnutrition, severity of malnutrition, and type of dietary deficiency are variable; hence it is difficult to generalize regarding effects on cellular growth of muscle. It seems clear, however, that whether estimated from cadaver analysis (Waterlow, 1956), biopsy specimen (Waterlow and Mendes, 1957), limb measurements (Standard *et al.*, 1959), or creatinine excretion (Alleyne, 1968; Standard *et al.*, 1959), muscle growth is more severely reduced than the reduction in total body weight would indicate. Loss of total body potassium (Alleyne, 1968) and muscle tissue protein (Waterlow, 1956; Graham *et al.*, 1969) can often be severe. Concomitantly, there is an increase in total body water when compared with body weight, which is due primarily to expansion of the extracellular fluid compartment (Alleyne, 1968; Graham *et al.*, 1969).

In a study of nine male Peruvian infants 5–30 months of age and suffering from severe malnutrition (Cheek *et al.*, 1970), gluteal muscle samples were analyzed for DNA and protein content, and creatinine excretion was used to assess muscle mass. The most striking change was a reduction of the protein/DNA ratio, which indicated a loss of "cell mass." There was only a slight reduction in muscle nuclear number when compared with infants of a similar height. These data conflict with data cited elsewhere in this chapter, which have demonstrated decreased DNA content in other tissues with severe infantile malnutrition. There are several possible explanations for this discrepancy. First, the calculation of nuclear number depends on the DNA concentration of the biopsy specimen, which may not reflect DNA content in muscles throughout the body. Second, the calculation depends on the assumption that the factor for conversion of creatinine excretion to muscle mass is the same in normal and malnourished children. This has not been entirely substantiated; indeed, Alleyne's work (1968) suggests that the factor of 20 may be in error for malnourished children. Finally, DNA replication and cell division are time-related phenomena and are logically better compared with controls of the same chronological age rather than the same height, as has been done in this study. Since these Peruvian children were retarded in growth, we might expect reductions in nuclear number if the comparisons were made against an age baseline.

Lee *et al.* (1974) studied six children age 4–14 years prior and after human growth hormone (HGH) administration for 12 months. Muscle biopsies were taken before and after treatment. HGH treatment significantly increased the growth rate of the group when the paired *t*-test was used for analysis. The amount of DNA within the muscle mass (measured by creatinine excretion) was significantly low for chronological age. After treatment, four of the six patients had an increase in cell number greater than the normal growth-curve slope. Cell size (protein/DNA) decreased after HGH treatment. The authors suggest that HGH stimulates DNA replication to a greater extent than does protein synthesis.

These studies, even though unique because of the great difficulty involved, have to be looked upon with a critical eye. Until further studies substantiate the valddity of the factor of 20 for creatinine excretion conversion to muscle mass, or until the DNA content of an entire muscle is measured, we must conclude that the extent of the effects of malnutrition on muscle nuclear number have not yet been completely delineated.

3.5. Other Tissues

The normal cellular growth patterns for most of the organs of the rat have been worked out. In general, weight and protein continue to increase until about 100 days of age. By contrast, DNA reaches a maximum before this in all organs. The time at which it does so varies with the particular organ. In brain and lung, DNA reaches a maximum at about 21 days of life; in liver, spleen, and kidney, at about 40 days of age; in submaxillary gland, at about 45 days of age; and in heart, at about 65 days of age. Malnutrition during the period of hyperplastic growth results in a reduced number of cells in all these organs.

In the human, cellular growth patterns have been studied in various organs during normal fetal development. The data indicate that total cell number, as measured by total organ DNA content, increases in all organs from 13 weeks of gestation until term. Cell size, as measured either by weight/DNA or protein/DNA ratios, remains unchanged throughout gestation in heart, kidney, spleen, thyroid, thymus, esophagus, stomach, large and small intestines, and tongue. In brain, lungs, liver, adrenal gland, and diaphragm, cell size increases slowly from the beginning of the seventh month of gestation until term.

More limited data during the first year of life demonstrate that cell number continues to increase rapidly in heart, liver, kidney, and spleen. Heart cell size begins to increase after 3 months of age, whereas in kidney, liver, and spleen, cell size does not change during the first year.

Children who died of marasmus during the first 2 years of life showed marked reductions in cell number in all organs studied. As described in a previous section, brain cell size was also reduced when the malnutrition extended into the second year. In contrast, cell size in the other organs was not significantly reduced, even if the malnutrition persisted beyond the first year.

4. References

Alfert, M., and Geschwindt, I. I., 1958, The development of polysomaty in rat liver, *Exp. Cell. Res.* **15**:230.

Alleyne, G. A. O., 1968, Studies on total body potassium in infantile malnutrition: The relation to body fluids, spaces, and urinary creatinine, *Clin. Sci.* **34**:199.

Altman, J., and Das, G., 1966, Autoradiographic and histological studies of postnatal neurogenesis. I. A longitudinal investigation of the kinetic,, migration and transformation of cells incorporating tritiated thymidine in infant rats with special reference to postnatal neurogenesis in some brain regions, *J. Comp. Neurol.* **126**:337.

Argyris, T. S., 1971, Adaptive effects of phenobarbital and high protein diet on liver growth in immature male rats, *Dev. Biol.* **25**:293.

Augustinsson, K. B., 1950, Acetylcholine esterase and cholinesterase, in: *The Enzymes: Chemistry and Mechanism of Action,* Vol. 1, Part 1 (J. B. Sumner and Karl Myrbäch, eds.), pp. 443–472, Academic Press, Inc., New York.

Bachrach, U., 1973, *Function of Naturally Occurring Polyamines,* Academic Press, Inc., New York.

Bagshaw, J. C., and Bond, B. H., 1974, Postnatal development of mouse liver: Increasing RNA polymerase activity and orotic acid incorporation, *Differentiation* **2**:269.

Bardon, A., and Paumůla, S., 1967, Ribonucleases in developing rat liver, *Acta Biochim. Pol.* **14**:341.

Beach, R. K., and Kostyo, J. L., 1968, Effect of growth hormone on the DNA content of muscles of young hypophysectomized rats, *Endocrinology* **82**:882, 1968.

Boivin, A., Vendrely, R., and Vendrely, C., 1948, L'Acide désoxyribonucléïque du noyau cellularie, dépositaire des caractères héréditaires; arguments d'ordre analytique, *C.R. Acad. Sci. Paris* **226**:1061.

Brasel, J. A., and Winick, M., 1972, Maternal nutrition and prenatal growth. Experimental studies of effects of maternal undernutrition on fetal and placental growth, *Arch. Dis. Child.* **47**:479.

Brasel, J. A., Ehrenkranz, R. A., and Winick, M., 1970, DNA polymerase activity in rat brain during ontogeny, *Dev. Biol.* **23**:424.

Bresnick, E., Sage, S., and Sandor, L. K., 1966, Ribonuclease activity in hepatic nuclei during development, *Biochim. Biophys. Acta* **114**:631.

Chase, H. P., Dorsey, J., and McKhann, G. M., 1967, The effect of malnutrition on the synthesis of a myelin lipid, *Pediatrics* **40**:551.

Cheek, D. B., 1968, Muscle cell growth in normal children, in: *Human Growth* (D. B. Cheek, ed.), pp. 337–351, Lea & Febiger, Philadelphia.

Cheek, D. B., 1971, Hormonal and nutritional factors influencing muscle cell growth, *J. Dent. Res.* **50**(suppl. 5):1385.

Cheek, D. B., and Hill, D. E., 1970, Muscle and liver cell growth: Role of hormones and nutritional factors, *Fed. Proc.* **29**:1503.

Cheek, D. B., Powell, C. K., and Scott, R. E., 1965, Growth of muscle mass and skeletal collagen in the rat. I. Normal growth, *Bull. Johns Hopkins Hosp.* **116**:378.

Cheek, D. B., Brasel, J. A., and Graystone, J. E., 1968, Muscle cell growth in rodents: Sex differences and the role of hormones, in: *Human Growth* (D. B. Cheek, ed.), pp. 306–325, Lea & Febiger, Philadelphia.

Cheek, D. B., Hill, D. E., Cordano, A., and Graham, C. G., 1970, Malnutrition in infancy: Changes in muscle and adipose tissue before and after rehabilitation, *Pediatr. Res.* **4**:135.

Cheek, D. B., Holt, A. B., Hill, D. E., and Talbert, J. L., 1971, Skeletal muscle cell mass and growth: The concept of the deoxyribonucleic acid unit, *Pediatr. Res.* **5**:312.

Chow, B. F., and Lee, C. F., 1964, Effect of dietary restriction of pregnant rats on body weight gain of the offspring, *J. Nutr.* **82**:10.

Cohen, S. S., 1971, *Introduction to the Polyamines,* Prentice-Hall, Inc., Englewood Cliffs, N.J.

Correy, E. L., 1935, Growth and glycogen content of the fetal liver and placenta, *Am. J. Physiol.* **112**:263.

Culley, W. J., and Lineberger, R., 1968, Effect of undernutrition on the size and composition of the rat brain, *J. Nutr.* **96**:375.

Dallman, P. R., 1971, Malnutrition: Incorporation of thymidine-^{3}H into nuclear and mitochondrial RNA, *J. Cell Biol.* **51**:549.

Dallman, P. R., and Manies, E. C., 1973a, Protein deficiency: Contrasting effects on DNA and RNA metabolism in rat liver, *J. Nutr.* **103**:1311.

Dallman, P. R., and Manies, E. C., 1973b, Protein deficiency: Turnover of protein and reutilization of amino acid in cell fractions of rat liver, *J. Nutr.* **103**:257.

Davies, A. S., 1972, Postnatal changes in the histochemical fibre types of porcine skeletal muscle, *J. Anat.* **113**:213.

Davison, A. N., and Dobbing, J., 1966, Myelination as a vulnerable period in brain development, *Br. Med. Bull.* **22**:40.
Dayton, D. H., Filer, L. J., and Canosa, C., 1969, Cellular changes in placentas of undernourished mothers in Guatemala, *Fed. Proc.* **28**:488.
Dobbing, J., and Sands, J., 1970, Timing of neuroblast multiplication in developing human brain, *Nature* **226**:639.
Dobbing, J., and Sands, J., 1973, Quantitative growth and development of human brain, *Arch. Dis. Child.* **48**:757.
Donaldson, H. H., 1924, *The Rat,* 2nd ed., Wistar Institute Press, Philadelphia.
Duckett, S., and Pearse, A. G. E., 1966, The chemo-architectronic patterns of the cerebral cortex of the embryonic and foetal human brain, in: *Proceedings of the 5th International Congress of Neuropathology,* Intl. Congr. Series 100, pp. 738–739, Excerpta Medica Foundation, Amsterdam.
Duckett, S., and Pearse, A. G. E., 1967, Monoamine oxidase cells in the developing human cortex, *Rev. Can. Biol.* **26**:173.
Duckett, S., and Pearse, A. G. E., 1968, The cells of Cajal-Retzius in the developing human brain, *J. Anat.* **102**:183.
Duckett, S., and Pearse, A. G. E., 1969, Histoenzymology of the developing human spinal cord, *Anat. Rec.* **163**:59.
Dunn, M. S., Murphy, E. A., and Rockland, L. B., 1947, Optimal growth of the rat, *Physiol. Rev.* **27**:72.
Elliott, D. A., and Cheek, D. B., 1968a, Muscle and liver cell growth in rats with hypoxia and reduced nutrition, in: *Human Growth* (D. B. Cheek, ed.), pp. 326–336, Lea & Febiger, Philadelphia.
Elliott, D. A., and Cheek, D. B., 1968b, Muscle growth in rats with exposure to hypoxia and food restriction, *J. Pediatr.* **69**:958.
Enesco, M., 1957, Increase in cell number and size and in extracellular space during postnatal growth of several organs of the albino rat, Ph.D. thesis, McGill University.
Enesco, M., and Leblond, C. P., 1962, Increase in cell number as a factor in the growth of the organs and tissues of the young male rat, *J. Embryol. Exp. Morphol.* **10**:530.
Enesco, M., and Puddy, D., 1964, Increase in the number of nuclei and weight in skeletal muscle of rats at various ages, *Am. J. Anat.* **114**:235.
Enwonwu, C. P., and Sreebny, L. M., 1971, Studies of hepatic lesions of experimental protein-calorie malnutrition in rats and immediate effects of refeeding on adequate protein diet, *J. Nutr.* **101**:501.
Epstein, C. J., 1967, Cell size, nuclear content and the development of polyploidy in the mammalian liver, *Proc. Natl. Acad. Sci. USA* **57**:327.
Ezekwe, M. O., and Martin, R. J., 1975, Cellular characteristics of skeletal muscle in selected strains of pigs and mice and the unselected controls, *Growth* **39**:95.
Faiferman, K., Hamilton, M. G., and Pogo, A. O., 1970, Nucleoplasmic ribonucleoprotein particles in rat liver. I. Selective degradation by nuclear nucleases, *Biochim. Biophys. Acta* **204**:550.
Fish, I., and Winick, M., 1969, Cellular growth in various regions of the developing rat brain, *Pediatr. Res.* **3**:407.
Fujii, T., and Villee, C. A., 1968, Comparison of ribonucleic acid metabolism in tissue of immature and adult rats, *Endocrinology* **82**:453.
Fukuda, M., and Sibatani, A., 1953, Biochemical studies on the number and composition of liver cells in postnatal growth of the rat, *J. Biochem.* **40**:95.
Geschwindt, I. I., and Li, C. H., 1949, Nucleic acid content of fetal rat liver, *J. Biol. Chem.* **180**:467.
Girija, N. A., Pradhan, D. S., and Sreenivasan, A., 1965, Effect of protein depletion on ribonucleic acid metabolism in rat liver, *Indian J. Biochem.* **2**:85.
Gluck, L., Talner, N. J., Stern, H., Gardner, T. H., and Kulovich, M. V., 1964, Experimental cardiac hypertrophy: Concentrations of RNA in the ventricles, *Science* **144**:1244.

Gordon, E. E., Kowalski, K., and Fritts, M., 1966, Muscle proteins and DNA in rat quadriceps during growth, *Am. J. Physiol.* **210:**1033.

Graham, C. C., Cordano, A., Blizzard, R. M., and Cheek, D. B., 1969, Infantile malnutrition: Changes in body composition during rehabilitation, *Pediatr. Res.* **3:**579.

Graystone, J. E., 1968, Creatinine excretion during growth, in: *Human Growth* (D. B. Cheek, ed.), pp. 182–197, Lea & Febiger, Philadelphia.

Graystone, J., and Cheek, D. B., 1969, The effects of reduced calorie intake and increased calorie intake (insulin induced) on the cell growth of muscle, liver, and cerebrum and on skeletal collagen in the post-weaning rat, *Pediatr. Res.* **3:**66.

Hegarty, P. V. J., Gundlach, L. C., and Allen, C. E., 1973, Comparative growth of porcine skeletal muscle using an indirect prediction of muscle fiber number, *Growth* **37:**333.

Hill, D. E., Holt, A. B., Parra, A., and Cheek, D. B., 1970, The influence of protein calorie versus calorie restriction on the body composition and cellular growth of muscle and liver in weanling rats, *Johns Hopkins Med. J.* **127:**146.

Howarth, R. E., 1972, Influence of dietary protein on rat skeletal muscle growth, *J. Nutr.* **102:**37.

Howarth, R. E., and Baldwin, R. L., 1971, Synthesis and accumulation of protein and nucleic acid in rat gastrocnemius muscles during normal growth, restricted growth, and recovery from restricted growth, *J. Nutr.* **101:**477.

Hubbard, R. W., Smoake, J. A., Matthew, W. T., Linduska, J. D., and Bowers, W. D., 1974, The effect of growth and endurance training on the protein and DNA content of rat soleus, plantaris and gastrocnemius muscles, *Growth* **38:**171.

Jackson, C. M., and Steward, C. A., 1920, The effects of inanition on the ultimate size of the body and of the various organs of the albino rat, *J. Exp. Zool.* **30:**97.

Jacobj, W., 1925, Über das rhythmische Wachstrum der Zellen durch Verdopplung ihres volumens, *Arch. Entwicklungsmech. Org.* **106:**124.

Jacobj, W., 1935, Die Zellkerngrosse beim Menschen, *Z. Mikrosk.-Anat. Forsch.* **38:**161.

Jasper, H. G., and Brasel, J. A., 1974, Rat liver DNA synthesis during the catch-up growth of nutritional rehabilitation, *J. Nutr.* **104:**405.

Johns, J. T., and Bergen, W. G., 1976, Growth in sheep: pre- and post-weaning hormone changes and muscle and liver development, *J. Anim. Sci.* **142:**192.

Jollie, W. P., 1964, Radioautographic observations on variations in desoxyribonucleic acid synthesis in rat placenta with increasing gestational age, *Am. J. Anat.* **114:**161.

Karp, R., Brasel, J. A., and Winick, M., 1971, Compensatory kidney growth after uninephrectomy in adult and infant rats, *Am. J. Dis. Child.* **121:**186.

Kennedy, G. C., 1957, The development with age of hypothalamic restrain upon the appetite of the rat, *J. Endocrinol.* **16:**9.

Klemperer, H. G., and Haynes, G. R., 1968, Thymidine kinase in rat liver during development, *Biochem. J.* **108:**541.

Kosterlitz, H. W., 1947, The effects of changes in dietary protein on the composition and structure of the liver cell, *J. Physiol.* **106:**194.

Kraft, N., and Shortman, K., 1970, A suggested control function for the animal tissue ribonuclease–ribonuclease inhibitor system, based on studies of isolated cells and phytohemagglutinin-transformed lymphocytes, *Biochim. Biophys. Acta* **217:**164.

Lash, T. W., Holtzer, H., and Swift, H., 1957, Regeneration of mature skeletal muscle, *Anat. Rec.* **128:**679.

Leduc, E. H., 1949, Mitotic activity in the liver of the mouse during inanition followed by refeeding with different levels of proteins, *Am. J. Anat.* **84:**397.

Lee, P. A., Blizzard, R. M., Cheek, D. B., and Holt, A. B., 1974, Growth and body composition in intrauterine growth retardation (IUGR) before and during human growth hormone administration, *Metabolism* **23:**913.

Lewis, C. G., and Winick, M., 1977, Pattern of cytoplasmic RNA in brain and liver of immature rats, *Proc. Soc. Exp. Biol. Med.* **156:**158.

Lewis, C. G., and Winick, M., 1978, Studies on ribosomal RNA synthesis *in vivo* in rat liver during short-term protein malnutrition, *J. Nutr.* **108:**329.

Logan, J. E., Mannell, W. A., and Rossiter, R. J., 1952, Chemical studies of peripheral nerve during Wallerian degeneration, *J. Biochim.* **51**:482.
McCance, R. A., and Widdowson, E. M., 1962, Nutrition and growth, *Proc. Roy. Soc. London Ser B* **156**:326.
Machovich, R., and Greengard, O., 1972, Thymidine kinase in rat tissues during growth and differentiation, *Biochim. Biophys. Acta* **286**:375.
Mannen, C. A., and Russell, D. H., 1977, Ornithine decarboxylase may function as an initiation factor for RNA polymerase, I, *Science* **195**:505.
Mariani, A., Miglaccio, . A., Spadoni, M. A., and Ticca, M., 1966, Amino acid activation in the liver of growing rats maintained with normal and with protein deficient diets, *J. Nutr.* **90**:25.
Miller, S. A., 1969, Protein metabolism during growth and development, in: *Mammalian Protein Metabolism* (H. N. Munro, ed.), pp. 183–233, Academic Press, Inc., New York, 1969.
Millward, D. J., Garlick, P. J., Stewart, R. J. C., Nnanyelugo, D. O., and Waterlow, J. C., 1975, Skeletal muscle growth and protein turnover, *Biochem. J.* **150**:235.
Mirsky, A. E., and Ris, H., 1949, Variable and constant components of chromosomes, *Nature* **163**:666.
Morpurgo, B., 1889, Sur les procès physiologiques de néoformation cellulaire durant l'inanition aigüe de l'organism, *Arch. Ital. Biol.* **11**:118.
Moss, F. P., 1968, The relationship between the dimensions of the fibers and the number of nuclei during restricted growth, degrowth, and compensatory growth, *Am. J. Anat.* **122**:565.
Munro, H. N., and Gray, J. A. M., 1969, The nucleic acid content of skeletal muscle and liver in mammals of different body size, *Comp. Biochem. Physiol.* **28**:897.
Munro, H. N., Naismith, D. J., and Wikramanayake, T. W., 1953, The influence of energy intake on ribonucleic acid metabolism, *Biochem. J.* **54**:198.
Nachmansohn, D., 1952, Chemical mechanisms of nerve activity, in: *Modern Trends in Physiology and Biochemistry* (E. S. G. Barron, ed.), pp. 229–276, Academic Press, Inc., New York.
Najdal, C., and Zajdela, F., 1966a, Polypoïdie somatique dans le foie de rat. I. Le rôle des cellules binuclées dans la genèse des cellules polypoïdes, *Exp. Cell. Res.* **42**:99.
Najdal, C., and Zajdela, F., 1966b, Polypoïdie dans le foie de rat. II. Le rôle de l'hypophyse et de la carence proteïque, *Exp. Cell Res.* **42**:117.
Naora, H. J., 1957, Microspectrophotometry of cell nuclei stained with the Feulgen reaction. IV. Formation of tetraploid nuclei in rat liver cells during postnatal growth, *J. Biophys. Biochem. Cytol.* **3**:949.
Nnanyelugo, D. O., 1976, Changes in RNA content during chronic protein deprivation in the hind limbs of rats, *Nutr. Rep. Int.* **14**:209.
Novello, F., and Stirpe, F., 1970, Simultaneous assay of RNA polymerase I and II in nuclei isolated from resting and growing rat liver with the use of α-amanitin, *FEBS Lett.* **8**:57.
Oliver, I. T., and Blumer, W. F. C., 1964, Metabolism of nucleic acids during liver maturation in the neonatal rat, *Biochem. J.* **91**:559.
Otten, J., 1968, Rat liver polyribosomes during the first days of life, *Growth* **32**:95.
Platt, B. S., 1962, Proteins in nutrition, *Proc. Roy. Soc. London B* **156**:337.
Platt, B. S., Heard, C. R. C., and Stewart, R. J. C., 1964, Experimental protein calorie deficiency, in: *Mammalian Protein Metabolism,* Vol. 2 (H. R. Munro and J. B. Allison, eds.), pp. 445–521, Academic Press, Inc., New York.
Quirin-Stricker, C., and Mandel, P., 1968, Étude du renouvellement du RNA des polysomes, du RNA de transfert et du RNA "messager" dans le foie de rat soumis à une jeûne proteïque, *Bull. Soc. Chim. Biol.* **50**:31.
Rahman, V. E., Cerney, E. A., and Peraino, C., 1969, Studies on rat liver ribonuclease in developing 2-acetylamino-fluorene fed and partially hepatectomized rats, *Biochim. Biophys. Acta* **178**:68.
Rosso, P., 1975, Maternal malnutrition and placental transfer of α-amino isobutyric acid, *Science* **187**:648.
Rosso, P., and Winick, M., 1975, Effects of early undernutrition and subsequent refeeding on alkaline ribonuclease activity of rat cerebrum and liver, *J. Nutr.* **105**:1104.

Rosso, P., Hormazabal, J., and Winick, M., 1970, Changes in brain weight, cholesterol, phospholipid and DNA content in marasmic children, *Am. J. Clin. Nutr.* **23**:1275.
Rosso, P., Nelson, M., and Winick, M., 1973, Changes in cellular RNA content and alkaline ribonuclease activity in rat liver during development, *Growth* **37**:143.
Rozovski, S. J., Winick, M., and Rosso, P., 1976, Adaptive changes in polyamine metabolism during malnutrition and refeeding, *Fed. Proc.* **35**:341.
Rozovski, S. J., Rosso, P., and Winick, M., 1978, The effect of malnutrition and rehabilitation on the metabolism of polyamines in rat liver, *J. Nutr.* **108**:1680.
Russell, D. H., 1973, *Polyamines in Normal and Neoplastic Growth,* Raven Press, New York.
Sereni, F., Sereni, L. P., Tomasi, V., and Barnabei, O., 1967, Synthesis and breakdown of rat liver RNA during the neonatal period, *Arch. Biochem. Biophys.* **121**:251.
Shaw, C., and Fillios, L. C., 1968, RNA polymerase activities and other aspects of hepatic protein synthesis during early protein depletion in the rat, *J. Nutr.* **96**:327.
Smith, C. A., 1947, Effects of maternal undernutrition upon the newborn infant in Holland, 1944–45, *J. Pediatr.* **30**:229.
Srivasta, U., Vu M.-L., and Goswami, T., 1974, Maternal dietary deficiency and cellular development of progeny in the rat, *J. Nutr.* **104**:512.
Standard, K. S., Wills, V. G., and Waterlow, J. C., 1959, Indirect indicators of muscle mass in malnourished infants, *Am. J. Clin. Nutr.* **7**:271.
Staun, H., 1963, Various factors affecting number and size of muscle fibers in the pig, *Acta Agric. Scand.* **13**:293.
Staun, H., 1972, The nutritional and genetic influence on number and size of muscle fibers and their response to carcass quality in pigs, *World Rev. Anim. Prod.* **8**:3.
Stickland, N. C., and Goldspink, C., 1973, A possible indicator muscle for the fibre content and growth characteristics of porcine muscle, *Anim. Prod.* **16**:135.
Swartz, F. J., 1956, The development in the human liver of multiple deoxyribose nucleic acid (DNA) classes and their relationship to the age of the individual, *Chromosome* **8**:52.
Swatland, H. J., 1976, Effect of growth and plane of nutrition on apparent muscle fiber numbers in the pig, *Growth* **40**:285.
Swift, H., 1950, The deoxyribonucleic acid content of animal nuclei, *Physiol. Zool.* **23**:169.
Swift, H., 1953, Quantitative aspects of nuclear nucleoproteins, *Int. Rev. Cytol.* **2**:1.
Tagliamante, B., Benedetti, C. P., and Spadoni, M. A., 1972, Effect of refeeding on liver thymidine kinase activity of food restricted suckling rats, *Nutr. Rep. Int.* **5**:305.
Thomson, R. Y., Heagy, F. C., Hutchison, W. C., and Davidson, J. N., 1953, The deoxyribonucleic acid content of the rat cell nucleus and its use in expressing the results of tissue analysis, with particular reference to the composition of liver tissue, *Biochem. J.* **53**:460.
Trenkle, A., 1974, Hormonal and nutritional interrelationships and their effects on skeletal muscle, *J. Anim. Sci.* **38**:1142.
Velasco, E. G., Brasel, J. A., Sigulem, D. M., Rosso, P., and Winick, M., 1973, Effects of vascular insufficiency on placental ribonuclease activity in the rat, *J. Nutr.* **103**:213.
Velasco, E. G., Rosso, P., Brasel, J. A., and Winick, M., 1975, Activity of alkaline RNase in placentas of malnourished women, *J. Obstet. Gynecol.* **123**:637.
Wannemacher, R. W., Jr., Cooper, W. K., and Yatvin, M. B., 1968, The regulation of protein synthesis in the liver of rats, *Biochem. J.* **107**:615.
Wasserman, M., Rosso, P., and Rozovski, S. J., 1976, Activity of polyamine synthesizing enzymes and rate of RNA accumulation in the rat placenta during normal gestation, *IRCS Med. Sci.* **4**:73.
Waterlow, J. C., 1956, The protein content of liver and muscle as a measure of protein deficiency in human subjects, *West Indian Med. J.* **5**:167.
Waterlow, J. C., and Mendes, C. B., 1957, Composition of muscle in malnourished human infants, *Nature* **180**:1361.
Wheatley, D. N., 1972, Binucleation in mammalian liver, *Exp. Cell. Res.* **74**:455.
Widdowson, E. M., 1970, Reported at Symposium on Fetal Malnutrition, sponsored by the National Foundation—March of Dimes, New York, January.

Widdowson, E. M., and Crabb, D. E., 1976, Changes in organs of pigs in response to feeding for the first 24 hours, *Biol. Neonate* **28**:261.
Widdowson, E. M., and McCance, R. A., 1960, Some effects of accelerating growth. I. General somatic development, *Proc. R. Soc. London B* **152**:188.
Widdowson, E. M., Crabb, D. E., and Milner, R. D. G., 1972, Cellular development of some human organs before birth, *Arch. Dis. Child.* **47**:652.
Wigglesworth, J. S., 1964, Experimental growth retardation in the foetal rat, *J. Path. Bact.* **88**:1.
Winick, M., 1967, Cellular growth of the human placenta. III. Intrauterine growth failure, *J. Pediatr.* **71**:390.
Winick, M., 1968a, Changes in nucleic acid and protein content of the human brain during growth, *Pediatr. Res.* **2**:352.
Winick, M., 1968b, Cellular growth of the placenta as an indicator of abnormal fetal growth, in: *Diagnosis and Treatment of Fetal Disorders* (K. Adamsons, ed.), pp. 83–101, Springer-Verlag New York Inc., New York.
Winick, M., 1969, The effect of nutrition on cellular growth, in: *Symposia of the Swedish Nutrition Foundation VII, Nutrition in Preschool Age* (G. Blix, ed.), pp. 33–41, Almqvist and Wiksell, Uppsala.
Winick, M., 1970a, Cellular growth in intrauterine malnutrition, *Pediatr. Clin. North Am.* **17**:69.
Winick, M., 1970b, Cellular growth of the fetus and placenta, in: *Fetal Growth and Development* (H. Waisman and G. Kerr, eds.), pp. 19–27, McGraw-Hill Book Company, New York.
Winick, M., 1970c, Nutrition and nerve cell growth, *Fed. Proc.* **29**:1510.
Winick, M., 1970d, Nutrition and mental development, *Med. Clin. North Am.* **54**:1413.
Winick, M., 1970e, Fetal malnutrition and growth processes, *Hosp. Pract.* **5**(May):33.
Winick, M., 1971, Cellular changes during placental and fetal growth, *Am. J. Obstet. Gynecol.* **109**:166.
Winick, M., and Noble, A., 1965, Quantitative changes in DNA, RNA and protein during prenatal and postnatal growth in the rat. *Dev. Biol.* **12**:451.
Winick, M., and Noble, A., 1966a, Quantitative changes in ribonucleic acids and protein during normal growth of rat placenta, *Nature* **212**:34.
Winick, M., and Noble, A., 1966b, Cellular response in rats during malnutrition at various ages, *J. Nutr.* **89**:300.
Winick, M., and Noble, A., 1967, Cellular response with increased feeding in neonatal rats, *J. Nutr.* **91**:179.
Winick, M., and Rosso, P., 1969a, Head circumference and cellular growth of the brain in normal and marasmic children. *J. Pediatr.* **74**:774.
Winick, M., and Rosso, P., 1969b, The effects of severe early malnutrition on cellular growth of human brain, *Pediatr. Res.* **3**:181.
Winick, M., and Rosso, P., 1970, *Proceedings of the Eighth International Congress on Nutrition,* Int. Congr. Ser. 213, pp. 531–538, Excerpta Medica, Amsterdam.
Winick, M., Coscia, A., and Noble, A., 1967, Cellular growth in human placenta. I. Normal placental growth, *Pediatrics* **39**:248.
Winick, M., Fish, I., and Rosso, P., 1968, Cellular recovery in rat tissues after a brief period of neonatal malnutrition, *J. Nutr.* **95**:623.
Winick, M., Velasco, E., and Rosso, P., 1969, DNA content of placenta and fetal brain, *Pan Am. Health Organ. Sci. Publ.* **185**:9.
Winick, M., Rosso, P., and Waterlow, J., 1970, Cellular growth of cerebrum, cerebellum, and brain stem in normal and marasmic children, *Exp. Neurol.* **26**:393.
Young, V. R., 1974, Regulation of protein synthesis and skeletal muscle growth, *J. Anim. Sci.* **38**:1054.
Zamenhof, S., Van Marthens, F., and Margolis, F. L., 1968, DNA (cell number) and protein in neonatal brain: Alteration by maternal dietary protein restriction, *Science* **160**:322.
Zeman, F. S., and Stanbrough, E. C., 1969, Effect of maternal protein deficiency on cellular development on the fetal rat, *J. Nutr.* **99**:274.

4

Nutrition and Brain Neurotransmitters

Edith Liwszyc Cohen and Richard Jay Wurtman

1. Introduction

This chapter describes the effects of nutrients on the synthesis of neurotransmitters in the CNS and on some of the functions thought to be mediated by the neurons that release these neurotransmitters. Neurotransmitters are compounds that the body uses to transmit information from one neuron to another or to muscle or glandular cells. The amount of a particular neurotransmitter present in the brain is determined by (1) the number of neurons that contain this neurotransmitter, (2) the number of synaptic boutons or terminals (the intracellular loci of most of the transmitter molecules) that are present in the neuron's axon or axonal branches, and (3) the number of neurotransmitter molecules present in each bouton. The numbers of CNS neurons and boutons probably can be affected by the nutritional state in developing animals; they appear to be fixed, however, in the adult (except, perhaps, for a small proportion of neurons that may be able to "sprout" even in adult animals). Some of the studies described in this chapter deal with chronic malnutrition in young animals and may therefore reflect changes in the numbers of neurons or synaptic boutons. However, the main focus of the following discussion is the acute and short-term control by nutritional factors of the numbers of neurotransmitter molecules present in boutons (and, thus, in neurons) in mature and

Abbreviations used in this chapter: DA, dopamine; NE, norepinephrine; 5-HT, serotonin; ACh, acetylcholine; GABA, gamma-aminobutyric acid; CAT, choline acetyltransferase; AChE, acetylcholinesterase; DOPA, 3,4-dihydroxyphenylalanine; AAAD, aromatic L-amino acid decarboxylase; VMA, 3-methoxy-4-hydroxy-mandelic acid; HVA, homovanillic acid; BBB, blood-brain barrier; 5-HIAA, 5-hydroxyindoleacetic acid; and 5-HTP, 5-hydroxytryptophan.

Edith Liwszyc Cohen and Richard Jay Wurtman • Laboratory of Neuroendocrine Regulation, Department of Nutrition and Food Science, Massachusetts Institute of Technology, Cambridge, Massachusetts.

developing organisms. (For a complete discussion of the effects of nutritional state on the brain, the reader is referred to Wurtman and Wurtman, 1977).

The formation of neurotransmitters in the mammalian brain is influenced both by acute changes in nutritional state (e.g., meal eating) and by the amounts of particular nutrients that are provided over longer periods of time. Chronic underavailability of certain nutrients can adversely affect brain function, even in mature organisms. This fact has been recognized for some time, but only recently have investigators established that short-term changes in nutrient consumption can also affect the levels of some brain neurotransmitters. During the past few years, abundant evidence has indicated that food consumption can, by changing the composition of the plasma, control the brain concentrations of compounds that are the precursors of such neurotransmitters as serotonin and acetylcholine. Moreover, the enzymes determining the rates at which these precursors (tryptophan and choline, respectively) are converted to their transmitters are normally less than fully saturated with their substrates; thus, the precursor levels control the rates at which these neurotransmitters are formed. This acute dietary control of neurotransmitter synthesis is extraordinary, inasmuch as it had previously been universally accepted that the brain is "protected" from the vagaries of food-related changes in blood composition. One challenging task now facing nutritionists and brain scientists alike is to discover the utility to the animal or human of coupling brain composition—and neuronal activity—to what one eats.

2. *Basic Neurobiology*

2.1. *Neurons, Neurotransmitters, and Brain*

Neurotransmitters are the chemical signals by which most nerve cells, or neurons, transmit messages to the cells that they innervate. Neurotransmitter molecules are released from the neuron's terminals and from the swellings (boutons) along its axonal tree, in response to a wave of depolarization that begins in the cell body (perikaryon) of the neuron and travels unidirectionally down the axon, often for a remarkably long distance (Fig. 1). In general, the release of the neurotransmitter occurs at a specific anatomic locus—the synapse (Fig. 2)—where the presynaptic, or *transmitting,* neuron makes anatomic contact with the cells that it innervates. [Some recent evidence suggests that one brain neurotransmitter, serotonin, may also be released at sites not associated with synapses (Descarries *et al.,* 1975; Chan-Palay, 1976).] After diffusing across the synaptic cleft, the neurotransmitter interacts with specific receptors, usually on the perikaryal or dendritic (receiving) surfaces of the postsynaptic cell; this interaction, in turn, increases or decreases the flux of particular ions across the cell's outer membrane, thereby changing the voltage across this membrane. A given neuron in the CNS may receive synaptic inputs from as many as 10,000 other neurons, utilizing as many different neurotransmitter molecules as are currently recognized (i.e., about 10 compounds). If, at

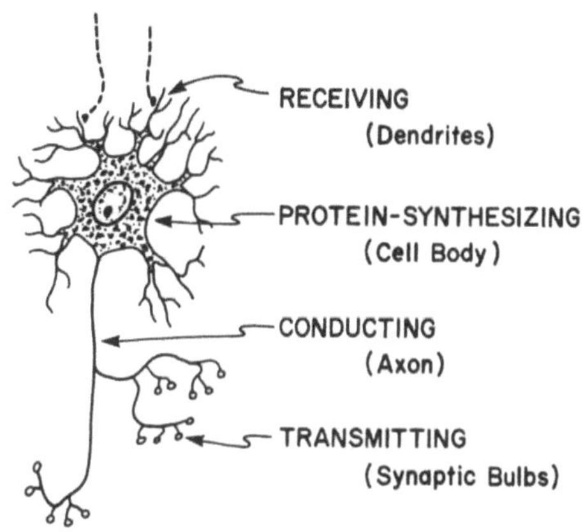

Fig. 1. Schematic diagram of interneuron in mammalian brain. (From Wurtman and Fernstrom, 1974.)

a given time, enough of the cell's postsynaptic receptors are bombarded by excitatory transmitters (those which elevate the cell's potential from its resting state, −70 mV, toward neutrality), the neuron depolarizes, generating an action potential that is conducted along its axonal tree and thereby causing its boutons to discharge its own neurotransmitter, as described above. Once a neurotransmitter molecule is released into a synaptic cleft, it is rapidly inactivated, either by enzymatic mechanisms that change its structure (e.g., acetylcholinesterase, which hydrolyzes acetylcholine) or by reuptake, an energy-requiring process by which it reenters the presynaptic neurons from which it had previously been released (e.g., norepinephrine, serotonin).

The neurons of the mammalian organism can be divided into two major categories: those whose terminals lie within the brain or spinal cord (CNS) and those whose terminals innervate other tissues and organs (peripheral nervous system, PNS). To demonstrate that a particular compound functions

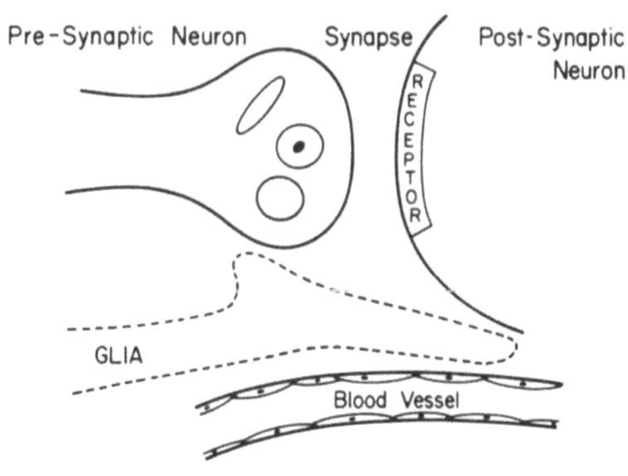

Fig. 2. Schematic diagram of a synapse.

as the transmitter of the latter peripheral neurons is relatively easy, but to "prove" this function for putative CNS neurotransmitters is difficult, if not impossible. Nonetheless, such proof requires satisfaction of the following criteria: (1) that the putative transmitter is synthesized (and stored) in the presynaptic neuron; (2) that the candidate transmitter is released from the presynaptic neuron when that neuron is depolarized; (3) that applying the putative transmitter to the postsynaptic membrane produces physiological effects similar to those caused by stimulating the presynaptic nerve (presumably because the membrane contains receptors for the transmitter); and (4) that drugs having the ability to block either the synthesis of the transmitter or its postsynaptic receptors, or to prolong its presence in the synaptic cleft, cause predicted effects on neurotransmission across that synapse (i.e., they must decrease, block, or potentiate it, respectively).

Based on the criteria above, acetylcholine and norepinephrine have been shown to function as neurotransmitter substances at peripheral synapses: acetylcholine is released at the neuromuscular junction, at all preganglionic autonomic synapses, and at parasympathetic postganglionic synapses; norepinephrine is released at postganglionic sympathetic synapses. As indicated above, one cannot actually "prove" that a compound functions as a central neurotransmitter—because, in part, one cannot yet collect fluid from within the synaptic cleft to show that its content of a putative transmitter increases when the presynaptic nerve fires. Nevertheless, indirect evidence indicates that a number of substances function as neurotransmitters in the brain. These are (listed in descending order of substantiation by experimental proof): dopamine (DA), norepinephrine (NE), serotonin (5-hydroxytryptamine, 5-HT), acetylcholine (ACh), gamma-aminobutyric acid (GABA), glycine, the acidic amino acids glutamate and aspartate, epinephrine, and perhaps some peptides. The structures of some of these molecules are diagrammed in this chapter.

Evidence of a central neurotransmitter function is the most extensively documented for the monoamines DA, NE (both catecholamines), and 5-HT (an indoleamine). This situation reflects (1) the availability of sensitive histochemical fluorescence techniques through which the monoamines can be identified within specific neurons; (2) the development of drugs that act with relatively great specificity to influence particular processes involved in their synthesis, metabolism, etc.; and (3) the relative ease with which an isotopic label can be introduced into the brain stores of these compounds. (The study of "catecholaminergic" and cholinergic synapses in the CNS is also facilitated by the existence of easily accessible peripheral synapses that utilize these transmitters.) In contrast, the acquisition of "proof" that glycine, aspartate, and glutamate function as CNS neurotransmitters proceeds with painstaking slowness because these compounds also subserve numerous other functions in brain cells, which are unrelated to neurotransmission and therefore complicate the investigation. The neuronal populations utilizing specific neurotransmitters have been described in detail for DA, NE, 5-HT (Dahlstrom and Fuxe, 1964; Fuxe, 1965a,b), and ACh (Shute and Lewis, 1967; Lewis and Shute, 1967; Lewis *et al.*, 1967). It should be noted that the majority of cells in the brain are not neurons at all but rather neuroglia, which "support," and possibly nourish, the neurons. The neuroglia consist of oligodendrocytes (which synthesize the myelin sheath that envelopes neuronal axons and facilitates conduction of the electrical impulse), astrocytes (which probably insulate brain capillaries and synaptic regions), and microglia (which may function as phagocytes).

2.2. *Biosynthesis and Metabolism of Major Brain Neurotransmitters*

The following discussion considers only the neurotransmitters about which most information is available (i.e., acetylcholine, the catecholamines, and serotonin).

2.2.1. Acetylcholine

Acetycholine (Fig. 3) is synthesized in cholinergic neurons from two precursors, choline and acetyl-CoA, in a reaction catalyzed by the enzyme choline acetyltransferase (CAT) (Nachmansohn and Machado, 1943). Analysis of pinched-off nerve endings (synaptosomes) has shown that in the nerve terminal, ACh is found both in the cytoplasm and in synaptic vesicles (the spherical structures characterizing the presynaptic region) (Whittaker, 1972). It remains to be determined whether the vesicles are used only to store the neurotransmitter or whether they are actually involved in quantal release of ACh. Such quantal release has been demonstrated at several peripheral synapses (Katz, 1966), but not, to date, in brain. After release into the synapse and interaction with postsynaptic receptors, the ACh is inactivated by hydrolysis to choline and acetate in a reaction catalyzed by the enzyme acetylcholinesterase (AChE). Some investigators have attempted to map the central cholinergic synapses by using histochemical stains for this enzyme. The utility of this approach is limited, however, because the enzyme is probably present at other loci, even those lacking synapses.

No currently available drugs can selectively suppress brain ACh synthesis *in vivo;* however, a number of agents do affect the action and metabolism of ACh in the brain. The most widely used of these are atropine, which blocks muscarinic cholinergic receptors, and physostigmine, which inhibits the action of the degradative enzyme AChE. The use of these drugs in experiments on nutritional control of brain ACh is discussed below.

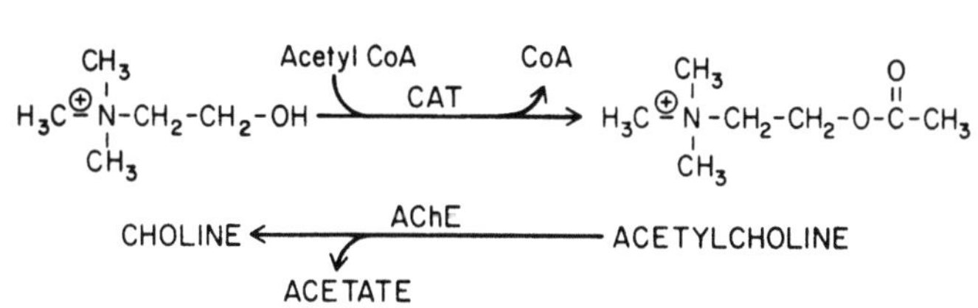

Fig. 3. Synthesis and catabolism of acetylcholine. CAT, choline acetyltransferase; AChE, acetylcholinesterase.

2.2.2. Catecholamines—Dopamine and Norepinephrine

The biochemistry of the catecholamines (Fig. 4) has been reviewed (Wurtman, 1966). The initial step in their biosynthesis involves the hydroxylation of tyrosine in the meta position to form 3,4-dihydroxyphenylalanine (DOPA) (Nagatsu *et al.*, 1964). This reaction is catalyzed by tyrosine hydroxylase, a stereospecific and substrate-specific enzyme that requires molecular O_2, Fe^{2+}, and a tetrahydropteridine cofactor for optimal activity. (Tyrosine hydroxylase is present only within those cells that make catecholamines—e.g., certain brain neurons, postganglionic sympathetic neurons, and adrenomedullary chromaffin cells.) Moreover, as described below, the *rate* at which tyrosine hydroxylation occurs *in vivo* seems to determine the overall rate of DA, and perhaps NE, synthesis. The activity of tyrosine hydroxylase (but not that of the analogous enzymes synthesizing ACh or 5-HT) is subject to considerable end-product inhibition by its neurotransmitter product.

The second step in catecholamine biosynthesis is the decarboxylation of the catechol amino acid DOPA to form the neurotransmitter amine DA. This reaction is mediated by a relatively nonspecific and widely distributed enzyme, aromatic L-amino acid decarboxylase (AAAD) (Lovenberg *et al.*, 1962); it requires pyridoxal phosphate (vitamin B_6) as a cofactor. NE-synthesizing cells

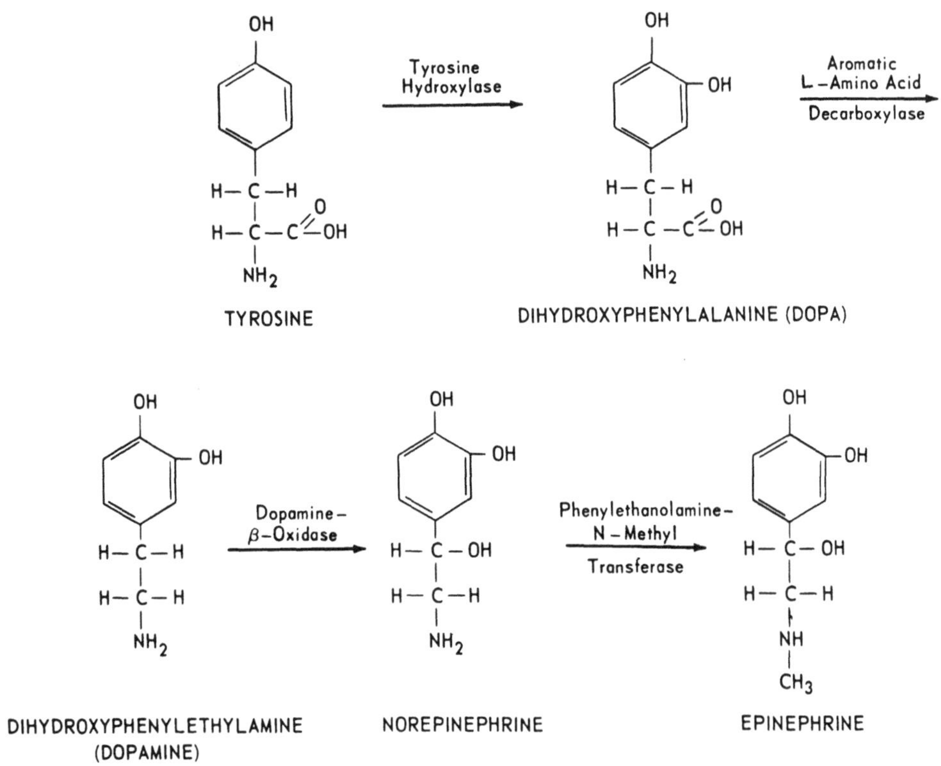

Fig. 4. Biosynthesis of catecholamines.

also contain a copper-containing enzyme, dopamine-beta-hydroxylase, which catalyzes the side-chain hydroxylation of DA to form NE (Levin *et al.*, 1960). The cofactors needed for this reaction are molecular oxygen and ascorbic acid.

The process by which released catecholamines are inactivated is largely nonenzymatic (i.e, they are physically removed from the synaptic cleft by reuptake into the presynaptic neurons.) Catecholamines may be metabolized within these presynaptic neurons and in most cells by the enzyme monoamine oxidase (MAO); they are also subject to enzymatic *O*-methylation (Axelrod, 1957), catalyzed by catechol-*O*-methyltransferase. The major metabolites resulting from the actions of these two enzymes are 3-methoxy-4-hydroxymandelic acid (VMA) and methoxyhydroxyphenyl glycol from NE and homovanillic acid (HVA) from DA. In general, only the acidic or neutral metabolites of neurotransmitters (and not the neurotransmitters themselves) are able to leave the brain and enter the circulation; in other words, the blood–brain barrier (BBB), which retards the entry of bases and hydroxylated compounds into the brain, works in both directions. In studies on humans, urinary concentrations of catecholamine and 5-HT metabolites have been used as indirect indices of the turnover of these neurotransmitters in the brain. Unfortunately, the utility of such data is sorely compromised by the fact that major peripheral sources of the catecholamines and 5-HT also exist; as a result, urinary metabolite levels primarily reflect monoamine metabolism outside the brain. For example, most of the major 5-HT metabolite (5-hydroxyindoleacetic acid, 5-HIAA) in urine derives from the large nonneuronal 5-HT pool that is synthesized in intestinal enterochromaffin cells and stored in blood platelets (Garattini and Valzelli, 1965).

The pharmacology of catecholaminergic neurons has been studied extensively, and many drugs act with some specificity to influence particular steps in catecholamine synthesis, metabolism, or action (Moskowitz and Wurtman, 1975). Such drugs have come into wide use in the treatment of schizophrenia (e.g., DA receptor-blocking agents), depression (drugs that block NE or 5-HT reuptake), Parkinson's disease (the DA precursor L-DOPA or drugs that stimulate DA receptors), hypertension (drugs that block peripheral noradrenergic transmission or modulate central noradrenergic mechanisms), angina pectoris (drugs that block cardiac beta-noradrenergic receptors), and, most recently, neuroendocrine disturbances such as galactorrhea and amenorrhea (treated by dopaminergic agonists that suppress prolactin secretion).

2.2.3. Serotonin

The pathways of serotonin (Fig. 5) and dopamine biosynthesis exhibit several similarities. The first step in 5-HT formation involves the hydroxylation of the essential amino acid tryptophan in the 5-position to form 5-hydroxytryptophan (5-HTP). This reaction is catalyzed by tryptophan hydroxylase (Lovenberg *et al.*, 1968), an enzyme that, like tyrosine hydroxylase, requires molecular O_2, Fe^{2+}, and a reduced pteridine cofactor for optimal activity. The decarboxylation of 5-HTP to form 5-HT is probably catalyzed by the same

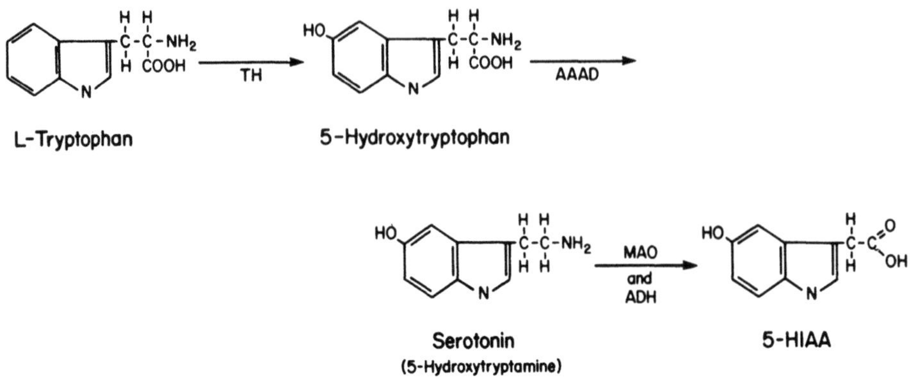

Fig. 5. Synthesis of serotonin and its metabolite 5-hydroxyindoleacetic acid (5-HIAA) from tryptophan in mammalian brain. TH, tryptophan hydroxylase; AAAD, aromatic L-amino acid decarboxylase; MAO, monoamine oxidase; ADH, aldehyde dehydrogenase.

enzyme (AAAD) as that which decarboxylates DOPA to form DA (Lovenberg *et al.*, 1962). Serotonin is catabolized to form 5-HIAA by enzymatic (MAO) deamination, followed by oxidation of the intermediate aldehyde. Its physiologic inactivation probably is accomplished by reuptake (i.e., removal from the synaptic cleft by entry into its cell of origin).

As was mentioned earlier, less is known about the natural history of brain 5-HT than about the catecholamines because of the lack of an easily accessible peripheral synapse that utilizes this neurotransmitter. A number of drugs affect 5-HT synthesis, reuptake, or metabolism, or interact with 5-HT receptors (Cooper *et al.*, 1974).

2.3. Relation between Neurotransmitter Level and Release

Since mammalian neurons presumably fulfill their communicative function by releasing neurotransmitter molecules into synapitc clefts—a process that cannot be monitored directly—it becomes important to determine whether the number of these molecules released when the neuron is depolarized bears any simple relationship to the amount of neurotransmitter present in the bouton or terminal. This knowledge is especially relevant in attempting to judge the physiological significance of nutritionally induced changes in transmitter level. In terms of the quantal theory, any change in the amount of neurotransmitter released by neuronal depolarization could result either from changes in the numbers of quanta released per impulse or from changes in the number of neurotransmitter molecules contained in each quantum.

To examine the relationship between the level of neurotransmitter in the brain neuron and the quantity that it releases, investigators must use indirect means. Some such strategies include: (1) measuring the amount of the neurotransmitter (or its metabolites) entering the CSF after various nutritional manipulations [for example, investigators in our laboratory have recently found (unpublished observations) that tryptophan administration, which increases brain 5-HT levels in cats, also increases the amount of 5-HT released into the perfused CSF of the cat]; (2) monitoring the spontaneous electrical activity of brain neurons whose transmitter levels have been increased (or decreased) as a result of a dietary manipulation, or similar monitoring of other neurons known to receive synaptic inputs from the diet-dependent neurons; (3) determining whether the effects on the electrical activity of neuron B caused by stimulating neuron A (electrically, or by providing sensory input) are altered when the neurotransmitter content of neuron A has been raised or lowered; and (4) examining the effects of changing neurotransmitter levels on physiological functions (e.g., temperature regulation) or behavioral states (e.g., locomotor activity) that are thought to involve the affected neurons.

A study modeled after this last strategy found that the chronic consumption of a corn diet (which is low in tryptophan) both depresses brain 5-HT levels (Fernstrom and Wurtman, 1971c) and modifies a behavioral function (pain sensitivity) previously shown to be mediated by serotoninergic neurons (Lytle *et al.*, 1975). Similar studies involving thermoregulation and *d*-amphetamine have shown that tyrosine-induced changes in brain DA *synthesis* also affect the *release* of the transmitter. In these studies, rats given *d*-amphetamine were placed in a cold environment (4–10°C), where they became hypothermic. [Control animals not receiving the drugs failed to do so (Yehuda and Wurtman, 1972a).] This hypothermic response of rats to *d*-amphetamine was shown to be mediated by the drug-induced release of DA from brain neurons (Yehuda and Wurtman, 1972b). When animals were also treated in such a way as to lower brain tyrosine levels (by being given other neutral amino acids, such as leucine or valine, which compete with tyrosine for uptake into the brain (Chiel and Wurtman, 1976), catechol synthesis was suppressed—because the substrate saturation of tyrosine hydroxylase was diminished (Gibson and Wurtman, 1978)—and *d*-amphetamine was found to have much less effect on body temperature (Chiel and Wurtman, 1976). Hence, the availability of tyrosine to the brain influences both the *synthesis* and *release* of DA.

Examining the relationship between levels and release of neurotransmitters is considerably easier in the case of peripheral synapses. For example, one can: (1) measure the ACh (or choline) released, spontaneously or after electrical stimulation, from *in vitro* preparations of cholinergically innervated smooth or skeletal muscle; (2) measure the postsynaptic contractile and bioelectric phenomena produced (e.g., in a preparation of smooth or skeletal muscle) when a presynaptic cholinergic nerve is caused to release its neurotransmitter; and (3) assay an enzyme (e.g., tyrosine hydroxylase in adrenal medulla) or a compound present within postsynaptic cells, whose activity or rate of synthesis is known to be controlled by ACh released presynaptically. As described below, the last method has also been used to study certain brain regions.

3. Nutrition and the Brain

3.1. Effect of Diet on the Brain

An adequate or inadequate nutritional state at various stages of the life cycle can affect the development and function of the CNS. The parameters affected can include virtually all of those that have thus far been examined: the total number of brain cells, the size of the cells, the number of synaptic connections, the composition of brain lipids, the rate of protein synthesis, the activity of enzymes, and the amount of various neurotransmitter substances. The relationships between nutrition, cellular growth, and biochemical maturation are discussed in detail in other chapters of this volume. In this chapter we concentrate on how long- and short-term nutritional inputs can affect the biochemistry of the brain by controlling the synthesis and levels of its most characteristic and probably most important constituents: its neurotransmitters.

There are, theoretically, several ways by which variations in nutritional state might result in changes in brain neurotransmitter levels. First, chronic malnutrition in the developing animal might affect the absolute and relative numbers of particular groups of neurons in the brain, thereby decreasing the number of cellular units capable of synthesizing and storing particular neurotransmitters. If, for example, severe malnutrition slows the division of 5-HT-producing cells and causes animals to have fewer than normal of these cells, one should anticipate that their brains will contain less tryptophan hydroxylase than those of well-fed littermates and will synthesize and store less 5-HT (even though the amount of 5-HT present in each serotoninergic neuron might be normal). It should be noted that the reduction in brain NE found in postnatally undernourished rats is associated with an *increase* in brain tyrosine hydroxylase activity (Shoemaker and Wurtman, 1973); this finding presumably reflects the operation of a feedback mechanism through which the noradrenergic neuron "tries" to compensate for precursor deficiency. Hence, measurements of the enzymes that synthesize brain neurotransmitters do not necessarily provide reliable indices of either the number of cells making this transmitter or the amounts of transmitter actually being produced.

A second effect from severe malnutrition is the possible suppressed arborization of growing neurons and the impaired formation of boutons and terminals (the loci of most neurotransmitter molecules), as well as of synapses. [Some histological evidence is now available in support of impaired synaptogenesis (Shoemaker and Bloom, 1977).] The effects of such impairments might or might not be permanent, but, in either case, they probably could produce the kinds of biochemical changes already described.

Nutritional state might also affect the brain levels of cofactors necessary for the optimal activity of enzymes involved in neurotransmitter synthesis (or catabolism). An inadequate supply of a cofactor can have the same effect on the neurotransmitter level as inadequate enzymatic activity. For example, pyridoxine deficiency, by decreasing the activity of AAAD, can suppress brain 5-HT synthesis (Le Blancq and Dakshinamurti, 1975).

Finally, and most important in the present context, variations in the availability to neurons of circulating compounds that are the precursors of the neurotransmitters can have dramatic effects on the synthesis and levels of these neurotransmitters. Thus, when excess quantities of the biosynthetic enzymes and their cofactors are present in the neurons (as is normally the case for ACh, 5-HT, and, under certain conditions, the catecholamines), the availability of the substrate for the rate-limiting enzyme will determine how much neurotransmitter is actually synthesized. Furthermore, provided that the rate of neurotransmitter degradation remains more or less constant, a change in its rate of synthesis will be reflected in a change in the levels of the neurotransmitter within the brain.

3.2. Protein Malnutrition

The biochemical composition of the brain can be altered by long-term consumption of protein-poor diets, especially during the early stages of development. Early malnutrition in experimental animals can be achieved by various methods, which include: (1) restricting the animal's access to its mother during lactation (Eayrs and Horn, 1955), (2) increasing the litter size (Widdowson and McCance, 1960), (3) decreasing the protein content of the maternal diet (Barnes *et al.*, 1966), (4) decreasing the amount of food available to the mother (Chow and Lee, 1964), and (5) combining these four methods. The period of deprivation may start before or during gestation or during lactation. Some of these methods have been used to study the effect of early protein malnutrition on the levels of brain neurotransmitters and on the activities of enzymes involved in neurotransmitter metabolism. These studies are summarized in Table I. Because the brain is an enormously heterogeneous structure with respect to cell types, and because nutritional state may affect some cell populations more than others, the levels of transmitters (or the activities of enzymes) known to be confined to only one cell type should be expressed *per brain,* and not simply per gram of tissue (or per protein, RNA, or DNA content); otherwise, one is in danger of dividing apples (e.g., the DA in a relatively small number of brain neurons) by bananas (e.g., the protein content or weight of the glia). Not all investigators express their data in this way, as can be seen from the following discussion of the results shown in Table I.

In one study, suckling rats were placed in litters of 16, rather than the usual 6 or 8, to generate postnatal malnutrition by limiting the accessibility to the mother; this treatment reportedly depressed brain NE and 5-HT levels and concentrations (Sereni *et al.*, 1966). In another experiment, the mothers were malnourished during pregnancy and/or lactation by feeding them a protein-deficient diet (8% rather than 24% protein); this treatment reduced the DA and NE in the brains of the offspring (Shoemaker and Wurtman, 1971). The neurotransmitter levels could be restored by nutritional rehabilitation (Shoemaker and Wurtman, 1973; Shoemaker *et al.*, 1974). Brain catecholamine levels were also found to be reduced in offspring of rats eating 20% wheat gluten during

Table I. Malnutrition and Brain Neurotransmitters

Reference	Experimental preparation	Findings
Sereni *et al.* (1966)	Rat; postnatal malnutrition; large litter size	Brain NE and 5-HT lowered
Shoemaker and Wurtman (1971, 1973); Shoemaker *et al.* (1974)	Rat; pre- and postnatal malnutrition; mothers fed low-protein diet (8%)	Brain DA and NE lowered; Tyr hydroxylase activity elevated; Tyr concentration unchanged; neurotransmitter levels restored with nutritional rehabilitation
Lee and Dubos (1972)	Rat; pre- and postnatal malnutrition; mothers fed 20% wheat gluten	Brain Da and NE lowered
Shoemaker and Wurtman (1973)	*Macaca mulatta*; prenatal malnutrition; mothers fed 2.5 g of protein/day	Pontine stem NE lowered
Sobotka *et al.* (1974)	Rat; postnatal malnutrition; mothers fed low-protein diet (12%)	Brain-stem 5-HT and 5-HIAA increased
Stern *et al.* (1974)	Rat; malnutrition	NE and 5-HT increased
Stern *et al.* (1975)	Rat; malnutrition; foot shock stress	5-HT, 5-HIAA, and NE lowered
Dickerson and Pao (1974)	Rat; pre- and postnatal maternal protein deficiency	5-HT and NE concentrations unchanged
Bernal *et al.* (1974)	Rat; intrauterine growth retardation; ligation of blood vessels	GABA concentration of brain synaptosomes increased
Rajalakshmi *et al.* (1974)	Rat; postweaning protein deficiency; preweaning malnutrition by increased litter size	Brain ACh lowered by postweaning malnutrition
Fernstrom and Wurtman (1971c)	Rat; postweaning malnutrition; corn diet	Brain Try and 5-HT reduced; fall in 5-HT partially reduced by Try supplement in diet
Enwonwu and Worthington (1973, 1974)	Rat; fed diets containing 0.5% or 18% lactalbumin for 5–8 weeks	Brain His, histamine, and homocarnosine increased; restored to normal within 1 week of protein refeeding
Hoeldtke and Wurtman (1973)	Human; kwashiorkor	Urinary DA and VMA decreased; NE unchanged
Sharma *et al.* (1968)	Human; kwashiorkor	Urinary 5-HIAA and serum 5-HT increased

pregnancy and lactation (Lee and Dubos, 1972) and in fetal offspring of *Macaca mulatta* given 2.5 g of protein/day during pregnancy (Shoemaker and Wurtman, 1973). Other investigators have reported increased brain concentrations of NE and 5-HT (Stern *et al.*, 1974), or no changes (Dickerson and Pao, 1974), in these neurotransmitters in brains of malnourished rat pups. A regional analysis of brains in one experiment showed an increase in brainstem content and concentration of 5-HT and 5-HIAA in rats whose mothers consumed a 12% casein diet during lactation (Sobotka *et al.*, 1974).

In an experiment testing intrauterine growth retardation (by means of ligation of the vessels of one uterine horn), brain concentrations of GABA in the synaptosomes were found to be three times higher among the growth-retarded rats than among the controls (Bernal *et al.*, 1974).

When weanling rats consumed a 5% protein diet for 5 weeks (compared to 20% protein), they had lowered brain ACh concentrations (Rajalakshmi *et al.*, 1974).

The effect of chronically consuming a tryptophan-poor diet (such as corn, which contains 5.6% protein) on brain 5-HT has been examined extensively in weanling rats; animals on such a diet were compared to littermates that were either pair-fed the same diet fortified with tryptophan or given 16.2% casein diet (by pair-feeding or *ad libitum*) (Fernstrom and Wurtman, 1971c). The corn-feeding lowered brain 5-HT; moreover, the following results showed that this reduction did indeed result from a decrease in the availability of its precursor, tryptophan.

1. The fall in brain 5-HT caused by a corn diet was partially blocked by fortifying the diet with tryptophan (Fernstrom and Wurtman, 1971c).

2. Acute administration of a single large dose of tryptophan to the corn-fed rats elevated brain 5-HT and 5-HIAA levels to at least the same levels as the controls; hence, all necessary enzymes and cofactors were available (Fernstrom and Hirsch, 1975).

3. Acute removal of tryptophan from the diet, by allowing normal rats to consume a single tryptophan-deficient meal, caused a similar dramatic reduction in brain 5-HT (and brain 5-HIAA) within 2 hr of food consumption (Biggio *et al.*, 1974).

4. In corn-fed rats, the rate of 5-HT synthesis, which is reduced to about one-third of that found in well-nourished animals, was rapidly increased by an intraperitoneal injection of tryptophan (Fernstrom and Hirsch, 1977).

Although histamine has not been shown to function as a neurotransmitter, it is present in nerve terminals (Snyder *et al.*, 1974) and may thus have some role in the transmission of nerve impulses. As with 5-HT and tryptophan, there is a very strong correlation between brain histamine levels and the levels of its amino acid precursor, histidine. Postweaning protein malnutrition, which increases brain histidine levels fivefold, also increases brain histamine levels threefold (Enwonwu and Worthington, 1973, 1974). Thus, histamine synthesis is probably also under precursor control.

The changes in brain neurotransmitter levels that result from experimental malnutrition seem to accompany changes in the activities of enzymes associated with their synthesis or catabolism. Thus, undernutrition in early life is associated with reductions in brain CAT and dopamine-beta-oxidase activities (Shoemaker *et al.*, 1974) and with the elevation in tyrosine hydroxylase activity described above (Shoemaker and Wurtman, 1971; Shoemaker *et al.*, 1974). AChE activity may also increase (Adlard and Dobbing, 1971; Im *et al.*, 1973); however, one group of investigators noted decreased cerebellar AChE activity as a result of neonatal malnutrition (Sobotka *et al.*, 1974). In a study of the developmental pattern of AAAD, Hernandez (1973) found that enzyme activity in the brains of rats raised in litters of 16 rose more slowly than in control animals raised in litters of 6, and it never attained the maximum activity found in the controls.

Studies on the effects of malnutrition on transmitters in human brain are few and, of necessity, indirect. The excretion of DA and VMA (the NE metabolite) in the urine is reduced in kwashiorkor (Hoeldtke and Wurtman, 1973). Although the decrease in DA may be artifactual [i.e., reflecting the change from a DOPA-containing cereal diet (Hoeldtke and Wurtman, 1974)], the reduced urinary VMA levels probably are an accurate reflection of endogenous NE metabolism. This NE could arise, however, from the adrenal gland or the sympathetic nerves, as well as from the brain. The urinary output of 5-HIAA and the serum concentration of 5-HT are elevated in children with kwashiorkor, especially if the disease is accompanied by steatorrhea (Sharma *et al.*, 1968). These findings cannot be reliably interpreted until the relation between amino acid metabolism and gut 5-HT synthesis is explored.

Since monoaminergic neurotransmitters are unable to leave the brain without first being metabolized by oxidative deamination, nutritionally induced changes in the levels of the neurotransmitters within blood or urine probably reflect washout from peripheral structures. The acidic and neutral transmitter metabolites in the urine originate from both the CNS and peripheral structures; hence, reductions in these levels could tell something about brain metabolism.

In this section we have reviewed some studies on the effects of prenatal, preweaning, and postweaning malnutrition and protein deficiency on brain neurotransmitters and on some of the enzymes associated with their biosynthesis and metabolism. Because chronic protein–calorie malnutrition is a rather crude experimental technique, it does not facilitate identifying the specific biochemical mechanism by which diet affects brain neurotransmitter levels and enzyme activities. Moreover, changes in transmitter levels are especially difficult to interpret because of the many concurrent changes taking place in malnourished animals, such as changes in brain weight. As described below, acute nutritional inputs also affect, preferentially, the production of several transmitters in experimental animals and, probably, in humans. Studies of these relationships can yield detailed information on both nutrient metabolism and neurochemical control mechanisms. They can also provide genuinely surprising insights on the extent to which brain function passively reflects peripheral metabolism. The next section describes how cofactor availability can modulate neurotransmitter synthesis; the section that follows it describes how precursor availability and—in some cases—the consumption of individual meals can control neurotransmitter synthesis.

3.3. Cofactor Availability

A number of cofactors, described above, are needed for the biosynthesis of the catecholamines, 5-HT, and ACh. These include, among others, pyridoxal phosphate (for the enzyme AAAD, which decarboxylates DOPA to form DA and 5-HTP to form 5-HT), tetrahydrobiopterin (for tyrosine hydroxylase and tryptophan hydroxylase), ascorbic acid (for dopamine-β-hydroxylase), and molecular oxygen (for all the oxidase enzymes above).There is growing evidence that pathologic reductions in the availability of these cofactors can suppress neurotransmitter formation. Thus, pyridoxine deficiency reduced brain 5-HT synthesis (Le Blancq and Dakshinamurti, 1975) by a mechanism independent of tryptophan levels or tryptophan hydroxylase activity; oxygen deficiency slowed catecholamine synthesis (Davis and Carlsson, 1973); and scurvy affected NE turnover (Thoa *et al.*, 1966). No information seems to be available, however, on the possibility that physiological variations in cofactor availability *normally* affect transmitter production. The levels of pyridoxal phosphate required for optimal *in vitro* activity of AAAD are higher than levels thought to exist in brain *in vivo* (Lloyd and Hornykiewicz, 1970); hence, enzyme activity *in vivo* may be limited by the availability of dietary vitamin B_6. However, it remains to be shown that brain pyridoxal phosphate levels normally vary, or that such variations cause parallel changes in monoamine production. The amount of tetrahydrobiopterin that has access to the tyrosine hydroxylase molecule would probably influence the rate of tyrosine hydroxylation *in vivo*. This access is known to be inhibited by DA and NE; indeed, it apparently constitutes the mechanism by which the amines exert end-product inhibition over catecholamine synthesis. Kettler *et al.* (1974) reported that administration of tetrahydrobiopterin directly to the brain could stimulate DA synthesis in normal animals. It should be determined whether levels of tetrahydrobiopterin in brain vary normally (perhaps in response to the consumption of certain types of diets) and, if so, whether this affects DA synthesis.

3.4. Precursor Availability

If the enzymes that synthesize the neurotransmitters require higher concentrations of their substrates than those normally found, and if these concentrations normally do undergo significant variations *in vivo*, then neurotransmitter synthesis (and, consequently, neurotransmitter levels) may be controlled physiologically by the concentrations of their precurrors. Moreover, if the precursor levels in brain are linked to their concentrations in the plasma, and if plasma concentrations, in turn, depend on nutritional state, it becomes possible for food consumption to exert major control over brain function by controlling transmitter synthesis. These hypothetical relationships do, in fact, obtain *in vivo*.

1. The consumption of a particular meal does change the concentrations of tryptophan, choline, and tyrosine in the plasma; it also changes plasma concentrations of other substances (the neutral amino acids) that compete with tryptophan and tyrosine for entry into the brain.

2. To be fully saturated, the transport systems that mediate the uptakes of tryptophan, choline, and tyrosine into brain cells (*all* brain cells, not only those that utilize these compounds for neurotransmitter synthesis) require

much higher concentrations of these amino acids than are normally present in the blood; hence food-induced changes in these plasma concentrations can and do cause predictable changes in their brain levels.

3. The enzyme that determines the overall rate of 5-HT synthesis (tryptophan hydroxylase) is not fully saturated with its substrate (tryptophan) *in vivo*; hence, as brain tryptophan levels change, so too does the rate at which brain neurons synthesize 5-HT. CAT is similarly unsaturated *in vivo,* and brain choline levels similarly control ACh synthesis. Tyrosine hydroxylase is probably also not fully saturated, and thus its activity *in vivo* is also subject to some control by brain tyrosine levels. Nevertheless, as described below, other mechanisms involving feedback loops probably exert a more significant control over catecholamine synthesis (under most conditions) than does precursor availability.

3.4.1. Tryptophan

Tryptophan, the precursor of 5-HT, is an essential amino acid—that is, it cannot be synthesized by the mammalian organism and must be obtained through the diet. The plasma concentrations of most amino acids, including tryptophan, undergo diurnal variations, both in rats (Fernstrom *et al.,* 1971) and in humans (Wurtman *et al.,* 1968). (The phases for the tryptophan rhythms differ between rats and humans, probably because rats eat nocturnally.) The K_m of tryptophan hydroxylase for tryptophan, as measured *in vitro,* is quite high—3×10^{-4} M (Lovenberg *et al.,* 1968)—whereas the concentrations of tryptophan in the brain are rather low—about $4–8 \times 10^{-5}$ M (McKean *et al.,* 1968; Fernstrom and Wurtman, 1974). [The *in vivo* K_m is estimated at 6×10^{-5} M (Carlsson *et al.,* 1972).] Thus, theoretically, the enzyme is unsaturated with its substrate *in vivo,* and the rate of its products' synthesis can change as a function of substrate availability.

Plasma tryptophan, brain tryptophan, and brain 5-HT levels all undergo diurnal changes in concentration that are related to food consumption. When rats were injected intraperitoneally with a low dose of tryptophan (12.5 mg/kg) at a time of day when blood tryptophan levels are lowest, blood tryptophan concentrations rose to a level not greater than the normal diurnal peak, and brain 5-HT levels also rose significantly (Fernstrom and Wurtman, 1971a). Varying the tryptophan dose gave a dose-response curve in which brain 5-HT levels reached a plateau at about the same time that the daily plasma and brain tryptophan concentrations normally peak.

The injection of insulin or the feeding of a high-carbohydrate diet raised the plasma tryptophan levels in rats, while lowering the plasma concentrations of most other amino acids (Fernstrom and Wurtman, 1971b, 1972a). Consequently, the pancreatic hormone also raised brain tryptophan and 5-HT levels. On the other hand, consumption of a high-protein diet, which increases plasma tryptophan levels but raises the plasma levels of the other neutral amino acids even more, had no effect on (or even lowered) brain concentrations of tryptophan and 5-HT (Fernstrom and Wurtman, 1972b). Consumption of a diet not containing the five neutral amino acids sharing the same carrier system as tryptophan caused a large increase in brain tryptophan, 5-HT, and 5-HIAA levels. This increase did not occur, however, if other amino acids (which do not compete with tryptophan for brain entry) were removed from the diet.

Based on these data, a model was developed by Fernstrom and Wurtman (1972b) for describing the nutritional control of brain 5-HT levels in rats (Fig. 6). This model proposes that these concentrations are physiologically regulated by the ratio of the concentrations of plasma tryptophan to the sum of the concentrations of the competing neutral amino acids (tyrosine, phenylalanine, leucine, isoleucine, and valine). This model applies to all the amino acids transported by the same carrier system into the brain. Thus, the ratio of *any* neutral amino acid to the sum of all the other neutral amino acids determines that amino acid's level in the brain—even if it is a drug that is not normally found in the circulation (such as L-DOPA, alpha-methyl-DOPA, or *p*-chlorophenylalanine).

Another aspect of the model is the relationship between the binding of tryptophan to serum albumin and its availability to the brain. Tryptophan is unique among the amino acids in that only about 10–20% of the circulating indole is in the "free" form; the rest is bound to serum albumin (McMenamy and Oncley, 1958). This fact has led some investigators to speculate that *free* circulating tryptophan might be the biologically important moiety in determining brain tryptophan levels (Knott and Curzon, 1972). Studies from our labo-

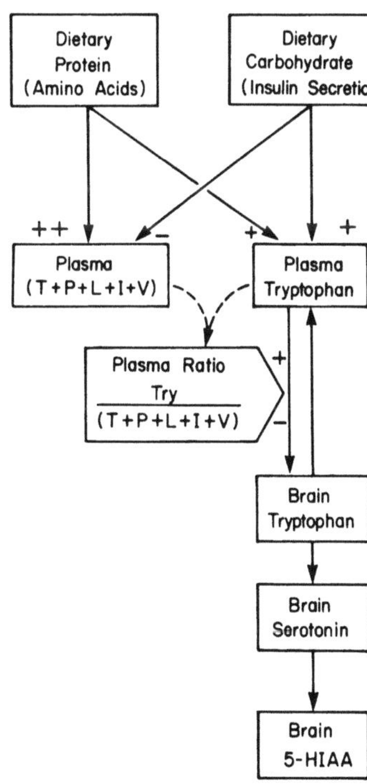

Fig. 6. Proposed sequence describing diet-induced changes in brain serotonin concentration in the rat. The ratio of tryptophan to tyrosine (T) plus phenylalanine (P) plus leucine (L) plus isoleucine (I) plus valine (V) in the plasma is thought to control the tryptophan concentration in the brain. (From Fernstrom and Wurtman, 1972b; copyright 1972 by the American Association for the Advancement of Science.)

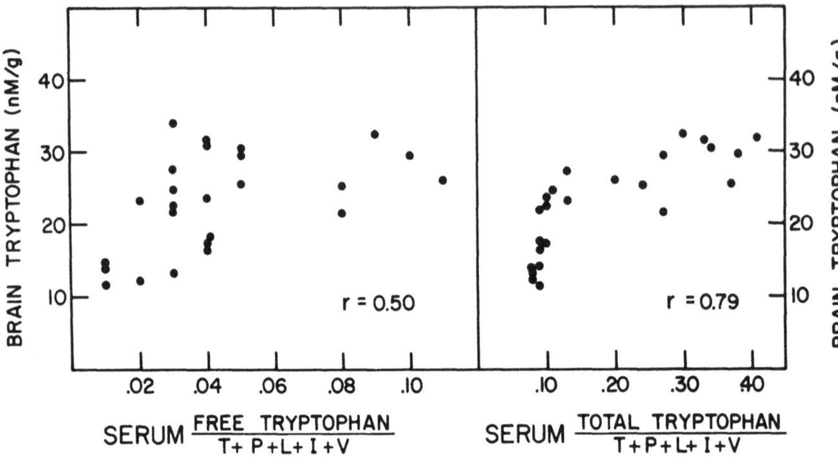

Fig. 7. Correlations between brain tryptophan and the ratios of both serum free and total tryptophan to the sum of the other neutral amino acids that compete with it for uptake in individual rats consuming single meals. Groups of fasting rats were given free access to one of the following diets and killed 2 hr later: carbohydrates + 0% fat, carbohydrates + 40% fat, carbohydrates + large neutral amino acids + 0% fat, and carbohydrates + large neutral amino acids + 40% fat.

ratory have shown that this is not the case. For example, treatments that lowered free tryptophan levels but raised its total (albumin-bound level in the plasma (e.g., the administration of insulin or the feeding of a carbohydrate meal, which, by reducing plasma concentrations of nonesterified fatty acids, increased the affinity of albumin for tryptophan) *increased* brain tryptophan and 5-HT levels (Madras *et al.*, 1973). Conversely, the addition of fat to the diet, which caused dose-dependent increases in serum nonesterified fatty acids, and therefore increased free tryptophan levels in serum, had no effect either on serum *total* tryptophan or brain tryptophan (Madras *et al.*, 1974). Brain tryptophan levels under all nutritional states correlated far better with the ratio of *total* serum tryptophan to the concentrations of the other neutral amino acids than with the corresponding ratio of *free* serum tryptophan (Fig. 7). Thus, the amount of tryptophan and the synthesis of 5-HT in the brain are directly related to the exact composition of each meal.

Ashley and Anderson (1975) have presented evidence that the brain utilizes the "information" that it obtains by coupling 5-HT synthesis to plasma amino acid concentrations in order to control some aspects of feeding behavior. The various proportions or varieties of proteins that animals chose to ingest over a 28-day experimental period correlated best with the effects of the diets on the plasma tryptophan/competing neutral amino acid ratio.

3.4.2. Acetylcholine

The K_m's of CAT, in homogenates of rat brain, for choline and for acetyl-CoA [400 μM and 18 μM, respectively (White and Wu, 1973)] are considerably higher than the *in vivo* concentrations of these precursors [approximately 37 μM for choline (Stavinoha and Weintraub, 1974; Cohen and Wurtman, 1975, 1976) and 7–11 μM for acetyl-CoA (Sollenberg, 1970)]. Therefore, as with brain 5-HT, changes in the brain choline (or acetyl-CoA) levels might be expected to affect the synthesis and levels of the neurotransmitter. In fact, the administration of choline by either injection or dietary means does elevate brain ACh levels (Cohen and Wurtman, 1975, 1976). The availability of acetyl-CoA may also affect brain ACh levels (Heinrich *et al.*, 1973).

Accurate measurement of brain choline and ACh concentrations was only recently made possible by the adoption of focused microwave irradiation as the means of killing the experimental animal; this treatment simultaneously inactivates brain enzymes, thereby preventing artifactual postmortem changes in brain choline and ACh levels. The development of new and sensitive radioisotopic assay methods (compiled by Hanin, 1974) has also had a major impact in allowing choline–ACh relationships to be examined.

It is widely believed (Ansell and Spanner, 1971)—but not by all (Kewitz and Pleul, 1976)—that the mammalian brain is incapable of *de novo* choline synthesis and must therefore depend on circulating free and covalently bound (e.g., as lecithin) choline for its supply of this ACh precursor. The choline in the circulation, in turn, derives mainly from two sources—synthesis by the liver and dietary ingestion. In fact, a significant fraction of plasma choline (estimated at about half) is apparently of dietary origin (Wise and Elwyn, 1965; Hanin and Schuberth, 1974). Intravenously administered labeled choline is very rapidly incorporated into brain ACh (Schuberth *et al.*, 1970; Haubrich *et al.*, 1972; Aquilonius *et al.*, 1973; Saelens *et al.*, 1973; Jenden *et al.*, 1974). *In vitro* studies of brain synaptosomal preparations have demonstrated the existence of two distinct uptake mechanisms for choline (Yamamura and Snyder, 1972; Haga and Noda, 1973)—i.e., a low-affinity system and a sodium-dependent high-affinity system [which is associated specifically with the terminals of cholinergic neurons (Kuhar *et al.*, 1973; Sorimachi and Kataoka, 1974)]. The mechanism by which choline is transported into the brain (through the BBB) apparently utilizes a low-affinity system and is not the same mechanism as that operating at nerve terminals (and, *in vitro,* in synaptosomes). The choline-specific transport system has a K_m of 0.22 mM and V_{max} of 6 nmol/min/g (Pardridge and Oldendorf, 1977). Thus, the K_m of the BBB for choline is much higher than that of either the high- or low-affinity transport systems reported for rat brain synaptosomes (1 μM and 90 μM, respectively; Yamamura and Snyder, 1972). Furthermore, since blood choline concentrations are on the order of 10 μM, the BBB choline carrier is not saturated *in vivo;* this is true even at plasma choline concentrations that are much higher than those normally found *in vivo* (Freeman *et al.*, 1975). Pardridge and Oldendorf (1977) calculated the rate of choline influx through the BBB to be 0.3 nmol/min/g; this rate is much lower than that calculated for transport through brain cell membranes and therefore probably determines the influx of circulating choline into the brain. Thus, one could predict, from the low affinity of the BBB for choline (i.e., a relatively high K_m of 0.22 mM) that plasma choline levels would control brain choline concentrations. Studies from our laboratory have shown that brain choline and, consequently, ACh concentrations do indeed depend on circulating choline levels (Cohen and Wurtman, 1975, 1976).

The mechanisms by which covalently bound choline (e.g., lecithin) travels to the brain have not been studied extensively, although Illingworth and Portman (1972) have shown that lysolecithin serves as a precursor for brain ACh in squirrel monkeys.

In initial experiments we showed that, when rats were injected intraperitoneally with a saline solution containing choline chloride (60 mg/kg, or about one-third of the daily dietary choline intake of these rats), serum and brain choline concentrations rose dramatically; this effect was soon followed by a significant increase in brain ACh levels (Fig. 8; Cohen and Wurtman, 1975). Other investigators (Haubrich *et al.*, 1975) have since confirmed this finding by the use of a higher choline dose (100 mg/kg). Furthermore, choline administered to guinea pigs by intracarotid perfusion (Haubrich *et al.*, 1974) or to rats by continuous infusion (Haubrich *et al.*, 1975; Racagni *et al.*, 1975) increased brain ACh levels.

In dietary studies with choline, Nagler *et al.*, (1968) found that the concentration of brain ACh was decreased by 30–35% when weanling rats were subjected to choline deficiency for 5 days. In our own experiments with adult rats fed a choline-deficient diet, serum choline levels were reduced by 30% after 10 days on the test diets (unpublished observations). We have further

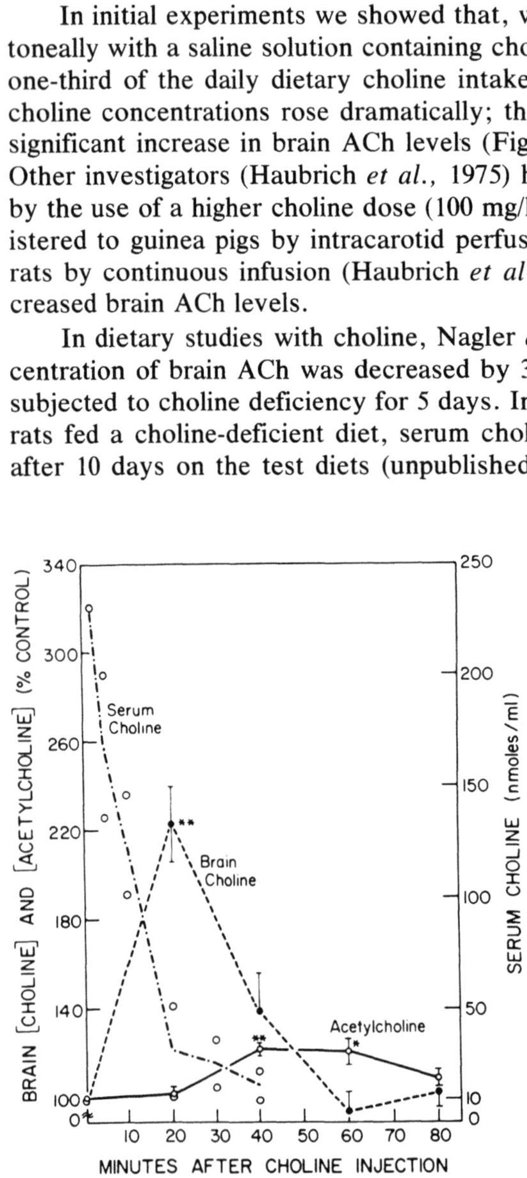

Fig. 8. Time course of the response of serum choline, brain choline, and brain acetylcholine to choline administered by injection. Groups of 5–9 male rats (150–200 g) received choline chloride (60 mg/kg, intraperitoneally) in saline (0.9% NaCl) or the diluent alone. At various intervals after the injection, the animals were killed by microwave irradiation of the head, and whole brain choline and ACh concentrations were measured. Data for brain choline and ACh levels are expressed as a percent of control mean. Bars indicate the standard error of the mean. Groups of 2–3 rats were injected as described above and killed by decapitation at various intervals after injection. Blood was collected from the cervical wound and serum choline concentrations were measured. Data for serum choline levels are expressed as nmol/ml. Open circles indicate the range of values for serum choline levels. Differences from corresponding concentrations in rats injected with saline are indicated by * ($p < 0.01$) and ** ($p < 0.001$).

observed that the chronic consumption of various amounts of choline, either from the food or the drinking water, caused parallel changes in serum choline and in choline and ACh concentrations in several brain regions [Fig. 9 (Cohen and Wurtman, 1976; Cohen, 1976)]. Thus, plasma choline levels are, contrary to earlier beliefs, normally quite variable.

An increase in brain choline and ACh levels could be induced by varying the choline concentration of the drinking water for only 3 days (Cohen, 1976) or even by administering a single dose of choline via stomach tube (Ulus *et al.*, 1977). Consumption of choline (15 mg/ml) in the drinking water for only 3 days increased serum choline levels by over 300% (unpublished observations).

The increases in brain ACh caused by ingestion of choline or by administration of physostigmine (to inhibit the AChE-catalyzed breakdown of ACh) were additive; this finding was interpreted as indicating that the mechanism by which choline ingestion affects tissue ACh levels involves accelerated *synthesis* of the neurotransmitter and not changes in its degradation (Cohen and Wurtman, 1976).

The elevations in brain ACh that follow choline administration were observed in all the examined brain regions, i.e., the caudate nucleus (a region rich in cholinergic interneurons), the cortex, and the hippocampus (a region rich in cholinergic nerve terminals) (Cohen and Wurtman, 1976; Hirsch *et al.*, 1977; Table II). The observation that ACh levels rise significantly within the hippocampus, as elsewhere in the brain (e.g., caudate nucleus), following choline administration indicates that the increase is not confined to cholinergic cell bodies, but must also take place at the neuronal loci (the boutons and terminals) from which ACh can be released.

Changes in neurotransmitter levels are physiologically significant only if they result in changes in the amounts of neurotransmitter actually released at the synapse. This chapter has already covered this point and listed some strategies for estimating neurotransmitter release after various treatments. One of these strategies has recently been used to determine whether choline-induced increases in tissue ACh levels are associated with changes in the release of the neurotransmitter. The problem was approached by examining both the enzyme tyrosine hydroxylase in the adrenal gland (this enzyme is involved in catecholamine biosynthesis and confined to cells that synthesize catecholamines) and the brain cells known to receive cholinergic synapses.

Rats received a single dose of choline by stomach tube, and ACh levels in the adrenal gland were markedly elevated for 1–8 hr; 24 hr after choline administration, the activity of adrenal tyrosine hydroxylase rose and remained elevated for 48 hr (Fig. 10) (Ulus *et al.*, 1977). This increase was not observed if rats received by stomach tube compounds that were not ACh precursors. It also failed to occur if an adrenal had been denervated prior to choline administration. Hence, the effect on tyrosine hydroxylase results not from a direct action of choline on adrenal chromaffin cells, but from the release of ACh, formed presynaptically from the administered choline.

Choline administration (100 mg/kg, intraperitoneally) also activates ty-

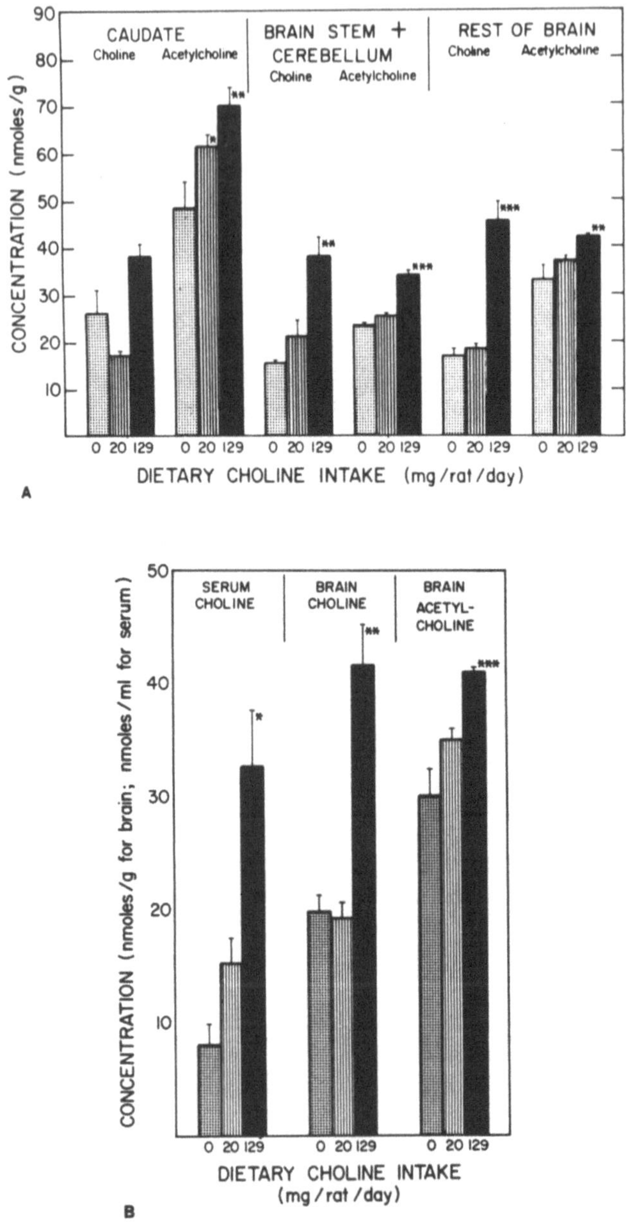

Fig. 9. Effect of dietary choline content on serum choline levels and on choline and acetylcholine concentrations in various brain regions. Groups of 8–10 male rats (each weighing 90 g) consumed diets containing an average of 0, 20, or 129 mg of choline per day for 11 days (choline chloride was added to the choline-deficient diet). Five to seven rats from each group were killed by microwave irradiation of the head, and their brains were dissected into regions and assayed for choline and ACh (A). Whole brain choline and ACh levels were calculated from the regional values; the remaining animals killed by decapitation and their sera were assayed for choline (B). Columns represent the mean concentrations; vertical lines represent standard errors of the means. Differences from corresponding concentrations in rats consuming no choline are indicated by $*p < 0.05$, $**p < 0.01$, and $***p < 0.001$. (From Cohen and Wurtman, 1976; copyright 1976 by the American Association for the Advancement of Science.)

Table II. Effect of Choline Chloride Administration on the Concentrations of Choline and Acetylcholine in Rat Hippocampus and Caudate Nuclei[a]

Group	Choline (nmol/g)	ACh (nmol/g)
Control		
Hippocampus	27.29 ± 3.00	14.23 ± 2.95
Caudate nuclei	34.14 ± 2.31	48.70 ± 2.00
20 min after choline chloride		
Hippocampus	42.60 ± 1.70[b]	26.50 ± 2.70[c]
Caudate nuclei	57.47 ± 5.50[b]	59.15 ± 2.45[c]
40 min after choline chloride		
Hippocampus	37.48 ± 2.99	29.60 ± 3.04[c]
Caudate nuclei	45.04 ± 3.53[d]	63.16 ± 3.00[c]

[a] Groups of 10 rats received choline chloride (60 mg/kg, intraperitoneally) or its diluent (water) and were killed 20 or 40 min after injection. Data are given as means ± standard errors of the means. (From Hirsch *et al.*, 1977.)
[b] $p < 0.01$ differs from control.
[c] $p < 0.02$ differs from control.
[d] $p <$ differs from control.

rosine hydroxylase in the caudate nucleus of the brain (Ulus and Wurtman, 1976)—an area rich in both cholinergic neurons and dopaminergic nerve terminals; this effect peaks 2 hr after choline administration and is blocked by atropine pretreatment (atropine blocks muscarinic cholinergic receptors). Therefore, it most likely reflects a receptor-mediated activation of existing molecules of tyrosine hydroxylase (Ulus and Wurtman, 1976).

These experiments strongly suggest that ACh release *is* affected by intraneuronal ACh levels, and, consequently, by the amount of choline in the diet.

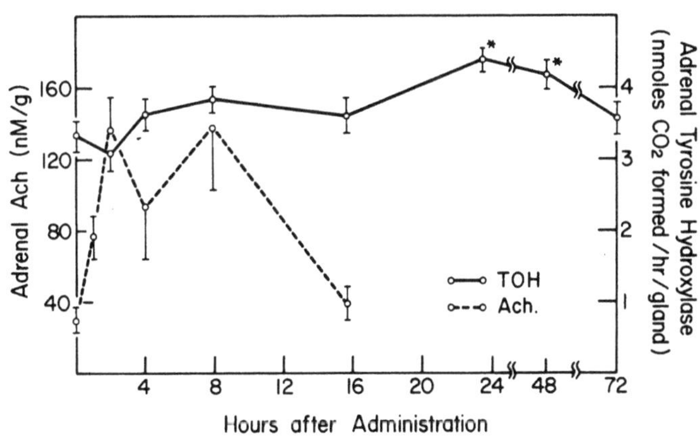

Fig. 10. Effect of choline chloride (20 mmol/kg, by stomach tube) on adrenal ACh content and tyrosine hydroxylase activity. Each point represents the mean ± standard error of the mean of 4–6 adrenals; similar data were obtained in a second experiment using 5–6 rats per group. *$p < 0.05$ compared with tyrosine hydroxylase activity in adrenals from control animals. (From Ulus *et al.*, 1977.)

It remains to be shown that choline in its usual dietary form (lecithins) is equally effective in modifying cholinergic functions.

3.4.3. Catecholamines

The effect of diet on brain catecholamine concentrations is not as readily apparent as it is with those of 5-HT and ACh. For example, when rats receive a large dose of tyrosine, no acute changes occur in brain DA or NE levels. This observation was initially interpreted to mean that tyrosine concentration does not significantly affect the rate of synthesis of the catecholamine neurotransmitters in the brain. However, tyrosine hydroxylase, like tryptophan hydroxylase, might not be saturated *in vivo* [its K_m for tyrosine is estimated to be 0.1 mM and brain tyrosine concentrations vary between 0.06 and 0.18 mM (Gibson and Wurtman, 1978)]; moreover, plasma and brain tyrosine levels normally vary within a broad dynamic range (i.e., they rise dramatically after a high-protein meal, probably because both the tyrosine and phenylalanine in the protein contribute to plasma tyrosine). Hence, it seemed likely that, under at least some circumstances, one could demonstrate precursor control of catecholamine synthesis. The paradigm initially used for this purpose involved giving a drug (RO4-4602) that blocks the decarboxylation of newly formed brain DOPA, and measuring the rate of accumulation of this intermediate in animals so treated as to raise or lower brain tyrosine levels (i.e., by giving them tyrosine or other neutral amino acids—which lower brain tyrosine by competing for uptake—or meals of varying composition). We found that treatments elevating brain tyrosine levels *increased* DOPA synthesis, whereas treatments lowering or maintaining brain tyrosine levels had, likewise, lowering or maintenance effects on brain DOPA levels (Wurtman *et al.*, 1974). Thus, an excellent correlation was obtained between brain tyrosine levels and the rate of synthesis of the catechol DOPA.

In other experiments, the relationship between brain tyrosine level and brain catechol*amine* production was examined by giving animals a MAO inhibitor and measuring the rates of DA or NE accumulation, or by giving animals probenecid—which blocks the efflux of HVA from the brain—and measuring its rate of accumulation. In both studies, poor correlations were noted between brain tyrosine concentrations and DA or HVA accumulation. However, when animals received yet another treatment (administration of the drug haloperidol, which blocks the effects of DA on its receptors and thereby accelerates DA synthesis), an excellent correlation was obtained between brain tyrosine concentrations and DA synthesis (as estimated by measuring HVA levels in the caudate nucleus) (M. C. Scally and R. J. Wurtman, submitted for publication).

Furthermore, the study of Chiel and Wurtman (1976) has shown that amino acid treatments that lowered brain tyrosine concentrations also reduced the effectiveness of *d*-amphetamine, which acts by liberating newly formed DA into brain synapses.

These varied findings are interpreted as follows. Catecholamine synthesis within the dopaminergic neurons of the brain depends on both the activity of tyrosine hydroxylase and the availability of the precursor tyrosine. When tyrosine levels are elevated, DA synthesis is accelerated and more neurotransmitter is released into synapses to interact with DA receptors. This increased interaction rapidly activates feedback mechanisms, which effectively decrease the activity of tyrosine hydroxylase, thereby restoring DA synthesis to normal. These feedback mechanisms were prevented from operating in two of the paradigms above: when the decarboxylation step was inhibited (thus reducing DA synthesis) and when DA receptors were blocked (thus preventing receptor interaction). Therefore, in these two experiments, it was possible to observe the tyrosine-induced increase in the synthesis rate of the neurotransmitters. However, in experiments involving pretreatment with MAO inhibitors or probenecid, the feedback loop could still operate. Hence, tyrosine availability is probably an important factor in situations in which catecholamine synthesis rates are elevated, or in catecholamine neurons that lack the postulated feedback loops (e.g., perhaps those terminating in the hypothalamus).

4. *Implications of Precursor Control of Brain Neurotransmitter Synthesis and Release*

One implication of the fact that the synthesis of several neurotransmitters is under nutritional control is that it may make possible the development of new modes of therapy for disease states involving neurons that utilize these transmitters. It seems likely that the diet—or even particular nutrients—can be used as a drug for treating diseases related to serotoninergic, cholinergic, and, possibly, catecholaminergic brain neurons; even more likely, foods or their constituents can be used to potentiate the effects of "true" drugs that act by increasing or decreasing chemical transmission across synapses. For example, several neurologic and psychiatric disorders are currently thought to involve a relative insufficiency of acetylcholine in the brain; these include, among others, tardive dyskinesia, Huntington's disease, mania, and even some forms of schizophrenia. It is at least possible that choline administration might, by increasing brain acetylcholine levels, modify some of the manifestations of these diseases. [In fact, oral choline administration has already been claimed to be beneficial to patients with some of these disorders (Barbeau *et al.*, 1979).] It seems safe to suggest that numerous additional uses will be found for the strange propensity of certain brain neurons to vary their activity passively according to the composition of the blood—and the diet.

5. *References*

Adlard, B. P., and Dobbing, J., 1971, Elevated acetylcholinesterase activity in adult rat brain after undernutrition in early life, *Brain Res.* **30**:198.

Ansell, G. B., and Spanner, S., 1971, Studies on the origin of choline in the brain of the rat, *Biochem. J.* **122**:741.

Aquilonius, S.-M., Flentge, F., Schuberth, J., Sparf, B., and Sundwall, A., 1973, Synthesis of acetylcholine in different compartments of brain nerve terminals *in vivo* as studied by the incorporation of choline from plasma and the effect of pentobarbitol on this process, *J. Neurochem.* **20:**1509.

Ashley, D. V., and Anderson, G. H., 1975, Correlation between the plasma tryptophan to neutral amino acid ratio and protein intake in the self-selecting weanling rat, *J. Nutr.* **105:**1412.

Axelrod, J., 1957, The *O*-methylation of epinephrine and other catechols *in vitro* and *in vivo*, *Science* **126:**400.

Barbeau, A., Growden, J. H., and Wurtman, R. J., 1979, in: *Choline and Lecithin in Brain Disorders,* Vol. 5, *Nutrition and the Brain* (R. J. Wurtman and J. J. Wurtman, eds.), Raven Press, New York.

Barnes, R. H., Cunnold, S. R., Zimmerman, R. R., Simons, H., MacLeod, R. B., and Krook, L., 1966, Influence of nutritional deprivations on early life on learning behavior of rats as measured by performance in a water-maze, *J. Nutr.* **89:**399.

Bernal, A., Morales, M., Feria-Velasco, A., Chew, S., and Rosado, A., 1974, Effect of intra-uterine growth retardation on the biochemical maturation of brain synaptosomes in the rat, *J. Nutr.* **104:**1157.

Biggio, G., Fadda, F., Fanni, P., Tagliamonte, A., and Gessa, G. L., 1974, Rapid depletion of serum tryptophan, brain tryptophan, serotonin and 5-hydroxyindoleacetic acid by a tryptophan-free diet, *Life Sci.* **14:**1321.

Carlsson, A., Kehr, W., Lindqvist, M., Magnusson, R., and Atack, C. V., 1972, Regulation of monoamine metabolism in the central nervous system, *Pharmacol. Rev.* **24:**384.

Chan-Palay, V., 1976, Serotonin axons in the supra- and subependymal plexuses and in the leptomeninges; their roles in local alterations of cerebrospinal fluid and vasomotor activity, *Brain Res.* **102:**103.

Chiel, H. J., and Wurtman, R. J., 1976, Suppression of ampetamine-induced hypothermia by the neutral amino acid valine, *Psychopharmacol. Commun.* **2**(3):207–217.

Chow, B. F., and Lee, C. J., 1964, Effect of dietary restriction of pregnant rats on body weight gain of the offspring, *J. Nutr.* **82:**10.

Cohen, E. L., 1976, *In vivo* studies of the effect of the availability of choline on the biosynthesis and content of acetylcholine in brain, Ph.D. thesis, Massachusetts Institute of Technology.

Cohen, E. L., and Wurtman, R. J., 1975, Brain acetylcholine: increase after systemic choline administration, *Life Sci.* **16:**1095.

Cohen, E. L., and Wurtman, R. J., 1976, Brain acetylcholine: control by dietary choline, *Science* **191:**561.

Cooper, J. R., Bloom, F. E., and Roth, R. H., 1974, *The Biochemical Basis of Neuropharmacology,* 2nd ed., Oxford University Press, Oxford.

Dahlstrom, A., and Fuxe, K., 1964, Evidence for the existence of monoamine-containing neurons in the central nervous system. I. Demonstration of monoamines in the cell bodies of brain stem neurons, *Acta Physiol. Scand.* **62**(suppl. 232):1.

Davis, J. N., and Carlsson, A., 1973, The effect of hypoxia on monoamine synthesis, levels and metabolism in rat brain, *J. Neurochem.* **21:**783.

Descarries, L., Beaudet, A., and Watkins, K. C., 1975, Serotonin nerve terminals in adult rat neocortex, *Brain Res.* **100:**563.

Dickerson, J. W., and Pao, S. K., 1974, Effect of pre- and post-natal maternal protein deficiency on free amino acids and amines of rat brain, *Biol. Neonate* **25:**114.

Eayrs, J. T., and Horn, G., 1955, The development of cerebral cortex in hypothyroid and starved rats, *Anat. Rec.* **121:**53.

Enwonwu, C., and Worthington, B., 1973, Accumulation of histidine, 3-methylhistidine, and homocarnosine in the brains of protein-calorie deficient monkeys, *J. Neurochem.* **21:**799.

Enwonwu, C. O., and Worthington, B. S., 1974, Concentrations of histamine in brain of guinea pig and rat during dietary protein malnutrition, *Biochem. J.* **144:**601.

Fernstrom, J. D., and Hirsch, M. J., 1975, Rapid repletion of brain serotonin in malnourished corn-fed rats following *l*-tryptophan injection, *Life Sci.* **17:**455.

Fernstrom, J. D., and Hirsch, M. J., 1977, Brain serotonin synthesis: Reduction in corn-malnourished rats, *J. Neurochem.* **28**:877–879.

Fernstrom, J. D., and Wurtman, R. J., 1971a, Brain serotonin content: Physiological dependence on plasma tryptophan levels, *Science* **173**:149.

Fernstrom, J. D., and Wurtman, R. J., 1971b, Brain serotonin content: Increase following ingestion of carbohydrate diet, *Science* **174**:1023.

Fernstrom, J. D., and Wurtman, R. J., 1971c, Effect of chronic corn consumption on serotonin content of rat brain, *Nature, New Biol.* **234**:62.

Fernstrom, J. D., and Wurtman, R. J., 1972a, Elevation of plasma tryptophan by insulin in the rat, *Metabolism* **21**:337.

Fernstrom, J. D., and Wurtman, R. J., 1972b, Brain serotonin content: Physiological regulation by plasma neutral amino acids, *Science* **178**:414.

Fernstrom, J. D., and Wurtman, R. J., 1974, Control of brain serotonin levels by the diet, in: *Advances in Psychopharmacology: Serotonin—New Vistas* (E. Costa and M. Sandler, eds.), Vol. 11, pp. 133–142, Raven Press, New York.

Fernstrom, J. D., Larin, F., and Wurtman, R. J., 1971, Daily variations in the concentrations of individual amino acids in rat plasma, *Life Sci.* **10**:935.

Freeman, J. J., Choi, R. L., and Jenden, D. J., 1975, Plasma choline: its turnover and exchange with brain choline, *J. Neurochem.* **24**:729.

Fuxe, K., 1965a, Evidence for the existence of monoamine neurons in the central nervous system. III. The monoamine nerve terminals, *Z. Zellforsch. Mikroskop. Anat. Abt. Histochem.* **65**:573.

Fuxe, K., 1965b, Evidence for the existence of monoamine neurons in the central nervous system. IV. The distribution of monoamine nerve terminals in the central nervous system, *Acta Physiol. Scand.* **64**(suppl. 247):39.

Garattini, S., and Valzelli, L., 1965, *Serotonin,* Elsevier North-Holland, Inc., New York.

Gibson, C. J., and Wurtman, R. J., 1978, Physiological control of brain norepinephrine synthesis by brain tyrosine concentration, *Life Sci.* **22**:1399–1406.

Haga, T., and Noda, H., 1973, Choline uptake systems of rat brain synaptosomes, *Biochim. Biophys. Acta* **291**:564.

Hanin, I. (ed.), 1974, *Handbook of Chemical Methods for the Assay of Acetylcholine and Choline,* Raven Press, New York.

Hanin, I., and Schuberth, J., 1974, Labelling of acetylcholine in the brain of mice fed on a diet containing deuterium labelled choline: Studies utilizing gas chromatography–mass spectrometry, *J. Neurochem.* **23**:819.

Haubrich, D. R., Reid, W. D., and Gillette, J. R., 1972, Acetylcholine formation in mouse brain and effect of cholinergic drugs, *Nature, New Biol.* **238**:88.

Haubrich, D. R., Wang, P. F. L., and Wedeking, P. W., 1974, Role of choline in biosynthesis of acetylcholine, *Fed. Proc.* **33**:477.

Haubrich, D. R., Wang, P. F. L., Clody, D. E., and Wedeking, P. W., 1975, Increase in rat brain acetylcholine induced by choline or deanol, *Life Sci.* **17**:975.

Heinrich, C. P., Stadler, H. and Weiser, W., 1973, The effect of thiamine deficiency on the acetylcoenzyme A and acetylcholine levels in the rat brain, *J. Neurochem.* **21**:1273.

Hernandez, R. J., 1973, Developmental pattern of the serotonin synthesizing enzyme in the brain of postnatally malnourished rats, *Experientia* **29**:1487.

Hirsch, M. J., Growdon, J. H., and Wurtman, R. J., 1977, Increase in hippocampal acetylcholine following choline administration, *Brain Res.* **332**:383–385.

Hoeldtke, R. D., and Wurtman, R. J., 1973, The excretion of catecholamines and catecholamine metabolites in kwashiorkor, *Am. J. Clin. Nutr.* **26**:205.

Hoeldtke, R. D., and Wurtman, R. J., 1974, Cereal ingestion and catecholamine excretion, *Metabolism* **23**:33.

Illingworth, D. R., and Portman, O. W., 1972, The uptake and metabolism of plasma lysophosphatidylcholine *in vivo* by the brain of squirrel monkeys, *Biochem. J.* **130**:557.

Im, H. S., Barnes, R. H., Levitsky, D. A., and Pond, W. G., 1973, Postnatal malnutrition and regional cholinesterase activities in brain of pigs, *Brain Res.* **63**:461.

Jenden, D. J., Choi, L., Silverman, R. W., Steinborn, J. A., Roch, M., and Booth, R. A., 1974, Acetylcholine turnover estimation in brain by gas chromatography mass spectrometry, *Life Sci.* **14**:55.

Katz, B., 1966, *Nerve, Muscle, and Synapse*, McGraw-Hill Book Company, New York.

Kettler, R., Bartholini, G., and Pletscher, A., 1974, *In vivo* enhancement of tyrosine hydroxylation in rat striatum by tetrahydrobiopterin, *Nature* **249**:476.

Kewitz, H., and Pleul, O., 1976, Synthesis of choline from ethanolamine in rat brain, *Proc. Natl. Acad. Sci. USA* **73**:2181.

Knott, P. J., and Curzon, G., 1972, Free tryptophan in plasma and brain tryptophan metabolism, *Nature* **239**:452.

Kuhar, M. J., Sethy, V. H., Roth, R. H., and Aghajanian, G. K., 1973, Choline: Selective accumulation by central cholinergic neurons, *J. Neurochem.* **20**:581.

Le Blancq, W. D., and Dakshinamurti, K., 1975, Non-parallel changes in brain monoamines in the pyridoxine-deficient rat, *Can. Fed. Biol. Soc.* **18**:32.

Lee, C.-J., and Dubos, R., 1972, Lasting biological effects of early environmental influences, *J. Exp. Med.* **136**:1031.

Levin, E. Y., Levenberg, B., and Kaufman, S., 1960, The enzymatic conversion of 3,4-dihydroxyphenylethylamine to norepinephrine, *J. Biol. Chem.* **235**:2080.

Lewis, P. R., and Shute, C. C. D., 1967, The cholinergic limbic system: projections of hippocampal formation, medial cortex, nuclei of the ascending cholinergic reticular system, and the subformical organ and supraoptic crest, *Brain* **90**:521.

Lewis, P. R., Shute, C. C. D., and Silver, A., 1967, Confirmation from choline acetylase of a massive cholinergic innervation to the rat hippocampus, *J. Physiol. (London)* **191**:215.

Lloyd, K. G., and Hornykiewicz, O., 1970, Occurrence and distribution of L-DOPA decarboxylase in the human brain, *Brain Res.* **22**:426.

Lovenberg, W., Weissbach, H., and Udenfriend, S., 1962, Aromatic *l*-amino acid decarboxylase, *J. Biol. Chem.* **237**:89.

Lovenberg, W., Jequier, E., and Sjoerdsma, A., 1968, A tryptophan hydroxylation in mammalian systems, *Adv. Pharmacol.* **6A**:21.

Lytle, L. D., Messing, R. B., Fisher, L., and Phebus, L., 1975, Effects of long-term corn consumption on brain serotonin and the response to electric shock, *Science* **190**:692.

Madras, B. K., Cohen, E. L., Fernstrom, J. D., Larin, F., Munro, H. N., and Wurtman, R. J., 1973, Dietary carbohydrate increases brain tryptophan and decreases free plasma tryptophan, *Nature* **244**:34.

Madras, B. K., Cohen, E. L., Munro, H. N., and Wurtman, R. J., 1974, Elevation of serum free tryptophan, but not brain tryptophan, by serum nonesterified fatty acids, *Adv. Biochem. Psychopharmacol.* **11**:143.

McKean, C. M., Boggs, D. E., and Peterson, N. A., 1968, The influence of high phenylalanine and tyrosine on the concentrations of essential amino acids in brain, *J. Neurochem.* **15**:235.

McMenamy, R. H., and Oncley, J. L., 1958, Specific binding of tryptophan to serum albumin, *J. Biol. Chem.* **233**:1436.

Moskowitz, M. A., and Wurtman, R. J., 1975, Catecholamines and neurologic diseases, *New Engl. J. Med.* **293**:274 (Part I), **293**:332 (Part II).

Nachmansohn, D., and Machado, A. L., 1943, The formation of acetylcholine. A new enzyme: "choline acetylase," *J. Neurochem.* **6**:397.

Nagatsu, T., Levitt, M., and Udenfriend, S., 1964, Tyrosine hydroxylase: The initial step in norepinephrine biosynthesis, *J. Biol. Chem.* **239**:2910.

Nagler, A. L., Dettbarn, W.-D., Seifter, E., and Levenson, S. M., 1968, Tissue levels of acetylcholine and acetylcholinesterase in weanling rats subjected to acute choline deficiency, *J. Nutr.* **94**:13.

Pardridge, W. M., and Oldendorf, W. H., 1977, Transport of metabolic substrates through the blood–brain barrier, *J. Neurochem.* **28**:5–12.

Racagni, G., Trabucchi, M., and Cheney, D. L., 1975, Steady-state concentrations of choline and acetylcholine in rat brain parts during a constant rate infusion of deuterated choline, *Naunyn-Schmiedebergs Arch. Exp. Pathol. Pharmakol.* **290**:99.

Rajalakshmi, R., Kulkarni, A. B., and Ramakrishnan, C. V., 1974, Effects of pre-weaning and post-weaning undernutrition on acetylcholine levels in rat brain, *J. Neurochem.* **23:**119.

Saelens, J. K., Simke, J. P., Allen, M. P., and Conroy, C. A., 1973, Some of the dynamics of choline and acetylcholine metabolism in rat brain, *Arch. Int. Pharmacodyn.* **203:**305.

Schuberth, J., Sparf, B., and Sundwall, A., 1970, On the turnover of acetylcholine in nerve endings of mouse brain *in vivo, J. Neurochem.* **17:**461.

Sereni, F., Principi, N., Perletti, L., and Piceni Sereni, L., 1966, Undernutrition and the developing rat brain. I. Influence on acetylcholinesterase and succinic acid dehydrogenase activities and on norepinephrine and 5-OH-tryptamine tissue concentrations, *Biol. Neonate* **10:**254.

Sharma, N. L., Pathak, A. K., and Bhargava, K. P., 1968, Serotonin metabolism in kwashiorkor, *Indian Pediatr.* **5:**261.

Shoemaker, W. J., and Bloom, F. E., 1977, The effect of undernutrition on brain morphology, in: *Nutrition and the Brain,* Vol. II (R. J. Wurtman and J. J. Wurtman, eds.), Raven Press, New York.

Shoemaker, W. J., and Wurtman, R. J., 1971, Perinatal undernutrition: Accumulation of catecholamines in rat brain, *Science* **171:**1017.

Shoemaker, W. J., and Wurtman, R. J., 1973, The effect of perinatal undernutrition on the metabolism of cetacholamines in the rat brain, *J. Nutr.* **103:**1537.

Shoemaker, W. J., Coyle, J., and Bloom, F. E., 1974, Perinatal undernutrition: Effect on neurotransmitters, transmitter synthetic enzymes, and the number of synaptic connections, *Trans. Am. Soc. Neurochem.* **4:**100.

Shute, C. C. D., and Lewis, P. R., 1967, The ascending cholinergic reticular system: Neocortical, olfactory and subcortical projections, *Brain* **90:**497.

Snyder, S. H., Brown, B., and Kuhar, M. J., 1974, The subsynaptosomal localization of histamine histidine decarboxylase and histamine methyltransferase in rat hypothalamus, *J. Neurochem.* **23:**37.

Sobotka, T. J., Cook, M. P., and Brodie, R. E., 1974, Neonatal malnutrition: neurochemical, hormonal and behavioral manifestations, *Brain Res.* **65:**443.

Sollenberg, J., 1970, Determination of acetyl coenzyme A, in: *Drugs and Cholinergic Mechanisms in the CNS* (E. Heilbronn and A. Winter, eds.), pp. 27–32, Forsvarets Forskningsanstalt, Stockholm.

Sorimachi, M., and Kataoka, K., 1974, Choline uptake by nerve terminals: A sensitive and specific marker of cholinergic innervation, *Brain Res.* **72:**350.

Stavinoha, W. B., and Weintraub, S. T., 1974, Choline content of rat brain, *Science* **183:**964.

Stern, W. C., Forbes, W. B., Resnick, O., and Morgane, P. J., 1974, Seizure susceptibility and brain amine levels following protein malnutrition during development in the rat, *Brain Res.* **79:**375.

Stern, W. C., Morgane, P. J., Miller, M., and Resnick, O., 1975, Protein malnutrition in rats: response of brain amines and behavior to foot shock stress, *Exp. Neurol.* **47:**56.

Thoa, N. B., Wurtman, R. J., and Axelrod, J., 1966, A deficient binding mechanism for norepinephrine in hearts of scorbutic guinea pigs, *Proc. Soc. Exp. Biol. Med.* **121:**267.

Ulus, I., and Wurtman, R. J., 1976, Choline administration: activation of tyrosine hydroxylase in dopaminergic neurons of rat brain, *Science,* **194:**1060.

Ulus, I., Hirsch, M. J., and Wurtman, R. J., 1977, Transsynaptic induction of adrenomedullary tyrosine hydroxylase activity by choline: evidence that choline administration increases cholinergic transmission, *Proc. Natl. Acad. Sci. USA* **74:**798–800.

White, H. L., and Wu, J. C., 1973, Kinetics of choline acetyltransferases (EC 2.3.1.6) from human and other mammalian central and peripheral nervous tissues, *J. Neurochem.* **20:**297.

Whittaker, V. P., 1972, The storage and release of acetylcholine, *Biochem. J.* **128:**73.

Widdowson, E. M., and McCance, R. A., 1960, Some effects of accelerating growth. I. General somatic development, *Proc. Roy. Soc. (London)* **B152:**188.

Wise, E. M., and Elwyn, D., 1965, Rates of reactions involved in phosphatide synthesis and small intestine of intact rats, *J. Biol. Chem.* **240:**1537.

Wurtman, R. J., 1966, *Catecholamines,* Little, Brown and Company, Boston.

Wurtman, R. J., and Fernstrom, J. D., 1974, Effects of diet on brain neurotransmitters, *Nutr. Rev.* **32:**193.
Wurtman, R. J., and Wurtman, J. J. (eds.), 1977, *Nutrition and the Brain,* Vols. I and II, Raven Press, New York.
Wurtman, R. J., Rose, C. M., Chou, C., and Larin, F., 1968, Daily rhythms in the concentrations of various amino acids in human plasma, *New Engl. J. Med.* **279:**171.
Wurtman, R. J., Larin, F., Mostafapour, S., and Fernstrom, J. D., 1974, Brain catechol synthesis: Control by brain tyrosine concentration, *Science* **185:**183.
Yamamura, H. I., and Snyder, S. H., 1972, Choline: High-affinity uptake by rat brain synaptosomes, *Science* **78:**626.
Yehuda, S., and Wurtman, R. J., 1972a, The effects of *d*-amphetamine and related drugs on colonic temperatures of rats kept at various ambient temperatures, *Life Sci.* **11:**851.
Yehuda, S., and Wurtman, R. J., 1972b, Release of brain dopamine as the probable mechanism for the hypothermic effect of D-amphetamine, *Nature* **240:**477.

5

Nutrition and Pregnancy

Pedro Rosso and Carolyn Cramoy

1. Maternal–Fetal Exchange

1.1. General Considerations

Fetal growth largely depends on an adequate supply of oxygen and nutrients. The changes that occur in the maternal organism during pregnancy, the so-called maternal adaptive mechanisms, are the most part directly or indirectly involved with the oxygen and nutrient needs of the fetus. The control of the maternal adaptive changes is not well understood. It seems clear, however, that to a large extent they are not under the direct influence of the fetus. Experiments have shown that both the rat and human placentae may remain attached to the uterus and, apparently, maintain an adequate secretion of hormones even after removal of the fetus (Bourdel and Jacquot, 1959; Friedman *et al.*, 1969). The sequence of the maternal adaptive changes also suggest the influence of placental hormones rather than fetal needs. For example, the maximum rate of expansion of blood volume and maternal weight gain takes place during midgestation, 10 weeks before maximum placental growth and 15 weeks before maximum fetal growth (Fig. 1). Thus, pregnancy seems to have two phases: a "maternal phase," which occurs during the first half of gestation and prepares the maternal organism for future fetal demands, and a "fetal phase" of maximum fetal requirements, occurring during the second half of gestation.

A sequence of events such as this suggests a complicated mechanism of feedback regulation. The initial stimulus would be provided by the conceptus and would induce, simultaneously, increased food intake and changes necessary to manage the greater influx of nutrients, and oxygen consumption, including expansion of blood volume and increased levels of certain plasma

Pedro Rosso and Carolyn Cramoy • Department of Pediatrics and Institute of Human Nutrition, Columbia University College of Physicians and Surgeons, New York, New York.

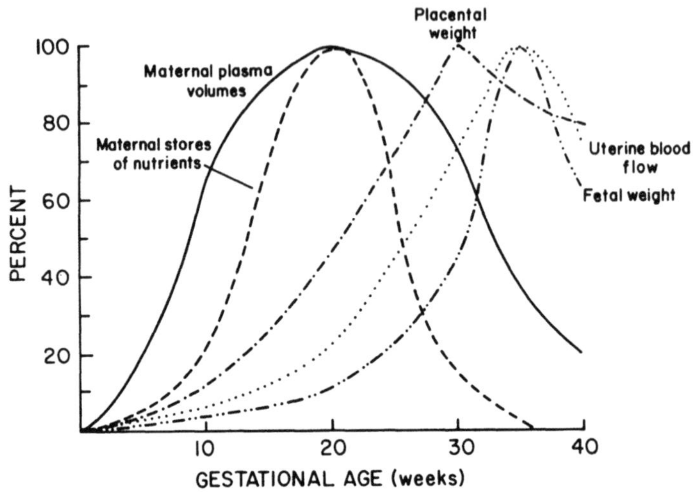

Fig. 1. Gestational changes in maternal stores, plasma volume, placental weight, uterine blood flow, and fetal weight in human pregnancy. Values are expressed as percent of maximal changes. Data from Hytten and Leitch (1971).

carriers. At the same time, the uterine vascularization would expand to provide the conceptus with an adequate quantity of oxygen and nutrients.

It is generally assumed that nutrients are distributed among tissues according to metabolic needs. Since the conceptus would have a higher metabolic rate than the maternal tissues, it would receive more nutrients per unit of body mass than would the mother. Moreover, it would be able to compete with the mother for the available nutrients (Hammond, 1944). However, as will be discussed in other sections of this chapter, many other factors besides the metabolic rate of the tissues may influence the way in which the available nutrients are divided between the mother and the conceptus. Among these factors, the most important one is probably the adequacy of the maternal adaptive mechanisms involved in the organization of a supply line to the conceptus.

1.2. Maternal Physiologic Changes

1.2.1. Alimentary Functions

After a period of a few weeks, in which food intake tends to decrease and in which many women complain of nausea and vomiting, appetite increases to a peak during midpregnancy and subsequently declines as pregnancy proceeds.

Although quantitative estimates of food intake are difficult to make and in many cases unreliable, the available data shows that American and Western European middle-class women increase their total caloric daily intake approximately 200 kcal over prepregnancy levels early in the second trimester of pregnancy and maintain a slightly lower, although still elevated intake during the third trimester (Beal, 1971; Lunell *et al.*, 1969).

A similar pattern of change in food intake during pregnancy has been described in the rat. In this species greater food ingestion is noticeable on day 5 of pregnancy, proportionally earlier than in humans, and reaches a peak at day 12 (Slonaker, 1924–1925). After day 12, food ingestion declines and in some animals it may reach prepregnancy levels on the last day of gestation.

Paradoxically, in spite of the higher food intake, the functions required to digest food appear to be depressed in pregnant women. For example, gastric secretion, as reflected by acid production, has been found to be lower during pregnancy when measured after fasting (Mehrotra *et al.*, 1969), after a water or saline meal (Hunt and Murray, 1958), or after histamine stimulation (Murray *et al.*, 1957). Pepsin secretion also seems to be lower (Gryboski and Spiro, 1956). In addition, gastric tone and motility are reduced with a consequent delay in gastric emptying time (Hansen, 1937). The reduced gastric motility may reflect an overall sluggishness of the gastrointestinal tract which is present in the small and probably also large intestines (Parry *et al.*, 1970).

Although the idea has gained some acceptance, there is no strong evidence that pregnant women have an increased ability to absorb nutrients. Iron and certain vitamins, such as B_{12} and folate, may be exceptions. It has been shown that the proportion of radioactive iron absorbed from a radioactively labeled dose increases twofold during gestation (Hahn *et al.*, 1951; Heinrich *et al.*, 1968). Similarly, the attained blood levels of B_{12} and folate after an oral dose of these vitamins is higher in pregnant than in nonpregnant women (Hellegers *et al.*, 1957; Chanarin *et al.*, 1959).

A great deal of attention has been centered on liver changes. In the rat, pregnancy induces a marked liver enlargement that reflects an increased deposition of protein associated with an increased DNA content (Kennedy *et al.*, 1958). Humans do not have liver changes of this nature. Some clinical indicators of liver dysfunction, however, such as a fall in the plasma albumin concentrations, elevated serum alkaline phosphatase activity, and elevated serum lipids, suggest a certain degree of hepatic participation. More specific tests of liver function, such as serum transaminase levels, prothrombin time, and cephalin and thymol turbidity, are usually within normal limits during gestation (Hytten and Lind, 1973).

1.2.2. Blood Volume and Hemodynamic Changes

The expansion of blood volume is one of the most interesting and early changes that occur in the mother. The increase is already apparent in humans at 10 weeks (Hytten and Paintin, 1963) and in the rat at 12 days (Knopp *et al.*, 1975). On the average, plasma volume increases 1250 ml above the nonpregnant level of 2600 ml in humans. A proportionally similar increase is seen in the rat.

Blood volume expansion is attained primarily through an expansion in the amount of intravascular fluids, and to a lesser degree, expansion in the circulating protein pool and red blood cells. This results in hemodilution reflected by a fall of the hemoglobin concentration, plasma proteins, and most serum electrolytes (Hytten and Lind, 1973). The erythrocyte count, for example, decreases from an average of about 4.5 million/μl to about 3.7 million/μl at about 30 weeks, rising slightly thereafter (Lundstrom, 1950; Darby *et al.*, 1953; Edgar and Rice, 1956). The concentration of hemoglobin tends to decrease even more from a normal range of 137–140 g/liter of whole blood to 110–210 g/liter of whole blood (Darby *et al.*, 1953; Edgar and Rice, 1956; Ventura and Klopper, 1951a).

In contrast to erythrocytes, the number of leukocytes increases from about 7200/μl in nonpregnant women to about 10,350/μl near term (Andrews and Bonsnes, 1951). The leukocytosis is due mainly to a 71% increase in the number of neutrophil polymorphonuclear cells, while lymphocytes rise only 15%. No significant change occurs in the number of monocytes and eosinophiles.

Serum concentration of protein decreases markedly during pregnancy from approximately 7 g/liter to 62 g/liter at term (Van Studnitz, 1955; DeAlvarez *et al.*, 1961; Rebound *et al.*, 1967). The fall occurs mainly in the first 3 months of pregnancy and levels off thereafter.

The individual protein fractions, with few exceptions, follow the same general pattern. Albumin concentration falls rapidly in the first 3 months, but it continues to fall, more slowly, until late pregnancy. The overall fall from prepregnancy levels of 37.0 g/liter is approximately 10 g/liter (Von Studnitz, 1955; DeAlvarez *et al.*, 1961; Reboud *et al.*, 1967). In contrast, the α- and β-globulin fractions rise progressively (Von Studnitz, 1955; DeAlvarez *et al.*, 1961; MacGillivray and Tovey, 1957). The 30% rise of the β-globin fraction is especially significant, since most of the carrier proteins used by nutrients belong to this fraction.

Serum concentrations of electrolytes such as sodium, potassium, magnesium, and chloride also tend to fall moderately during pregnancy (Herbinger and Wichmann, 1967; Michel, 1971; Newman, 1957).

Teleologically, the expansion in blood volume can be explained as fulfilling the need to transport more oxygen and nutrients to the fetus and also to compensate for the volume of blood "trapped" by the increased uterine circulation. The exact mechanisms by which the blood expansion is achieved are still obscure.

In addition to the changes in the volume of circulating blood, there are several hemodynamic changes, most of them secondary to the expanded volume, such as increased cardiac output, increased heart rate, reduced blood pressure, and increased peripheral blood flow (Walters and Lim, 1975).

1.2.3. Other Functional Changes

Changes involving other systems include an elevation in the ventilation rate throughout pregnancy from about 7 liters/min to about 10 liters/min. This increase is achieved almost entirely by an increase in tidal volume (the volume of air inspired and expired with each breath) (Hytten and Lind, 1973). Also, there is a rapid rise, early in pregnancy, in the renal plasma flow from about 500 ml/min to approximately 700 ml/min. Concomitant with the increase in renal plasma flow, glomerular filtration rate rises from 90 ml/min to 140–150 ml/min. Since this rise is proportionally larger than the rise in renal plasma flow, the proportion of the plasma flow that is filtered is also raised (Hytten and Lind, 1973).

1.3. Maternal Metabolic Changes

1.3.1. Body Composition and Deposit of Maternal Stores

The increase in food intake during the second trimester seems to be disproportionately high in relation to fetal needs. As a consequence, the maternal body receives an excess of nutrients that are deposited in various tissues. The largest portion of the "stored" nutrients is made up of lipids. Thus, the maternal weight gain not directly attributable to changes in the size of the conceptus or the organs related to the reproductive process is mostly fat and the fluids retained in the expanded plasma volume.

In humans, total weight gain during pregnancy has been measured in several series. For Western European women it has been found to be approximately 12.5 kg, with a wide margin of variation, ranging from −10 kg to + 30 kg (Hytten and Leitch, 1971). This wide range reflects mainly prepregnancy maternal weight since, generally, lean women tend to gain weight, whereas overweight women do the opposite (Beal, 1971).

The rate of maternal weight gain is not uniform, and most pregnant women gain more weight between weeks 17 and 20 than at any other period in pregnancy. As shown in Table I, there is a rapid increase in the rate of maternal weight gain between weeks 10 and 20, a more gradual increase between weeks 20 and 30, and a slight decrease during the last 10 weeks of gestation.

Table I. Components of Weight Gain during Pregnancy[a]

	Increase in weight up to:			
Component	10 weeks	20 weeks	30 weeks	40 weeks
Fetus	5	300	1,500	3,400
Placenta	20	170	430	650
Amniotic fluid	30	350	750	800
Uterus	140	320	600	970
Mammary gland	45	180	360	405
Blood	100	600	1,300	1,250
Extravascular fluid	0	30	80	1,680
"Stores"	310	2,050	3,480	3,345
	650	4,000	8,500	12,500

[a] Adapted from Hytten and Leitch (1971).

Based on analysis of body composition, it has been determined that a total of 925 g of protein is deposited in the conceptus (Hytten and Leitch, 1971). This quantity represents an average daily retention of 3.3 g of protein. Because of the changing rates of fetal growth during gestation, the daily retention will be only 0.6 g during the first 10 weeks and 1.8, 4.8, and 6.1 g/day during successive 10-week periods. In apparent contrast with the theoretical estimates, however, balance studies have shown an average daily protein retention of 6.25 g (Calloway, 1974). Such an excess over the quantity of protein retained by the fetus can be construed to reflect a significant protein accumulation in maternal tissues. However, there is no direct evidence that such accumulation takes place, and the excess protein retention of the metabolic studies has been interpreted as a methodological artifact (Hytten and Leitch, 1971). The main theoretical argument against pregnancy deposits of protein is based on the fact that protein accumulation will determine a proportional retention of water. Thus, it is reasoned that a gain of 1600 g of protein, which is the approximate net accumulation suggested by the balance studies, would produce a retention of 6900 ml of water. Since there is no evidence for a maternal fluid retention of such magnitude, it has been concluded, by exclusion, that the approximate 3300 g of dry weight that constitutes the maternal stores is made up almost entirely of fat. However, total body potassium determinations and K^+-retention studies in adolescent girls (King *et al.*, 1973) and mature pregnant women (King *et al.*, 1976) demonstrate an accumulation of potassium throughout gestation that can be explained only by an increase in lean body tissues and therefore by a protein accumulation. If these studies are proven to be correct, only 50% of the maternal stores would be fat and the rest lean tissue.

Paradoxically, although it is widely accepted that maternal fat is accumulated during pregnancy, body composition studies do not support such an assumption (Seitchik *et al.*, 1963; McCartney *et al.*, 1959). These inconclusive results are thought to reflect methodological problems, since changes in skinfold thickness clearly demonstrate increments in the subcutaneous adipose tissue fat in several regions of the body (Hytten and Leitch, 1971).

In the rat, changes in body composition associated with pregnancy are well documented. Both lipid and protein retention in the carcass begin during the first week of pregnancy and continue until near term, when most of the excess protein is removed (Beaton *et al.*, 1954). In contrast, an excess of fat persists after delivery (Naismith, 1966). No changes in the cellularity of the adipose tissue have been found, indicating that fat is accumulated by an increase in cell size and not in cell number (Knopp *et al.*, 1970).

1.3.2. Carbohydrate Metabolism

The plasma concentration of glucose decreases steadily during the course of pregnancy, reaching the lowest level, approximately 68 mg/100 ml, near term (Hytten and Lind, 1973). One of the main characteristics of carbohydrate metabolism during pregnancy is the progressive increase in resistance of the maternal tissues to the hypoglycemic effect of insulin, in spite of the fact that pregnant women release significantly greater amounts of insulin than do non-pregnant ones (Yen, 1973). Further, the amounts of insulin released after intravenous glucose administration increases as pregnancy approaches term (Spellacy, 1971).

The changes in carbohydrate metabolism are still poorly understood. The assumption is that various factors are involved, including the hypoglycemic effect of the continuous glucose extraction by the conceptus and the elevation in the plasma concentration of several hormones responsible for insulin, such as glucocorticoids, estrogen, progesterone, and somatomammotropin (Spellacy, 1975).

1.3.3. Lipid Metabolism

The concentration of total lipids in the serum rises progressively during pregnancy from about 6 g/liter to about 10 g/liter. This rise reflects a higher concentration of each major fraction (Von Studnitz, 1955; DeAlvarez *et al.*, 1959).

Cholesterol concentration falls during the first trimester, then begins a rapid rise to approximately 250–290 mg/100 ml (Von Studnitz, 1955; Green, 1966). The change probably reflects metabolic adjustment rather than dietetic changes, since neither a vegetarian diet nor a low-cholesterol diet influence the rise (Green, 1966; Mullick *et al.*, 1964). Further evidence of the metabolic change is the fact that insulin or glucose administration, which in nonpregnant women can reduce cholesterol levels, raises the levels even further (Dannenburg and Burt, 1965).

The individual phospholipids have specific patterns of change. For example, lysolecithin curves show a progressive, although minor, decrease, while cephalins and sphingomyelins increase moderately. Lecithins increase from 2.2 to 3.5 mmol/liter (Svanborg and Vikrot, 1965). The mechanisms and metabolic significance of these change are still unknown.

Levels of free fatty acids (FFA) increase from 600–700 μeq/liter to 900–1000 μeq/liter. The rise largely reflects changes in saturated fatty acids, with a changing proportion of saturated to unsaturated. Individual FFAs have their own characteristic patterns, with an increase in the proportion of palmitic and stearic and a decrease in that of oleic, linoleic, and arachidonic (Bottigioni *et al.*, 1966).

The gestational increase in maternal plasma levels of the major lipid fractions suggest marked modifications in lipid metabolism. The nature of these modifications and their implications are still poorly understood. The most likely mechanism for the elevated plasma lipid concentration is an overproduction of hepatic triglycerides associated with an increased mobilization of fatty acids from the adipose tissues.

Because of its connection with carbohydrate metabolism, the progressive raise in the levels of FFA could be interpreted as resulting from the lower levels of glucose. In support of this idea, intravenous administration of glucose in pregnant women decreased by about 500 μeq/liter, the plasma concentration of FFA (Burt *et al.*, 1969). The explanation seems unlikely, however, considering the marked disproportion between the slight reduction in plasma glucose level and the marked elevation in FFA levels.

In the rat, it has been shown that the concentration of triglycerides rises to a maximum at days 17–19 of pregnancy and then rapidly declines to near prepregnancy levels at parturition. It has been suggested that the higher triglyceride concentration might be related to the decrease in lipoprotein lipase activity in adipose tissue (Otway and Robinson, 1968). However, hypertriglyceridemia occurs before lipoprotein lipase activity of adipose tissue falls below the nonpregnant levels. Further, although plasma triglyceride concentration is inversely related to enzyme activity between 12 and 20 days of gestation, lipoprotein lipase activity is always equal to or above that in nonpregnant rats (Hamosh *et al.*, 1970; Knopp *et al.*, 1975). These results suggest that other factors, including enhanced hepatic synthesis, are responsible for the elevation in plasma triglycerides. Also in the rat, it has been shown that there is a heightened availability of FFA within adipose tissue during late pregnancy (Knopp *et al.*, 1970). This coincides with increased splitting of stored triglycerides. The enhanced lipolysis is independent of glucose availability.

1.3.4. Amino Acid and Protein Metabolism

Pregnant women have a marked increase in the urinary excretion of amino acids, especially threonine, serine, alanine, histidine, glycine, and glutamine (Hytten and Cheyne, 1972). The reason for the increased excretion is not clear. It has been suggested that elevated cortisol secretion may play a role, since administration of ACTH and cortisol to nonpregnant patients has been shown to induce similar changes (Zinneman *et al.*, 1963).

Current knowledge on other changes of amino acid metabolism during pregnancy is scarce and fragmentary. For example, the pattern of changes of plasma amino acids is still inadequately defined. The general trend seems to be toward a lower total concentration of amino acids (Christensen *et al.*, 1957). In particular taurine, serine, proline, glycine, valine, tyrosine, ornithine, lysine, and arginine are consistently reduced, whereas histidine and glutamic acid levels remain unchanged. The levels of all amino acids tend to rise during the last month of pregnancy (Hytten and Leitch, 1971).

In the rat some amino acids, such as alanine, increase markedly during the last week of gestation, whereas others, such as glycine, serine, and proline, remain unchanged. The rest of the amino acids, however, show a downward trend, especially lysine, which decreases by about 40% (Southgate, 1971). The progressive reduction in plasma concentration of amino acids may reflect a considerable amino acid drain by the conceptus. Indirect evidence supporting this assumption is provided by fasting experiments done in rats at 17–19 days of gestation. In pregnant rats, fasting induces a marked hypoglycemia compared with nonpregnant animals. The pregnant animals also had a concomitant hyperketonemia. Hypoglycemia in starving pregnant rats is secondary to a poor gluconeogenic response caused by substrate unavailability (Freinkel, 1972). This has been demonstrated by showing that the liver of pregnant rats has an enhanced gluconeogenic capacity near term and by suppressing the hypoglycemic response of fasting with the administration of exogenous alanine. The role of the fetus in the reduced availability of substrates has been shown in experiments in which pregnant rats whose fetuses had been removed (leaving the placentas attached) did not develop hypoglycemia after starvation (Freinkel, 1972). Starvation experiments done in humans during the first trimester of pregnant have also shown a more rapid and profound reduction in gluconeogenic amino acids, especially alanine (Felig *et al.*, 1972).

Protein utilization increases during pregnancy. This phenomenon is apparently mediated by decreased urea synthesis and decreased activity of enzymes involved in amino acid catabolism (Beaton *et al.*, 1954; Naismith, 1973).

1.3.5. Control of Maternal Metabolic Adaptations

It has been suggested, largely based on observations made in the rat, that maternal metabolic changes, especially those affecting amino acids and lipids, may reflect "anabolic" and "catabolic" phases of pregnancy (Naismith, 1966; Knopp *et al.*, 1973) (see Fig. 2). The anabolic phase, when protein and fat stores are deposited, would comprise the first 2 weeks of gestation in the rat. During this period of pregnancy the fetus is still relatively small, and therefore its total demands are minimal. The catabolic phase would take place during the last third of pregnancy, when fetal demands are the highest. During the catabolic phase the mother would mobilize some of her extra stores of muscle protein and fat to ensure that fetal needs are met. Fat storage during the anabolic phase would reflect an increased conversion of glucose into fatty acids from the adipose tissue (Knopp *et al.*, 1973). In contrast, during the catabolic phase, adipose tissue fatty acid formation would decline to one-third of nonpregnancy levels, and maternal fat stores would be increasingly mobilized as free fatty acids (Knopp *et al.*, 1973; Hamosh *et al.*, 1970). The changes in adipose tissue metabolism would be mainly caused by maternal hyperphagia and the peripheral response of adipose tissue to insulin (Knopp *et al.*, 1973). Thus, hyperinsulinism and excess food intake would promote the fat storage that takes place during the first 2 weeks. Later, because of diminished tissue responsiveness to insulin, maternal fat storage would decline. The anabolic phase of protein metabolism, instead, would be caused by increased levels of progesterone and the effect mediated through a reduced secretion of corticosteroids (Naismith, 1966). In support of such a possibility, it has been shown that progesterone administration in the rat causes a significant reduction in the weight of the adrenal glands and lowers the plasma concentration of corticosteroids (Naismith and Fears, 1972).

The catabolic phase of pregnancy would be regulated by levels of estrogen.

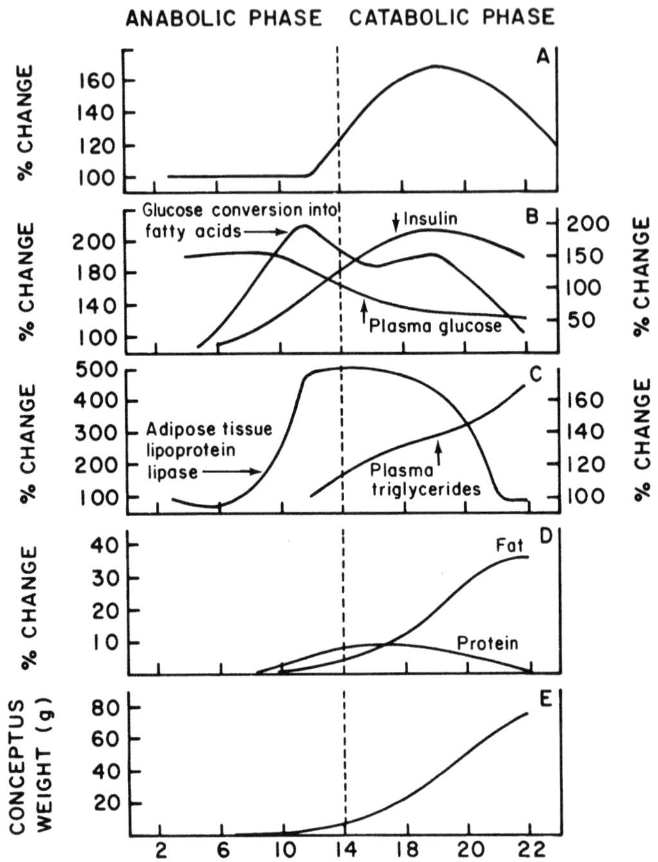

Fig. 2. Anabolic and catabolic phases of pregnancy in the rat. Values expressed as percent of prepregnancy levels. A, food intake; B, carbohydreate metabolism; C, lipid metabolism; D, maternal body composition (carcass); E, fetal weight (absolute values). Data from Naismith (1966) and Knopp *et al.* (1973).

Increased secretion of estrogen by the fetoplacental unit during the last part of pregnancy would stimulate, by influencing ACTH secretion, the synthesis of corticosteroids by the maternal adrenal gland. The corticosteroids would, in turn, increase maternal muscle protein catabolism (Naismith, 1969). There is evidence that supports this possibility. The increase in plasma levels of estriol and corticosteroids during gestation parallels increments in the weight of the conceptus (Ratanasopa *et al.*, 1967; Bayliss *et al.*, 1955). In the rat, administration of estradiol benzoate enhances synthesis and secretion of ACTH (Gemzell, 1952), while injections of cortisone cause a loss of protein from the carcass and a gain in liver protein (Goodlad and Munro, 1959). As previously discussed, the same changes in the distribution of body protein have been seen to occur during the last week of pregnancy in the rat.

1.4. Growth of the Conceptus

1.4.1. Fetal Growth

1.4.1a. Somatic Growth. Information of prenatal physical growth in humans derives from cross-sectional studies, based on samples obtained in abortions, premature deliveries, and term (Lubchenko *et al.,* 1963; Hendricks, 1964; Usher and McLean, 1974; Hytten and Leitch, 1971). The methodological problems derived from the use of such materials are obvious. More recently, ultrasound scanning has made it possible to monitor fetal growth longitudinally as reflected by changes in the biparietal diameter (Campbell, 1974).

Cross-sectional studies indicate that fetal growth, determined by changes in body weight and length, has a sigmoid curve, with two areas of relatively slow growth separated by one of rapid growth (Fig. 3). The first period of slow fetal growth occurs before week 14 of pregnancy. The average normal weight at 10 weeks is about 5 g. After week 14 the phase of rapid, linear, growth begins and lasts until approximately week 34 of gestation. During this 20-week period, fetal weight increases from approximately 50 g to 2500 g. After week 34 of gestation the weight curve begins to fall off, and at term the average birth weight is usually 3300 g.

The gestational differences in the rate of fetal growth are apparent when prenatal growth is expressed as growth velocity, or g/day. Such a plotting shows that the fetus gains approximately 5 g/day at 16 weeks (Usher and McLean, 1974). After 37 weeks there is a rapid fall of growth velocity that lasts until the first week of postnatal life. The slowing of fetal growth rate near term is also apparent in other mammalian species, such as rat, (Rosso, 1975) sheep, (Cloete, 1939) guinea pig (Draper, 1920), rabbit (Rosahn and Greene, 1936), and rhesus monkey (Cheek, 1975). The cause of this preterm reduction in the rate of fetal growth is still unclear, although the subsequent postnatal reacceleration of growth, after regular feeding begins, suggests that it may reflect a limited availability of nutrients.

Fetal needs are determined by the increments in fetal weight due to synthesis of new tissue and by the current size of the fetus. Thus, fetal requirements would be maximum near term.

Although the general growth of the fetus reflects growth of the different organs and tissues, as has been seen in postnatal life, each organ has its own characteristic pattern of development. In the mammalian species the brain is the organ that develops earliest. This is reflected by the greater brain weight/body weight ratio of the neonates compared with adults.

1.4.1b. Cellular Growth. The organ whose prenatal growth has been most extensively studied is the brain and its different regions. Available data show that DNA content of the whole brain increases linearly until birth and continues to increase at a slower rate during the first few months of postnatal life (Dobbing and Sands, 1973).

The curve for DNA content in fetal brain shows two different slopes, with the steeper one up to 20 weeks of gestation (Fig. 4). The interpretation of the

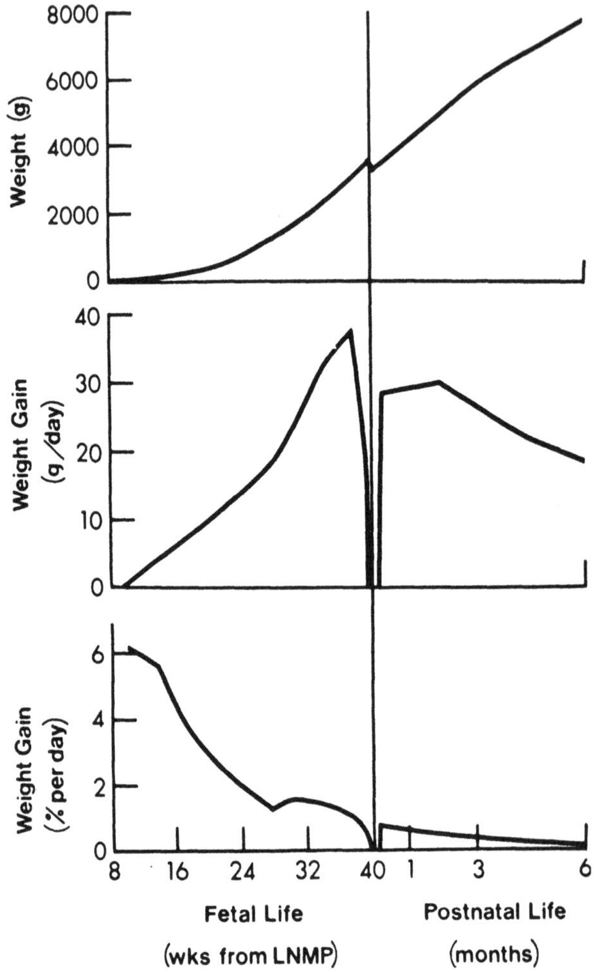

Fig. 3. Changes in body weight of the human fetus during normal gestation. From Usher and McLean (1974).

authors is that the first peak corresponds to neuronal division. Therefore, the bulk of DNA increments during late gestation, and postnatally, would represent glial proliferation (Dobbing and Sands, 1973).

Changes in brain weight and protein content compared with DNA content reflect the different phases of cellular growth. In human brain protein content continues to increase for several years after birth, indicating that hypertrophic growth occurs mainly postnatally (Dobbing and Sands, 1973; Winick, 1968).

Prenatal cellular growth of the cerebrum in the rhesus monkey is remarkably similar to that in humans. However, cerebellum has a proportionally faster prenatal rate of cell division in the monkey (Cheek, 1975). Other parameters, such as increase in cerebral cholesterol concentration (a parameter of myelination), are also similar in man and monkey, suggesting that some primates may provide valuable models for a better understanding of normal and abnormal events of human brain growth (Cheek, 1975).

During prenatal life, brain growth is mainly proliferative; therefore, increases in DNA content are proportional to protein content and to weight changes. Increases in brain weight, in turn, are reflected by increases in head circumference. This relationship determines a linear correlation between DNA content and head circumference during gestation and the first year of life. Thus, DNA content of a fetal brain can be calculated, with an acceptable degree of accuracy, from head circumference (Winick and Rosso, 1969). A similar relationship of total brain DNA to head circumference also exists in

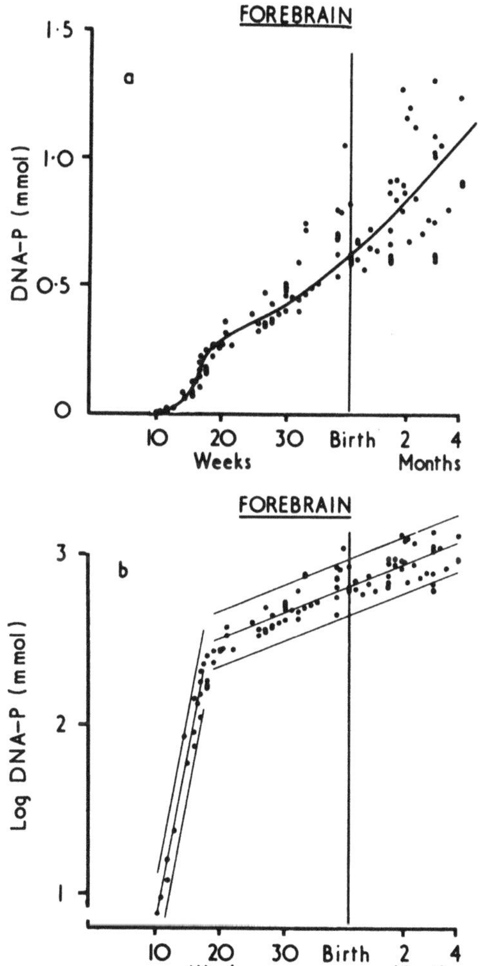

Fig. 4. DNA-P content of the human forebrain during prenatal and early postnatal life. From Dobbing and Sands (1973).

the rhesus monkey but is best fitted by a quadratic expression (Holt *et al.*, 1975).

Human cellular growth patterns have also been studied in several organs other than brain during normal fetal development (Winick *et al.*, 1972). However, there are growth curves available only for heart, liver, kidney, and gastrocnemius muscles (Widdowson *et al.*, 1972). In all these organs there is a marked linear increase in DNA content that is greater between 13 and 25 weeks of gestation than at later ages. Up to 25 weeks of gestation, DNA content of each organ approximately doubles every week. In kidney and heart the protein/DNA ratio increases slowly up to 30 weeks of intrauterine life and rapidly thereafter. In contrast, protein/DNA ratio in liver remains almost constant throughout gestation (Fig. 5). The difference in protein/DNA ratio in these organs reflect diversities of timing for the different phases of cellular

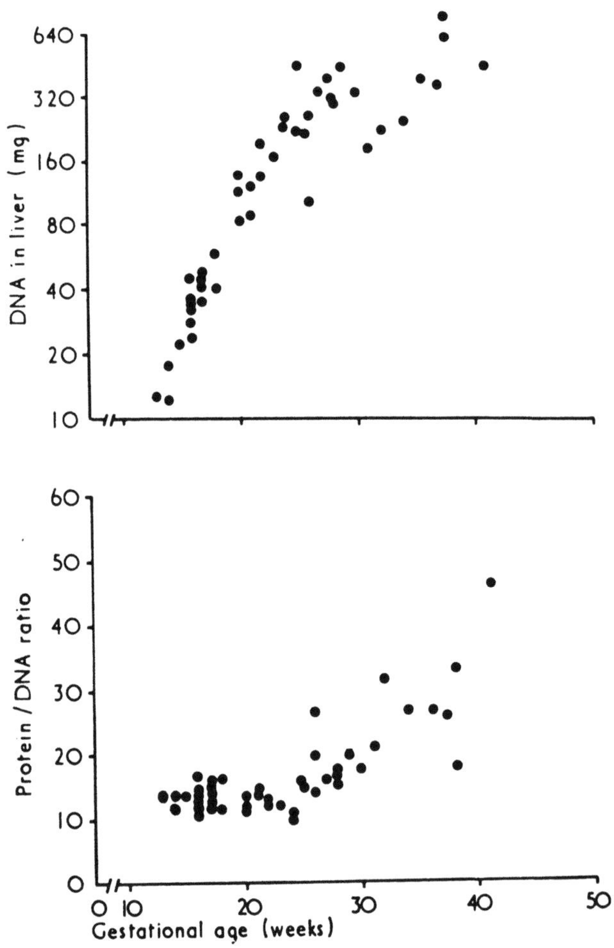

Fig. 5. DNA content and protein/DNA ratio of human liver at different gestational ages. From Widdowson *et al.* (1972).

growth. The combination of simultaneous proliferative and hypertrophic growth in different organs found in the human fetus has also been described in the monkey (Cheek, 1975). In contrast, in the rat, prenatal growth is almost exclusively hyperplastic (Winick and Noble, 1965).

Growth of muscle, as reflected by changes in DNA content of the gastrocnemii, also increases more rapidly before week 25 of gestation than thereafter. Protein/DNA ratio doubles between weeks 15 and 30 and triples during the last week of gestation. No data are available for prenatal human growth or changes in the total muscle mass of the body. In rhesus monkey, the percentage of body weight represented by skeletal muscle has been determined by careful dissection of the muscle. It was found that during the entire intrauterine life and until 4 days postnatally, skeletal muscle accounted for 20–25% of body weight (Beatly and Bocek, 1975). In the macaque there is good agreement between values of DNA content and protein content in various muscles. If the same relationship holds in the human fetus, data from gastrocnemii would reflect changes in total muscle mass.

1.4.1c. Body Composition. Water content of the fetus decreases throughout development from 92–94% of the body weight in a 10-g fetus to 70–72% in a term fetus (Widdowson and Spray, 1951; Kelly *et al.*, 1951). Fetal lipid content changes are opposite to those of water content. In fetuses of up to 500 g, lipids are almost exclusively structural ones. In larger fetuses, fat is accumulated in the various depots at an increasing rate during the last third of pregnancy. Available data on term babies suggest a proportion of body fat ranging from 15 to 30% (Widdowson and Spray, 1951; Kelly *et al.*, 1951) (Fig. 6).

Deposit of nitrogen has a sigmoid curve similar to the general curve of fetal growth. The nitrogen content increases continuously from about 1% in the 10-g fetuses to approximately 2.5% in the term fetuses. The decrease in the slope of the nitrogen curve near term is due to the rapidly increasing

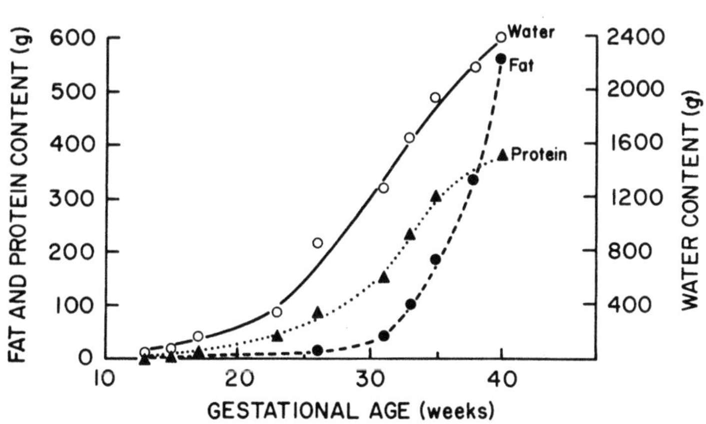

Fig. 6. Changes in body composition of the human fetus. Drawn using data from Widdowson and Spray (1951).

proportion of body fat, which exceeds the increase of nitrogen content (Widdowson and Spray, 1951; Kelly *et al.*, 1951) (Fig. 6).

Total ash largely consits of calcium, magnesium, sodium potassium, phosphorus, chlorine, iron, copper, zinc, and iodine. The quantity of these elements in the fetal body increases linearly during gestation. Calcium content, for example, increases from approximately 1 g at 18–20 weeks of gestation, to 10 g at 30 weeks and 28 g at 40 weeks. Similarly, reflecting the progressive calcification of the bones, the concentration of calcium in the body increases progressively from 2.1 g/kg of fat-free body weight during the first trimester of pregnancy to 9.6 g/kg of fat-free weight near term (Widdowson and Spray, 1951; Kelly *et al.*, 1951). Changes in the body concentration of other minerals and electrolytes are shown in Table II.

1.4.2. Placental Growth

1.4.2a. Changes in Placental Weight. Three days after fertilization, the human blastocyst begins to differentiate into a small area, destined to form the embryo, and a larger one that will become the trophoblast. When, on the fourth or fifth day after ovulation, the embryo, in the morula stage, finally enters the uterine cavity, the endometrium is about 5 mm in thickness. The morula then lies free within the uterine cavity for about 2 or 3 days. After this time, the trophoblast starts to penetrate the endometrium and by day 11 or 12 the blastocyst is completely embedded in the endometrium (Wynn, 1975). As gestation proceeds, the trophoblast differentiates into a syncitial layer (syncytiotrophoblast) and a cellular layer (cytotrophoblast). The projections of the trophoblast into the uterine mucosa develop later into villi. Vascularization of the villi and further branching into secondary and tertiary villi, especially the latter, create the principal structures of exchange of the human placenta. The villi increase in number, and the circumference and thickness of the placenta increase until about the fourth month. Thereafter, while there is no further increase in thickness, there is a continued increase in circumference. As the placenta matures, septa develop from the uterine mucosa and project into the intervillous space, dividing the placenta into 10–30 cotyledons.

As shown in Fig. 7, the human placental membrane consists of three layers of cells: the trophoblast, in which, as mentioned above, the cytotrophoblast and the syncytiotrophoblast can be recognized; connective tissue; and endothelial cells of the fetal placental villi. The cytotrophoblast is particularly prominent during the first trimester of pregnancy. Thereafter, the number of its cells decreases, but it persists until term and gives rise to the syncytiotrophoblast (Tao and Hertig, 1965). The syncytiotrophoblast is thick during early development and it has small cytoplasmic projections or microvilli. The nuclei of the early syncytiotrophoblast are large and evenly spaced. The cytoplasm contains numerous mitochondria and vacuoles. As gestation proceeds, the cells become progressively thinner. In some regions they tend to clump in knots and in other areas the thickness decreases markedly to form "vasculosyncytial membranes" (Fox, 1967).

Table II. Chemical Composition of the Body of the Developing Fetus[a]

Body weight (g)	Approximate fetal age (weeks)	Per kg of whole body		Per kg of fat-free body tissues										
		Water (g)	Fat (g)	Water (g)	N (g)	Ca (g)	P (g)	Mg (g)	Na (meq)	K (meq)	Cl (meq)	Fe (mg)	Cu (mg)	Zn (mg)
300	13	900	5	906	10	3.0	2.0	0.10	20	40	81	—	—	—
100	15	890	5	894	10	3.0	2.0	0.10	100	40	70	50	—	—
200	17	885	5	889	14	4.0	3.0	0.15	100	40	70	50	3.5	18
500	23	880	6	885	14	4.4	3.0	0.20	100	44	66	56	3.5	18
1000	26	860	10	869	14	6.1	3.4	0.22	90	44	66	65	3.5	18
1500	31	847	23	867	17	6.8	3.8	0.24	85	44	66	68	3.8	18
2000	33	810	50	853	20	7.9	4.3	0.24	85	44	63	84	4.2	18
2500	35	776	74	838	21	9.0	4.8	0.25	85	48	56	95	4.3	18
3000	38	727	120	826	21	9.5	5.3	0.27	90	49	55	95	4.5	18
3500	40	686	160	816	21	10.2	5.8	0.27	95	51	54	95	4.8	18

[a] Adapted from Widdowson and Spray (1951).

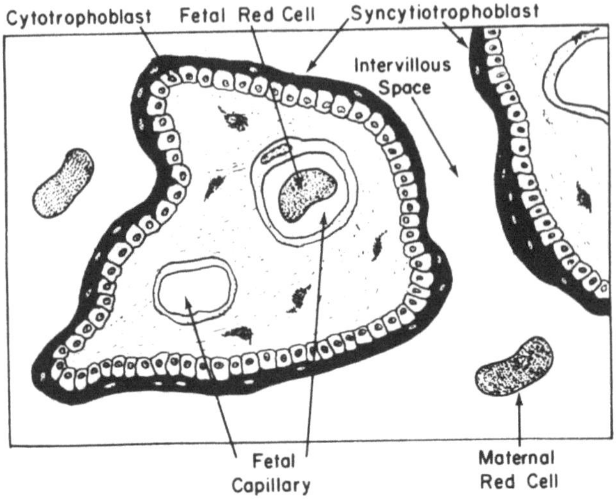

Fig. 7. Schematic section of human placenta villous.

As pregnancy advances, the terminal villi form an increasing number of subdivisions, with an increase in the ratio of surface area to volume. The villi decrease in cross-sectional area and the number of capillaries increase (Lister, 1964). After about 35 weeks, the basement membrane of the trophoblast and the endothelial cells increase in thickness (McKay *et al.*, 1958). Some maternal and fetal vessels are obliterated and fibrin is deposited in the intervillous space.

The area of placental exchange, measured by various morphometric techniques, has been reported to vary from about 6 to 15 m^2 (Clavero and Botella-Llusia, 1963; Aherne and Dunhill, 1966a; Dodds, 1922). The villous surface area of the normal human placenta increases from about 4.8 m^2 to about 11 m^2 at 40 weeks. This increase in accompanied by a proportional increase in the surface area of the fetal placental capillaries (Aherne and Dunhill, 1966a,b).

The difficulties of determining the characteristics of placental growth are even greater than those of determining fetal growth. Besides the methodological errors of the cross-sectional studies or the difficulties of applying ultrasound techniques, it is still unclear how the organ should be prepared for adequate weighing. For example, one of the main difficulties in this respect is how much blood does the placenta contain. These types of problems are responsible for some of the differences in the different series published. In the two largest series of placental weight available, this organ was weighed with the membranes and the umbilical cord attached (McKeown and Record, 1953; Hosemann, 1949). Thus, the values are not strictly placental weight. The data show that the curve of placental growth, as reflected by changes in placental weight, is sigmoid, as in the fetus. In these studies, the weight of the placenta at term was reported to be approximately 650 g. Other series, including a smaller number of cases, have found that after removing the cord and the membranes and after a partial blood drainage the weight of the placenta at term is only 450 g (Winick *et al.*, 1967). This figure is probably a more accurate estimate of true placental weight.

1.4.2b. Cellular Growth. The study of the cellular growth of the placenta has received increasing attention in the recent years. Normal placental growth was studied first in the rat and subsequently in the human, the sheep, and the rhesus monkey (Winick *et al.*, 1967; Kulhaneck *et al.*, 1974; Hill, 1975).

In the rat, placental weight increases linearly until approximately day 20 of gestation (length of pregnancy in the rat is 21 days) and then falls off slightly. Protein and RNA content also continue to increase until term, whereas DNA fails to increase after the seventeenth day of gestation (Winick *et al.*, 1967). The leveling off of placental DNA content of this time reflects cessation of DNA synthesis. This interpretation is supported by radioautographic and [^{14}C]thymidine incorporation studies (Winick *et al.*, 1967; Winick and Noble, 1966a). In rat placenta, cell division stops around the seventeenth day, and the rest of the growth is due to an increase in protoplasmic elements without further cell division. Thus, just as in other organs of the rat, three phases of cellular growth may be described in placenta. From the tenth day until about the sixteenth day of gestation, DNA synthesis and net protein synthesis increase proportionally, reflecting cell-number increments, while cell size remains unchanged. This is the hyperplastic or proliferative phase of placental growth. From the sixteenth to the eighteenth days the rate of cell division slows down while protein accumulation continues, indicating that hyperplasia and hypertrophy are occuring together. Finally, around the seventeenth day, cell division stops while weight and protein still continue to increase. The protein/DNA ratio increases, indicating that cell size increment, or hypertrophy, is the major mechanism for placental growth at this time. Although the exact timing of these events is not so clear as with rats, available data indicate that human placenta grows in a qualitatively similar manner. Weight, protein, and RNA content increases until term. DNA, however, ceases to increase after the placenta reaches about 300 g. This corresponds to a fetal weight of about 2400 g or a gestational age of 34–36 weeks. Thus, as demonstrated in the rat, cell division also ceases before term in human placenta (Winick *et al.*, 1967). This indicates that a constant number of placental cells sustain the rapid increase in fetal weight of the last weeks of pregnancy.

In vitro studies have demonstrated that after week 34 of pregnancy, there is a shift in the pathway of glucose metabolism away from the hexose monophosphate shunt. This finding is consistent with a reduction in nucleic acid synthesis, thus supporting the evidence that DNA synthesis ceases at this time (Beaconsfield *et al.*, 1964).

Sheep placenta has a pattern of growth comparable to rat and human placenta (Kulhaneck *et al.*, 1974). However, in rhesus monkey, total placental DNA content continues to increase until term (Hill, 1975). There are other significant differences between human placental cellular growth and that of the rhesus monkey. For example, placental RNA content in the macaque increases linearly until term but the slope of the curve is not as steep as with DNA (Hill, 1975). Thus, while in human and rat placenta the RNA/DNA ratio increases during the last half of pregnancy, in the rhesus monkeys it decreases. In contrast, changes in protein/DNA ratio are similar in both species.

1.4.2c. Endocrine Functions. The placenta, besides its role in gas exchange and fetal nutrition and excretion, is also a unique endocrine organ able to synthesize a variety of steroid hormones, such as estrogens and progesterone, and polypeptide hormones, such as chorionic gonadotrophin and chorionic somatomammotropin. In addition, the production of a number of other polypeptide hormones has been attributed to the placenta: human chorionic thyrotropin, adrenocorticotropic hormone, melanocyte-stimulating hormone, oxytocin, insulin, relaxin, and various pressor factors. The endocrine function of the placenta has been the subject of several recent reviews (Hytten and Leitch, 1971; Moghissi and Hafez, 1974; Thau and Lanman, 1975).

Steroid hormones. In the nonpregnant female, estrogens act as specialized growth hormones for the female reproductive organs: fallopian tubes, uterus, cervix, vagina, and breasts.

Recent evidence indicates that placental growth is also influenced by estrogens. Increase in placental weight following ovariectomy in rats and rabbits and elimination of the increase by estrogen treatment suggest that estrogens inhibits placental growth, and thereby fetal growth in these species (Abdul-Karim *et al.*, 1971; Csapo *et al.*, 1974).

Under the combined influence of estrogens and progesterone the endometrium proliferates and the endometrial glands grow, providing a suitable medium for the maintenance of the fertilized egg before nidation and an appropriate surface for blastocyst implantation. A number of more generalized metabolic effects have been attributed to estrogen: expansion of blood volume (Friedlander *et al.*, 1936), proliferation of leukocytes (Cruickshank *et al.*, 1970), and increased secretion of certain β-globulins (Dowling *et al.*, 1956; Carruthers *et al.*, 1966). It is conceivable, however, that estrogens may have other metabolic functions, such as those discussed in previous sections, mediated by its influence on corticosteroid secretion.

Plasma levels of estriol increase steadily during pregnancy with a steep rise, from 4.5 mg/liter to 14 mg/liter, between weeks 18 and 35 of pregnancy and a subsequent plateau (Tulchinsky and Kovenman, 1971).

Estriol is the main metabolite of estrogen found in urine, and its rate of excretion parallels changes in estrogen plasma levels (Klopper and Billewicz, 1963). The excretion of estriol is widely used to monitor fetal well-being since, in certain conditions, such as preeclampsia, anencephaly, threatened abortion, and others, it may have diagnostic and therapeutic value.

In the cycling human female, progesterone synthesis is predominantly postovulatory when, together with estrogens, it induces the secretory phase of the uterine glands in preparation for ovum implantation. The general metabolic effect of progesterone acts in some way to prevent abortion induced by premature uterine contractions (Csapo and Wood, 1968). In several species progesterone is also involved in some aspects of maternal behavior, such as hair pulling, nest building, and mothering (Thau and Lanman, 1975). The progesterone concentration in maternal blood rises during pregnancy to values ranging from 11 to 32 mg/100 ml of plasma, as compared with nonpregnancy levels, varying from 0.1 to 2 mg/100 of plasma (Johansson, 1969; LeMaire *et al.*, 1970).

Protein hormones. Human placental gonadotropins (HCG) have been found in several primates, including the macaque. There is also evidence that gonadotropins originating in the uterus or the placenta exist in other species (Thau and Lanman, 1975). HCG is a glycoprotein and in humans it is secreted mainly during the first half of gestation, with a peak rate of secretion at 60 days of gestation. The action of HCG is somewhat similar to that of the luteinizing hormone (Thau and Lanman, 1975). The stimulating effect on the early corpus luteum may be needed to maintain function until the time when the corpus luteum is no longer essential for the maintenance of pregnancy. It has recently been found that HCG inhibits the *in vitro* lymphocyte-stimulating effect of phytohemagglutinin (Adcock *et al.*, 1973). This finding has been interpreted by some as indicating a partial suppression of immunity and as affording the means to the human fetus, which is immunologically foreign to the mother, of avoiding rejection as a foreign graft.

Human chorionic somatomammotropin (HCS) is a polypeptide with prolactin-like and growth-hormone-like activities. The mammotropic activities of HCS are reflected in its promotion of crop sac growth in the pigeon and of milk production in the pseudopregnant rabbit (Thau and Lanman, 1975). Human growth-hormone-like effects have been demonstrated in patients with hypopituitarism. These studies showed reduced urinary excretion of nitrogen, potassium, and phosphorus, while urinary loss of calcium increased. However, HCS is relatively low in growth-promoting activity, having less than $\frac{1}{100}$ the biological activity of HGH (Thau and Lanman, 1975).

Two opposing effects of HCS on carbohydrate metabolism have been described. On the one hand, HCS is diabetogenic, producing insulin antagonism, as does human growth hormone. On the other hand, it also promotes the secretion of insulin (Samaan *et al.*, 1968). Plasma-free fatty acids are also increased by HCS (Thau and Lanman, 1975). Many of these effects are similar to changes which occur during human pregnancy. It is difficult, however, to attribute any of them to HCS independent of concurrently acting hormones, particularly growth hormone, estrogens, and corticosteroids.

HCS becomes detectable in maternal plasma after the first month of pregnancy and subsequently rises to values of about 9 mg/ml at term (Genazzani *et al.*, 1972), when its production rate has been estimated at about 1 g/day (Kaplan *et al.*, 1968). These values are higher than those of any other human polypeptide hormone.

The mechanisms controlling HCS secretion are not clearly defined, although there is evidence that secretion can be influenced by maternal events. For example, fasting for up to 72 hr during the first half of pregnancy induces a marked elevation in plasma levels of HCS (Tyson *et al.*, 1971). These changes in HCS secretion support the idea that the feto-placental unit may react to conditions leading to hypoglycemia, by using HCS as a mobilizer of maternal lipid and protein stores.

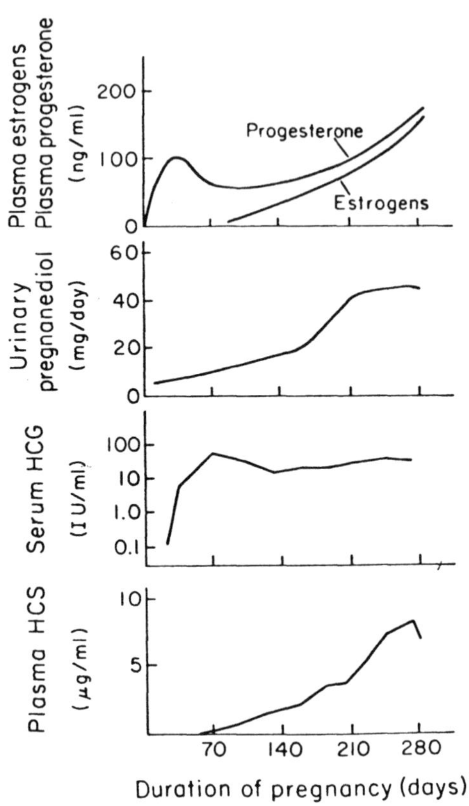

Fig. 8. Plasma hormone levels and urinary pregravidiol during normal human pregnancy. From Thau and Lanman (1975).

Changes in plasma levels of placental steroid and protein hormones during pregnancy are presented in Fig. 8.

1.5. Transfer of Nutrients into the Conceptus

All the substrates, cofactors, energy, and oxygen needed for fetal growth must be transported into the fetus via a supply line whose major components are the maternal blood, which serves as a vehicle for the nutrients; the uterine circulation; and the placenta.

1.5.1. Uterine Vascularization and Circulation

In humans, as well as other mammalian species, uterine vascularization and blood flow increases during gestation manyfold over prepregnancy levels. For example, in the sheep, uterine blood flow represents approximately 2% of the cardiac output before pregnancy and 20% at term (Dilts *et al.,* 1969). As first described in the rabbit (Barcroft and Ratschild, 1932), the expansion of uterine circulation precedes by a considerable margin of time the period of rapid fetal growth that occurs during the last third of gestation (Fig. 1). This changing relationship between uterine blood perfusion and fetal growth is best reflected by differences in the rate of perfusion per unit of fetal mass during the first and second halves of gestation. In the sheep, up to day 60 of gestation (normal gestation = 147 days), this rate averages more than 650 ml/kg/min. From day 90 of gestation until term, however, the rate of uterine blood flow averages less than 250 ml/kg/min (Huckabee *et al.*, 1961, 1972). More recent studies in thc sheep have demonstrated that blood flow to the myometrium and endometrium follows this general pattern of high perfusion per gram of tissue in early pregnancy and of subsequent decline (Rosenfeld *et al.*, 1974). In contrast, placental blood perfusion increases in proportion to the growth of the placental cotyledons up to day 80 of gestation and continues to increase after placental weight declines in the last third of pregnancy. Thus, placental blood flow per gram of placental cotyledons increased from a minimum of approximately 0.4 ml/min at 90 days to a maximum of 3 ml/min near term.

Owing to differences in the growth curves of the different components of the pregnant uterus, the distribution of blood flow within the uterus varies considerably from the nonpregnant to the pregnant state and from early to late pregnancy (Rosenfeld *et al.*, 1974). The myometrium, endometrium, and caruncles of the nonpregnant uterus each receive approximately one-third of the uterine blood flow. In early pregnancy approximately 50% of the uterine blood flow goes to the endometrium and only 27% to the sites of implantation. Then there is a rapid shift of flow to the placenta, which receives approximately 64% of the uterine blood flow at 140 days of gestation. The measuremnt of uterine blood flow in humans is an indirect and rather inaccurate procedure. The combination of available data from studies done by different authors at different gestational ages (Assali *et al.*, 1960; Romney *et al.*, 1955; Metcalfe *et al.*, 1955; Blechner *et al.*, 1974) indicate a marked increase from approximately 50 ml/min at 10 weeks to 490 ml at term. The reported values at term range from 175 ml to 840 ml. Such a degree of variability probably reflects the methodological difficulties of the procedure. Based on data on O_2 and CO_2 exchange, it has been calculated that approximately 75% of the total uterine blood flow enters the intervillous space and therefore exchanges with fetal blood (Bartels *et al.*, 1962). Studies using ^{133}Xe clearance support this estimate (Jansson, 1969).

As in pregnant sheep, in humans uterine blood flow per kilogram of uterus and its content is higher during early pregnancy than at any other time and remains remarkably constant during the last half of pregnancy. Assuming that approximately 75% of the total uterine blood flow enters the intervillous space, the blood flow to the placenta is the highest during the first 10 weeks, it declines until the 20th week, and increases moderately until term. In contrast, placental blood flow per kilogram of fetus decreases progressively during gestation (Fig. 9). Thus, in the sheep and in humans, distribution of uterine blood flow per kilogram of conceptus seems to have a remarkably similar pattern.

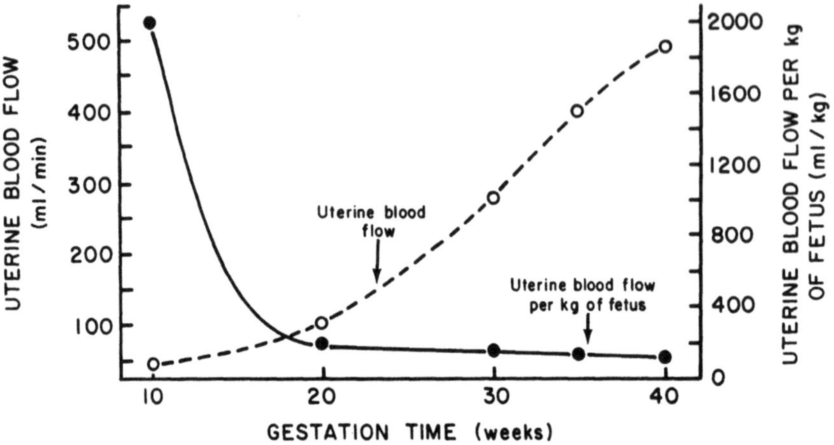

Fig. 9. Uterine blood flow per organ and per fetal body weight during normal human pregnancy. Data from Hytten and Leitch (1971).

It has been speculated that the decreasing availability of nutrients and oxygen per unit of body mass, determined by the relative decrease in the uterine circulation, may be a limiting factor for the metabolic needs of the fetus near term. The evidence supporting such a concept, however, is controversial.

The expansion of the uterine circulation seems to be under the hormonal regulation of estrogens and progesterone. In the nonpregnant female of several species, estrogen administration causes a marked increase in the rate of uterine blood flow (Kalman, 1958; Dickson *et al.*, 1969; Greiss and Anderson, 1970) and, at least in the sheep, a decrease in the arteriovenous oxygen difference (Huckabee *et al.*, 1970). Increased uterine blood flow and reduced arteriovenous oxygen difference are also found during the follicular phase of the estrous cycle. Progesterone administration to nonpregnant castrated sheep does not seem to have any direct effect on uterine blood flow, except to block the vasodilation response of estrogen administration (Caton, 1974). Assuming that the effects of these hormones in the nonpregnant and in the pregnant ewe are similar, it seems reasonable to attribute the high rates of uterine blood flow in early pregnancy to a predominance of estrogen effects. Similarly, the proportionally smaller rate of increase of uterine blood perfusion during the second half of pregnancy would reflect a relative increase in progesterone levels. Since estrogen and progesterone are synthesized by the feto-placental unit, it is conceivable that they may be used by the fetus as a mechanism to increase its availability of oxygen and nutrients. There is no evidence, however, of the existence of a feedback mechanism of this nature.

1.5.2. Placental Transfer Mechanisms

The placenta is part of a system for the exchange of substances between the mother and fetus. The systems has three major components: (1) the uterine circulation and the placental blood flow on the maternal side, (2) the placenta, and (3) the placental blood flow in the fetal side.

The data on maternal–fetal transfer of substances has been extensively reviewed (Longo, 1972). Although present knowledge is still limited, there is evidence that a series of different mechanisms are used, including the following: simple diffusion, facilitated diffusion, active transport, pinocytosis, and bulk flow.

1.5.2a. Simple Diffusion. Simple diffusion represents the movement of a molecular species across the membrane by random thermal motion from an area of high concentration to one of low concentration. It is a passive process involving no energy consumption of work by the membrane. Although simple diffusion usually occurs because of chemical gradients, charged ionic species may move in response to electrochemical gradients. The net quantity of molecules transferred by simple diffusion is directly proportional to the concentration and/or electrochemical difference across the membrane; hence simple difusion continues until uniform concentration or electrochemical equilibrium is established. Besides the concentration or electrochemical gradient on each side of the membrane, simple diffusion is also influenced by certain characteristics of the membrane, such as thickness, area, and diffusibility, and the physicochemical characteristics of the molecules. Among the latter, molecular size, electrical charge, and lipid solubility are the most important (Fig. 10).

1.5.2b. Facilitated Diffusion. The mechanisms of facilitated diffusion are still elusive. Like simple diffusion, it occurs only when a concentration or electrochemical gradient exists across a membrane, and it continues until an equilibrium is reached. It differs, however, from simple diffusion in that the rate of transfer is faster, becomes constant at high concentrations, and can be interfered by molecules with similar spacial configuration. These characteristics suggest the presence of certain "transfer sites" or "carriers" that can react or combine with the molecules being transferred.

Like simple diffusion, facilitated diffusion does not occur against concentration or electrochemical gradients and does not require energy (Fig. 10).

1.5.2c. Active Transport. This type of transfer mechanism requires energy consumption, and can proceed against oncentration or electrochemical gradients. The mechanisms of active transport have not been established yet, but, again, the most accepted hypothesis involves the presence of membrane "carriers" that would combine with the substrate. Either the carrier or the carrier–substrate complex would undergo endergonic chemical changes linked to energy availability provided by adenosine triphosphate molecules.

The presence of the "carrier" would determine certain of the characteristics of this transport mechanism, such as reduced rate of transfer at high concentration, and competition by molecules with similar special configuration. Other characteristics, such as inhibition during anaerobic conditions or by metabolic inhibitors, reveal the energy requirement of the system (Fig. 10).

1.5.2d. Ultrafiltration. Under certain conditions of hydrostatic and osmotic pressure, when fluids cross a semipermeable membrane carry with them solutes (solvent drag). Although there is no firm evidence that this type of transfer mechanism takes place in placenta, it is considered to be responsible for the rapid movement of water among amniotic fluid, fetus, and mother.

1.5.2e. Pinocytosis. In this mechanism, plasma membrane invaginate, engulfing solute and water and creating small vacuoles than can cross the cells and discharge their contents on the opposite side. Pinocytosis has been observed in the placenta, and histochemical studies have demonstrated a high protein content in the vacuoles. The process is still poorly understood. Histochemical studies have demonstrated that pinocytotic vesicles are surrounded by high concentrations of ATPase, suggesting energy requirements, but the need of energy for pinocytosis has not been substantiated by more direct studies (Fig. 10).

1.5.3. Transfer of Specific Substances

1.5.3a. Gases. The respiratory gases O_2, CO_2, and CO and the metabolically inert gases are presumed to cross the placenta by simple diffusion (Barron, 1951–1952; Longo *et al.*, 1967; Meschia *et al.*, 1967). In a review of this subject, Metcalfe *et al.* (1967) pointed out that placental transfer of oxygen is influenced by several factors, including the placental diffusing capacity, the uterine and umbilical arterial O_2 tensions, the characteristics of the maternal and fetal oxyhemoglobin saturation curves, the maternal and fetal placental hemoglobin flow rates, the pattern of maternal to fetal blood flows, and the amount of CO_2 exchanged. Each of these determinants of O_2 exchange is, in turn, a function of other factors.

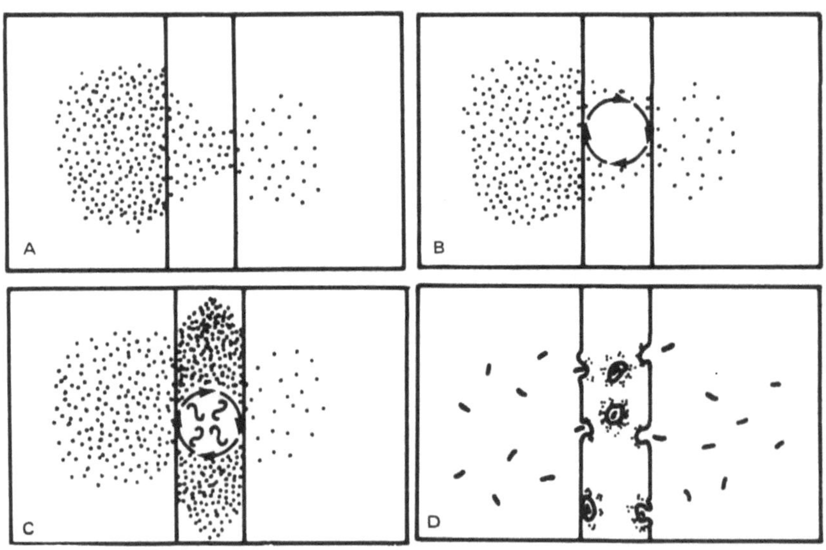

Fig. 10. Transfer mechanisms of the human placenta. A, simple diffusion; B, facilitated diffusion; C, active transfer; D, pinocytosis.

An uninterrupted O_2 supply is vital for fetal survival. The rate of fetal oxygen utilization has been estimated to be 13.5–15 ml/min (for a fetus near term) while the supply of fetal oxygen in reserve is estimated to be about 30 ml. Thus, oxygen deprivation would cause irreversible brain damage in a few minutes.

1.5.3b. Carbohydrates. Glucose and other carbohydrates cross the placenta by facilitated diffusion (Widdas, 1952). The mean maternal glucose concentration is about 90–100 mg/100 ml, while the mean fetal concentration is about 70–75 mg/100 ml (Zuspan *et al.*, 1966; Paterson *et al.*, 1967). Besides fetal consumption, placental metabolism of glucose would account for part of the maternal-to-fetal glucose concentration difference (Davies, 1955).

As is characteristic in facilitated diffusion, the placental transfer of glucose is stereospecific. For example, D-glucose is transferred much more rapidly than L-glucose (Longo and Kleinzeller, 1970) or than fructose (a ketohexose with the same molecular weight as glucose) (Chinard *et al.*, 1956). Competitive inhibition has also been demonstrated among various hexoses (Colbert *et al.*, 1958; Ely, 1966).

The possible role of maternal insulin in the rate of uptake of glucose by placental tissues and its eventual transfer into the fetus has received some attention. The available evidence, however, is conflicting. Some authors have reported an increased glucose uptake by human placental slices in the presence of insulin (Villee, 1953; Litonjua, 1967), while others have been unable to observe such an effect (Szabo and Grimaldi, 1970).

The human placenta contains a large amount of glycogen, most of which is probably synthesized from maternal glucose. Placentas of diabetic women or alloxan diabetic rats have increased glycogen concentration (Heijkenskjold and Gemzell, 1957). In contrast, placental glycogen concentration is reduced in malnourished rats (Rosso *et al.*, 1976).

It has been shown that placental glycogen concentration is highest at about 8 weeks and that it declines throughout pregnancy with the lowest concentration near term (Ville, 1953). Glycogen concentration is higher in the decidua than in the fetal side (Loveland *et al.*, 1931; Huggett and Morrison, 1955); however, [^{14}C]glucose given intravenously in the rabbit is incorporated more rapidly into the glycogen of the trophoblast (Huggett and Morrison, 1955). There seems to be a continuous exchange among maternal glucose, placental glycogen, and fetal glucose. The fact that, during the last third of gestation, the decrease in placental glycogen is accompanied by a reciprocal increase in the glycogen of the fetal liver suggests the presence of a regulatory mechanism coordinating these changes. However, there is no adequate explanation yet of the physiological role of placental glycogen. For example, in the rabbit, placental glycogen concentration was found to be independent of sudden alterations in maternal blood glucose (Lockhead and Cramer, 1908; Huggett, 1929).

In ungulates, fructose concentration in fetal blood is higher than that of glucose (Bacon and Bell, 1948). In these species the placenta contains only small amounts of glycogen. Fructose is also present, although in small quantities, in human fetal blood; its metabolic role is unknown (Hagerman and Villee, 1952).

1.5.3c. Lipids. Free fatty acids (FFA), such as linoleic, margaric, palmitic, and stearic, and acetate are transferred across the placentas of the rabbit, the guinea pig, and subhuman primates by simple diffusion (Van Duyne *et al.*, 1962; Hershfield and Nemeth, 1968; Postman *et al.*, 1969). Placental transfer of FFA by simple diffusion has also been observed *in vitro* in perfused human placentas (Szabo *et al.*, 1969). In contrast, the sheep placenta seems to be relatively impermeable to these substances (James *et al.*, 1971).

In the guinea pig placenta, there is no significant difference in the rate of transfer when fatty acids are given singly or in groups (Kayden *et al.*, 1969). Further, rate of transfer of fatty acids is not influenced by the degree of saturation, the length of the chain, or whether the chain is odd or even (Kayden *et al.*, 1969).

In humans the levels of free fatty acids are higher in the maternal plasma, and their concentration is closely correlated with fetal plasma concentration (Whaley *et al.*, 1968). Analysis of the subcutaneous fat of human fetuses and newborns suggests that during the second trimester of pregnancy the transfer of fatty acids accounts for most of the fat deposited in the fetus. Near term, however, the rate of fat deposit would exceed the capacity for placental transfer and lipids derived from *de novo* synthesis in the fetal tissues would constitute the majority of the lipids deposited at term (Hirsch *et al.*, 1960; Dancis *et al.*, 1973). In contrast, in the guinea pig 10–20% of [^{14}C]palmitate injected into the mother near term can be recovered in lipids of the fetus. Based on these data and the rate of transport of palmitate in the maternal blood, it has been suggested that placental transport of free fatty acids can account for almost the entire accumulation of lipids near term (Bohmer and Havel, 1975). Similar conclusion have been drawn from quantitative studies on the maternal–fetal transfer of FFA in the rat near term (Hummel *et al.*, 1975). Also, in the guinea pig it has been shown that fetal liver during late gestation maintains a high rate of synthesis of triglycerides and very low density lipoprotein used for plasma transport of the triglycerides (Bohmer and Havel, 1975).

There is little information on the transfer rate of triglycerides. In the guinea pig placenta they appear to cross very slowly (Kayden *et al.*, 1969).

Maternal cholesterol also crosses the placental slowly compared with free fatty acids. Fetal plasma cholesterol concentration is lower than that of maternal blood, and in the rat, maternal cholesterol accounts for only 10–20% of the total (Goldwater and Stetten, 1947; Gelfand *et al.*, 1960). One of the reasons for the lower cholesterol concentration in the fetus would be a reduced concentration of the β-lipoprotein to which plasma cholesterol is bound. In the fetus the plasma concentration of this protein is only 30% of the maternal concentration.

Transfer mechanisms for phospholipids are still poorly understood, although they seem to be metabolized during transfer with hydrolysis of the phosphate group and subsequent resynthesis in the placenta (Popjak, 1954; Popjak and Beeckmans, 1950; Biezenski, 1969).

1.5.3d. Amino Acids. Amino acids are transferred across the placenta by active transport. Numerous studies support this assumption. Fetal blood levels of amino acids are higher than maternal levels. In humans the ratio between the two concentrations varies from 1.2 to 4.0 for the different amino acids (Glendening *et al.*, 1961; Ghadimi and Pecora, 1964; Van Slyke and Meyer, 1913; Young and Prenton, 1969). Thus, net transfer toward the fetus is against a gradient. The participation of a carrier in amino acid transfer is suggested by experiments showing that natural L-amino acids are transferred more rapidly than the D forms (Pape *et al.*, 1957; Mischel, 1963; Reynolds and Young, 1971). In addition, a decreased transfer rate of amino acids at high concentrations or during simultaneous infusion of two amino acids has been described (Dancis *et al.*, 1958; Dancis and Shafran, 1958). These results suggest saturation phenomena and competition for transport sites, respectively. The energy dependency of placental amino acid transport has been demonstrated *in vitro* using different metabolic inhibitors (Longo *et al.*, 1973; Smith *et al.*, 1973).

In vitro studies with human placental fragments (Enders *et al.*, 1976) have shown that neutral amino acids are transported by systems similar to those described in other tissues (Christensen, 1973). These systems have been designated A, L, and ASC. Transport by the A system is Na^+-dependent and very sensitive to changes in pH and temperature of the extracellular fluid. This system also shows strong mutual inhibition among the amino acids that use it preferentially, such as glycine, alanine, serine, and proline, and the nonmetabolizable amino acid α-amino isobutyric acid (AIB) and *N*-methyl AIB. In human placenta, *N*-methyl alanine, threonine, glutamine, alanine, serine, glycine, proline, and AIB have also been found to use this system preferentially (Miller and Berndt, 1974; Enders *et al.*, 1976). The L system, in contrast, is independent of Na^+ and less sensitive to pH, temperature, and the presence of metabolic inhibitors. In human placenta the less polar branched-chain amino acids, such as isoleucine, valine, and phenylalanine, preferentially use this system. The ASC system is also Na^+-sensitive, and it can be distinguished from the A system by the fact that it is not inhibited by *N*-methyl AIB. Amino acids using this system are alanine and serine and, in human placenta, threonine and glutamine. It has been shown that preincubation of human placental fragments increases the rate of *in vitro* transport by system A (Smith *et al.*, 1973). This increase is blocked by cycloheximide, puromycin, or actinomycin D, and metabolic inhibitors such as dinitrophenol and cyanide (Smith and Depper, 1974; Gusseck *et al.*, 1975).

Placental transfer of most amino acids would consist of three steps: (1) uptake from the maternal circulation, (2) placental concentration, and (3) release into the fetal circulation. Only the first step would require energy consumption in order to transport the amino acid into the cell against a considerable concentration gradient. The third step would be a mere passive diffusion of amino acids from the high placental concentration into the fetal plasma.

Evidence supporting the existence of this three-step transfer mechanism have been described in the guinea pig (Hill and Young, 1973) (see Fig. 11).

There are data suggesting that some amino acids may not necessarily follow a three-step mechanism. For example, in contrast with the guinea pig placenta, it has been shown that glutamate crosses the placenta of the macaca mulatta at an extremely slow rate (Stegink *et al.,* 1975). Similarly, in human placenta, *in vivo* experiments done at 16–22 weeks of pregnancy have shown that both cysteine and cystine levels in maternal plasma are higher than in fetal plasma. It was also found that after an intravenous load of these amino acids, they were transported into the fetal blood less rapidly than other amino acids and that the fetal blood concentration never rose above maternal levels (Gaull *et al.,* 1973).

1.5.3e. Proteins. The general assumption is that proteins, especially large ones, cross the human placenta by pinocytosis. There are several facts, however, which suggest that other factors, in addition to the simple engulfment process of pinocytosis, may be involved especially for smaller protein. Several plasma proteins, such as albumin, *l*-acid glycoprotein, transferrin, gamma globulin, and fibrinogen can cross the placenta (Gitlin and Biasucci, 1969). The rate of transfer of these protein fractions is considerably slower than that of amino acids and is influenced by the concentration gradient between the maternal and fetal plasma and by the molecular weight (Gitlin *et al.,* 1964). It has been calculated that the rate of transfer of most proteins is inversely proportional to the square root of the molecular weight (Gitlin *et al.,* 1964). IgG gamma globulin is one of the known exceptions to this rule, and in the rhesus monkey, for example, is transported faster than albumin, in spite of the fact that its molecular weight is twice that of albumin (Baugham *et al.,* 1958).

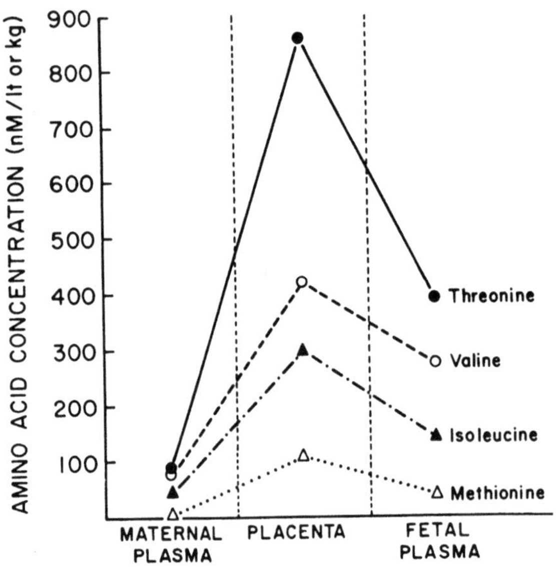

Fig. 11. Concentration of certain amino acids in maternal plasma, placental tissue, and fetal plasma in the guinea pig. Data from Hill and Young (1973).

Because of its role in the passive immunity mechanisms of the neonate, transplacental passage of IgG gamma globulin has received considerable attention. It has been shown that IgG is transported by a carrier-mediated mechanism that can be saturated by high maternal concentrations of IgG (Gitlin and Koch, 1968). Such a system will undergo maturation during pregnancy, reaching in humans a significant capacity to transport IgG at approximately 22 weeks (Gitlin and Biasucci, 1969; Karte, 1969; Morphis and Gitlin, 1970).

As suggested by some clinical situations, the human placenta seems to be almost completely impermeable to some small and medium-size polypeptides, such as insulin, growth hormone, and thyroid-stiumulating hormones (Jost and Picon, 1970).

Transport of proteins into the fetus has marked species differences. In the rabbit, guinea pig, and probably the rat, maternal plasma proteins are secreted into the uterine cavity and then transported through the splanchnopleure into the yolk sac (Brarubell, 1954).

1.5.3f. Water-Soluble Vitamins. The placental mechanisms to transport water-soluble vitamins are still largely unknown; nevertheless, considering the fact that maternal plasma concentrations of most of these vitamins are lower than fetal plasma concentrations, it is conceivable that they are actively transported. In many cases, however, the degree of protein binding may be higher in fetal blood, resulting in a higher concentration of total vitamin yet in a reduced concentration of free vitamin. Thus, this fraction could be transferred by simple diffusion mechanisms.

Fetal blood concentration of thiamine is 1.7 times higher than that of maternal blood. It is not known, however, what proportion of the fetal levels represents thiamine pyrophosphate, the active form, or thiamine the precursor (Slobody *et al.*, 1949). In the guinea pig there is a maternal–fetal concentration gradient of thiamine, but transfer is relatively slow (Brink *et al.*, 1959).

Fetal free riboflavin concentration is approximately 5 times higher than the maternal concentration. Concentration of total riboflavin (including the bound fraction) is similar in maternal and fetal blood (Lust *et al.*, 1953).

Pyridoxal phosphate and its precursor, pyridoxamine phosphate, have been found to be 3.6 and 2.9 times higher in fetal cord blood than in maternal peripheral blood (Contractor and Shane, 1970). This finding suggests an active placental transport of the vitamin, although it is also conceivable that the vitamin is metabolized and rephosphorylated in the placenta (Contractor and Shane, 1971). Plasma levels of pyridoxal phosphate are lower in premature infants than are full-term ones, which suggests that the transport of pyridoxal phosphate becomes more active during the last trimester of pregnancy (Reinken and Marigold, 1973).

Cord blood concentrations of vitamin B_{12} are higher than maternal blood, but the difference varies in different series (Killander and Vahlquist, 1954; Okuda *et al.*, 1956; Boger *et al.*, 1957). Little is known about the characteristics of placental transfer in humans. In the mouse, B_{12} injected into the mother accumulates in the placenta before passing into the fetus (Ullberg *et al.*, 1967).

Levels of folate in cord blood and sera are 2.5 times higher than that of maternal blood (Braestrup, 1937; Manahan and Eastman, 1938; Snelling and Jackson, 1939). The transport mechanisms are still poorly understood. It has been shown in the rat that after [^{14}C]ascorbic acid is injected into the maternal circulation, the placental concentration of label is much higher than in fetal tissues. Such a concentration suggests the existence of active transport (Rosso and Norkus, 1976). Studies in guinea pig placenta indicate that only dehydroascorbic acid may cross the placenta and is rapidly converted into ascorbic acid by the fetal tissues. Ascorbic acid itself would cross slowly, if at all (Raiha, 1959). However, in human placenta perfused *in vitro* it has been shown that ascorbic acid crosses this organ against a concentration gradient and that both oxytocin and papaverine significantly inhibit its transfer (Hensleigh and Krantz, 1966).

1.5.3g. Lipid-Soluble Vitamins. Little is known about the transport mechanisms for lipid-soluble vitamins. Most of the fragmentary information consists of studies done in vitamin A. Vitamin A concentration in cord blood of full-term infants is about 50% lower than the average maternal concentration of the vitamin. There is no correlation, however, between maternal and fetal levels of vitamin A (Lund and Kimble, 1943). In contrast, there is a high correlation between maternal levels of carotene and the levels of carotene in the infant (Lund and Kimble, 1943; Barnes, 1951). Based on these observations it has been proposed that in humans transplacental passage of vitamin A accounts for only a small proportion of the fetal vitamin and that fetal synthesis of vitamin A from carotene accounts for the rest (Barnes, 1951). Transplacental passage of carotene would be accomplished by simple diffusion since it is not concentrated by the placenta (Barnes, 1951). Transfer of vitamin A across the placenta has also been demonstrated in the rat, and it has been suggested that in this species the rate of placental transfer is inversely proportional to the fat content of the organ (Williamson, 1948). It has been shown in the rat that both vitamin D_3 and 25-dehydroxycholecalciferol cross the placenta (Haddad *et al.*, 1971). There is no evidence, however, suggesting the type of transport mechanisms involved. In human plasma, concentration of vitamin D is usually slightly higher than fetal plasma concentration regardless of gestational age (Rosen *et al.*, 1974).

Maternal plasma levels of vitamin E are approximately five times higher than fetal levels (Varangot, 1942; Straumfjord and Quaife, 1946), which suggests that either simple or facilitated diffusion may be the transport mechanisms involved. A rapid passage of vitamin E from maternal to fetal blood has been shown in rats (Sternberg and Pascoe-Dawson, 1959).

Very little is known about vitmain K transfer except that it may cross the rat placenta (Dam *et al.*, 1955).

1.5.3h. Minerals. It is generally assumed that most of the electrolytes and minerals cross the placenta by simple diffusion or "solvent drag." There is evidence, however, that some of the minerals, such as Ca^{2+}, P, Zn, and iron, may use active transport. In the guinea pig and the primate (Dancis and Money, 1960; Battaglia *et al.*, 1968), placental transfer of sodium is greatly dependent on the nature and condition of the membrane rather than on the rate of blood flow. The transfer of sodium has therefore been characterized as "membrane-limited." It is also known that sodium net transport in materno-fetal and fetomaternal direction is similar in the isolated, artificially perfused guinea pig placenta. This finding suggests that there are no active mechanism involved in sodium transfer (Schroder *et al.*, 1972).

The mechanisms of potassium transfer are less clear. It is known to readily cross the guinea pig and rat placentas (D'Silva and Harrison, 1953), and it has nearly similar concentrations in fetal and maternal blood in humans and dogs (Earle *et al.*, 1951; Serrano *et al.*, 1964). Deprivation studies in the rat have shown that plasma concentration of potassium in the fetus is maintained even when the maternal concentration decreases as much as 50% (Stewart and Welt, 1961; Dancis and Springer, 1970). The maintenance of higher concentrations of potassium in the fetus of the deprived animals is compatible with the existence of an active transport mechanism.

Calcium concentration in humans is higher in the cord blood than the maternal blood. The higher concentration includes both the protein-bound and the ultrafiltrable fractions (Delivoria-Papadopoulos *et al.*, 1967). In the guinea pig and the sheep, net transfer of calcium to the fetus continues against very high gradients (Bowden and Wolkoff, 1967; Twardock and Austin, 1970).

Kinetic analysis of calcium transport across the sheep placenta in a chronic preparation has shown that daily transport rates of calcium from mother to fetus were 215 mg/kg of fetus and from fetus to mother 12 mg/kg of fetal weight (Ramberg *et al.*, 1973). In a similar preparation in primate, values obtained were 391 and 326 mg/kg fetal weight, respectively (McDonald *et al.*, 1965). Differences in these rates were explained as due to histological characteristics of the placenta in both species.

Recent experiments have shown that human placental plasma membranes vesicles can accumulate Ca^{2+}, reaching concentrations that are 24-fold higher than the medium. This process would be dependent on ATP hydrolysis by the placental Ca^{2+}-ATPase (Shami *et al.*, 1975). As suggested by studies on the guinea pig showing that fetal plasma concentration of phosphorus is, approximately, twice the maternal concentration, it is possible that placental transfer of phosphorus also requires energy consumption (Fuchs and Fuchs, 1956).

Recent studies have shown that, in contrast with other minerals, such as cobalt, bromine, rubidium, and gold, with similar maternal and cord blood concentrations, the concentration of zinc is almost twice as high in the cord blood than the maternal blood (Alexiou *et al.*, 1976). As previously discussed, such a concentration gradient is compatible with the existence of active placental mechanisms for the transfer of zinc.

Placental transfer of iron has been investigated in several species (Baker and Morgan, 1973; Majumdar and Wadsworth, 1974). The available evidence suggests that the first step involves dissociation of iron from the maternal transferrin–iron complex, passage through the placenta, probably bound to a 4000- to 4700-molecular-weight carrier (Larkin *et al.*, 1970), and association with fetal transferrin molecules on the fetal side. In the guinea pig placenta, low oxygen tension and substances such as sodium arsenite, sodium cyanide, or dinitrophenol markedly reduce the rate of iron transfer (Wong and Morgan, 1953). This is clear evidence of an active, energy-requiring transport mechanism for iron. The rat and the guinea pig have hemochorial placentae that allow direct contact between the maternal plasma transferrin and the trophoblast cells. In contrast, in species such as the cat and the dog, the maternal capillary endothelium separates the maternal blood from the trophoblast cells. In the cat, transfer of ^{59}Fe into the fetus is very slow and only small amounts of plasma transferrin are taken up by the placenta (Baker and Morgan, 1973). This has been interpreted as an indication that another source of iron, such as maternal erythrocytes, may provide most of the fetal supply of iron in the cat.

1.5.4. Changes in Placental Transfer during Pregnancy

During the last half of pregnancy, the rate of fetal growth greatly exceeds the rate of placental growth. This phenomenon is reflected by a marked increase of the ratio of fetal weight to placental weight. The obvious functional implications of such disproportionate growth is that a relatively smaller placenta must sustain a rapidly growing fetus. Under these circumstances the only way in which the fetal needs can be met is by increasing the rate of transfer of nutrients (i.e., near term more molecules must cross the placenta per unit of placental mass per minute than at midgestation). Until recently, this interesting phenomenon has not been explicity recognized. Theoretically, an increased placental capacity to transfer nutrients can be achieved by increasing placental blood perfusion in both maternal and fetal sides, by increasing the area of exchange and the permeability of the placenta membrane or by increasing the efficiency of the active mechanisms. The latter would require either an increase in the number of carriers, or an increase in the availability of energy, or both.

Studies on the rate of sodium exchange across the placenta in humans and other species (Flexner and Gellhorn, 1942; Flexner *et al.*, 1948) have shown a progressive increase in the efficiency of the exchange as pregnancy progresses, reaching a peak near term and then rapidly falling. In the human placenta the permeability to sodium increases from about 0.3 mg/g of placenta/hr at 9 weeks to 7 mg/g/hr at 36 weeks, and then it decreases to 4 mg/g/hr at 40 weeks. These functional changes have been explained as a progressive vascularization and thinning of the barrier between the maternal and fetal circulation with a concomitant increase in the surface area of the membrane and a reduced distance through which sodium must diffuse. A comparative study of placental transfer among several species has been interpreted as supporting the hypothesis (Flexner and Gellhorn, 1942). For example, the ungulate placenta has several cellular layers between the maternal and fetal circulations. In contrast, in the guinea pig and human placentas the trophoblast is in direct contact with maternal blood. The exchange rate of sodium in the latter is considerably faster than in the ungulate.

Increased placental transfer during gestation has also been described for phosphorus in experiments done in guinea pigs. Between days 30 and 65 of gestation, the amount of phosphorus taken up by the fetus increases from less than 0.5 mg/24 hr to 27 mg/24 hr, while the amount of phosphorus retained by the placenta is nearly constant (Fuchs and Fuchs, 1956).

Although an increased placental transfer of substances such as sodium or phosphorus may be accomplished by changes in the thickness of the placental membrane or the rate of blood perfusion, the gestational changes in the rate of iron transfer suggest that more sophisticated mechanisms are also operating with greater efficiency. In experiments done in the guinea pig it was found that placental transfer of ^{59}Fe increased markedly after day 25 of pregnancy, to reach a maximum at about day 54 and then decreased markedly by day 65 (Wong and Morgan, 1953). Interestingly enough, these changes were disproportionately high when compared with changes in fetal growth, suggesting that the fetus was receiving an apparent excess of iron until few days before delivery.

More recently, the rate of placental transfer of radioactive-labeled nutrients such as glucose and ascorbic acid and nutrient analogs such as α-amino isobutyric acid (AIB) were studied in rat. In these experiments, the labeled compounds were injected into the maternal circulation and the concentration of label measured in placentas and fetuses removed 10 min later. It was found that the concentration of AIB in placental tissues increased severalfold between days 14 and 21, suggesting higher capacity of the placenta to take up and concentrate this substance. Similarly, the concentration of AIB in fetal tissues also increased proportionally to the placental changes, indicating that the quantity of AIB transported into the fetus is proportionally higher near term than in previous days (Rosso, 1975). A similar pattern of changes in the rate of placental transfer was found for glucose. In contrast, the increase in the rate of transfer of ascorbic acid was proportional to the increment in fetal weight (Rosso and Norkus, 1976). Thus, the fetal concentration of label 10 min after maternal injection of [^{14}C]ascorbic acid remained constant between 15 and 21 days of gestation.

The apparent excess in the quantity of glucose and AIB transported into the fetal rat during the last third of pregnancy, or the excess of iron transported into the guinea pig, suggests that there may be a "safety margin" for certain substances. This concept of "safety factor" for the fetus, expressed as the ratio of placental transfer to fetal requirements, was introduced after the first studies on sodium exchange during gestation were completed (Flexner *et al.*, 1948). Based on the values obtained at term, it was calculated that the exchange rate was 1000 times the rate of accumulation of sodium by the fetus. The teleological interpretation of the findings is that during the period of maximum fetal growth, the fetus receives certain nutrients in excess of its needs as a way to prevent the possibility of shortages. It is hard, however, because of current ignorance on fetal metabolism and requirements, to understand the metabolic implications of the apparent disproportion and therefore to accept

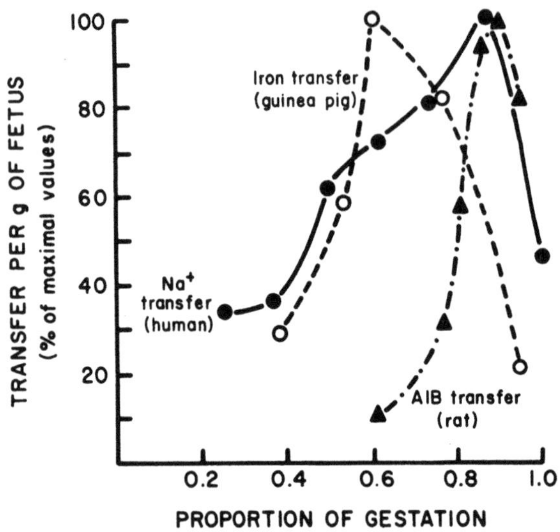

Fig. 12. Rates of placental transfer of sodium, iron, and α-amino isobutyric acid (AIB) at different times of gestation in human, rat, and guinea pig. Data from Flexner *et al.* (1948), Wong and Morgan (1953), and Rosso (1975).

it as a safety factor. It will be interesting to determine if other substances whose rate of placental transfer increase near term, but which are not directly related to fetal growth, such as IgG, are also transferred in apparent excess during a certain period of gestation. The rate of transfer of various nutrients during gestation in three different species are compared in Fig. 12.

Besides the changes in sodium transfer discussed above, there are some indirect indications that the human placenta also may increase the concentrations of nutrients in the fetus during part of the last trimester of pregnancy. For example, the plasma concentration of vitamin B_6 and folate are low in premature infants compared with term newborns (Dawson, 1972; Reinken and Marigold, 1973). Also, premature infants are more prone to develop deficiency of iron, suggesting that the stores of this mineral were lower than in more mature babies. Deficiency of riboflavin and lower plasma levels of vitamin B_{12} also have been described in premature infants (Killander and Vahlquist, 1954; Cole *et al.*, 1974). Further, concentrations of amino acids in the umbilical vein are higher in babies born between 34 and 38 weeks than in full-term babies (Young and Prenton, 1969).

2. *Nutrition and Growth of the Human Conceptus*

2.1. *Maternal Weight Gain and Birth Weight*

In the sixteenth, seventeenth, and eighteenth centuries, great attention was paid to the diets of pregnant women. It was recognized that the fetus needed nutrients and that since only the mother could provide them, it was important that the maternal diet be adequate. At the same time, overeating was considered conducive to a large series of complications, such as "plethora of humors," "fetal outgrow of the womb," and abortions (Eschelman, 1975).

The idea that pregnant women should not overeat is also a recurrent theme in the treatises of midwifery and obstetrics of the nineteenth century. In these treatises, however, only secondary importance is given to maternal diet. It is usually discussed in chapters devoted to the "hygiene of pregnancy," and the recommendations are a mixture of empiricism and mythology. As in the previous centuries, the emphasis is on moderation, light meals, and low carbohydrate intake. During the last half of the century, the prevalent idea was that overeating may cause larger babies and, as a consequence, more difficult labors. In an age in which maternal mortality was extremely high and cesarean sections a desperate alternative, limitation of fetal size was an understandable goal. This constitutes the background of the first scientific study on pregnancy and diet (Prochownik, 1901). Published in 1901, it reports the effect of a restricted diet throughout pregnancy on the birth weight of 48 full-term babies. The average weight of the males was 2906 g and the females 2735 g. The normal standards for this population were 3333 g and 3250 g for males and females, respectively.

In 1916, a classic study, apparently neglected by the following decades, demonstrated that maternal nutritional status has a profound influence on birth weight and the outcome of pregnancy (Smith, 1916). A group of women of similar height (5 ft 2 inches) from London and Dublin were divided into three nutrition groups according to their height/weight ratios. Then, based on dietary history, height/weight ratio, and physical examination, they were further divided in smaller groups. The assessment of the maternal nutritional status was considered to be more accurate in the smaller groups and categorized as poor, average, or good. Women considered to have poor nutritional status (either because of low height/weight ratio or low ratio combined with poor history and signs of undernutrition) had a higher incidence of stillbirth babies than those in the other groups. Similarly, they had a higher incidence of prematurity and of term babies weighing less than 6 lb (2.7 kg).

Accepting the assumption that weight gain reflected maternal nutritional status during pregnancy and that, in turn, nutritional status influenced fetal growth, a study was conducted in 1923 in 150 women (Davis, 1923). The population consisted of primiparae and multiparae in similar proportions, presumably all middle-class white. Maternal height was not controlled. It was found that birth weight increased with increasing pregnancy weight gain from 3144 g in those women gaining 7 kg to 3620 g in those gaining 13.6 kg. With few exceptions (Toombs, 1931) this observation has been corroborated by the majority of subsequent studies (Slemons and Fagan, 1927; Cummings, 1934; Kerr, 1943; Bailey and Kurland, 1945; Ademowore *et al.*, 1972).

One of the largest series in which the correlation between maternal body weight and birth weight was analyzed, included 6675 white and 5236 black women from Baltimore, Md. (Eastman and Jackson, 1968). The data show, once again, that an increase in weight gain during pregnancy is associated with a parallel increase in birth weight and a progressive decrease in the number of infants under 2501 g. Increased prepregnancy weight is also associated with increased birth weight and reduced incidence of low-birth-weight babies.

Weight gain and prepregnancy weight were found to act independently of each other. When they varied in the same direction, their effects were additive, whereas change in opposite directions tended to neutralize their separate effects. Thus, the largest babies were found when both prepregnancy weight and weight gain were high, smallest infants when both were low, and average-weight infants when one was high and the other was low (Table III). Further, based on the high incidence of low-birth-weight babies in women gaining less than 11 lb and the high rate of neonatal mortality of these infants, it was proposed that any woman who at week 20 of pregnancy has not gained at least 10 lb be regarded as "high-risk" case.

When maternal body weight is correlated with birth weight, there are two uncontrolled variables involved in the correlation: maternal caloric intake and maternal height. Although extreme variations in body weight reflect overall caloric intake, maternal height is probably the most important determinant of total body weight.

The specific influence on birth weight of maternal body size and caloric intake was investigated in a study, carried out in Aberdeen, Scotland, on 489 primigravidas. This study found that caloric intake was less among women of the lower than the upper social classes. It was also less among short women than among tall women, and this relationship was maintained regardless of whether the mother was underweight or overweight. Finally, caloric intake was directly related to birth weight. This correlation disappeared, however, when women of the same height were compared. These results were interpreted as indicating that maternal body weight or "body size" was the antecedent factor. Thus, larger and taller women would require more calories and, independently, deliver larger babies (Thomson, 1958, 1959a,b).

The conclusions reached in the Aberdeen study contrast with a more recent multivariate analysis of birth weight conducted in New York City in a low-socioeconomic black population (Rush *et al.*, 1972). This study shows that maternal weight gain during pregnancy correlates with caloric intake and birth weight independent of "maternal size." The correlation was interpreted as indicating that caloric intake determines fetal size and birth weight.

The discrepancies with the Aberdeen data were explained by the fact that in the Scottish study, maternal size was calculated based on the midpregnancy

Table III. Joint Relationship of Prepregnancy Weight and Maternal Weight Gain during Pregnancy with Birth Weight in Singleton, Term Deliveries in a Group of White Women[a]

Prepregnancy weight (kg)	Weight gain (kg)	Mean birth weight (g)	Percentage of low-birth-weight babies
72.7	13.6	3831	0.0
72.7	5.0	3628	2.3
54.5	13.6	3453	1.5
54.5	5.0	3044	5.8

[a] Adapted from Eastman and Jackson (1968).

weight of the subjects. It is known that a great part of the maternal pregnancy weight gain takes place between weeks 10 and 30 of gestation. Thus, weight at 20 weeks may have been influenced by recent caloric intake; therefore, it was not an antecedent factor but an intervening variable.

The influence of the maternal nutritional status before pregnancy on birth weight has frequently been overlooked. Several studies have been done in which the nutritional intake during pregnancy is the only maternal variable that has been correlated with birth weight. The majority of these studies, carried out in middle-class American women, have failed to show any correlation between these two variables and have, erroneously, concluded that maternal nutritional status is not an important determinant of birth weight (Williams and Fralin, 1942; Speert *et al.,* 1951). The inadequacy of this approach is further demonstrated by longitudinal studies done on monthly food intake of pregnant women, showing that, either because of medical advice or by still-unknown biological mechanisms, overweight women tend to decrease their food intake while underweight women tend to increase it (Beal, 1971). Since, as previously discussed, overweight women are likely to have babies of normal weight even when on a restricted diet, the lack of correlation between caloric intake and birth weight seems predictable.

Similar studies done in women from low-socioeconomic groups, in which the pregnancy nutritional status was probably inadequate, shows that there is a highly significant correlation between food intake and birth weight (Burke *et al.,* 1943). These results agree with the finding that in underweight women an adequate gain has a positive effect on birth weight.

2.2. Food-Supplementation Studies

Starting over four decades ago, several attempts have been made to modify the outcome of pregnancy by nutrition interventions. Unfortunately, as pointed out in a review of this topic (Bergner and Susser, 1970), most of these studies are plagued by flaws in their scientific design. Some of these flaws are serious enough to make the validity of the results questionable. Among the various methodological problems, the size of the sample, the assessment of the maternal nutritional status, and the type and amount of supplement given are probably the most critical. In some early studies the type of supplement was apparently decided on an empirical basis. In others there are more scientific approaches, based on nutritional concepts prevailing at that time, which are questionable in light of current knowledge.

The first supplementation study to be conducted on a large scale was carried out in clinics of the London area between 1938 and 1939 (People's League of Health, 1942). The aim of the study was to determine whether supplementation of the diet with vitamins and minerals would improve the course of pregnancy and labor and benefit the newborn child. The choice of the supplement was based on the results of a preliminary survey, carried out among 1000 women, indicating that inadequate intake of vitamins and minerals was the main nutritional problem. A total of 5022 women were included in the study and no significant differences were found between the supplemented and unsupplemented groups in terms of birth weight or incidence of prematurity.

In 1941, a food supplementation study was done in Montreal, Canada, on patients attending a prenatal clinic (Ebbs *et al.*, 1941). Based on their dietary histories, women were divided into three groups. One group taking a "poor diet" was used as control and given a placebo.

A second group, also taking a "poor diet," was given dietary counseling and food supplements plus minerals and vitamins. A third group, found to have an acceptable, although not optimal diet, was given only dietary advice. No difference in the birth weight of the babies born in the different groups was found, although the control group had a higher rate of prematurity. This study is a good example of the type of methodological problems mentioned before: nutritional status of the mother is classified solely on the information provided by one dietary history, noncompliance is not adequately controlled, and the size of the sample is probably too small to expect a statistically significant effect. However, the basic problem is the inadequate criteria to assess maternal nutritional status before and after food supplementation. Since the nutritional status of the mother is the intervening variable in birth weight, the improvement of maternal nutritional status is the only criterion that can reveal if the supplement had any effect.

The type and quantity of the supplement provided is obviously crucial. For example, it is conceivable that the food supplementation study done in London may have focused on the wrong nutrients. Although minerals and vitamins appeared to be the nutrients most often deficient in the average diet of the London population, current knowledge suggests that energy and protein intake are the major factors influencing pregnancy outcome. Other food supplementation studies, carried out in Scotland (Cameron and Graham, 1944), Chicago (Dieckmann *et al.*, 1944), and Philadelphia (Tompkins *et al.*, 1955) between 1943 and 1953, used either minerals and vitamins or protein concentrate, and the results did not indicate any major effect on the outcome of pregnancy in the treated groups. The study done in Philadelphia is probably the most carefully designed and conducted (Tompkins *et al.*, 1955). Patients registering for prenatal care before or at week 16 of pregnancy and without any previous or present serious illness were assigned to four groups kept equal in respect to age, parity, and race. One group received a protein concentrate as a supplement, a second group a polyvitamin capsule, a third group both the protein and vitamin, and a fourth group only the dietetic counseling given to all groups. The incidence of babies weighing less than 5.5 lb (2475 g) at birth was similar in all groups. However, when babies born after 39 weeks of gestation and weighing between 5.1 and 5.5 lb were excluded from the premature category and included in the analysis, the difference between the protein/vitamin-supplemented group and the controls became statistically significant.

More recent and, generally, better designed studies have failed totally to clarify this issue. Some have demonstrated a clear beneficial effect of food or nutrient supplementation, while other have been less conclusive. The first type of study is exemplified by a food supplementation program in Montreal, Canada, and by studies in India and Guatemala.

The Montreal study (Higgins *et al.*, 1973) was carried out between 1963 and 1971 on low-income women, many of them unmarried, attending the clinics of an obstetric hospital. Based on dietary history, women were provided with food supplements that increased their average daily caloric intake from 2251 kcal to 2782 kcal and protein intake from 68 g to 100 g. Active counseling, including at least one home visit, was also provided. The results obtained in these women were compared with those from women of similar background provided only with dietetic counseling and with those of private patients. In the food-supplement group the percentage of premature birth was 6.7%, similar to that of private patients (6.5%) and lower than nonsupplemented women from a similar socioeconomic background (9.0%).

More significantly, prematurity and perinatal mortality rates were also significantly lower in the group that received food supplements and counseling than in women who did not receive services.

The average length of the dietary services was 18.7 weeks, and duration influenced birth weight. Thus, average birth weight in mothers receiving counseling and food supplements for less than 15 weeks was 3178 g, whereas in those receiving it for more than 28 weeks it was 3438 g. After birth weight was adjusted for other variables, it was still found to be higher in the latter group (3329 g versus 3254 g).

In the studies carried out in India (Iyengar, 1967), a group of 13 women from a low socioeconomic background were admitted to the hospital approximately 4 weeks before term and given a daily diet containing 2100 kcal and 60 g of protein. The mean birth weight in these women was 3028 g. A similar birth weight (3028 g) was found in 12 women from a comparable background also admitted into the hospital 4 weeks before term and given the same diet plus 100 g of skimmed milk daily. In contrast, the mean birth weight in a group of 26 women not admitted was 2704 g. Besides receiving a better diet, the hospitalized women had the opportunity to rest for longer periods of time. This may have helped to increase the effect of the diet because of the saving of energy otherwise consumed in physical activity.

The Guatemalan study (Lechtig *et al.*, 1975) was conducted in four rural villages where there was a very low average income and poor sanitary conditions. These populations had very high rates of perinatal and infant mortality, high prevalence of infantile malnutrition, and a high incidence of intrauterine infections. The average diet consisted largely of corn and beans, with animal protein forming 12% of the total protein intake. The mean daily intake of calories and protein during pregnancy was about 1500 kcal and 40 g, respectively. These women were probably affected by chronic malnutrition during childhood, as reflected by an average height of 149 cm. The mean maternal weight at the end of the first trimester was 49 kg. In a normal population the average weight gain at this state of pregnancy is approximately 1.5 kg; thus the prepregnancy weight/height ratio of the population was probably adequate. The mean weight gain during gestation, however, was only 7 kg. Two types of supplements were distributed: one containing protein and calories and the other containing only calories. The caloric density of the latter, however, was approximately one-third that of the protein–calorie supplement. Attendance at the supplementation center was voluntary, and this resulted in a wide range of supplement intake during pregnancy. A total of 405 cases were included in the study. In both supplemented groups, birth weight showed a consistent association with the number of calories received during pregnancy. No apparent effect was determined by the presence of protein. The mean birth weight in the women supplemented with more than 20,000 kcal was 3105 g, whereas in those that received less than 20,000 kcal it was 2994 g.

More significantly, the proportion of low-birth-weight babies in the high-supplement group was only 9%, compared with 19% in the low-supplement group (Fig. 13). The relationship between caloric supplementation and birth weight was not due to maternal height, parity, or gestational age.

In contrast with the preceding results, two recent nutrient-supplementation studies conducted in the United States have failed to show conclusive effects of the intervention on birth weight. The subjects of both studies were women from low socioeconomic groups, mostly black, from New York and Philadelphia. The supplements used were protein, vitamins, and minerals. In the study carried out in Philadelphia (Osofsky, 1975), a group of 118 low-income women without history of past or present serious illness and registered in a prenatal clinic before the week 28 of pregnancy were compared with 122 similar individuals provided with a protein–mineral supplement. Both groups were studied sequentially. Nutritional assessments were carried out up to four

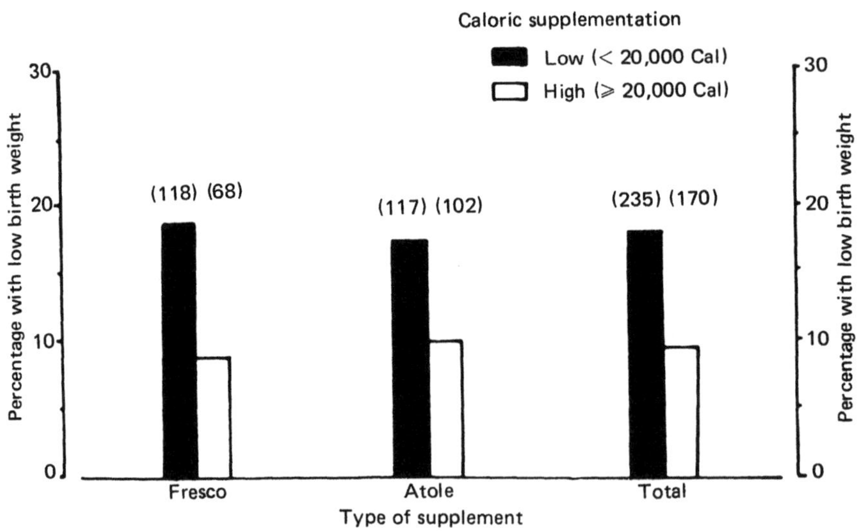

Fig. 13. Relationship between supplemented calories during pregnancy and proportion of low-birth-weight (2500 g) babies. Numbers in parentheses indicate number of cases. From Lechtig *et al.* (1975).

times in some subjects and various maternal parameters were monitored. After birth, infants were measured and their gestational age carefully assessed.

At the beginning of the study, the average weight of the subjects was 131 lb, and their daily caloric and protein intakes were 1,918 kcal and 71.3 g, respectively. The supplemented group, on the average, ate only half of the recommended amount of supplement. Still their intake was significantly higher in protein, iron, and calcium. No major significant difference in the outcome of pregnancy were found. However, both groups gained the same amount of weight (27 lb). This suggests a similar cumulative intake in controls and supplemented women.

In the study carried out in New York [Susser, 1977 (personal communication)], the population was selected from black women attending a prenatal clinic. Patients enrolled had less than 30 weeks of gestation and met any of the following criteria: protein intake less than 50 g in the preceding 24 hr, prepregnant weight less than 110 lb, and/or low weight gain and prior low-birth-weight infant. Women were assigned to a supplement group (receiving two 8-oz cans of a beverage containing 40 g of animal protein, 470 cal, and minerals and vitamins) a complement group (also receiving 8-oz cans of beverage containing 6 g of animal protein, 322 cal, and the same amount of minerals and vitamins), or a control group receiving regular clinical care (including mineral and vitamin supplements). Approximately 1000 women, evenly distributed among the three groups, were included. No significant differences in birth weight, nor in other fetal dimensions, were found among the three treatment groups.

All available explanations of the absence of the effect in the New York City study have been tested, including distribution of confounding factors that might have been present despite the randomization of the design, the degree to which supplementation actually took place, and the degree to which supplementation might have been substituted for the regular diet. Furthermore, a more detailed analysis demonstrated an excess of premature deliveries, prior to 35 weeks of gestation, among the women in the high-protein supplement group and a consequent proportional excess of neonatal mortality in that group. Even more striking were the findings that over and above the excess of premature deliveries, the high-protein supplement led to fetal growth retardation prior to 35 weeks of gestation.

The effect of the high-protein supplementation was not adverse throughout gestation. When only term pregnancies were compared, there was some indication of increased birth weight. The difference, however, was not statistically significant.

When results were analyzed according to different risk conditions, significant increases in birth weight due to the supplement were found among women with low-pregnancy weight and among smokers. Nutritional supplementation produced no effect among nonsmokers and women with a prepregnancy weight of more than 110 lb.

Both U.S. studies were well designed and adequately controlled. The only possible explanation for the negative results obtained is that in spite of low socioeconomic conditions, the majority of the women had an adequate nutritional status before the intervention. This conclusion seems to conflict with the dietary histories, which show an average caloric intake in the Philadelphia group 300 cal lower than the recommended dietary allowance and a low protein intake in the New York test groups. However, the weight gain reported in the Philadelphia study was optimum, suggesting that in spite of apparent deficiency, overall caloric intake during pregnancy was adequate.

The conclusion that can be drawn from these supplementation studies, is that, while in the low socioeconomic groups of developing countries the majority of women suffer some degree of undernutrition both before and during pregnancy, only a minority of women from analogous socioeconomic strata are undernourished in the more industrialized countries. The data, rather than being used as a case against dietary counseling and food supplementation during pregnancy, should be interpreted as demonstrating a need to refine evaluation processes and to concentrate available resources on the group that is underweight and fails to gain weight properly during pregnancy or that shows other evidence of nutritional abnormalities.

2.3. *Effect of Famine*

World Wars I and II created situations of inadequate nutrition or famine in different populations for different lengths of time. Such tragic conditions made it possible to study the effects of undernutrition, associated with a stressful situation, on birth weight. The first observations, published in the German literature between 1915 and 1918, described no significant changes in birth weight or body length in areas of Germany and the Austrian Empire afflicted by the food shortage caused by the blockade (Hytten and Leitch, 1971).

In contrast, the more reliable information collected during World War II does show an effect of war conditions and famine on pregnancy outcome. Birth weights recorded in Wuppertal, Germany, for the period 1937–1948 reveal an abrupt fall to 185 g below the prewar levels in 1945, a year of acute food shortage. After the abrupt fall there was a steady rise, coincidental with improved living conditions, reaching prewar levels in 1948 (Bergner and Susser, 1970).

The effects of famine on pregnancy have been demonstrated by data collected during the Dutch famine (Stein *et al.*, 1975) and the siege of Leningrad (Antonov, 1947).

During the winter months of 1944–1945, as a result of a food embargo, the population of western Holland had its average daily ration suddenly reduced from 1800 cal to 600 cal. The period of famine lasted approximately 28 weeks, and several women were affected during the last two trimesters of pregnancy. In these women, mean birth weight fell 327 g, or 9%, compared with the prefamine mean value (Fig. 14). Besides affecting fetal growth, famine also had a distinct effect on mortality rates, as indicated by a higher incidence of stillbirth and neonatal mortality in the affected groups. Although the in-

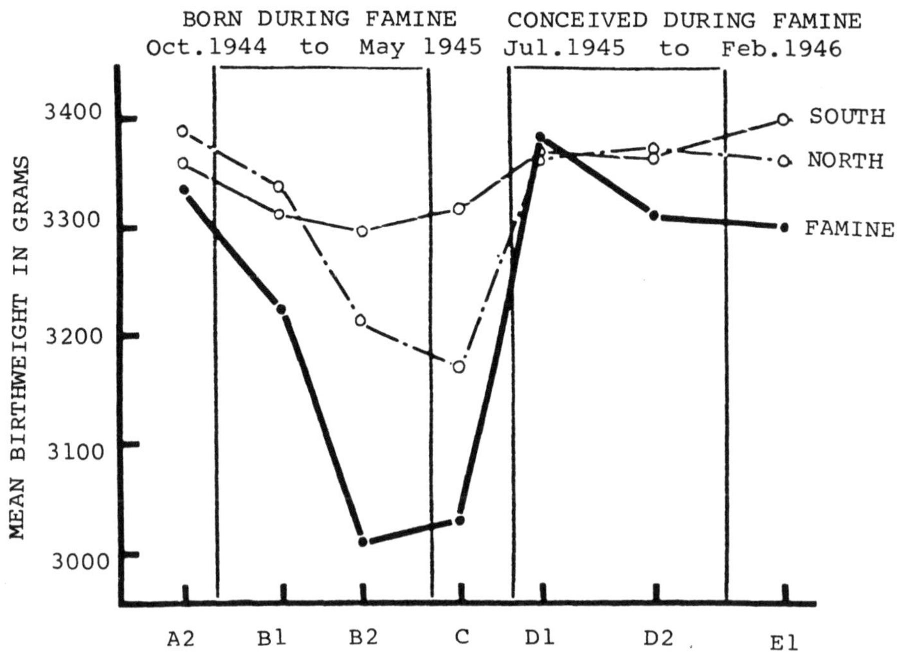

Fig. 14. Effect of famine on mean birth weight. South and North represent control areas. Group A2 was conceived and born before the famine. Group B1 was exposed to the famine for the third trimester. Group B2 was exposed for the second as well as the third trimesters. Group C was exposed during the middle 6 months. Group D1 was exposed to famine during the first and second trimesters of gestation, and D2 was exposed only during the first. E1 group was conceived and born after the famine. From Stein *et al.* (1975).

creased number of stillbirths was clearly associated with exposure to famine during the first trimester of pregnancy, it was not possible to separate the effect of prenatal famine exposure on neonatal mortality from the effect of other general wartime factors that occurred during the siege of Leningrad, in which famine conditions persisted for almost 6 months, the average fall in birth weight was 529 g for boys and 542 g for girls. No adequate information on the mortality rates of these babies is available for comparison with the Dutch situation.

2.4. Maternal–Fetal Exchange and Maternal Nutritional Status

The evidence discussed in preceding sections suggests a connection between the quantity of maternal stores, mainly fat, at a given time and the rate of fetal growth. Thus, obese women with an excess of stores are apparently able to sustain a normal rate of fetal growth even when their food intake during pregnancy is very low and they are losing weight. In contrast, underweight women, whose nutrient stores are presumably small, are unable to maintain a normal rate of fetal growth unless they increase their food intake appreciably. These observations seem to suggest that a restricted maternal diet during pregnancy would affect the fetus only when a critical level of maternal stores has been reached. An analysis of the data from the Dutch famine suggest that fetal growth retardation occurs when the mother, after consuming or failing to deposit the pregnancy stores, begins to use her own prepregnancy body stores. Although no information in this respect has been provided, it seems reasonable to assume that after several months of food rationing, the prepregnancy weight of the average Dutch women in 1944–1945 was around 55 kg. During the prefamine period, and after several months of food rationing, the postpartum maternal weight 9–10 days after delivery was 59 kg. Assuming that, after a loss of at least 1 liter of extra fluid accumulated during pregnancy, the mother still has 3 kg of extra weight due to uterus, mammary glands, expanded blood volume, and tissue fluid, the amount of maternal stores present at the end of pregnancy in the average Dutch woman was approximately 1 kg. The mean birth weight of babies born from this group of women was still normal (3338 g). As shown in Table IV, however, the group exposed to famine during the third trimester had a probable loss of prepregnancy stores of approximately 0.4 kg and a concomitant drop in birth weight. The phenomenon became more evident in the group exposed to famine during the second and third trimesters. The net loss of prepregnancy stores in these mothers was approximately 1.5 kg (approximately 3% of the initial body weight) and the fetal weight 3011 g (9.6% lower than prefamine levels). The women that conceived their babies after the famine, when food supplies were again adequate, had normal or perhaps slightly above normal stores, and the fetal weight was again in the usual range of 3300 g.

The data shown in Table IV, demonstrating that a significant degree of fetal growth retardation can be produced under conditions in which the mother is losing only 1.5 kg of her prepregnancy weight, are striking. Any adult woman can lose by starvation 10–20% of her body weight without endangering her life. Why during pregnancy is the mother not making those extra stores available to the fetus? Obviously in this situation fetal growth retardation is not caused by maternal depletion but rather by lack of access to the maternal

Table IV. Effect of Severe Maternal Undernutrition at Different Gestational Ages on Maternal Postpartum Weight, Maternal Stores, and Mean Birth Weight[a]

	Postpartum weight (kg)	Maternal pregnancy stores[b] (kg)	Mean birth weight (g)
Prefamine	59	+1.0	3338
Famine period			
3rd trimester	57.6	−0.4	3220
2nd and 3rd trimesters	56.5	−1.5	3011
1st and 2nd trimesters	61.0	+3.0	3370
1st trimester	61.5	+3.5	3312
Postfamine	62.0	+4.0	3308

[a] Adapted from Stein *et al.* (1975). Some figures taken from graphs.
[b] Maternal stores calculated by subtracting to the postpartum weight, 3 kg (reflecting other maternal components, such as enlarged uterus, mammary glands, etc.) and assuming a pregavid weight of 55 kg.

prepregnancy stores. This possibility contrasts with current ideas that the division of nutrients between the mother and the conceptus is mainly regulated by metabolic needs. This hypothesis has been directly tested only once, however (Frazer and Huggett, 1970). In this study, a group of pregnant rats were either underfed by an overall reduction in food intake or "overfed" by growth hormone administration along with a standard diet fed *ad libitum*. The results showed that changes in maternal body weight were proportional to the weight of the conceptus and to the levels of nutrient intake. These findings were interpreted to indicate that a certain minimum amount of nutrients is always supplied to the fetus and that any intake above this is distributed so that three-fourths goes to the conceptus and the remaining one-fourth goes to the mother.

Data from this study show, however, that the body weight of rats fed 7 g of diet per day, the most restricted group, was similar to prepregnancy levels. In contrast, the weight of the conceptus in these animals was markedly lower than controls. These results would indicate that if the fetus is indeed able to compete with the mother for available nutrients, the competition is restricted to those nutrients ingested in excess and stored during pregnancy and does not include those present in maternal tissues at conception.

A review of the data available in the literature demonstrates that in various mammalian species in which maternal malnutrition has been induced during pregnancy the loss of maternal body weight tends to be proportionally less than the reduction in fetal body weight (Rosso, 1977a). In the rat, a 50% reduction in dietary intake throughout pregnancy produces a 10% loss of maternal body weight and a 25% reduction of birth weight. When diets are 75% restricted, maternal losses in body weight range between 26 and 36%, compared with initial body weight, while birth weight is reduced by more than 50% (Berg, 1965). Also in the rat, a 4% casein diet throughout the entire pregnancy causes a 20% loss in the initial body weight of the mother and a 40% loss in the average weight of the fetus (Hastings-Roberts and Zeman, 1977).

In the guinea pig, either a caloric or a protein restriction during the last half of pregnancy produces a significant increase in fetal mortality and approximately a 25% reduction in birth weight with no significant changes in maternal body weight (Young and Widdowson, 1975). In the pig, a severe protein deficiency maintained throughout pregnancy causes a 17% loss of maternal body weight, compared with prepregnancy weight, and a 33% deficit in birth weight (Pond *et al.*, 1969). There is evidence that in certain subhuman primates the same phenomenon of maternal sparing occurs. For example, the mean body weight of rhesus monkeys fed throughout gestation a diet containing only 25% as much protein as a control diet drops 0.4 kg compared with prepregnancy values (Kohrs *et al.*, 1976). This amounts to approximately a 7% loss of prepregnancy body weight. The mean birth weight of the neonates, however, was reduced approximately 15%. Such a level of restriction produces some maternal deaths and a very high incidence of fetal losses due to abortion, stillbirth, and perinatal mortality. In other experiments, also in rhesus monkeys (Riopelle *et al.*, 1975) no significant differences in birth weight were found in groups fed 1, 2, or 4 g of protein/kg/day throughout most of pregnancy, although average weights tended to be lower in the low-protein groups. The total number of cases analyzed (approximately 15 animals in each group) is too small, however, to draw any definitive conclusion on this respect. Significantly, the incidence of abortions, prematures, and stillbirths was higher in the protein-restricted animals. Maternal body weight after delivery was 18% higher than prepregnancy weight in the control group (4 g of protein/kg), 12% higher in the group consuming 2 g/kg, and 1% lower in the most restricted group. Thus, as in the human situation, no significant growth retardation seems to occur in the fetal monkey unless the maternal prepregnancy stores are significantly affected. The proportional effect of malnutrition on maternal and fetal weight in various mammals is shown on Table V.

Only a limited amount of information is available on changes in maternal body composition during dietary restriction. In a study done in rats fed different levels of protein, it was found that a diet containing approximately 11% casein produces a 5% loss in protein content of the liver at term (Naismith, 1969). In these animals there was a 15% reduction in protein content in the fetus, consistent with the idea that the mother is proportionally spared by malnutrition. However, this difference was found to be statistically nonsignificant.

More recent studies on body composition of rats fed a 5% casein diet during pregnancy have shown that, compared with nonpregnant animals fed a similar diet, the percentage of protein in the carcass does not change significantly in these animals, but the percentage of body fat is higher (Morgan, 1975). Since, in animals fed a 5% casein diet, birth weight is reduced approximately 25%, it seems unquestionable that the fetus is disproportionally more affected than the mother. Similarly, a 50% dietary restriction in the rat produces loss of fat and lean dry matter similar to nonpregnant pair-fed rats (Lederman and Rosso, 1978). In these animals individual fetal weight was reduced by 20%. Given the body composition of the newborn rats, such a loss may represent 30% or more of the expected dry weight at term. Thus, even in extreme conditions, the fetal rat is not able to induce extra mobilization of maternal stores.

The information discussed above suggests that in several mammalian species the pregnant mother is able to compartmentalize available nutrients. The mechanisms for this effect would operate through an inadequate maternal adaptation to pregnancy and, ultimately, by a reduced maternal supply line. The successful completion of each of the various maternal adaptive changes requires an adequate intake of nutrients and/or an adequate nutritional status. For example, if the mother is not adequately nourished, she may not expand the blood volume and the uterine circulation to the needed limits. As a consequence, placental growth and fetal growth will be impaired. This hypothesis is supported by experimental data showing a direct correlation in pregnant rats between maternal body weight and plasma volume. In rats fed a 6% casein diet or a restricted diet both maternal body weight and plasma volume are significantly reduced at term when compared with a well-fed control group

Table V. Effect of Malnutrition during Pregnancy on Maternal and Fetal Body Weight in Some Mammalian Species[a]

Species	Maternal weight (g)			Newborn weight (g)			Source
	Prepregnancy	Postpartum	Postpartum after malnutrition	Normal	After malnutrition	Type of restriction	
Rat	215	+28 g (+13%)[b]	−22 g (−10%)	3.88[b]	−0.45 g (−12%)	50% dietary restriction	Berg (1965)
	215	+28 g (+13%)[b]	−68 g (−31.6%)	3.88[b]	−1.99 g (−51%)	75% dietary restriction	Berg (1965)
	240	+40 g (+16.6%)	−18 g (−7.5%)	5.50	−1.0 g (−19%)	6% casein diet	Rosso, P. (unpublished)
Guinea pig	—	900	−93 g (−10%)	78.5	−21 g (−28%)	Low protein (30% of control)	Young and Widdowson (1975)
	—	900	−105 g (−11.6%)	78.5	−17 g (−22%)	60% dietary restriction	Young and Widdowson (1975)
Pig	128,000	+21,000 (+16.4%)	−21,000 (−16%)	1120	−370 g (−33%)	Low-protein diet	Pond *et al.* (1969)
Macaca mulatta	5800[c]	+1400 g (+24%)	−400 g (−6.9%)	450	−66 g (−14.6%)	Low-protein diet	Kohrs *et al.* (1976)
	—	+1000 g	— (−1.0%)	492[c]	−52 g (−10.6%)[d]	Low-protein diet	Riopelle *et al.* (1975)

[a] Values represent averages.
[b] Day 20 of gestation.
[c] Data taken from a graph.
[d] Difference nonsignificant.

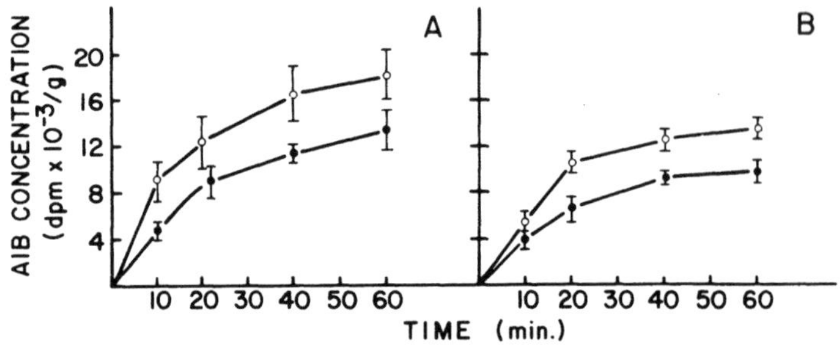

Fig. 15. α-Amino isobutyric acid (AIB) concentration in rat fetuses after injection of labeled AIB in control (○) and protein-restricted (•) mothers at days 20 (A) and 21 (B) of gestation. From Rosso (1977b).

(Rosso, 1977a). More direct evidence that maternal fetal exchange of nutrients is altered by malnutrition has been provided by studies in which transfer of α-amino isobutyric acid (AIB) and glucose were compared in control and protein-restricted rats. Rats fed a 6% or a 27% casein diet were compared (Rosso 1977b,c). In these experiments radioactively labeled AIB was injected in the maternal circulation at days 20 and 21 of gestation. Samples of maternal blood and placentas and fetuses were removed at various time intervals. It was found that, in malnourished animals, AIB had a lower rate of disappearance from the maternal plasma, stayed longer in the placenta, and was transported in reduced amounts to the fetus (Fig. 15). Maternal transfer of glucose and of α-methyl-D-glucopyranoside (AMG), a nonmetabolizable glucose analog, was also significantly decreased. In contrast with results obtained in rats, placental transfer studies done in protein- or protein/calorie-restricted guinea pigs have shown that the amount of AIB transported into the fetus per gram of fetal weight was similar in controls and protein/calorie-restricted animals and increased in the protein-restricted ones. A possible interpretation of the apparent discrepancy between these two studies is that the guinea pig is more sensitive than the rat to nutrient deprivation. Thus, shortly after the onset of maternal malnutrition the nutrient supply to the fetus would be drastically reduced. The finding that the restricted guinea pigs had a 50% fetal mortality supports such an assumption. After fetal mass is so drastically reduced, transfer would continue at a reduced rate proportional to fetal weight or, for unknown reasons, even relatively higher near term in the protein-restricted animals.

The reduced maternal-fetal transfer of nutrients in the rat may be caused by a reduced blood perfusion or a reduced placental transfer or both. While the rate of blood perfusion in malnourished animals or humans has not been measured yet, there is evidence that the placenta may be affected. Rats fed a low-protein diet from day 6 of pregnancy may have smaller placentas at term, with a 25% lower protein and DNA content than placentas from a control group. In contrast, RNA content in these placentas has been found to be increased at day 13 of gestation (Winick, 1969) and reduced at term (Rosso *et al.*, 1976). Studies of the morphologic characteristics of the placentas of women from low socioeconomic groups, presumably malnourished, have indicated that these placentas are smaller and have reduced villous surface and reduced number of cotyledons (Murthy *et al.*, 1976; Laga *et al.*, 1972a). Placental DNA content has been found to be significantly reduced in a population of poor women (Dayton *et al.*, 1969). Other studies done in similar populations have also found lower average values for DNA content compared with placentas delivered by women that presumably are better nourished. However, the difference between groups was not statistically significant (Winick, 1969; Laga *et al.*, 1972b). The apparent discrepancies between those studies may be explained by factors such as sample size and differences in the nutritional status of the groups being compared. Total protein content has also been reported to be slightly but not significantly decreased in placentas from a malnourished population (Laga *et al.*, 1972b). These results suggest that a moderate degree of maternal undernutrition, probably a caloric deficit, produces only a minor interference with proliferative and hypertrophic phases of placental growth. Some parameters of RNA metabolism, however, seem to be more affected by a deficit of nutrients than DNA and protein content (Laga *et al.*, 1972b). For example, the polysome/monosome ratio has been found to be 50% lower in placentas from malnourished women. Polysomes consist of a strand of messenger RNA with ribosomes attached and are the basic units for protein synthesis. Two types of polysomes exist in the cell, those attached to the secretory membrane of the endoplasmic reticulum, or "bound polysomes," and those not associated to such structure, or "free polysomes." For both populations of women the "bound polysomes" averaged 21% of the total number of ribosomes. This finding suggests a similar capacity to synthesize export proteins, conceivably peptide hormones, in the low socioeconomic, presumably undernourished group. Further, in spite of the higher percentage of polysomal disaggregation, suggesting a reduced organ capacity to synthesize protein, cell-free protein synthesis per milligram of ribosomal RNA was similar in the malnourished and well-nourished women. Thus, when total capacity of the placenta for protein synthesis, obtained by multiplying *in vitro* amino acid transportation per milligram of RNA content, was compared, it was found to be similar in both populations. This finding would suggest that in spite of the reduced polysome/monosome ratio, maternal malnutrition would not reduce the overall metabolic efficiency of the organ. Further evidence of abnormal RNA metabolism has been provided by studies demonstrating elevated alkaline ribonuclease activity (RNase) in the placentas of malnourished women when compared with a well-nourished population (Velasco *et al.*, 1975). The cellular role of this enzyme is still poorly understood. High levels of RNase are usually associated with an increased rate of RNA turnover (Girija *et al.*, 1965; Quinn-Stricker and Mandel, 1968). A low-protein diet or dispensary restriction has been shown to cause elevation of RNase in liver and brain in the growing rat (Rosso, 1975). A protein-calorie supplementation during pregnancy has been found to reduce placental levels of RNase activity (Lechtig *et al.*, 1977), suggesting that nutritional status may influence placental metabolism. Some of the biochemical changes induced by malnutrition in the placenta are listed in Table VI.

Table VI. Placental Changes Reported in Women Presumably Affected by Some Degree of Malnutrition

Morphological changes
1. Reduced weight
2. Reduced number of cotyledons
3. Reduced villous surface
Biochemical changes
1. Reduced DNA content
2. Lower polysome/monosome ratio
3. Increased alkaline ribonuclease activity

The picture of maternal-fetal exchange that emerges from the data discussed here is far more complicated than the current idea that the main factor of maternal fetal exchange is the metabolic rate of maternal and fetal tissue. In fact, contrary to the idea of fetal parasitism, there seem to be feedback mechanisms operating in the mother that reduce the maternal supply line to the fetus when nutrients are in short supply.

From a teleological point of view, the possibility that the mother, rather than the fetus, is proportionally spared during malnutrition seems to be a more plausible situation than the traditional one based on the "fetal parasite" concept. In the mammalian species, pregnancy is only the first stage of the reproductive cycle. Lactation, beginning shortly after delivery, is in most species equally essential for the survival of the young and more demanding, from a nutritional point of view, on the mother. Thus, it would seem a rather inadequate evolutionary adaptation if after a shortage of food a depleted mother would deliver a normal baby that she is unable to care for. However, if a moderately malnourished mother must care for a runt, chances are that the mother will recover when food becomes available and conceive again. For the survival of species, this seems to be a better solution.

2.5. Postnatal Consequences of Maternal Malnutrition

Studies on perinatal mortality have shown that the death rate is lowest in babies whose birth weight is 3500–4000 g and that mortality increases progressively as birth weight increases or decreases. Perinatal mortality in babies heavier than 4000 g is generally associated with complications arising from more difficult deliveries. Mortality in lighter babies reflects a wider spectrum of postdelivery problems, with those babies weighing less than 2500 g accounting for the highest percentage of deaths.

The effect of maternal malnutrition in a single case may not be apparent. If a baby weighs 3100 g instead of 3300 g, he will be considered normal, and probably his perinatal period and subsequent growth will also be normal. Statistically, however, the 200-g loss in body weight might double the risk of perinatal mortality (Bergner and Susser, 1970). This higher risk increases disproportionally in smaller babies.

When a high percentage of a population is receiving substandard diets before and during pregnancy the average birth weight decreases. Compared with a better nourished population, this will determine a shift to the left of a normal distribution curve, with a resulting higher percentage of babies weighing less than 2500 g (Mata *et al.*, 1975; Rosso and Luke, 1978). This excess of low-birth-weight babies will determine a disproportionate increase in the perinatal mortality rate. This problem is evident when well-nourished and poorly nourished populations are compared (WHO, 1965). As previously discussed, socioeconomic differences imply also differences in other maternal parameters that can influence birth weight, maternal height, for example. However, as demonstrated by the Guatemalan experience, an improvement in maternal nutrition status induces a dramatic drop in the percentage of low-birth-weight babies (Lechtig *et al.*, 1975).

In spite of its obvious implications, the relationship between maternal nutrition and perinatal mortality is a relatively new concept. The traditional focus has been on the possibility that maternal malnutrition may induce permanent negative changes. Such negative changes may not be necessarily associated with fetal growth retardation. For example, there is evidence that ketonemia during prolonged periods of gestation, for example obese women losing weight, may affect psychomotor development (Churchill and Berendes, 1969). Although these studies have been interpreted as an indication that ketone bodies per se may have a harmful effect, it is conceivable that other associated factors may also play a role. For example, besides ketonemia, a fasting woman is likely to have marked changes in the plasma concentration of amino acids and lower levels of glycemia.

The possible harmful effect of prenatal malnutrition on brain growth and subsequent mental capacity has been the focus of numerous studies. The concern stems from observation in experimental animals (mainly the rat) of a deficit in brain growth associated with maternal malnutrition and subsequent altered behavior (see the next section). In addition, human studies also indicate that postnatal undernutrition retards brain growth and induces a permanent, although to a large extent reversible, deficit in mental performance (Winick and Rosso, 1975).

The investigation of whether maternal malnutrition induces subsequent lasting effects on mental performance is an extremely difficult task because of the influence of postnatal factors. Women who are malnourished during pregnancy are generally poor and, in many cases, likely to offer their children an environment that is also emotionally and intellectually deprived. A deprived environment is known to have a negative effect on mental development.

Some of the postnatal environmental variables that can negatively affect development were not present in the majority of the Dutch population exposed to famine. This fact made the Dutch famine a unique source of information. Using data collected in the 18-year-old male population at the age of military induction, it was possible to establish that the subjects presumably exposed to famine were similar to those not exposed in terms of physical growth and mental capacity. Except for the higher neonatal mortality in this group, there were no indications that other factors may have created a selective process by which only the less affected individuals reached military induction (Stein *et al.*, 1975).

Unfortunately, there are no behavioral data on these individuals. Although the background data does not suggest the possibility of any serious behavioral problems, it does not rule out the possibility that famine may have induced changes in this area. As will be discussed later, the most striking differences between prenatally malnourished and control rats are behavioral ones and not performance.

Again, it must be emphasized that the implications of prenatal undernutrition in a baby born in the low socioeconomic group of a developing country probably differs from that of the Dutch situation. In Holland, a previously well-nourished population was acutely undernourished for a brief period of time and then allowed again to eat *ad libitum* and to reorganize for the surviving babies the stimulatory environment of a highly educated and technological society. In the low-income groups of the less industrialized nations, a baby born with a similar deficit in brain growth is likely to suffer a superimposed deficit caused by postnatal undernutrition and the negative influences of a deprived environment.

The only nutrition intervention study presently available in which the effect of nutrient supplementation on maternal nutritional status and the central nervous system were determined was carried out in a group of low-income black women in Philadelphia (Osofsky, 1975). The subjects were randomly assigned to two groups, one of which was provided with a protein-mineral supplement from the time of enrollment, before week 28 of gestation, until term. The babies were assessed on the third day of life using Brazelton Neonatal Behavioral Assessment measures. Two significant relationships were found with the nutrition intervention and none with the variables considered most important developmentally. As discussed before, however, it is unlikely that these women suffered from undernutrition during pregnancy, since both controls and supplemented groups had a similar rate of weight gain. Further, since tests were performed during the newborn period, their implications for later development are uncertain.

In contrast with the previous study, an observational prospective study, part of the "Collaborative Perinatal Study" of the National Institute of Neurological Diseases and Stroke, found an inverse correlation between the percentage of infants with abnormal physical and mental development at 1 year of age and maternal weight gain during pregnancy (Singer *et al.*, 1968). The population included in this study was largely urban and from a low socioeconomic strata. Although the differences between groups are highly significant, they cannot be accepted as proof that nutritional inadequacies during pregnancy are recognizable for the retarded physical and mental development observed. It is conceivable that women who failed to gain adequate weight were also less likely to provide after birth the environment conducive to optimal development of the infants. Thus, the lower performances may reflect postnatal rather than prenatal events.

As previously mentioned, maternal undernutrition is associated with a higher incidence of low-birth-weight infants. Many studies have shown that low birth weight is antecedent of reduced mental competence, and that the lower the birth weight, the greater the subsequent deficit in mental capacity (Benton, 1940; Drillien, 1970; Wiener, 1970). In babies with birth weight under 1500 g, it is conceivable that perinatal complications resulting in brain damage, such as hypoxia or hypoglycemia, play a role in their later lower performances. This possibility is supported by the fact that this group also has the highest incidence of cerebral palsy and other neurological sequelae. In babies with more moderate reductions in birth weight, the situation is less clear. Two groups of babies can be found in these categories, prematures and small-for-gestational age. The premature group is more susceptible to perinatal complications associated with subsequent neurological abnormalities, and therefore their subsequent lower mental capacity may reflect perinatal injuries. In contrast, the small-for-gestational-age babies are less afflicted by perinatal complications; thus their subsequent lower mental capacity may preferentially reflect prenatal events. A recent prospective (Fitzhardinge and Stevens, 1972) study has shown that cerebral palsy was uncommon in this group, but there was a high incidence of minimal brain dysfunction characterized by hyperactivity, a short attention span, poor fine coordination, and hyperreflexia. The average IQs were lower than those of the general population, and the school performance was poor.

From the conflicting body of evidence discussed here, it is hard to draw any conclusions about the effects of maternal malnutrition per se on subsequent mental capacity. It is hoped that future studies may help to resolve this important question.

The only significant difference so far discovered in the Dutch famine study between the prenatally affected populations and the nonaffected ones is a higher incidence of obesity in the male population (Ravelli *et al.,* 1976). No data are available for the female population. The effect was limited to those exposed to famine during the first trimester of pregnancy. It has been speculated that a reduced maternal plasma level of glucose may have had a lasting effect on the centers regulating food intake. There is no evidence, however, to substantiate such a hypothesis.

3. Nutrition and Growth of the Conceptus in Animal Models

3.1. Fetal Growth

Reduced caloric or protein intake throughout the entire pregnancy, or during the last half of pregnancy, reduces the rate of fetal growth and/or increases fetal mortality in all the eutherian mammals in which the phenomenon has been studied (Rosso, 1977a). However, most of the current information on the characteristics and postnatal consequences of fetal growth retardation associated with maternal malnutrition is based on studies conducted on the rat. In this species either a 50% overall reduction of food intake or a decrease in protein content of the diet to 5–6% from early pregnancy or throughout pregnancy causes a 20–30% reduction in birth weight (Chow and Lee, 1964; Zeman, 1967; Berg, 1965). Less severe restrictions of dietary protein, or limiting the protein restriction to the last week of pregnancy, causes a fall in birth weight of a smaller magnitude (Naismith, 1969, 1973). The growth-retarding effect of maternal protein restriction on the fetus is already noticeable at day 12 of pregnancy and becomes progressively more severe during the last week of pregnancy (Rosso, 1977a).

Since in the newborn rat the quantity of adipose tissue is very low compared with other species (Naismith, 1966), the reduced body weight can be equated to a proportional degree of growth retardation. Growth retardation is also reflected in reduced body length and reduced weight of every major organ, including brain, liver, kidney, spleen, and thymus (Zamenhoff *et al.*, 1968; Zeman, 1967; Kenney, 1969).

Although the organ-weight reduction is roughly proportional to the reduction of body weight, some organs tend to be less affected than others; among these, brain seems to be the organ least affected. In the rat, most studies show brain weight of prenatally malnourished pups to be 15–23% less than that of control animals (Zamenhoff *et al.*, 1968; Zeman, 1967). Such a percent reduction is usually similar to or less than the reported percent reduction in birth weight. Other organs whose weight is increased relative to body weight are lungs, heart, and thymus. In contrast, liver, kidney, and spleen weight is decreased relative to body weight (Zeman, 1967; Kenney, 1969). The differential effect of prenatal malnutrition on the weight of the organs can be explained by their different patterns of cellular growth (Winick and Noble, 1966b).

The smaller intrauterine malnourished pups seem to have a high rate of neonatal mortality compared with control pups. Such a high mortality is probably associated with a reduced metabolic capacity to withstand the stress of birth and the period that follows birth. It has been demonstrated in the rabbit that intrauterine malnourished pups have a reduced fat content and a markedly reduced concentration of glycogen in liver (Hafez *et al.*, 1967). These changes suggest an increased susceptibility of the malnourished pup to the complications derived from the combination of fasting and reduced environmental temperature in the period following birth.

No congenital abnormalities have been associated with either overall dietary restriction or protein restriction alone. However, a reduction in litter size has been reported, its magnitude depending on the intensity and the time of pregnancy at which the restriction is imposed. Most authors report no significant differences in litter size between restricted and nonrestricted animals when a 50% overall restriction or a low-casein diet is used after day 5 of pregnancy. Restriction imposed earlier, however, almost invariably reduces litter size, especially when a protein-free diet, is being used. The mechanism for such an effect is not well understood. However, it is conceivable that it may have hormonal mediation, since injection of estrogen and progesterone reduces the effect (Hazelwood and Nelson, 1965).

In the rat, prenatal growth is almost exclusively proliferative; thus the fetal growth retardation caused by maternal restriction of either calories or protein implies a reduced cell number in every organ. This fact is reflected by a reduced DNA content in all organs of the affected pups (Winick, 1969; Zeman, 1970).

DNA content of the brain in the newborns of restricted mothers has been found to be reduced 10–15% when compared to a control group (dams fed 20 or 25% protein diets) (Zamenhoff *et al.*, 1968; Winick, 1969; Zeman, 1970). A proportionally similar deficit in brain DNA content has also been described in the guinea pig after an overall dietary restriction (Chase *et al.*, 1971) and in the miniature pig after maternal protein deprivation (Tumbleson *et al.*, 1972).

Reduced DNA content of the brain has been the focus of a considerable number of studies. Autoradiographic methods have made it possible to determine that at day 16 of gestation, fetuses from protein-restricted mothers have an overall reduction in the rate of cell division in the brain. The magnitude of this effect has been shown to follow a well-defined, regional pattern, probably determined by different rates of cell division in the various areas of the brain (Winick and Rosso, 1975). For example, in regions where at this time the rate of cell division is relatively slow, such as the cerebrum white and gray matter, the number of cells incorporating labeled thymidine is only moderately reduced compared with control fetuses. However, areas adjacent to the third ventricle and subiculum, where proportionally more cells are dividing at this age, show great reductions in the rate of labeled thymidine incorporation. Finally, the cerebellum and the area adjacent to the lateral ventricle had the most marked reduction in the number of dividing cells when compared to control samples. These data indicate that during maternal malnutrition the areas of the fetal brain most affected are those with a high rate of cell division. A similar phenomenon has been described in rats suffering postnatal malnutrition from birth until 21 days of age (Winick and Rosso, 1975).

The reduced rate of cell division in the brain, produced by maternal malnutrition, does not seem to be caused only by a lack of substrate. Other factors, such as reversible alteration in the mechanisms controlling cell division, are probably involved. This possibility is supported by a study in which explants of neonatal mouse cerebellum from animals whose mothers received an 8% casein diet during gestation continued to show generalized neuronal retardation *in vitro* in spite of the fact that the culture medium was similar to that of the control animals (Allerand, 1972).

In addition to the cellular changes, there is also evidence that maternal protein restriction in the rat causes changes in the accumulation of norepinephrine and dopamine of the brain. In contrast, the activity of tyrosine hydroxylase, the enzyme that converts tyrosine into dihydroxyphenylalanine (DOPA), is increased (Shoemaker and Wurtman, 1971).

The effect of prenatal malnutrition on other organs is still a relatively unexplored area. It has been shown that progeny of rats fed a 6% casein diet throughout pregnancy have smaller kidneys, with a reduced number of less differentiated glomeruli and with fewer collecting ducts. Proximal tubules are also shortened and less convoluted (Zeman, 1968).

Histological and functional changes in the intestines of the prenatally malnourished rats have also been reported. Smaller intestinal diameter and shortening of the jejunal villi, with histological evidence of a decreased absorption of protein and fat by jejunal enterocytes, were the main changes reported. In addition, cells in which the absorptive defect was most evident had a decreased content of cytoplasmic organelles (Loh *et al.*, 1971). In contrast with the rat, the offspring of protein-deprived guinea pigs have been described as having intestinal villi that are longer, narrower, and more numerous than the controls. In addition, these animals showed a more efficient absorption of [^{14}C]oleic acid and [^{14}C]triolein than did control young (Zeman and Widdowson, 1974).

Considering the relatively minor deficit in its DNA content, the lungs seem to be, together with brain, one of the organs least affected by prenatal deficiency in the rat. There is evidence, however, that lung metabolism is significantly altered (Hawrylewicz *et al.*, 1975). Progeny of rats fed a 10% casein diet throughout pregnancy were compared with those of a control group fed a 27% casein diet. At birth the respiratory control index of mitochondria from deficient rats was significantly reduced. A similar reduction was also found in the amount of ADP utilized. This would be indirect evidence of a reduced rate of ATP synthesis in the lungs.

3.2. *Postnatal Consequences of Maternal Malnutrition*

Progency of rats fed either a low-protein or a caloric-restricted diet, or both, remain smaller throughout life. However, when adults, the deficit in the body weight is proportionally smaller than at birth (Chow and Rider, 1973). Data from different studies show the deficit in body weight to be approximately 20% at 21 days of age (Zeman, 1970), 12% at 76 days of age (Caldwell and Churchill, 1967), and only 8–10% in older males (Chow and Rider, 1973). The reduced body weight reflects a reduced body size, with shortening of the bones. Studies on the skeletal development of fetuses from rats fed a low-casein diet during pregnancy (Shroder and Zeman, 1973b) demonstrate that these animals have retarded ossification already apparent at day 17 of gestation. At this age the retarded ossification was estimated to be a developmental lag of 24 hours. Such a developmental lag becomes progressively more severe postnatally, with a 4-day lag in bone center development during the first weeks of life. The rate of bone elongation is also significantly slower in these animals (Shrader and Zeman, 1973b). At 65 and 90 days of age, tibial elongation per day is similar in control and prenatally restricted animals, while in the latter group body weight is increasing proportionally faster. Such a disproportion between body-weight increments and bone elongation in growing animals suggests a specific lesion either in the skeleton or in the mechanisms that modulate skeletal growth.

Another explanation of the lasting deficit in growth of the offspring of malnourished mothers has been the possibility of a permanent metabolic alteration. Such a metabolic abnormality, involving an abnormal utilization of absorbed amino acids, has been described in animals malnourished during gestation (McLeod *et al.*, 1972) and also in combined pre- and postnatal malnutrition (Lee and Chow, 1965, 1968).

Considerable interest has been focused on the possible consequences of the cellular deficit of the brain induced by prenatal malnutrition. Offspring from protein-restricted rats reared in litters of four pups, a situation that accelerates growth in normal animals, still have a reduced DNA content of the brain at 21 days of age (Zeman, 1970). Since no significant cell division is expected to occur in brain after this age, one would predict a lasting deficit in brain DNA. However, no differences in DNA content of the brain between controls and the progeny of restricted mothers has been found in 6-month-old animals (Stephan, 1971). A shorter period of rehabilitation also induces partial recovery of DNA deficit in some regions of the brain of the guinea pig. In this species DNA content reaches control levels in the cortex but remains 17% below control content in cerebellum (Chase, 1973). The implications of such apparent recovery of the prenatally induced cell deficit are unclear.

Brain is made up of a heterogeneous cell population that can be broadly categorized as belonging to two basic groups, neurons and glial cells. Each type of cell has its own unique timing for proliferative growth. In the rat, neurons demonstrate proliferative growth, restricted almost exclusively to prenatal life (Altman and Das, 1966). By contrast, the bulk of glial cells is still actively dividing for some time after birth. DNA polymerase studies in the rat would indicate that the peak of glial cell division occurs at day 10 of postnatal life (Brasel *et al.*, 1970). Therefore, it is conceivable that the 10–15% reduction in brain DNA content at birth caused by maternal protein malnutrition represents a reduction in neuronal cells. Since at birth these cells have already ceased to divide, they would not be affected by dietary rehabilitation. Furthermore, postnatal nutritional recovery would accelerate proliferative growth of glial cells. Histological studies done in progeny of protein restricted rats support this hypothesis by demonstrating a larger proportion of glial cells during postnatal recovery (Siassi and Siassi, 1973).

Long-lasting changes associated with maternal malnutrition have also been reported in kidneys. Pups from restricted mothers were reared in litters of four animals in an attempt to reverse the changes present at birth. At day 21 of age it was found that a partial recovery in DNA content of the kidneys had occurred, but the cellularity was restricted to existing nephrons and to a lengthening of the proximal convoluted tubes (Allen and Zeman, 1973a). Thus, in the rat, prenatal growth retardation due to maternal malnutrition determined a permanent deficit in the number of nephrons in the kidneys. Such a deficit probably explains the functional changes in the kidney reported in these animals. At 6 days of age the prenatally malnourished pups have a reduced response to either water or osmotic diuresis when compared to the progeny of rats fed a 25% casein diet (Hall and Zeman, 1968). The renal function is still altered when they are 22 days old, demonstrating a reduced glomerular filtration rate consistent with the reduced number of nephrons (Allen and Zeman, 1973b).

In contrast, spleen and thymus are able to attain a weight similar to control animals within a few days. Further, no abnormalities have been found in the capacity of the prenatally malnourished rat to develop certain types of immune responses later in life (Kenney, 1969).

There is evidence that the endocrine system of prenatally malnourished rats may also be altered. Comparing the *in vitro* rate of growth hormone synthesis and release in pituitaries of 36-day-old male control and prenatally restricted animals, it was found that these functions were significantly reduced in the pituitaries of deficient animals (Shroder and Zeman, 1973a). It has also been reported that plasma levels and pituitary concentration of growth hormones are lower in progeny of rats fed a 50% overall restricted diet (Stephen *et al.*, 1971).

Further support to the possibility that altered endocrine function may be one of the mechanisms by which postnatal recovery of deficient animals is prevented is given by data showing that administration of growth hormone to prenatally malnourished pups increases epiphyseal width, tibial length, and body weight (Zeman *et al.*, 1973). These findings suggest that bone from prenatally restricted animals may have a normal growth potential.

4. *Effects of Specific Deficiencies on Fetal Growth*

4.1. *Vitamin A*

Serum vitamin A levels decrease during the first trimester of pregnancy and then rise until shortly before term. Serum levels fall immediately prior to and during delivery (Gal and Parkinson, 1974). All these changes are very moderate.

Deficiency of vitamin A in rats has been shown to have profound effects on reproductive efficiency (Takahashi *et al.*, 1975). Severe deficiency will prevent conception. A diet containing only retinoic acid will sustain pregnancy through day 14 of gestation, and then fetal death and resorption begin. Those few pups which may reach term are stillborn or die shortly after birth (Takahashi *et al.*, 1975). In addition to a high fetal mortality, vitamin A deficiency in the rat, as well as in other animals, produces severe congenital malformations of the skeleton and other organs (Giroud, 1968). No conclusive evidence linking congenital malformations and vitamin A deficiency in humans has yet been provided.

Large doses of vitamin A also have a teratogenic effect in animals, resulting in cleft palate and other craniofacial changes. The existence of a similar problem in humans is still a matter of active research. Gal *et al.* (1972) reported high vitamin A concentrations in the maternal blood and fetal liver of malformed fetuses. Renal anomalies similar to those seen in animals have been reported in an infant whose mother ingested large doses of vitamin A during pregnancy (Bernhardt and Dorsey, 1974).

The recommended dietary allowance (RDA) for vitamin A during pregnancy is 1000 RE (5000 IU), an increase of 200 RE above the nonpregnant recommendation to allow fetal storage of the vitamin without compromising maternal liver stores.

4.2. Vitamin D, Calcium, and Phosphorus

During pregnancy there is a progressive net retention of calcium, most of the calcium being deposited in the fetus, approximately 30 g (Widdowson and Spray, 1951). Maternal mean bone mineral content at term does not differ from normal prepregnancy values (Christiansen *et al.*, 1976). Fetal content of both calcium and phosphorus is directly related to fetal weight. The total fetal accumulation of calcium amounts to only 2.5% of maternal body calcium.

Maternal serum calcium levels drop steadily until the middle of the last trimester of pregnancy, after which there is a slight rise (Michel, 1971; Mull and Bill, 1934; Newman, 1947, 1957). These changes are believed to reflect a decrease in protein-bound calcium resulting from the relative hypoalbuminemia of pregnancy. The pattern of calcium level parallels that of albumin in maternal serum. It is generally agreed that the level of ionic calcium does not change during pregnancy (Andersch and Obertst, 1936; Kerr *et al.*, 1962; Reitz *et al.*, 1972), although one study does report a progressive decline in ionized calcium (Tan *et al.*, 1972).

It has been postulated that estrogen tends to inhibit bone resorption, leading to increased parathyroid hormone (PTH) levels in plasma. Plasma calcitonin (CT) levels are also elevated, probably to protect against excessive maternal bone resorption while allowing increased intestinal absorption. Human chorionic somatomammotropin (HCS) apparently increases the rate of bone turnover (Hearey and Skillman, 1971; Pitkin, 1975a). No changes in maternal plasma levels of vitamin D with increasing length of gestation have been described. Supplementation with calcium and vitamin D does not significantly affect maternal serum levels of calcium (Newman, 1953, 1956).

Fetal serum calcium levels increase through gestation with the full term being hypercalcemic with respect to maternal levels (Reitz *et al.*, 1972; Delivoria-Papadopoulos *et al.*, 1967; Tan and Raman, 1972; Mull, 1936; Thalme, 1966; Crawford, 1965; Armstrong *et al.*, 1970). In contrast, fetal serum phosphorus concentration falls throughout gestation, reaching levels of approximately 6 mg/100 ml at term (Mull, 1936). Fetal PTH levels in cord blood are approximately one-fourth of the maternal levels and within or below the nonpregnant adult range (Reitz *et al.*, 1972; Pitkin, 1975a). CT levels, on the other hand, are significantly higher than maternal levels or the adult nonpregnant levels (Samaan *et al.*, 1973; Hesch *et al.*, 1973). This balance of PTH to CT is believed to favor skeletal growth and is probably maintained by the active transport of calcium from the maternal circulation.

Newborns experience a drop in ionic serum calcium and a rise in serum phosphorus during the first 3 or 4 days of life. Serum calcium then rises by the end of the first week to a point approximately 10 mg/100 ml below cord levels. The changes probably reflect a gradual adaptation of the infants' endocrine system to extrauterine life as PTH production increases and CT output decreases (Denzer *et al.*, 1939; Todd *et al.*, 1939; Khattab and Forfar, 1970). Some infants fail to regulate its calcium level during the first days of life, and calcemia accompanied by tetany results. It has been shown that these hypocalcemic episodes are significantly more frequent in infants born in winter and spring and that they are also related to parity, maternal age, and lower social class (Roberts *et al.*, 1973).

A recent study found that 56% of infants who suffered neonatal tetany later showed severe enamel hypoplasia of the deciduous teeth. Histological examination indicated a prolonged disturbance of enamel formation in the 3 months prior to birth. Also, an inverse relationship was shown between mean daily hours of bright sunlight in each calendar month and the incidence of neonatal tetany 3 months later (Purvis *et al.*, 1973).

Rosen *et al.* (1974) found that hypocalcemic premature infants fell into two groups, those in which maternal and neonatal levels of vitamin D were strikingly low and those in which there was no correlation between calcium levels and vitamin D levels. Dietary intake of vitamin D was greater in the latter group and in controls than in the former group. This evidence points to vitamin D deficiency as being only one of the causes of neonatal hypocalcemia. Other associated conditions include maternal hyperparathyroidism, diabetes, untreated coeliac disease, "placental insufficiency," and delivery by cesarian section (Roberts *et al.*, 1973). The high phosphorus load of cow's milk and cow's milk-based formulas increases the incidence of hypocalcemic convulsion in infants at risk (Gardner, 1952; Oppe and Redstone, 1972; Cockburn *et al.*, 1973).

In a study of three racial groups in England, Watney *et al.* (1971) found that mean serum calcium was lower and mean serum phosphorus proportionally higher in Asian mothers than in Caucasian and West Indian mothers. Although only minor, nonsignificant, differences were found in cord blood calcium and phosphorus levels, significant differences among groups in calcium plasma levels were found at 6 days of age. At this time, 33% of the infants in the Asian group had serum calcium levels of 8 mg/100 ml or less, whereas in the Caucasians and West Indian infants, these values were found in only 19% and 7% of the infants, respectively. Dietary history revealed no difference in vitamin D intake among the groups. However, the intake of Asian mothers was particularly deficient in calcium. The ratio of calcium to phytates in the Asian diet was 5:1, versus a ratio of 17:1 in the West Indian diet. Six-day serum calcium was lower in infants whose mothers admitted not taking prescribed vitamin supplements containing 400 IU of vitamin D and 250 μg of calcium.

A case of fetal rickets has been reported by Russell and Hill (1974). The mother had low serum of vitamin D. Although the mother showed no signs of osteomalacia during pregnancy, she did develop the disease some months after delivery. Moncrieff and Fadahunsi (1974) also reported a case of neonatal rickets in the infant of a mother suffering from osteomalacia during pregnancy.

There is some nonconclusive evidence that maternal hypervitaminosis D is associated with severe infantile hypercalcemia, craniofacial abnormalities, supravalvular aortic and pulmonic stenosis, hypertension, mental retardation, and nephrocalcinosis (Friedman and Roberts, 1966; Gal *et al.*, 1972; Nanda, 1974).

The RDA for vitamin D during pregnancy is 400 IU, the same amount recommended for infant children and adolescents. The RDAs for calcium and phosphorus are raised to 1200 mg/day, an increase of 400 mg/day above non-pregnant levels.

4.3. *Vitamin E*

Maternal serum levels of tocopherols rise throughout pregnancy to about 60% above nonpregnancy levels (Straumfjord and Quaife, 1946). Since most plasma vitamin E is carried by lipoproteins, this increase is probably associated with the normal hyperlipidemia of pregnancy (Pitkin, 1975b).

Body stores of α-tocopherol in preterm infants are disproportionately low when compared with full-term infants. The average store of a 1000-g infant is only 3 mg, while that of a 500-g infant is 20 mg (Dju *et al.*, 1952). A vitamin-E-responsive hemolytic anemia has been described in premature infants, usually at 6–10 weeks of age. Clinical signs include edema of legs, external genitalia, and eyelids; watery nasal discharge; tachypnea; and restlessness (Oski and Barness, 1963; Dallman, 1974). Prevention of the disease is possible through supplementation of the infants with 5–25 IU/day of α-tocopherol. Polyethylene glycol-1000-succinate may make supplementation with 5–10 IU/day adequate (Gross and Melhorn, 1974; Dallman, 1974).

Tateno and Ohshima (1973) reported a correlation between birth weight and cord maternal blood levels of vitamin E. However, failure to control for gestational age and other variables make this study difficult to interpret.

The RDA for vitamin E during pregnancy is 15 IU, an increase of 3 IU above nonpregnant levels, to allow for fetal storage of the vitamin.

4.4. *Vitamin K*

Vitamin K_2 is synthesized by bacteria in the human gut. Deficiency in adults is extremely rare, suggesting that intestinal production is adequate to meet the body's need for the vitamin.

The sterile gut of the newborn infant is unable to produce vitamin K, and hemmorrhagic disease may develop due to a postnatal deficiency. Premature infants are more likely to develop deficiency than term infants, owing to inadequate tissue stores (Schaffer, 1971). Parenteral K_1 is generally given to all newborns in the United States as a preventive measure. Studies have shown hat oral administration of K_1 to the mother during the last week or two of pregnancy will also help to prevent hemorrhagic disease and raise the stage 1 and stage 2 prothrombin levels in the newborn (Owen *et al.*, 1967). Parenteral administration of synthetic vitamin K, menadione, to the mother has been associated with hyperbilirubinemia and kernicterus of premature infants and severe hyperbilirubinemia in full terms (Lucey and Dolan, 1959).

4.5. Thiamine

Evidence suggests an increased demand for thiamine during pregnancy. In studies of the thiamine dose necessary to produce an excretion peak, Lockhardt *et al.* (1943) found the required dose to increase throughout pregnancy, reaching three times the nonpregnant dose near term. Lockhardt concluded that there is an increased requirement for thiamine during pregnancy. Siddal and Mull (1945) found that private patients consuming a well-balanced diet excreted an average of 200 μg/day or more of urinary thiamine in each trimester of pregnancy. Daily supplementation with 0.75 mg of thiamine doubled the rate of urinary excretion. However, based on the results of a study of Melnick (1942), which showed 200 μg/day to be the normal excretion level for an adult population, Siddal and Mull concluded that women receiving a well-balanced diet do not need thiamine supplementation during normal pregnancy. More recent studies have shown also that during pregnancy, blood thiamine levels are reduced, as are the levels of urinary excretion (Baker *et al.*, 1975).

Heller *et al.* (1974a) determined thiamine status of pregnant women by using the erythrocyte transketolase (ETK) activation test. ETK values after activation were significantly higher in pregnant women than in the general population. Values remained more or less constant throughout the course of pregnancy. These results were interpreted as indicating that approximately 25% of "representative Central European women" were thiamine-deficient. The incidence of complications of pregnancy among these women, however, was not significantly higher than that among women classified as nondeficient. In addition, thiamine status and clinical parameters such as number of previous pregnancies or abortions, concentration of hemoglobin urine analysis, and size of the baby were not significantly correlated.

King (1967) has reported cases of acute cardiac failure developing 2-4 days after birth. Mothers all had low thiamine intake during pregnancy but no overt signs of beriberi. Early treatment with 5-10 mg of thiamine/day was essential for recovery. Apparently, in these cases the babies were born with very low stores that were rapidly depleted before the intestinal flora could provide a minimum supply of the vitamin.

The RDA for thiamine is 5 mg/1000 kcal during the first 6 months of gestation and 0.6 mg/1000 kcal during the last trimester.

4.6. *Riboflavin*

Riboflavin requirements appear to be increased during pregnancy. A drop in fasting morning urinary riboflavin excretion has been noted during the last 4 months. Excretion rises sharply after delivery but drops again as lactation is established (Brezezinski *et al.*, 1952). The incidence of angular stomatitis, glossitis, and cheilosis in the mother would correlate with low plasma riboflavin concentration (Brezezinski *et al.*, 1952; Clarke, 1971).

Studies or riboflavin deficiency in human pregnancy have reported an increased incidence of hyperemesis gravidarum during the second half of pregnancy, increased incidence of prematurity, and increased incidence of unsuccessful lactation (Brezezinski *et al.*, 1947). However, since a diet low in riboflavin is usually low in many other essential nutrients as well, including animal protein, the significance of such a finding is uncertain. Using erythrocyte glutathione reductase activity to measure riboflavin status, 25% of a group of middle-class European women were found to be deficient during pregnancy, while 40% were deficient at term. This study did not find any relation between riboflavin deficiency and abortion, hydroamnios, preeclampsia, stillbirth, or birth weight or length (Heller *et al.*, 1974b).

Severe riboflavin deficiency causes malformations in the rat. Most malformations involve skeletal system abnormalities such as shortened bones, fused ribs, syndactylism, and cleft palate. Warkany and Schraffenberger (1944) concluded that the formation of the membranous skeleton which precedes the cartilaginous as well as the osseous skeleton is inhibited by the riboflavin deficiency. Failure of human studies to reflect deletereous effects of riboflavin deficiency on the fetus probably reflects a lesser degree of deficiency than in experimental models. Until the discovery of riboflavin antagonists, production of congenital malformations in rats required feeding of riboflavin-deficient diets to the mother for weeks before mating. In these severe dietary deficiency experiments, the yield of malformed fetuses was only about one-third of the litter, while use of the antagonist galactoflavin increases the yield to almost 100% of offspring (Nelson *et al.*, 1956). Additionally, animal experiments have shown that riboflavin deficiency affects fetal growth only during organogenesis. Thus, the appearance of slight riboflavin deficiency late in human pregnancy would not present the same threat to the fetus that is seen in rats made deficient prior to or early in pregnancy.

During pregnancy, an additional 0.3 mg/day is added to the normal RDA for each maternal age group.

4.7. *Folate*

The daily requirements for folate are determined by the metabolic rate and the rate of cell synthesis. Therefore, folate requirements are higher during conditions which increase metabolic rate, such as infection or hyperthyroidism, and during conditions which increase cell synthesis, such as hemolytic anemia.

Serum folate levels are significantly lower than nonpregnant levels in the third trimester of pregnancy (Whalley *et al.*, 1969). However, the finding of significantly reduced serum and red cell folate levels in oral contraceptive users (Shajania *et al.*, 1969) indicates that reduced blood levels seen in pregnancy are also due, at least in part, to hormonal changes of pregnancy. The incidence of folate-responsive megaloblastic anemia is significantly increased during pregnancy, indicating a true increase in demand for the vitamin. Incidence runs about 2.5–5.0% in pregnant women in developed countries and is considerably higher in the developing countries. About 1 in 15 pregnant women show at least transitional cells suggestive of megaloblastosis (Knipscheer, 1975). Among a poor population in New York City, 16% of pregnant women were found to have low red cell folate levels (Herbert *et al.*, 1975).

Both red cell folate and serum folate levels are significantly lower in the first trimester of pregnancy in women who will develop megaloblastic anemia than in those who will remain normoblastic throughout pregnancy (Temerley *et al.*, 1968).

Supplementation of pregnant women with folic acid increases serum folic acid levels (Chisholm, 1966) and reduces the incidence of macrocytosis (Fleming *et al.*, 1974) but has not been shown to have any effect on hemoglobin levels (Chisholm, 1966; Fleming *et al.*, 1974; Iyengar and Rajalakshmi, 1975). Maternal problems associated with megaloblastic anemia include glossitis, ulcerations of the mucous membrane of the mouth and vagina, malabsorption, and steatorrhea (Knipscheer, 1975).

Cord blood values are not affected by maternal deficiency, and the ratio of cord to maternal level may be much greater than the normal 2.5 times (Avery and Ledger, 1970). Serum blood levels of folic acid in premature infants are comparable to full-term levels at birth, but in a series of 20 such infants, two-thirds showed evidence of folate deficiency at 2–3 months (Vanier and Tyas, 1967).

Experiments in which rats are maintained on a folate-deficient diet during pregnancy have produced multiple fetal anomalies, including malformations of the eye, palate, lip, gastrointestinal tract, aorta, central nervous system, kidneys, and skeleton (Asling *et al.*, 1955; Monie and Nelson, 1963; Armstrong and Monie, 1966). Most animal experiments have involved not only total dietary deprivation of folic acid, but also the use of a folate antagonist such as 9-methylpteroylglutamic acid.

In humans, treatment with folate antagonists (methotrexate, aminopterin, chlorambucil) during pregnancy will generally produce abortion. However, some cases of term infants with severe congenital anomalies associated with the use of these drugs during pregnancy have been described (Committee on Maternal Nutrition, 1970).

Naturally occurring folate deficiency in pregnant women has not been proven to have any adverse effect on pregnancy outcome. Correlations have been reported between red cell folate level and the incidence of malformations, small for gestational-age babies and third-trimester bleeding (Hibbard, 1975; Streiff and Little, 1967). Other studies, however, have found no correlation between the occurrence of megaloblastic anemia and malformations, prematurity, or perinatal mortality (Giles, 1966; Knipscheer, 1975; Pritchard *et al.*, 1970). The possible correlation between folate deficiency and abruptio placentae has been the greatest source of controversy. Some of the earliest studies in which such a correlation was described used FIGLU excretion as a measure of folate status (Hibbard and Hibbard, 1963). However, the reliability of this test has been questioned (Chanarin, 1969). A correlation between megaloblastic changes and abruptio placentae has been reported by some investigators (Coyle and Geoghegan, 1962; Hibbard and Hibbard, 1963; Varadi *et al.*, 1966), while others have been unable to find any such correlation (Whalley *et al.*, 1969).

Measuring serum and red cell folate levels, most investigators have found no relation between low levels and the occurrence of abruptio placentae (Alpern *et al.*, 1969; Hall, 1972; Whalley *et al.*, 1969). One study reported that 100% of abruptio placentae patients had low red cell folate levels, while none of the normal delivery patients had low levels (Streiff and Little, 1967). Hibbard (1975) reported a correlation between low red cell folate levels and the incidence of abruptio placenta. However, careful examination of the data reveals that this correlation only holds true when abruptio placenta patients not included in the original study group are added to the analysis.

In folate-supplementation studies, no effect on pregnancy outcome is generally seen (Cooper *et al.*, 1970; Fleming *et al.*, 1974). However, a study among Bantu women found that supplementation with folate significantly reduced the incidence of prematurity. The normal Bantu diet contains almost no folate (Brumslag *et al.*, 1970). Iyengar and Apte (1970) had similar findings in poor Indian women, reducing the incidence of prematurity from 34 to 18%.

It seems logical that deficiency of a vitamin as intimately involved in cell division as folic acid would have some ill effects on the outcome of pregnancy. The failure of studies to show such effects in humans seems to be a function of severity of deficiency. The fact that folic acid antagonists do produce abortions and malformations in humans highlights this point. The normal folate levels seen in newborns whose mothers suffered from low folate levels during pregnancy suggests that the basic needs of the fetus for this particular vitamin are met even when the maternal availability is low. It is conceivably, however, that under those conditions fetal stores of folate are low.

The RDA for folic acid is doubled during pregnancy to 800 μg/day.

4.8. *Vitamin B_{12}*

Serum B_{12} levels fall progessively during pregnancy and rise sharply immediately after delivery (Green *et al.*, 1975). However, normal adult stores of B_{12} are considered to be adequate for an average of 2000 days (Knipscheer, 1975). This, coupled with the findings of Metz *et al.* (1965) that supplementation with vitamin B_{12} would not raise blood levels during pregnancy, indicates that the fall in plasma concentration during pregnancy does not reflect solely an increased demand for the vitamin. Herbert (1970) has shown that B_{12} serum levels rise when folate supplementation is given to pregnant women with low folate levels. This may indicate that the fall in B_{12} level seen in normal pregnancy is a direct result of the fall in folate level which normally accompanies pregnancy

Deficiency of B_{12} due to dietary inadequacy is rare except in vegetarians consuming no animal products. In the United States most cases are due to impaired absorptions. Pernicious anemia is rare in women of childbearing age, and when it does occur it is generally accompanied by infertility (Ball and Giles, 1964). All the premature infants (gestational age 29–37 weeks) in a study group had normal B_{12} levels at birth. However, at 40 days of age, 50% of them had low values (Pathak *et al.*, 1972).

The RDA for vitamin B_{12} during pregnancy has been set at 4 μg/day based on an estimated fetal demand of 0.3 μg/day and an estimated increase in maternal demand of 0.33 μg/day. The RDA for nonpregnant females is 3.0 μg/day.

4.9. Vitamin B_6

Blood pyridoxine levels fall progressively during pregnancy to as low as 25% nonpregnant levels (Coursin and Brown, 1961). An early study (Coursin and Brown, 1961) reports the greatest fall in blood pyridoxine to be found in the pyridoxamine-5′-phosphate portion, with a much less drastic fall in the pyridoxal-5′-phosphate (PLP) level. However, a more recent study (Contractor and Shane, 1970) reports no significant fall in pyridoxamine-5′-phosphate level, but a significant drop in PLP levels. PLP represents the coenzyme form of vitamin B_6, activating well over 60 enzymes involved in reactions such as transamination, racemization, decarboxylation, cleavage, dehydration, and desulfhydration.

The gradual decrease in plasma pyridoxal phosphate throughout pregnancy is accompanied by increasing levels of urinary xanthurenic acid excretion after a tryptophan load (Hamfelt and Hahn, 1969). Estrogen administration to nonpregnant women brings about a rise in urinary tryptophan metabolites following a tryptophan load. It appears that certain progestogens can reduce this action of estrogen, with the result that some oral contraceptive agents (OCAs) have less effect on tryptophan metabolism than do others. Urinary excretion in both pregnant women and OCA users can be normalized by supplementation with pyridoxine HCl. In pregnancy a dose of 6–10 mg is required to obtain this effect (Coursin and Brown, 1961; Lumeng *et al.*, 1976).

Experimental evidence indicates that estrogen conjugates formed in the liver inhibit the activity of hepatic kynureninase by competing with pyridoxal phosphate for receptor sites of the apoenzyme. Elevated urinary excretion of *N*-methylnicotinamide indicates a probable increase in the rate of synthesis of nicotinic acid ribonucleotide from L-tryptophan in OCA users. Also, studies suggest that several pyridoxal phosphate-dependent metabolic pathways for amino acid catabolism are accelerated during pregnancy, increasing the possibility of development of a true B_6. The major urinary metabolite of vitamin

B_6 in humans is 4-pyridoxic acid. In pregnancy, urinary excretion of a 4-pyridoxic acid after a test dose of pyridoxine hydrochloride is lower than in normal nonpregnant women (Wachstein and Gudiatis, 1953). Some OCA users also excrete reduced levels of this metabolite (Rose *et al.*, 1972).

The question of whether or not the drop in blood pyridoxine levels and the apparent disturbance of tryptophan metabolism in pregnancy are a physiological adjustment to pregnancy remains unanswered. Many advocate the administration of 6–10 mg of supplemental pyridoxine to "normalize" pyridoxal phosphate levels and tryptophan load test results (Lumeng *et al.*, 1976; Coursin and Brown, 1961). Considerig the average 50% expansion of blood volume seen in pregnancy, such a "normalization" of blood pyridoxal phosphate levels would entail a 50% increase in the total amount of circulating coenzyme. Because pyridoxal phosphate in the blood exists principally bound to albumin and the concentration of this protein is reduced by hemodilution, it seems unlikely that blood pyridoxal phosphate levels should remain at or above nonpregnant levels throughout pregnancy

In B_6 deficiency in the nonpregnant state, the excretion of *N*-methylnicotinamide is reduced, indicating impairment of the pyridoxal phosphate synthesis of nicotinic acid from tryptophan. Yet, in pregnancy the excretion of this metabolite is elevated, apparently indicating increased synthesis. The fact that 4-pyridoxic acid excretion is reduced during pregnancy may indicate a slowed destruction of the vitamin, thus lowering the relative demand for the vitamin.

Fetal levels are raised by maternal supplementation with B_6 (Cleary *et al.*, 1975). Pyridoxal phosphate levels in premature infants weighing 1900–2500 g were found to be deficient, indicating, as with many of the vitamins, stepped-up transport during the last months of pregnancy (Reinken and Marigold, 1973).

In rats, pyridoxine deficiency during pregnancy has been reported to result in runting, neonatal death, low birth weight, hypoplasia of the thymus and spleen, deformities, and reduced immunological competence in the young (Davis, 1974; Robson and Schwarz, 1975).

Adult, nonpregnant volunteers placed on a B_6-deficient diet had developed personality changes, including irratibility, depression and loss of a sense of responsibility, filiform hypertrophy of the lingual papilla, aphthous stomatitis, nasolabial seborrhea, an acneform papular rash of the forehead, abnormal electroencephalograms, and alterations in tryptophan metabolism (Sauberlich *et al.*, 1970). Pyridoxine administration corrected all abnormalities which had developed. High intakes or protein hasten the unset of B_6 deficiency. A tryptophan load test in the vitamin-B_6-deficiency patient induces elevated urinary excretion of xanturenic acid, kynurenine, hydroxykynurenine, kynurenic acid, acetylkynurenine, and guinolinic acid.

Supplementation with vitamin B_6 has not been found to have a significant effect on any clinical complication of pregnancy except "poor appetite" (Hillman *et al.*, 1963). Low maternal blood levels of pyridoxine are not associated with neonatal clinical sequellae (Heller *et al.*, 1973).

A study of pyridoxine toxicity in rats reported no change in number of live fetuses, number of dead fetuses, number of resorption sights, fetal weight, or incidence of anomalies with doses of up to 80 mg/kg/day (Khera, 1976). Doses as large as 1 g/kg have been tolerated without ill effects in adult rats, rabbits, and dogs (Unna and Honig, 1968).

The RDA for vitamin B_6 during pregnancy and lactation is 2.5 mg/day, an increase of 0.5 mg/day over nonpregnant levels.

4.10. Vitamin C

It is generally agreed that serum vitamin C levels falls during pregnancy, (Javert and Stander, 1943; Martin *et al.*, 1957; Mason and Rivers, 1971; Baker *et al.*, 1975), although some investigators have reported no significant change (Hoch and Marrack, 1948; Dawson *et al.*, 1969). Women with an intake greater than 80 mg/day maintain essentially the same average serum vitamin level throughout pregnancy, while those with lower intakes show progressively lower levels in each trimester (Martin *et al.*, 1957).

Urinary excretion of a test dose of ascorbic acid is reduced during pregnancy and lactation (Toverud, 1939). If saturation is defined as excretion of at least 50% of a test dose within 12 hr of administration, the length of time required to achieve saturation with an intravenous dose of 300 mg of ascorbic acid per day is increased during pregnancy (Irwin and Hutchins, 1976).

The incidence of gingival changes during pregnancy is significantly related to serum vitamin C levels, although an incidence of 8% in those with high concentrations indicates that poor vitamin C nutriture is not the sole cause of this condition during pregnancy (Martin *et al.*, 1957).

Some researchers have reported a lowered serum vitamin C levels in patients with threatened abortion or a history of previous abortions (Javert and Stander, 1943; King, 1945). Others have been unable to find any such relation (Martin *et al.*, 1957; Vobecky *et al.*, 1974).

Prematurity is significantly increased in those women with vitamin C intakes of less than 20 mg/day or with low serum levels (Martin *et al.*, 1957; Wideman *et al.*, 1964). A relationship between serum ascorbic acid levels and premature rupture of the fetal membrane has been reported. However, the number of mothers involved is small, and further investigation is required (Wideman *et al.*, 1964).

Animal studies and clinical evidence indicate that large doses of vitamin C during pregnancy may result in an increased demand for vitamin C in the offspring (Cochrane, 1965; Norkus and Rosso, 1975). This effect is apparently caused by an prenatally induced increase in the rate of catabolism of the vitamin (Fig. 16).

The RDA for vitamin C during pregnancy is 60 mg/day, an increase of 15 mg/day over the nonpregnant level.

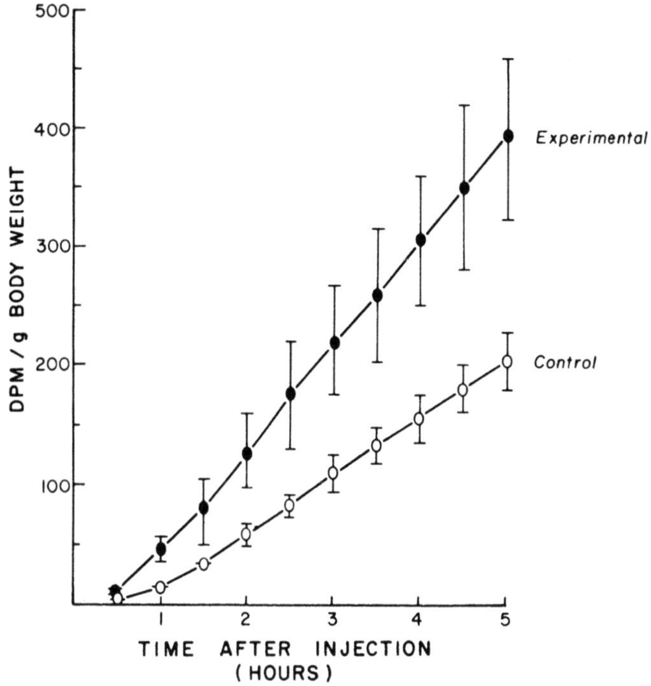

Fig. 16. Cumulative excretion of $^{14}CO_2$ following injection of [^{14}C]ascorbic acid in control pups and pups from mothers fed a high-vitamin-C diet during the last half of pregnancy. From Norkus and Rosso (1975).

4.11. Iron

Several changes are seen in the clinical parameters of iron metabolism during pregnancy. Hematocrit, hemoglobin concentration, and serum iron fall. Iron-binding capacity rises (Hytten and Lind, 1973).

Hemodilution accounts for a large part of the fall in hematocrit, hemoglobin concentration, and serum iron. Expansion of plasma volume occurs early in pregnancy, while expansion of red cell volume is greatest in the last trimester (Hytten and Leitch, 1971).

Evidence indicates that a true iron-deficiency state is common during pregnancy and that supplementation is generally necessary to prevent total depletion of maternal iron stores (Committee on Maternal Nutrition, 1970). One study reported that 67% of women not receiving iron and 46% of those receiving iron supplements had no stainable iron stores at the end of pregnancy (Fleming *et al.*, 1974). Iron stores are still lacking in the bone marrow 6 months postpartum in women who received no iron during pregnancy (Lowenstein *et al.*, 1962). Thus, the requirement for iron is greatly increased during pregnancy.

The average iron content of a single fetus at birth is 0.2–0.3 g (Widdowson and Spray, 1951). The expansion of maternal red cell volume during pregnancy requires approximately 0.5 g. Therefore, the total iron demand of pregnancy is close to 1 g. These additional demands of pregnancy cannot be met by prepregnancy iron stores, which average only 0.3 g among healthy college women (Scott and Pritchard, 1967). To meet the iron requirement and maintain maternal iron stores, the body must absorb approximately 7 mg of iron each day during the latter half of pregnancy.

The present uptake of iron from the intestine rises during pregnancy. Uptake of ^{59}Fe in one study rose from 32% prior to pregnancy to 75% at 6 months and 90% at term (Heinrich *et al.*, 1968; Hahn *et al.*, 1951).

A fall in hemoglobin concentration of about 0.5 g/100 ml is apparent by the third month. Minimum concentration is seen at 30–32 weeks, followed by a slight rise as red blood cell production increases and plasma volume levels off (Hytten and Leitch, 1971). The average hemoglobin concentration in nonpregnant women is 13.7–14.0 g/100 ml. A hemoglobin concentration of less than 11.0 g/100 ml is generally considered indicative of anemia in pregnancy (Committee on Maternal Nutrition, 1970). In pregnant women not receiving supplemental iron, the average hemoglobin level at term is 11.1 g/100 ml. In women receiving supplemental iron, the average is 12.3–12.4 g/100 ml. The average hemoglobin level in late pregnancy was 12.5 g in 21 women who received no iron supplements during pregnancy but demonstrated iron stores at the end of pregnancy (Pritchard *et al.*, 1970).

The increase in both plasma volume and total red cell volume during pregnancy is partly dependent on the availability of iron, with increases of both being significantly less in iron-deficient women than in those with iron reserves (Lawrence, 1962). This inability to expand plasma volume may mask a decrease in hemoglobin during pregnancy.

A hematocrit of 40–42% is normal in a nonpregnant woman. During pregnancy the hematocrit falls to an average of 43% (Hytten and Leitch, 1971). A rise is seen near term as red blood cell production increases.

Total iron binding capacity (TIBC) normally rises 15% in iron-sufficient pregnant women (Kitay and Harbort, 1975). A rise in TIBC has also been reported in women taking oral contraceptives (Mardell *et al.*, 1969). This indicates that at least part of the rise in TIBC seen in pregnancy is hormonal in origin. Rises in TIBC which accompany iron deficiency during pregnancy are much greater than those seen in the nonpregnant state (Kitay and Harbort, 1975).

Mean cell hemoglobin concentration varies little from the normal average of 32–33%. Levels may fall slightly in non-iron-supplemented women. Supplementation with large doses of iron may raise the level to 35% (Hytten and Leitch, 1971).

The mean hemoglobin concentration of the fetus is unaffected by maternal iron levels. However, infants of anemic mothers tend to develop anemia in the first year of life (Stevenson, 1973).

The practice of stripping the umbilical cord can deliver an additional 50–75 ml of blood containing 35–40 mg of iron to the newborn (Stevenson, 1973).

Reduced blood loss during delivery and a reduction in the number of women requiring transfusion during and after delivery are reported with iron supplementation during pregnancy (Wills *et al.*, 1947; Pritchard, 1959). Some studies indicate a decrease in mean birth weight at term in infants of women with iron deficiency (Chisholm, 1966; Whalley *et al.*, 1969), while others report no such correlation (Kitay and Harbort, 1975). An increase in the incidence of prematurity has been reported in iron deficiency (Scott, 1962), but no significant evidence exists linking iron deficiency with congenital anomalies.

During pregnancy the Food and Nutrition Board (1974) recommended supplementation of the diet with 30–60 mg of elemental iron because increased pregnancy demands cannot be met through normal dietary sources.

4.12. *Zinc*

Total body zinc increases by approximately 50 mg during pregnancy (Sanstead, 1973). However, maternal plasma and hair levels fall (Hambridge and Drogemueller, 1974; Widdowson *et al.*, 1974). The concentration of zinc in the serum of the fetus is very high at 20–26 weeks and falls toward term, although it is always higher than maternal levels (Berfenstam, 1949, 1952; Vikbladh, 1951).

Fetal liver concentrations declines during gestation in a manner similar to maternal plasma levels but remains higher than normal adult liver values. At term, 25% of fetal body zinc is found in the liver (Widdowson *et al.*, 1972).

Zinc deficiency is highly teratogenic in rats and birds, resulting in abnormalities of the skeletal and nervous systems (Turk *et al.*, 1959; Blamberg *et al.*, 1960; Hurley and Swenerton, 1966). It has been postulated that the incidence of malformations of the central nervous system is higher in areas where zinc deficiency is prevalent (Sever and Emanuel, 1973).

In the rat, very high doses of zinc have been shown to cause growth retardation and increased fetal death and resorption but no increase in external anatomical malformations (Schlicker and Cox, 1968).

The RDA for zinc during pregnancy is 20 mg/day, an increase of 5 mg above nonpregnant levels.

4.13. *Copper*

Maternal serum copper levels rise throughout pregnancy (DeJorge *et al.*, 1965; Do-Kumov, 1968). This rise is due, at least in part, to increasing estrogen levels. Estrogen therapy brings about a rise in serum copper levels of nonpregnant women (Russ and Raymunt, 1956; Von Studnitz and Berezin, 1958). The activity of ceruloplasmin, a copper-containing oxidase, almost doubles by late pregnancy (Hytten and Leitch, 1971).

The copper concentration in fetal serum is lower at term than at any other time of life (Widdowson *et al.*, 1974). Maternal concentration at this time is approximately five times higher than fetal concentration. This large difference reflects the high level of ceruloplasmin in maternal serum (Scheinberg *et al.*, 1954).

About 50% of the body copper of the fetus is found in the liver (Widdowson *et al.*, 1974). Much of this copper is associated with protein in the mitochondria to form "neonatal hepatic mitochondrocuprein," a Cu–protein complex unique to the fetus, containing 10 times as much copper as any other known protein (Porter, 1966). This is believed to be a storage compound and disappears soon after birth.

The copper concentration of the placenta remains constant throughout gestation, with an average value of 227 μg/100 ml (Poczekaj *et al.*, 1963). In multiple pregnancies, the placental copper content is significantly higher than in a single pregnancy (O'Leary *et al.*, 1966).

No correlation has been found between serum copper level and maternal age, race, parity, or birth weight. Several investigators have reported elevated serum copper or urinary copper in preeclampsia/eclampsia (O'Leary *et al.*, 1966).

No RDAs for copper have been established. In the nonpregnant adult, an intake of 2.0 mg/day is sufficient to maintain balance. This amount is generally obtained in the average mixed diet (Food and Nutrition Board, 1974).

4.14. Sodium

Total sodium can be expected to increase by approximately 900 mmol during pregnancy (Hytten and Leitch, 1971; Weir *et al.*, 1971). The increase in sodium requirement for pregnancy is estimated at 25 g total or an additional 4 meq/day (Pitkin *et al.*, 1972).

Exchangeable sodium per kilogram of body weight does not show any increase during normal pregnancy. Rather, the increase in total exchangeable sodium is due to an increase in total body fluid (Dieckmann and Pottinger, 1957; Plentl and Gray, 1959; Chesley, 1966).

The rate of aldosterone secretion increases significantly in pregnancy. Sodium restriction results in a greatly increased secretion rate, while large quantities of sodium will result in a secretion rate lower than that of normal pregnancy but still severalfold greater than that seen in the nonpregnant state (Jones *et al.*, 1959; Vande Wiele *et al.*, 1960; Watanabe *et al.*, 1963). Elevated plasma concentrations of renin, renin substrate, angiotensin II, and aldosterone are also seen in normal pregnancy (Weir *et al.*, 1970, 1971; Robertson *et al.*, 1971; Skinner *et al.*, 1972).

In the rat, degranulation of the juxtaglomerular cells and hypertrophy and hyperplasia of the zona glomerulosa of the adrenals to the point of exhaustion are associated with restriction of sodium during pregnancy. Additionally, low sodium intake results in a less expected increase in "effective blood volume" and in small litters and fetuses (Wardlaw and Pike, 1963).

Current evidence indicates that sodium restriction does not help to prevent toxemia of pregnancy or to alleviate the disease once it has developed (Committee on Maternal Nutrition, 1970; Lindheimer and Katz, 1973).

5. References

Abdul-Karim, R. W., Nesbitt, R. E., Jr., Drucker, M., and Rizk, P. T., 1971, The regulatory effect of estrogens on fetal growth, I. Placental and fetal body weights, *Am. J. Obstet. Gynecol.* **109**:656.

Adcock, E. W., III, Teasdale, F., August, C. S., Cox, S., Meschia, G., Battaglia, F. C., and Naughton, M. A., 1973, Human chorionic gonadotropin: Its possible role in maternal lymphocyte suppression, *Science* **181**:845.

Ademowore, A. S., Conrey, N. G., and Kime, J. S., 1972, Relationship of maternal nutrition and weight gain to new born birthweight, *Obstet. Gynecol.* **39**:460.

Aherne, W., and Dunhill, M. S., 1966a, Quantitative aspects of placental structure, *J. Pathol. Bacteriol.* **91**:123.

Aherne, W., and Dunhill, M. S., 1966b, Morphometry of the human placenta, *Br. Med. Bull.* **22**:5.

Alexiou, D., Grimanis, A. P., Grimani, M., Papaevangeleou, G., and Papadatos, C., 1976, Concentration of zinc, cobalt, bromine, rubidium and gold in maternal and cord blood serum, *Biol. Neonate* **29**:191.

Alfonso, J. F., and DeAlvarez, R. R., 1963, Further starch jelfractionation of new protein zones in pregnancy, *Am. J. Obstet. Gynecol.* **86**:815.

Allen, L. H., and Zeman, F. J., 1973a, Influence of increased postnatal nutrient intake on kidney cellular development in progeny of protein-deficient rats, *J. Nutr.* **103**:929.

Allen, L. H., and Zeman, F. J., 1973b, Kidney function in the progeny of protein-deficient rats, *J. Nutr.* **103**:1467.

Allerand, C. D., 1972, Effect of prenatal protein deprivation on neonatal'cerebellar development *in vitro*, *Nature* **239**:157.

Alpern, J. B., Haggard, M. E., and McGanity, W. J., 1969, Folic acid, pregnancy and abruptio placenta, *Am. J. Clin. Nutr.*, **22**:354.

Altman, J., and Das, G. D., 1966, Autoradiographic and histological studies of postnatal neurogenesis, *J. Comp. Neurol.* **126**:337.

Aly, H. E., Donald, E. A., and Simpson, M. W. H., 1971, Oral contraceptives and vitamin B_6 metabolism, *Am. J. Clin. Nutr.* **24**:297.

Andersch, M., and Obertst, F. W., 1936, Filterable serum calcium in late pregnant and parturient women and in the newborn, *J. Clin. Invest.* **15**:131.

Anderson, L. L., 1975, Embryonic and placental development during prolonged inanition in the pig, *Am. J. Physiol.*, **229**:1687.

Andrews, W. C., and Bonsnes, R. W., 1951, The leukocytes during pregnancy, *Am. J. Obstet. Gynecol.* **61**:1129.

Antonov, A. N., 1947, Children born during the siege of Leningrad in 1942, *J. Pediatr.* **30**:250.

Armstrong, R. C., and Monie, I. W., 1966, Congenital eye defects in rats following maternal folic-acid deficiency during pregnancy, *J. Embryol. Exp. Morphol.* **16**:531.

Armstrong, W. D., Singer, L., and Makowski, E. L., 1970, Placental transfer of fluoride and calcium, *Am. J. Obstet. Gynecol.* **107**:432.

Arora, D. J. S., and deLamirande, G., 1971, The influence of starvation on rat liver polysomes, *Can. J. Biochem.* **49**:1150.

Asling, C. W., Nelson, M. W., Wright, H. V., and Evans, H. M., 1955, Congenital skeletal abnormalities in fetal rats resulting from maternal pteroylglutamic acid deficiency during gestation, *Anat. Rec.* **121**:775.

Assali, N. S., Rauramo, L., and Peltonen, T., 1960, Measurement of uterine blood flow and oxygen consumption in early human pregnancy, *Am. J. Obstet. Gynec.* **79**:86.

Avery, B., and Ledger, W. J., 1970, Folic acid metabolism in well-nourished pregnant women, *Obstet. Gynecol.* **35**:616.

Bacon, J. S. D., and Bell, D. J., 1948, Fructose and glucose in the blood of the fetal sheep, *Biochem. J.* **42**:397.

Bailey, J. S., and Kurland, I. I., 1945, Relationship of maternal weight gain and weight of newborn infant, *Am. J. Obstet. Gynecol.* **50**:202.

Baker, E., and Morgan, E. H., 1973, Placental iron transfer in the cat, *J. Physiol. (London)* **232:**485.
Baker, H., Ziffer, H., Pasher, I., and Sobotka, H., 1958, A comparison of maternal and fetal folic acid and vitamin B_{12} at parturition, *Br. Med. J.* **2:**978.
Baker, H., Frank, O., Thomson, A. D., Langer, A., Munues, E., DeAngelis, B., and Kaminetzky, H. A., 1975, Vitamin profile of 174 mothers and newborns at parturition, *Am. J. Clin. Nutr.* **28:**59.
Ball, E. W., and Giles, C., 1964, Folic acid and vitamin B_{12} levels in pregnancy and their relation to megaloblastic anemia, *J. Clin. Pathol.* **17:**165.
Barcroft, J., and Ratschild, P., 1932, The volume of blood in the uterus during pregnancy, *J. Physiol.* **76:**447.
Barnes, A. C., 1951, The placental metabolism of vitamin A, *Am. J. Obstet. Gynecol.* **61:**368.
Barron, D. H., 1951–52, Some aspects of the transfer of oxygen across the synesmochorial placenta of the sheep, *Yale J. Biol. Med.* **24:**169.
Bartels, H., Moll, W., and Metcalfe, J., 1962, Physiology of gas exchange in the human placenta, *Am. J. Obstet. Gynecol.* **84:**1714.
Battaglia, F. C., Behrman, R. E., Meschia, G., Seeds, A. E., and Bruns, P. D., 1968, Clearance of inert molecules, Na and Cl ions across the primate placenta, *Am. J. Obstet. Gynecol.* **102:**1135.
Baugham, D. R., Hobbs, K. I., and Terry, R. J., 1958, Selective placental transfer of serum-proteins in the rhesus, *Lancet* **2:**351.
Bayliss, R. J. S., Browne, J. C. M., Round, B. P., and Steinbeck, A. W., 1955, Plasma 17-hydroxycortico steroids in pregnancy, *Lancet* **1:**62.
Beaconsfield, P., Ginsburg, J., and Jeacock, M., 1964, Glucose metabolism in the pentose phosphate pathway relative to nucleic acid and protein synthesis in the human placenta, *Dev. Med. Child Neurol.* **6:**469.
Beal, V. A., 1971, Nutritional studies during pregnancy, I. Changes in intake of calories, carbohydrates, fat protein, and calcium, *J. Am. Dietet. Assoc.* **58:**312.
Beatly, C. H., and Bocek, R. M., 1975, Metabolic aspects of fetal skeletal muscle, in: *Fetal and Postnatal Cellular Growth* (D. B. Cheek, ed.), pp. 257–268, John Wiley & Sons, Inc., New York.
Beaton, G. H., Beare, J., Ryu, M. H., and McHenry, E. W., 1954, Protein metabolism in the pregnant rat, *J. Nutr.* **54:**291.
Benton, A. L., 1940, Mental development of prematurely born children: A critical review of the literature, *Am. J. Orthopt.* **10:**719.
Berfenstam, R., 1949, Studies on carbonic anhydrase in premature infants, *Acta Paediatr. Scand.* **77:**124.
Berfenstam, R., 1952, A clinical and experimental investigation into the zinc content of plasma and blood corpuscles with special reference to infancy, *Acta Paediatr. Scand.* **87:**3.
Berg, B. N., 1965, Dietary restriction and reproduction in the rat, *J. Nutr.* **87:**344.
Bergner, L., and Susser, M. W., 1970, Low birth weight and prenatal nutrition: An interpretative review, *Pediatrics* **46:**946.
Bernhardt, I. R., and Dorsey, D. J., 1974, Hypervitaminosis A and congenital renal anomalies in a human infant, *Obstet. Gynecol.* **43:**750.
Biezenski, J. J., 1969, Role of placenta in fetal lipid metabolism, I. Injection of phospholipids double labeled with ^{14}C glycerol and 32-P into pregnant rabbits, *Am. J. Obstet. Gynecol.* **104:**1177.
Blamberg, D. L., Blackwood, V. B., Suppleee, W. C., and Combs, G. F., 1960, Effect of zinc deficiency in hen on hatchability and embryonic development, *Proc. Soc. Exp. Biol. Med.* **104:**217.
Blechner, J. N., Stenger, U. G., and Prystowsky, H., 1974, Uterine blood flow in women at term, *Am. J. Obstet. Gynecol.* **120:**633.
Boger, W. P., Bayne, G. M., Wright, L. D., and Beck, G. D., 1957, Differential serum vitamin B_{12} concentration in mothers and infants, *New Engl. J. Med.* **256:**1085.

Bohmer, T., and Havel, R. J., 1975, Genesis of fatty liver and hyperlipemia in the fetal guinea pig, *J. Lipid Res.* **16:**454.

Bothwell, T. H., Pribilla, W. F., Mebust, W., and Finch, C. A., 1958, Iron metabolism in the pregnant rabbit, iron transport across the placenta, *Am. J. Physiol.* **193:**615.

Bottiglioni, F., Flamigni, C., Caramazza, G., and Tirelli, R., 1966, Ricerche gas-chromatografiche negli acid grassi liberi del plasma nello stato puerperale, *Boll. Soc. Ital. Biol. Sper.* **42:**893.

Bourdel, G., and Jacquot, R., 1959, Rôle du placenta dans la faculté anabolisante des rattes gestantes, *Compt. Rend.* **242:**552.

Bowden, J. W., and Wolkoff, A. S., 1967, Fetal blood calcium responses to maternal calcium infusions in sheep, *Am. J. Obstet. Gynecol.* **99:**55.

Braestrup, P. W., 1937, Studies of latent scurvy in infants, II. Content of ascorbic acid in the blood serum of women in labour and in children at birth, *Acta Paediatr. Scand.* **19:**328.

Brarubell, F. W. R., 1954, Transport of proteins across the fetal membranes, *Cold Spring Harbor Symp. Quant. Biol.* **19:**71.

Brasel, J. A., Ehrenkranz, R. A., and Winick, M., 1970, DNA polymerase activity in rat brain during ontogeny, *Dev. Biol.* **23:**424.

Brezezinski, A., Bromberg, Y. M., and Braun, K., 1947, Riboflavin deficiency in pregnancy, its relationship in pregnancy, its relationship to course of pregnancy and to condition of fetus, *J. Obstet. Gynaecol. Br. Emp.* **54:**182.

Brezezinski, A., Bromberg, Y. M., and Braun, K., 1952, Riboflavin excretion during pregnancy and early lactation, *J. Lab. Clin. Med.* **39:**84.

Brink, C., Esila, P., Karvanen, M. J., and Laamanen, A., 1959, Transfer of thiamine across the placenta of guinea pig, *Acta Physiol. Scand.* **47:**375.

Brown, R. R., Thornton, M. J., and Price, J. M., 1961, The effect of vitamin supplementation on the urinary excretion of tryptophan metabolites by pregnant women, *J. Clin. Invest.* **40:**617.

Brumslag, N., Edelstein, T., and Metz, J., 1970, Reduction of incidence of prematurity by folic acid supplementation in pregnancy, *Br. Med. J.* **18:**16.

Burke, B. S., Beal, V. A., Kirkwood, S. B., and Stuart, H. C., 1943, The influence of nutrition during pregnancy upon the condition of the infant at birth, *J. Nutr.* **26:**569.

Burt, R. L., Leake, N. H., and Rhyme, A. L., 1969, Glucose tolerance during pregnancy and the puerperium, *Obstet. Gynecol.* **33:**634.

Caldwell, D. F., and Churchill, J. A., 1967, Learning ability in the progeny of rats administered a protein deficient diet during the second half of gestation, *Neurology* **17:**95.

Calloway, D. H., 1974, Nitrogen balance during pregnancy, in: *Nutrition and Fetal Development*, Vol. 2 (M. Winick, ed.), pp. 79–94, John Wiley & Sons, Inc., New York.

Cameron, C. S., and Graham, S., 1944, Antenatal diet and its influence on stillbirths and prematurity, *Glasgow Med. J.* **142:**1.

Campbell, S., 1974, Physical methods of assessing size at birth, in: *Size at Birth* (E. Wolvestholme, ed.), CIBA Foundation Symposium 27, p. 275, Elsevier/Excerpta Medica, Amsterdam.

Carruthers, M. E., Hobbs, C. B., and Warren, R. L., 1966, Raised serum copper and caerulophasmin levels in subjects taking oral contraceptives, *J. Clin. Path.* **19:**498.

Cartwright, G. E., and Wintrobe, M. M., 1964, Copper metabolism in normal subjects, *Am. J. Clin. Nutr.* **14:**224.

Caton, D., 1974, Uteroplacental circulation and its regulation, in: *The Placenta* (K. Moghissi and E. S. E. Hafez, eds.), p. 67, Charles C Thomas Publisher, Springfield, Ill.

Chanarin, I., 1969, *The Megaloblastic Anemias,* Blackwell Scientific Publications Ltd., Oxford.

Chanarin, I., MacGibbon, B. M., O'Sullivan, W. J., and Mollin, D. L., 1959, Folic acid deficiency in pregnancy, the pathogenesis of megaloblastic anemia of pregnancy, *Lancet* **2:**634.

Chase, H. P., 1973, The effect of intrauterine and postnatal undernutrition on normal brain in development, *Ann. N.Y. Acad. Sci.* **205:**231.

Chase, P. M., Dabiere, C. S., Welch, N. N., and O'Brien, D., 1971, Intrauterine undernutrition and brain development, *Pediatrics* **47:**491.

Cheek, D. B., 1975, The fetus, in: *Fetal and Postnatal Cellular Growth* (B. Cheek, ed.), pp. 3–22, John Wiley & Sons, Inc., New York.

Chesley, L. C., 1966, Sodium retention and pre-eclampsia, *Am. J. Obstet. Gynecol.* **95:**127.
Chinard, F. P., Danesino, V., Hartman, W. L., Huggett, A. St., G., Paul, W., and Reynolds, S. R. M., 1956, The transmission of hexoses across the placenta in the human and the rhesus monkey (*Macaca mulatta*), *J. Physiol. (London)* **132:**289.
Chisholm, M. A., 1966, A controlled clinical trial of prophylactic folic acid and iron in pregnancy, *J. Obstet. Gynaecol. Br. Commonw.* **73:**191.
Chow, B. F., and Lee, C. J., 1964, Effect of dietary restrictions of pregnant rats on body weight gain of the offspring, *J. Nutr.* **82:**10.
Chow, B. F., and Rider, A. A., 1973, Implications of the effects of maternal diets in various species, *J. Am. Sci.* **36:**167.
Christakis, G., ed., 1973, Nutritional assessment in health programs, *Am. J. Publ. Health,* **63**(Part 2).
Christensen, H. N., 1973, On the development of amino acid transport systems, *Fed. Proc.* **32:**19.
Christensen, P. J., Date, J. W., Schonheyder, F., and Volgvartzk, S., 1957, Amino acids in blood plasma and urine during pregnancy, *Scand. J. Clin. Lab. Invest.* **9:**54.
Christiansen, C., Rodbro, P., and Heinild, B., 1976, Unchanged total body calcium in normal human pregnancy, *Obstet. Gynecol.* **31:**712.
Churchill, J. A., and Berendes, H. W., 1969, Intelligence of children whose mothers had acetonemia during pregnancy, *Pan Am. Health Org. Sci. Publ. 185,* p. 30.
Cinnamen, A. D., and Beaton, J. R., 1970, Biochemical assessment of vitamin B_6 status in man, *Am. J. Clin. Nutr.* **23:**696.
Clarke, H. C., 1971, The riboflavin deficiency syndrome of pregnancy, *Surg. Forum* **22:**394.
Clavero, J. A., and Botella-Llusia, J., 1963, Measurement of the villus surface in normal and pathologic placentas, *Am. J. Obstet. Gynecol.* **86:**234.
Cleary, R. E., Lumeng, L., and Li, T. K., 1975, Maternal and fetal plasma levels of pyridoxal phosphate at term: adequacy of vitamin B_6 supplementation during pregnancy, *Am. J. Obstet. Gynecol.* **121:**25.
Cloete, J. H. L., 1939, Prenatal growth in the merino sheep onderstepoort, *J. Vet. Sci.* **13:**417.
Cochrane, W. A., 1965, Overnutrition in prenatal and neonatal life: a problem? *Can. Med. Assoc. J.* **93:**893.
Cockburn, F., Brown, J. K., Belton, N. R., and Forfar, J. O., 1973, Neonatal convulsions associated with primary disturbances of calcium, phosphorus and magnesium metabolism, *Arch. Dis. Child.* **48:**77.
Colbert, R. M., Calton, F. M., Dinda, R. E., and Davies, J., 1958, Competitive transfer of sorbose and glucose in placenta of rabbit, *Proc. Soc. Exp. Biol. Med.* **97:**867.
Cole, S. H., Lopez, R., and Cooperman, J. M., 1974, Quantitative estimation of riboflavin deficiency in low socio-economic pediatric population by a new method, *Pediatr. Res.* **8:**379/105.
Committee on Maternal Nutrition, Food and Nutrition Board, 1970, *Maternal Nutrition and the Course of Pregnancy,* National Academy of Sciences, U.S. Government Printing Office, Washington, D.C.
Contractor, S. F., and Shane, B., 1970, Blood and urine levels of vitamin B_6 in mother and fetus before and after loading the mother with vitamin B_6, *Am. J. Obstet. Gynecol.* **107:**635.
Contractor, S. F., and Shane, B., 1971, Metabolism of ^{14}C-pyridoxal in the pregnant rat, *Biochim. Biophys. Acta* **230:**127.
Coons, C. M., 1933, Dietary habits during pregnancy. Study on food intake of 15 women at various periods of pregnancy, *J. Am. Dietet. Assoc.* **9:**95.
Cooper, B. A., Cantile, G. S. D., and Brunton, L., 1970, The case for folic acid supplementation during pregnancy, *Am. J. Clin. Nutr.* **23:**848.
Coursin, D. B., and Brown, V. L., 1961, Changes in vitamin B_6 during pregnancy, *Am. J. Obstet. Gynecol.* **82:**1307.
Coyle, C., and Geoghegan, F., 1962, The problem of anemia in a Dublin maternity hospital, *Proc. Roy. Soc. Med.* **55:**764.
Crawford, J. S., 1965, Maternal and Cord Blood at Delivery, IV. Glucose, sodium, potassium, calcium and chloride, *Biol. Neonate* **8:**222.

Creasy, R. K., Barrett, C. T., DeSwiet, M., Kahampaa, K. V., and Rudolph, A. M., 1972, Experimental intrauterine growth retardation in the sheep, *Am. J. Obstet. Gynecol.* **112**:566.

Cruickshank, J. M., Morris, R., Butt, W. R., and Crook, A. C., 1970, The relationship of total and differential leukocyte counts with urinary oestrogen and plasma cortisol levels, *J. Obstet. Gynaecol. Br. Commonw.* **77**:734.

Csapo, A. P., and Wood, C., 1968, The endocrine control of the initiation of labour in the human, in: *Recent Advances in Endocrinology* (V. H. T. James, ed.), pp. 207–239, Little, Brown and Company, Boston.

Csapo, A. P., Dray, F., and Erdos, T., 1974, Estradiol 17 Beta: Inhibitor of placental growth, *Lancet* **2**:51.

Cummings, H. H., 1934, An interpretation of weight changes during pregnancy, *Am. J. Obstet. Gynecol.* **27**:808.

Dallman, P. R., 1974, Iron, vitamin E and folate in the preterm infant, *J. Pediatr.* **85**:742.

Dam, H., Prange, I., and Sondergaard, E., 1955, Deposition of injected massive doses of colloidal vitamin K_1 in chicks, after partial blockage of the reticulo-endothelial system and in pregnant rats, *Acta Pharmacol. Toxicol.* **11**:90.

Dancis, J., and Money, W. L., 1960, Transfer of sodium and iodoantipyrine across guinea pig placenta using an in situ perfusion technique, *Am. J. Obstet. Gynecol.* **80**:215.

Dancis, J., and Shafran, M., 1958, The origin of plasma proteins in the guinea pig fetus, *J. Clin. Invest.* **37**:1093.

Dancis, J., and Springer, D., 1970, Fetal homeostatis in maternal malnutrition: potassium and sodium deficiency in rats, *Pediatr. Res.* **4**:345.

Dancis, J., Olsen, G., and Folkart, G., 1958, Transfer of histidine and xylose across the placenta and into the red blood cell and amniotic fluids, *Am. J. Physiol.* **194**:44.

Dancis, J., Lind, J., Oratiz, M., Smolens, J., and Vara, P., 1961, Placental transfer of proteins in human gestation, *Am. J. Obstet. Gynecol.* **82**:167.

Dancis, J., Jansen, V., Kayden, H. J., Schneider, H., and Levitz, M., 1973, Transfer across perfused human placenta. II. Free fatty acids, *Pediatr. Res.* **7**:192.

Dancis, J., Jansen, V., Kayden, H. J., Bjornson, L., and Levitz, M., 1974, Transfer across perfused human placenta. III. Effect of chain length on transfer of free fatty acids, *Pediatr. Res.* **8**:796.

Dannenburg, W. N., and Burt, R. L., 1965, The effect of Insulin and glucose on plasma lipids during pregnancy and the puerperium, *Am. J. Obstet. Gynecol.* **92**:195.

Darby, W. J., McGanity, W. J., Martin, M. B., Brigforth, E., Denson, P. M., Kaser, M. M., Ogpe, P. J., Newbill, J. A., Stockwell, A., Ferguson, M. E., Touster, O., McLellan, G. S., Williams, C., and Cannon, R. O., 1953, The vanderbilt co-operative study of maternal and infant nutrition, IV. Dietary, laboratory, and physical findings in 2129 delivered pregnancies, *J. Nutr.* **51**:565.

Darby, W. J., McGanity, W. J., McLaren, D. S., Paton, D., Alemu, A. Z., and Medhew, A. McG., 1960, Bitot's spots and vitamin A deficiency, *Public Health Rep.* **75**:738.

Davies, J., 1955, Permeability of the rabbit placenta to glucose and fructose, *Am. J. Physiol.* **181**:532.

Davis, C. H., 1923, Weight in pregnancy, its value as a routine test, *Am. J. Obstet. Gynecol.* **6**:575.

Davis, S. D., 1974, Immunodeficiency and runting syndrome in rats from congenital pyridoxine deficiency, *Nature* **251**:548.

Dawson, E. B., Clark, R. R., and McCarthy, W. J., 1969, Plasma vitamins and trace metal changes during teen-age pregnancy, *Am. J. Obstet. Gynecol.* **104**:953.

Dawson, K. R., 1972, Folic acid and low birth weight infants, *Scot. Med. J.* **17**:371.

Dayton, D. H., Filer, L. J., and Canosa, C., 1969, Cellular changes in placentas of undernourished mothers in Guatemala, *Fed. Proc.* **28**:488.

DeAlvarez, R. R., Gaiser, D. F., Simkins, D. M., Smith, E. K., and Bratvold, G. E., 1959, Serial studies of serum lipids in normal human pregnancy, *Am. J. Obstet. Gynecol.* **77**:743.

DeAlvarez, R. R., Alfonso, J. F., and Sherrard, D. J., 1961, Serum protein fractionation in normal pregnancy, *Am. J. Obstet. Gynecol.* **82**:1096.

DeJorge, F. B., Delascio, D., and Antunes, M. L., 1965, Copper and copper oxidase concentrations in the blood serum of normal pregnant women, *Obstet. Gynecol.* **26**:225.
Delivoria-Papadopoulos, M., Battaglia, F. C., Bruns, P. D., and Meschia, G., 1967, Total protein bound and ultrafiltrable calcium in maternal and fetal plasma, *Am. J. Physiol.* **213**:363.
Denzer, B. S., Reiner, M., and Weiner, S. B., 1939, Serum calcium in the newborn, *Am. J. Dis. Child.* **57**:809.
Dickson, W. M., Bosc, M. J., and Locatelli, A., 1969, Effect of estrogen and progesterone on uterine blood flow of castrate sows, *Am. J. Physiol.* **217**:1431.
Dieckmann, W. J., and Pottinger, R., 1957, Total exchangeable sodium and space in normal and pre-eclamptic patients determined with sodium-22, *Am. J. Obstet. Gynecol.* **74**:816.
Dieckmann, W. J., Adain, F. L., Michael, H., Kiamen, S., Dunkle, F., Costin, M., Campbell, A., Wensley, A. C., and Lovang, E., 1944, Calcium, phosphorus, iron and nitrogen balances in pregnant women, *Am. J. Obstet. Gynecol.* **47**:357.
Dilts, P. V., Jr., and Brinkman, C. R., 1969, Uterine and systematic hemodynamic interrelationships and their response to hypoxia, *Am. J. Obstet. Gynecol.* **103**:138.
Dine, M. E., and Snyder, L. M., 1970, Iron deficiency and PMN segmentation, *New Engl. J. Med.* **282**:691.
Dju, M., Mason, K. E., and Filer, L. J., Jr., 1952, Vitamin E (tocopherol) in human fetuses and placenta, *Études Néonates* **50**:42.
Dobbing, J., and Sands, J., 1973, Quantitative growth and development of human brain, *Arch. Dis. Child.* **48**:757.
Dodds, G. S., 1922, The area of the chorionic villi in the full term placenta, *Anat. Rec.* **24**:287.
Do-Kumov, S. I., 1968, Serum copper and pregnancy, *Am. J. Obstet. Gynecol.* **101**:217.
Dowling, J. T., Freinkel, N., and Ingbar, S. H., 1956, Thyroxine-binding by sera of pregnant women, *J. Clin. Endocrinol.* **16**:288.
Draper, R. L., 1920, The prenatal growth of the guinea pig, *Anat. Rec.* **18**:369.
Drillien, C. M., 1970, The small-for-date infant: Etiology and prognosis, *Pediatr. Clin. North Am.* **17**:9.
D'Silva, J. L., and Harrison, R. J., 1953, The distribution of radioactive potassium in the uterus of pregnant rats and guinea pigs, *J. Emryol. Exp. Morphol.* **1**:357.
Duyne, C. M., Havel, R. J., and Felts, J. M., 1962, Placental transfer of palmitic acid-1-C^{14} in rabbits, *Am. J. Obstet. Gynecol.* **84**:1069.
Earle, D. P., Bakuru, H., and Hirsch, D., 1951, Plasma potassium level in newborn, *Proc. Soc. Exp. Biol. Med.* **76**:756.
Eastman, N. J., and Jackson, E., 1968, Weight relationship in pregnancy, I. The bearing of maternal weight gain and prepregnancy weight on birth weight in full term pregnancies, *Obstet. Gynecol.* **23**:1002.
Ebbs, J. H., Tisdall, F. F., and Scott, W. A., 1941, The influence of prenatal diet on the mother and child, *J. Nutr.* **22**:515.
Edgar, W., and Rice, H. W., 1956, Administration of iron in ante-natal clinics, *Lancet* **1**:599.
Ely, P. A., 1966, The placental transfer of hexose and polyols in the guinea pig as shown by umbilical perfusion of the placenta, *J. Physiol. (London)* **184**:255.
Enders, R. H., Judd, R. M., Donohue, T. M., and Smith, C. H., 1976, Placental amino acid uptake, III. Transport systems for neutral amino acids, *Am. J. Physiol.* **230**:706.
Eschelman, M. K., 1975, Diet during pregnancy in the sixteenth and seventeenth centuries, *J. Hist. Med.* **30**:23.
Fay, J., Cartwright, G. E., and Wintrobe, M. M., 1949, Studies on free erythrocyte protoporphyrin, serum iron, serum iron-binding capacity and plasma copper during normal pregnancy, *J. Clin. Invest.* **28**:487.
Felig, P., Kim, Y. J., and Lynch, V., 1972, Amino acid metabolism during starvation in human pregnancy, *J. Clin. Invest.* **51**:1195.
Fitzhardinge, P. M., and Stevens, E. M., 1972, The small-for-date infant. II. Neurological and intellectual sequelae, *Pediatrics* **50**:50.
Fleming, A. F., Martin, J. D., Hahnel, R., and Westlake, A. J., 1974, Effects of iron and folic

acid antenatal supplements on maternal haematology and fetal well-being, *Med. J. Aust.* **2**:429.

Flexner, L. B., and Gellhorn, A., 1942, The comparative physiology of placental transfer, *Am. J. Obstet. Gynecol.* **43**:965.

Flexner, L. B., Cowie, D. B., Hellman, L. M., Wilde, W. S., and Vosburgh, G. J., 1948, The permeability of the human placenta to sodium in normal and abnormal pregnancies and the supply of sodium to the human fetuses determined with radioactive sodium, *Am. J. Obstet. Gynecol.* **55**:469.

Food and Nutrition Board, 1974, Recommended Dietary Allowances, National Academy of Sciences, Washington, D.C.

Fox, H., 1967, Senescence of placental villi, *J. Obstet. Gynecol.* **74**:881.

Frazer, J. F. D., and Huggett, A. St., G., 1970, The partition of nutrients between mother and conceptuses in the pregnant rat, *J. Physiol.* **207**:783.

Freinkel, N., 1972, Accelerated starvation and the mechanisms for conservation of maternal nitrogen during pregnancy, *Isr. J. Med. Sci.* **8**:426.

Friedlander, M., Laskey, N., and Silbert, S., 1936, Effect of estrogenic substance on blood volume, *Endocrinology* **20**:329.

Friedman, S., Gans, B., Eckerlin, B., Goldman, J., Kaufman, H., and Rumny, M., 1969, Placental hormone activity after removal of the fetus in a case of advanced abdominal pregnancy, *J. Obstet. Gynaecol. Br. Commonw.* **76**:554.

Friedman, W. F., and Roberts, W. C., 1966, Vitamin D and the supervalvular aortic stenosis syndrome. The transplacental effect of vitamin D on the aorta of the rabbit, *Circulation* **34**:77.

Fuchs, F., and Fuchs, A. R., 1956, Studies on the placental transfer of phosphate in the guinea pig, *Acta Physiol. Scand.* **38**:379.

Gal, I, and Parkinson, C. E., 1974, Effects of nutrition and other factors on pregnant women's serum vitamin A levels, *Am. J. Clin. Nutr.* **27**:688.

Gal, I., Sharman, I. M., and Pryse-Davis, J., 1972, Vitamin A in relation to human congenital malformations, in: *Advances in Teratology*, Vol. 5 (D. H. M. Wollam, ed.), Chap. 6, Academic Press, Inc., New York.

Gardner, L. I., 1952, Tetany and parathyroid hyperplasia in the newborn infant: Influences of dietary phosphate load, *Pediatrics* **9**:534.

Gaull, G. E., Rahia, N. C., Saarikoski, S., and Sturman, J. A., 1973, Transfer of cyst(e)ine and methionine across the human placenta, *Pediatr. Res.* **7**:908.

Gelfand, M. M., Stream, G. J., Pavilanis, V., and Steinberg, J., 1960, Studies in placental permeability, *Amer. J. Obstet. Gynecol.* **79**:117.

Gemzell, C. A., 1952, Increase in the formation and secretion of ACTH in rats following the administration of oestradiol monobenzoate, *Acta Endocrinol.* **11**:221–228.

Genazzani, A. R., Pocola, F., Neri, P., and Fioretti, P., 1972, Human chorionic somatomammotropin (HCS): plasma levels in normal and pathological pregnancies and their correlation with placental function, *Acta Endocrinol.* **168**:1.

Ghadimi, H., and Pecora, P., 1964, Free amino acids of cord plasma as compared with maternal plasma during pregnancy, *Pediatrics* **33**:500.

Giles, C., 1966, An account of 335 cases of megaloblastic anemia of pregnancy and the puerperium, *J. Clin. Pathol.* **19**:1.

Girija, N. S., Pradham, D. S., and Sreenivasan, A., 1965, Effect of protein depletion on ribonucleic acid metabolism in rat liver, *Indian Biochem. J.* **2**:85.

Giroud, A., 1968, Nutrition of the embryo, *Fed. Proc.* **27**:163.

Gitlin, D., and Biasucci, A., 1969, Development of gamma G, gamma A, gamma M, beta IC-beta IA, CI esterase inhibitor, ceruloplasmin, transferrin, hemopexin, haptoglobin, fibrinogen, plasminogen, alpha 1-antitrypsin, orosomucoid, beta-lipoprotein, alpha 2-macroglobulin, and prealbumin in the human conceptus, *J. Clin. Invest.* **48**:1433.

Gitlin, D., and Koch, C., 1968, On the mechanism of maternofetal transfer of human albumin and gamma-G globulin in the mouse, *J. Clin. Invest.* **47**:1204.

Gitlin, D., Kumate, J., Urrusti, J., and Morales, C., 1964, The selectivity of the human placenta in the transfer of plasma proteins from mother to fetus, *J. Clin. Invest.* **43:**1938.

Glendening, M. B., Margolis, A. J., and Pape, E. W., 1961, Amino acid concentrations in fetal and maternal plasma, *Am. J. Obstet. Gynecol.* **81:**591.

Goldwater, W. H., and Stetten, DeW., Jr., 1947, Studies in fetal metabolism, *J. Biol. Chem.* **169:**722.

Goodlad, G. A. J., and Munro, H. N., 1959, Diet and the action of cortisone on protein metabolism, *Biochem. J.* **73:**343.

Green, J. G., 1966, Serum cholesterol changes in pregnancy, *Am. J. Obstet. Gynecol.* **95:**387.

Green, R., Colamn, N., and Metz, J., 1975, Comparison of results of microbiologic and radioisotopic assays for serum vitamin B_{12} during pregnancy, *Am. J. Obstet. Gynecol.* **122:**4.

Greiss, F. C., and Anderson, S. G., 1970, Effect of ovarian hormones on the uterine vascular bed, *Am. J. Obstet. Gynecol.* **107:**829.

Gross, S., and Melhorn, D. K., 1974, Vitamin E-dependent anemia in the premature infant, *J. Pediatr.* **85:**753.

Grossowicz, N., Aronovitch, M., Rachmilewitz, G., Izak, A., Sadovsky, A., and Bercovici, B., 1960, Folic and folinic acid in maternal and foetal blood, *Br. J. Haematol.* **6:**296.

Gruenwald, P., 1975, The supply line of the fetus, in: *The Placenta* (P. Gruenwald, ed.), pp. 1–17, University Park Press, Baltimore.

Gryboski, W. A., and Spiro, H. M., 1956, The effect of pregnancy on gastric secretion, *New Engl. J. Med.* **255:**1131.

Gusseck, D. J., Yuen, P., and Longo, L. D., 1975, Amino acid transport in placental slices. Mechanisms of increased accumulation by prolonged incubation, *Biochim. Biophys. Acta* **401:**278.

Haddad, J. G., Jr., Boissean, V., and Avioli, L. V., 1971, Placental transfer of vitamin D_3 and 25-hydroxycholecalciferol in the rat, *J. Lab. Clin. Med.* **77:**908.

Hafez, E. S. E., Lindsay, D. R., and Moustafa, L., 1967, Effect of feed intake of pregnant rabbits on nutritional reserves of neonates, *Am. J. Vet. Res.* **28:**1153.

Hagerman, D. D., 1962, Metabolism of tissues from pregnant diabetic rats *in vitro, Endocrinology* **70:**88.

Hagerman, D., D., and Villee, C. A., 1952, The transport of fructose by human placenta, *J. Clin. Invest.* **31:**911.

Hahn, P. F., Carothers, E. L., Darby, W. J., Martin, M., Sheppard, C. W., Cannon, R. O., Beam, A. S., Densen, P. M., Peterson, J. C., and McClellan, G. S., 1951, Iron metabolism in human pregnancy as studied with the radioactive isotope Fe^{59}, *Am. J. Obstet. Gynecol.* **61:**477.

Hall, C. A., 1971, Vitamin B_{12} binding proteins, *Ann. Int. Med.* **75:**297.

Hall, M. H., 1972, Folic acid deficiency and abruptio placentae, *J. Obstet. Gynaecol. Br. Commonw.* **79:**222.

Hall, S. M., and Zeman, F. J., 1968, Kidney function of the progeny of rats fed a low protein diet, *J. Nutr.* **95:**49.

Hambridge, K. M., and Drogmueller, W., 1974, Changes in plasma and hair concentration of zinc, copper, chromium and manganese during pregnancy, *Obstet. Gynecol.* **44:**666.

Hamfelt, A., and Hahn, L., 1969, Pyridoxal phosphate concentration in plasma and tryptophan load test during pregnancy, *Clin. Chim. Acta* **25:**91.

Hamfelt, A., and Tuvemo, T., 1972, Pyridoxal phosphate and folic acid concentration in blood and erythrocyte aspartate amino transferase activity during pregnancy, *Clin. Chim. Acta* **41:**287.

Hamil, B. L., Munks, B., Moyer, E. Z., Krucher, M., and Williams, H. H., 1947, Vitamin C in the blood and urine of the newborn and in cord and maternal blood, *Am. J. Dis. Child.* **74:**417.

Hammond, J., 1944, Physiological factors affecting birth weight, *Proc. Nutr. Soc.* **2:**8.

Hamosh, M., Clary, T. R., Chernick, S. S., and Scow, R. O., 1970, Lipoprotein lipase activity of adipose and mammary tissue and plasma triglyceride in pregnant and lactating rats, *Biochim. Biophys. Acta* **210:**473.

Hansen, A. E., Wiese, H. F., Adam, D. J. D., Boelsche, A. N., Haggard, M. E., Davis, H., Newsom, W. T., and Pesut, L., 1964, Influence of diet on blood serum lipids in pregnant and newborn infants, *Am. J. Clin. Nutr.* **15:**11.

Hansen, R., 1937, Zur Physiologie des Magens in der Schwangershaft, *Gynak* **61:**2306.

Hastings-Roberts, M. M., and Zeman, F., 1977, Effects of protein deficiency, pair-feeding, or diet supplementation on maternal, fetal and placental growth in rats, *J. Nutr.* **107:**973.

Hawrylewicz, C. J., Kissane, J. Q., Blair, W. H., and Heppner, C. A., 1975, Effect of maternal protein malnutrition on neonatal lung development and mitochondrial functions, *Nutr. Rep. Int.* **7:**253.

Hazelwood, R. L., and Nelson, M. M., 1965, Steroid maintenance of pregnancy in rats in the absence of dietary protein, *Endocrinology* **77:**999.

Hearey, R. P., and Skillman, T. G., 1971, Calcium metabolism in normal human pregnancy, *J. Clin. Endocrinol.* **33:**661.

Heijkenskjold, F., and Gemzell, C. A., 1957, Glycogen content in the placenta of diabetic mothers, *Acta Paediatr.* **46:**74.

Heinrich, H. C., Bartels, H., Heinisch, B., Hausmann, K., Kuse, R., Humke, W., and Mauss, H. J., 1968, Intestinale ^{59}Fe-resorption und prälatenter Eisenmangel während er Gravidität des Menschen, *Klin. Wochschr.* **46:**199.

Hellegers, A., Okuda, K., Nesbit, R. E. L., Smith, D. W., and Chow, B. F., 1957, Vitamin B_{12} absorption in pregnancy and in the newborn, *Am. J. Clin.* **5:**327.

Heller, S., Salkeld, R. M., and Korner, W. F., 1973, Vitamin B_6 status in pregnancy, *Am. J. Clin. Nutr.* **26:**1339.

Heller, S., Salkeld, R. M., and Korner, W. F., 1974a, Vitamin B_1 status in pregnancy, *Am. J. Clin. Nutr.* **27:**1221.

Heller, S., Salkeld, R. M., and Korner, W. F., 1974b, Riboflavin status in pregnancy, *Am. J. Clin. Nutr.* **27:**1225.

Hendricks, C. H., 1964, Patterns of fetal and placental growth, *Obstet. Gynecol.* **24:**357.

Hensleigh, P. A., and Krantz, K. E., 1966, Extracorporeal perfusion of the human placenta, I. Placental transfer of ascorbic acid, *Am. J. Obstet. Gynecol.* **96:**5.

Herbert, V., 1970, Drugs effective in megaloblastic anemias: vitamin B_{12} and folic acid, in: *The Pharmacological Basis of Therapeutics*, 4th ed. (L. S. Goodman and A. Gilman, eds.), pp. 1414–1444, Macmillan Publishing Co., Inc., New York.

Herbert, V., 1971, Recent developments in cobalamin metabolism, in: *The Cobolamins* (H. R. V. Arnstein and R. J. Wrighton, eds.), pp. 2–16, J.& A. Churchill Ltd., London.

Herbert, V., and Tisman, G., 1971, Iron deficiency and megaloblastoid marrow, *New Engl. J. Med.* **284:**448.

Herbert, V., and Zalusky, R., 1962, Interrelations of vitamin B_{12} and folic acid metabolism; folic acid clearance studies, *J. Clin. Invest.* **41:**1263.

Herbert, V., Colman, N., Spivack, M., Ocasio, E., Ghanta, V., Kimmel, K., Brenner, L., Freundlich, J., and Scott, J., 1975, Folic acid deficiency in the United States: Folate assays in a prenatal clinic, *Am. J. Obstet. Gynecol.* **123:**175.

Herbinger, W. von, and Wichmann, H., 1967, Die extrazellularen und intraerythrozytaren Electrolyte während der zweiten Schwangershaftshälfte, *Gynaecologia (Basel)* **163:**1.

Hershfield, M. S., and Nemeth, A. M., 1968, Placental transport of free palmitic and linoleic acids in the guinea pig, *J. Lipid Res.* **9:**460.

Hesch, R. D., Woodhead, S., Huefrer, M., and Waf, H., 1973, Gastrointestinal stimulation of calcitonin in adults and newborns, *Horm. Metab. Res.* **5:**235.

Hibbard, B. M., 1975, Folates and the fetus, *S. Afr. Med. J.* **49:**1223.

Hibbard, B. M., and Hibbard, E. D., 1963, Aetiological factors in abruptio placentae, *Br. Med. J.* **2:**1430.

Higgins, A. C., Crampton, E. W., and Moxley, J. E., 1973, Nutrition and the outcome of pregnancy, in: *Endocrinology* (R. O. Scow, ed.), pp. 1071–1077, Excerpta Medica/American Elsevier, New York.

Hill, D. E., 1975, Cellular growth of the rhesus monkey placenta, in: *Fetal and Postnatal Cellular Growth* (D. B. Cheek, ed.), pp. 283–288, John Wiley & Sons, Inc., New York.

Hill, D. E., Myers, R. E., Holt, A. B., Scott, R. E., and Cheek, D. B., 1971, Fetal growth retardation produced by experimental placental insufficiency in the rhesus monkey, II. Chemical composition of the brain, liver, muscle, carcass, *Biol. Neonate* **19:**68.

Hill, P. M. M., and Young, M., 1973, Net placental transfer of free amino acids against varying concentrations, *J. Physiol.* **235:**409.

Hillman, R. V., Coband, P. G., Nelson, D. E., Arpio, P. D., and Tufane, R. J., 1963, Pyridoxine supplementation during pregnancy, clinical and laboratory observations, *Am. J. Clin. Nutr.* **12:**427.

Hirsch, J., Farguhar, J., Ahreus, E. H., Jr., Peterson, M. L., and Stoffel, W., 1960, Studies of adipose tissues in man, a microtechnique for sampling and analysis, *Am. J. Clin. Nutr.* **8:**499.

Hoch, H., and Marrack, J. R., 1948, The composition of the blood of women during pregnancy and after delivery, *J. Obstet. Gynecol.* **55:**1.

Hodges, R. E., Hodd, J., Conham, J. E., Suaberlich, H. E., and Baker, E. M., 1971, Clinical manifestations of ascorbic acid deficiency in man, *Am. J. Clin. Nutr.* **24:**432.

Holt, A. B., Cheek, D. B., Mellits, D. E., and Hill, D. E., 1975, Brain size and the relation of the primate to the non primate, in: *Fetal and Postnatal Cellular Growth* (D. B. Cheek, ed.), pp. 23–44, John Wiley & Sons, Inc., New York.

Hosemann, H., 1949, Schwangerschaftsdauer und Gewicht der Placenta, *Arch. Gynaekol.* **176:**453.

Huckabee, W. E., Metcalfe, J., Prystowsky, H., and Barron, D. H., 1961, Blood flow and oxygen consumption of the pregnant uterus, *Am. J. Physiol.* **200:**274.

Huckabee, W. E., Crenshaw, C., Curet, L. B., Mann, L., and Barron, D. H., 1970, Effect of exogenous oestrogen on blood flow and oxygen consumption of the uterus of the nonpregnant ewe, *Q. J. Exp. Physiol.* **55:**16.

Huckabee, W. E., Crenshaw, C., Curet, L. B., Mann, L., and Barron, D. H., 1972, Uterine blood flow and oxygen consumption in the unrestrained pregnant ewe, *Q. J. Exp. Physiol.* **57:**12. **57:**12.

Huggett, A. St. G., 1929, Maternal control of placental glycogen, *J. Physiol. (London),* **67:**360.

Huggett, A. St. G., and Morrison, S. D., 1955, Placental glycogen in the rabbit, *J. Physiol. (London)* **129:**68P.

Hummel, L., Schinmeister, W., and Wagner, H., 1975, Quantitative evaluation of the maternal-fetal transfer of free fatty acids in the rat, *Biol. Neonate* **26:**263.

Hunt, J. N., and Murray, F. A., 1958, Gastric function in pregnancy, *J. Obstet. Gynaecol. Br. Emp.* **65:**78.

Hurley, L. S., and Swenerton, H., 1966, Congenital malformations resulting from zinc deficiency in rats, *Proc. Soc. Exp.* **123:**692.

Hytten, F. E., and Cheyne, G. A., 1969, The size and composition of the human pregnant uterus, *J. Obstet. Gynaecol. Br. Commonw.* **76:**400.

Hytten, F., and Cheyne, G., 1972, The aminoaciduria of pregnancy, *J. Obstet. Gynaecol. Br. Commonw.* **79:**424.

Hytten, F. E., and Leitch, I., 1971, *The Physiology of Human Pregnancy,* p. 332, Blackwell Scientific Publications Ltd., Oxford.

Hytten, F. E., and Lind, T., 1973, *Diagnostic Indices in Pregnancy,* CIBA-Geigy Ltd., Basel.

Hytten, F. E., and Paintin, D. B., 1963, Increase in plasma volume during pregnancy, *J. Obstet. Gynaecol. Br. Commonw.* **70:**402.

Irwin, M. I., and Hutchins, B. K., 1976, A prospect of research on vitamin C requirements of man, *J. Nutr.* **106:**821.

Iyengar, L., 1967, Effects of dietary supplements late in pregnancy on the expectant mother and her newborn, *Indian J. Med. Res.* **55:**85.

Iyengar, L., and Apte, S. V., 1970, Prophylaxis of anemia in pregnancy, *Am. J. Clin. Nutr.* **23:**725.

Iyengar, L., and Rajalakshmi, K., 1975, Effect of folic acid supplement on the birth weights of infants, *Am. J. Obstet. Gynecol.* **122:**332.

James, E., Meschia, G., and Battaglia, F. C., 1971, A-V differences of free fatty acids and glycerol in the bovine umbilical circulation, *Proc. Soc. Exp. Biol. Med.* **138:**823.

Jansson, J., 1969, Xenon clearance in the myometrium of pregnant and non-pregnant women, *Acta Obstet. Gynecol. Scand.* **48:**302.

Javert, C. T., and Stander, H. J., 1943, Plasma vitamin C and prothrombin concentration in pregnancy and in threatened, spontaneous and habitual abortion, *Surg. Gynecol. Obstet.* **76:**115.

Johansson, E. D. B., 1969, Progesterone levels in peripheral plasma during the luteal phase of the normal human menstrual cycle measured by a rapid competitive binding technique, *Acta Endocrinol.* **61:**592.

Jones, K. M., Lloyd-Jones, R., Riondel, A., Tait, J. E., Tait, S. A., Bulbrook, R. D., and Greenwood, F. C., 1959, Aldosterone secretion and metabolism in normal men and women and in pregnancy, *Acta Endocrinol.* **30:**321.

Jost, A., and Picon, L., 1970, Hormonal control of fetal development and metabolism, *Adv. Metab. Disord.* **4:**123.

Kalman, S. M., 1958, Effects of estrogens on uterine blood flow in the rat, *J. Pharmacol. Exp. Ther.* **124:**179.

Kaplan, S. L., Gurpide, E., Sciarra, J. J., and Grumbach, M. M., 1968, Metabolic clearance rate and production rate of chorionic growth hormone prolactin in late pregnancy, *J. Clin. Endocrinol. Metab.* **28:**1450.

Karte, H., 1969, The development of immuonproteins in the pre- and postnatal time, *Z. Klin. Chem.* **7:**204.

Kayden, H. J., Dancis, J., and Money, W. L., 1969, Transfer of lipids across the guinea pig placenta, *Am. J. Obstet. Gynecol.* **104:**564.

Kelly, H. J., Sloan, R. E., Hoffman, W., and Saunders, C., 1951, Accumulation of nitrogen and six minerals in the human fetus during gestation, *Hum. Biol.* **23:**61.

Kennedy, G. C., Pearle, W. M., and Parrot, D. M. V., 1958, Liver growth in the lactating rat, *J. Endocrinol.* **17:**158.

Kenney, M. A., 1969, Development of spleen and thymus in offspring of protein-deficient rats, *J. Nutr.* **98:**202.

Kerr, A., Jr., 1943, Weight gain in pregnancy and its relation to weight of infants and to length of labor, *Am. J. Obstet. Gynecol.* **45:**950.

Kerr, C., Loken, H. F., Glendening, M. B., Gordon, G. S., and Page, E. W., 1962, Calcium and phosphorus dynamics in pregnancy, *Amer. J. Obstet. Gynecol.* **83:**2.

Khattab, A. K., and Forfar, J. O., 1970, Interrelationships of calcium, phosphorus and glucose levels in mother and newborn infant, *Biol. Neonate* **15:**26.

Khattab, A. K., Nagdy, S. A., Moural, K. A. H., and El Fizghal, H. I., 1970, Foetal maternal ascorbic acid gradient in normal Egyptian subjects, *J. Trop. Pediatr.* **16:**112.

Khera, K. S., 1976, Teratogenicity study in rats given high doses of pyridoxine (vitamin B_6) during organogenesis, *Experientia* **31:**469.

Killander, A., and Vahlquist, B., 1954, B_{12}-vitamin-koncentrationen i serum fran fullgangna och prematurt fodda barn, *Nord. Med.* **51:**777.

King, E. Q., 1967, Acute cardiac failure in the newborn due to thiamine deficiency, *Exp. Med. Surg.* **25:**173.

King, J. C., Calloway, D. H., and Morgan, S., 1973, Nitrogen retention, total body ^{40}K, and weight gain in teenage pregnant girls, *J. Nutr.* **103:**772.

King, J. C., Alberts, J., and Kodama, A. M., 1976, Nitrogen and potassium retention in healthy adult pregnant women, *Fed. Proc.* **35:**597.

King, W. E., 1945, Vitamin studies in abortions, *Surg. Gynecol. Obstet.* **80:**139.

Kitay, D. Z., and Harbort, R. A., 1975, Iron and folic acid deficiency in pregnancy, *Clin. Perinatol.* **2:**255.

Klopper, A., and Billewicz, W., 1963, Urinary excretion of oestriol and pregnanediol during normal pregnancy, *J. Obstet. Gynaecol. Br. Commonw.* **70:**1024.

Knipscheer, R. J. U. L., 1975, Megaloblastic anemia in pregnancy and folate, in: *Aspects of Obstetrics Today* (T. K. A. B. Eskes, ed.), pp. 87–99, Excerpta Medica/American Elsevier, New York.

Knopp, R. H., Herrera, E., and Freinkel, N., 1970, Carbohydrate metabolism in pregnancy, VIII. Metabolism of adipose tissue isolated from fed and fasted pregnant rats during late gestation, *J. Clin. Invest.* **49:**1438.

Knopp, R. H., Sandek, C. D., and Avky, R. A., 1973, Two phases of adiopose tissue metabolism in pregnancy: Maternal adaptations for fetal growth, *Endocrinology* **92:**984.

Knopp, R. H., Boroush, M. A., and O'Sullivan, J. B., 1975, Lipid metabolism in pregnancy, II. Postheparin lipolytic activity and hypertriglyceridemia, *Metabolism* **24:**481.

Kohrs, M. B., Harper, A. E., and Kerr, G. R., 1976, Effects of a low-protein diet during pregnancy on the rhesus monkey, I. Reproductive efficiency, *Am. J. Clin. Nutr.* **29:**136.

Kulhaneck, J. F., Meschia, G., Makowski, E. L., and Battaglia, F. C., 1974, Changes in DNA content and urea permeability of the sheep placenta, *Am. J. Physiol.* **226:**1257.

Kunin, C. M., and Finland, M., 1961, Clinical pharmacology of the tetracyline antibiotics, *Clin. Pharmacol. Ther.* **2:**51.

Laga, E. M., Driscoll, S. G., Munro, H. N., 1972a, Comparison of placentas from two socio-economic groups, I. Morphometry, *Pediatrics* **50:**24.

Laga, E. M., Driscoll, S. G., and Munro, H. N., 1972b, Comparison of placentas from two socio-economic groups, II. Biochemical characteristics, *Pediatrics* **50:**33.

Larkin, E. C., Weintraub, L. R., and Crosby, W. H., 1970, Iron transport across the rabbit allantoic placenta, *Am. J. Physiol.* **218:**7.

Laurel, C. B., and Morgan, E., 1964, Iron exchange between transferrin and the placenta in the rat, *Acta Physiol. Scand.* **62:**271.

Lawrence, A. C., 1962, Iron status in pregnancy, *J. Obstet. Gynaecol. Br. Commonw.* **69:**29.

Lawrence, C., and Klipstein, F. A., 1967, Megaloblastic anemia of pregnancy in New York City, *Am. Intern. Med.* **66:**25.

Lechtig, A., Habicht, J. P., Delgado, H., Klein, R. E., Yarbrough, C., and Martorell, R., 1975, Effect of food supplementation during pregnancy on birthweight, *Pediatrics* **56:**508.

Lechtig, A., Rosso, P., Delgado, H., Bassi, J., Martorell, R., Yarborough, C., Winick, M., and Klein, R. E., 1977, Effects of moderate maternal malnutrition on the levels of alkaline ribonuclease activity of the human placenta, *Ecol. Food Nutr.* **6:**83.

Lederman, S. A., and Rosso, P., 1978, The effects of food restriction during pregnancy on maternal weight and body composition, *Fed. Proc.* **37:**1465.

Lee, C. J., and Chow, B. F., 1965, Protein metabolism in the offspring of the underfed mother rats, *J. Nutr.* **87:**439.

Lee, C. J., and Chow, B. F., 1968, Metabolism of proteins by progeny of underfed mother rats, *J. Nutr.* **94:**20.

LeMaire, W. J., Conly, P. W., Moffat, A., and Cleveland, W. W., 1970, Plasma progesterone secretion by the corpus luteum of term pregnancy, *Am. J. Obstet. Gynecol.* **108:**132.

Lindheimer, M. D., and Katz, A. I., 1973, Sodium and diurectics in pregnancy, *New Engl. J. Med.* **288:**891.

Lister, U. M., 1964, Structural changes in the capillaries of human chorionic villi occurring with age, *J. R. Microscop. Soc.* **83:**455.

Litonjua, A. D., 1967, Studies on the glucose metabolism of the human placenta. Effects of insulin, *Acta Med. Phillippina* **3:**247.

Lockhardt, H., Kirkwood, S., and Harns, R. S., 1943, Effect of pregnancy and puerperium on the thiamine status of women, *Am. J. Obstet. Gynecol.* **46:**358.

Lockhead, J., and Cramer, W., 1908, The glucogenic changes in the placenta and the fetus of the pregnant rabbit: A combination to the chemistry of growth, *Proc. R. Soc. (London)* **B80:**263.

Loh, K. R. W., Shrader, R. E., and Zeman, F. J., 1971, Effects of maternal protein deprivation on neonatal intestinal absorption in rats, *J. Nutr.* **101:**1663.

Longo, L. D., 1972, Disorders of placental transfer, in: *Pathophysiology of Gestation*, Vol. 2 (N. S. Assali and C. R. Brinkman, eds.), pp. 1–76, Academic Press, Inc., New York.

Longo, L. D., and Kleinzeller, A., 1970, Transport of monosaccharides by placental cells, *Fed. Proc.* **29:**802.

Longo, L. D., Power, G. G., and Forster, R. E., 1967, Respiratory function of the placenta as determined with carbon monoxide in sheep and dogs, *J. Clin. Invest.* **46:**812.

Longo, L. D., Yuen, P., and Gusseck, D. J., 1973, Anabolic glycogen-dependent transport of amino acids by the placenta, *Nature* **243:**531.

Loveland, G., Maurer, E. E., and Snyder, R. R., 1931, The diminution of the glycogen store of the rabbit placenta during the last third of pregnancy, *Anat. Rec.* **49:**265.

Low-Beer, T. S., McCarthy, C. F., Austad, W. I., Brzechwa-Audukiewicz, A., and Read, A. E., 1968, Serum vitamin B_{12} levels and vitamin B_{12} binding capacity in pregnant and non-pregnant Europeans and West Indians, *Br. Med. J.* **4:**160.

Lowenstein, L., Hsieh, Y. S., Brunton, L., Deleeuw, N. K., and Cooper, B. A., 1962, Nutritional deficiency and anemia in pregnancy, *Postgrad. Med. J.* **31:**72.

Lubchenko, L. O., Hanshan, C., Dressler, M., and Boyd, E., 1963, Intrauterine growth as estimated from liveborn birth-weight data at 24 to 42 weeks of gestation, *Pediatrics* **32:**793.

Lucey, J. F., and Dolan, R. G., 1959, Hyperbilirubinemia of newborn infants associated with parenteral administration of a vitamin K analog to the mothers, *Pediatrics* **23:**553.

Lumeng, L., Clearly, R. E., Wagner, R., Yu, P. L., and Li, T. K., 1976, Adequacy of vitamin B_6 supplementation during pregnancy: A prospective study, *Am. J. Clin. Nutr.* **29:**1376.

Lund, C. J., and Kimble, M. S., 1943, Plasma vitamin A and carotene of the newborn infant, *Am. J. Obstet Gynecol.* **46:**207.

Lunstrom, P., 1950, Studies on erythroid elements and serum iron in normal pregnancy, *Acta Med. Soc. Upsalien* **55:**1.

Lunell, N. O., Persson, B., and Sterky, G., 1969, Dietary habits during pregnancy: A pilot study, *Acta Obstet. Gynecol. Scand.* **48:**187.

Lust, J. E., Hagerman, D. D., and Villee, C. A., 1953, The transport of riboflavin by human placenta, *J. Clin. Invest.* **33:**38.

Majumdar, A. P. N., and Wadsworth, G. R., 1974, The influence of the level of protein in the diet of the pregnant mouse on transfer of iron to the foetus, *Nutr. Rep. Int.* **9:**47.

Manahan, C. P., and Eastman, N. J., 1938, The cevitamic acid content of fetal blood, *Bull. Johns Hopkins Hosp.* **62:**478.

Mardell, M., Symmons, C., and Zilva, J. F. A., 1969, A comparison of the effect of oral contraceptives, pregnancy and sex on iron metabolism, *J. Clin. Endocrinol.* **29:**1489.

Martin, M. P., Bridgforth, E., McGanity, W. J., and Darby, W. J., 1957, The Vanderbilt Cooperative study of maternal and infant nutrition, X. Absorbic acid, *J. Nutr.* **62:**201.

Mason, M., and Rivers, J., 1971, Plasma ascorbic acid levels in pregnancy, *Am. J. Obstet. Gynecol.* **109:**960.

Mata, L. J., Kronmal, R. A., Urrutia, J. J., and Garcia, B., 1975, Antenatal events and postnatal growth and survival of children. Prospective observation in a rural Guatemalan village, *Proc. Western Hemishpere Nutr. Congr. IV* (P. L. White and N. Selvey, eds.), pp. 107–116, Publishing Sciences Group, Inc., Boston, Mass.

McCartney, C. P., Pottinger, R. E., and Harrod, J. P., 1959, Alterations in body composition during pregnancy, *Am. J. Obstet. Gynecol.* **77:**1038.

McDonald, M. S., Hutchison, D. L., Helper, M., and Flynn, E., 1965, Movement of calcium in both directions across the primate placenta, *Proc. Soc. Exp. Biol. Med.* **119:**476.

MacGillivray, I., and Tovey, J. E., 1957, A study of the serum protein changes in pregnancy and toxaemia using paper electrophoresis, *J. Obstet. Gynaecol. Br. Emp.* **64:**361.

McKay, D. G., Hertig, W. T., Adams, E. C., and Richardson, M. V., 1958, Observations on the human placenta, *Obstet. Gynecol.* **12:**1.

McKeown, T., and Record, R. G., 1953, The influence of placental size on foetal growth according to sex and order of birth, *J. Endocrinol.* **10:**73.

McLaren, D. S., 1963, *Malnutrition and the Eye,* Academic Press, Inc., New York.

McLeod, K. I., Goldrick, R. B., and Whyte, H. M., 1972, The effect of maternal malnutrition on the progeny in the rat. Studies on the growth, body composition and organ cellularity in first and second generation progeny, *Aust. J. Exp. Biol. Med. Sci.* **50:**435.

Mehrotra, J., Jandon, J. L., and Thukla, R. C., 1969, Study of gastric secretion with reference to pregnancy in Indian women, *Indian J. Physiol. Pharmacol.* **13:**64.

Melnick, D., 1942, Vitamin B_1 (thiamine) requirement of man, *J. Nutr.* **24:**139.

Meschia, G., Battaglia, F. C., and Bruns, P. D., 1967, Theoretical and experimental study of transplacental diffusion, *J. Appl. Physiol.* **22:**1171.

Metcalfe, J., Rowney, S. L., Ramsey, L. H., Reid, D. E., and Burwell, C. S., 1955, Estimation of uterine blood flow in normal human pregnancy at term, *J. Clin. Invest.* **34:**1632.

Metcalfe, J., Bartels, H., and Moll, W., 1967, Exchange in the pregnant uterus, *Physiol. Rev.* **47:**782.

Metz, J., Festenstein, H., and Welsh, P., 1965, Effect of folic acid and vitamin B_{12} supplementation during pregnancy in a population subsisting on a suboptimal diet, *Am. J. Clin. Nutr.* **16:**472.

Michel, C. F., 1971, Der Serum-Magnesium-Gehalt in der Schawangerschaft und unter der Geburt im Vergleich zum Serum Kalzium, *Z. Geburtsh. Gynaekol.* **174:**276.

Miller, R. K., and Berndt, W. O., 1974, Characterization of neutral amino acid accumulation by human placental slices, *Am. J. Physiol.* **227:**1236.

Mischel, W., 1963, Der diaplazentare Transport stereoisomerer Aminosäuren, *Arch. Gynaekol.* **198:**181.

Moghissi, K. S., and Hafez, E. S. E. (eds.), 1974, *The Placenta,* Charles C Thomas, Publisher, Springfield, Ill.

Moncrieff, M., and Fadahunsi, T. O., 1974, Congenital rickets due to maternal vitamin D deficiency, *Arch. Dis. Child.* **49:**810.

Monie, I. W., and Nelson, M. M., 1963, Abnormalities of pulmonary and other vessels in rat fetuses from maternal pteroglutamic acid deficiency, *Anat. Rec.* **147:**397.

Morgan, B. O. G., 1975, Effects of prenatal and postnatal undernutrition on development in the rat, Ph.D. thesis, University of London.

Morgan, E. H., 1961, Plasma-iron and haemoglobin levels in pregnancy. The effect of oral iron, *Lancet* **1:**9.

Morphis, L. G., and Gitlin, D., 1970, Maturation of the maternofoetal transport system for human gamma-globulin in the mouse, *Nature* **228:**573.

Mull, J. W., 1936, Variations in serum calcium and phospholipids during pregnancy, II. The effect on the fetal circulation, *J. Clin. Invest.* **15:**513.

Mull, J. W., and Bill, A. H., 1934, Variations in serum calcium and phosphorus during pregnancy, I. Normal variations, *Am. J. Obstet. Gynecol.* **27:**510.

Mullick, S., Bagga, O. P., and DuMullick, V., 1964, Serum lipid studies in pregnancy, *Am. J. Obstet. Gynecol.* **89:**766.

Murray, F. A., Erskine, J. P., and Fielding, Y., 1957, Gastric secretion in pregnancy, *J. Obstet. Gynaecol. Br. Emp..* **64:**373.

Murthy, L. S., Agarural, K. N., and Khanna, S., 1976, Placental morphometric and morphologic alterations in maternal undernutrition, *Am. J. Obstet. Gynecol.* **124:**641.

Naismith, D. J., 1966, The requirement for protein, and the utilization of protein and calcium during pregnancy, *Metabolism* **15:**582.

Naismith, D. J., 1969, The foetus as a parasite, *Proc. Nutr. Soc.* **28:**25.

Naismith, D. J., 1973, Adaptations in the metabolism of protein during pregnancy and their nutritional implications, *Nutr. Rep. Int.* **7:**383.

Naismith, D. J., and Fears, R. B., 1972, Progesterone, the hormone of protein anabolism in early pregnancy, *Proc. Nutr. Soc.* **31:**79A.

Nanda, R., 1974, Effect of vitamin A on the potentiality of rat palatal processes to fuse *in vivo* and *in vitro, Cleft Palate J.* **11:**123.

Nelson, M. M., Baird, C. D. C., Wright, H. V., and Evans, H. M., 1956, Multiple congenital abnormalities in the rat resulting from riboflavin deficiency induced by the antimetabolite galactoflavin, *J. Nutr.* **58:**125.

Newman, R. L., 1947, Blood calcium: A normal curve for pregnancy, *Am. J. Obstet. Gynecol.* **53:**817.

Newman, R. L., 1953, Further observations on serum calcium and phosphorous in pregnancy, *Am. J. Obstet. Gynecol.* **65:**796.

Newman, R. L., 1956, Calcium and phosphorus balance in pregnancy, *Obstet. Gynecol.* **8:**561.

Newman, R. L., 1957, Serum electrolytes in pregnancy parturition and puerperium, *Obstet. Gynecol.* **10:**51.

Norkus, E. P., and Rosso, P., 1975, Changes in ascorbic acid metabolism of the offspring following high maternal intake of this vitamin in the pregnant guinea pig, *Ann. N.Y. Acad. Sci.* **258**:401.

Okuda, K., Helliger, A. E., and Chow, B. F., 1956, Vitamin B_{12} serum level and pregnancy, *Am. J. Clin. Nutr.* **4**:440.

O'Leary, J. A., Novalis, G. S., and Vosburgh, G. J., 1966, Maternal serum copper concentration in normal and abnormal gestations, *Obstet. Gynecol.* **28**:112.

Oppe, T. E., and Redstone, D., 1972, Clinical and chemical correlates in convulsions of the newborn, *Lancet* **1**:135.

Oski, F. A., and Barness, L. H., 1963, Vitamin E deficiency: a previously recognized cause of hemolytic anemia in the premature infant, *J. Pediatr.* **70**:211.

Osofsky, H. J., 1975, Relationships between prenatal medical and nutritional measures, pregnancy outcome, and early infant development in an urban poverty setting, I. The role of nutritional intake, *Am. J. Obstet. Gynecol.* **123**:682.

Otway, S., and Robinson, D. S., 1968, The significance of changes in tissue clearing-factor lipase activity in relation to the lipaemia of pregnancy, *Biochem. J.* **106**:677.

Owen, G. M., Nelson, C. E., Baker, G. L., Conner, W. E., and Jacobs, J. P., 1967, Use of vitamin K in pregnancy. Effect on serum bilirubin and plasma prothrombin in the newborn, *Am. J. Obstet. Gynecol.* **99**:368.

Paintin, D. B., Thompson, A. M., and Hytten, F. E., 1966, Iron and the haemoglobin level in pregnancy, *J. Obstet. Gynecol. Br. Commonw.* **73**:181.

Pape, E. W., Glendening, M. B., Margolis, A., and Harper, H. A., 1957, Transfer of D-and L-histidine across the human placenta, *Am. J. Obstet. Gynecol.* **73**:589.

Parry, E., Shields, R., and Tumbull, A., 1970, The effect of pregnancy on the colonic absorption of sodium, potassium and water, *J. Obstet. Gynecol. Br. Commonw.* **77**:616.

Paterson, P., Phillips, L., and Wood, C., 1967, Relationship between maternal and fetal blood glucose during labor, *Am. J. Obstet. Gynecol.* **98**:938.

Pathak, A., Godwin, H. A., and Prudent, C. M., 1972, Vitamin B_{12} and folic acid values in premature infants, *Pediatrics* **50**:584.

People's League of Health, 1942, Nutrition of expectant and nursing mothers, *Lancet* **2**:10.

Pitkin, R. M., 1975a, Calcium metabolism in pregnancy: A review, *Am. J. Obstet. Gynecol.* **121**:724.

Pitkin, R. M., 1975b, Vitamins and minerals in pregnancy, *Clin. Perinatol.* **2**:221.

Pitkin, R. M., Kaminetsky, H. A., Newton, M., and Pritchard, J. A., 1972, Maternal nutrition. A selective review of clinical topics, *J. Obstet. Gynecol.* **40**:773.

Plentl, A. A., and Gray, M. J., 1959, Total body water, sodium space and total exchangeable sodium in normal and toxemic pregnant women, *Am. J. Obstet. Gynecol.* **78**:472.

Poczekaj, J., Hejduk, J., and Chodera, A., 1963, Behavior of copper in trophoblast and in placenta at term, *Gynaecologia* **155**:155.

Pond, W. G., Strachan, D. N., Sinha, Y. N., Walker, E. F., Jr., Dunn, J. A., and Barnes, P. H., 1969, Effect of protein deprivation of swine during all or part of gestation on birth weight, postnatal growth rate and nucleic acid content or brain and muscle of progeny, *J. Nutr.* **99**:61.

Popjak, G., 1954, The origin of fetal lipids, 1954, *Cold Spring Harbor Symp. Quant. Biol.* **19**:200.

Popjak, G., and Beeckmans, M. L., 1950, Are phospholipids transmitted through the placenta? *Biochem. J.* **46**:99.

Porter, H., 1966, The tissue copper proteins: cerebrocuprein, erythrocuprein, hepatocuprein, and neonatal hepatic mitochondrocuprein, in: *Biochemistry of Copper* (J. Peisach, P. Aisen, and W. E. Blumberg, eds.), pp. 159–174, Academic Press, Inc., New York.

Postman, O. W., Behrman, R. E., and Soltys, P., 1969, Transfer of free fatty acids across the primate placenta, *Am. J. Physiol.* **216**:143.

Pritchard, J. A., 1959, Anemia in obstetrics and gynecology: An evaluation of therapy with parenteral iron, *Am. J. Obstet. Gynecol.* **77**:74.

Pritchard, J. A., and Scott, D. E., 1974, Effect of maternal anemia on fetal growth and development, in: *Birth Defects and Fetal Development: Endocrine and Metabolic Factors* (K. S. Moghissi, ed.), pp. 77–88, Charles C Thomas, Publisher, Springfield, Ill.

Pritchard, J. A., Scott, D. E., Whalley, P. J., and Haling, R. F., 1970, Infants of mothers with magaboloblastic anemia due to folate deficiency, *J. Am. Med. Assoc.* **211:**1982.
Prochownik, L., 1901, Über Ernährungscuren in der Schwangerschaft, *Ther. Monat.* **15:**446.
Purvis, R. J., MacKay, G. S., Coobburn, F., Barrie, W. J., Wilkinson, E. M., Balton, N. R., and Forfar, J. O., 1973, Enamel hypoplasia of the teeth associated with neonatal tetany: A manifestation of maternal vitamin D deficiency, *Lancet* **2:**811.
Quinn-Stricker, C., and Mandel, P., 1968, Étude du renouvellement du RNA des polysomes, du RNA de transfert, et du RNA "messager" dans le foie de rat soumis à une jeûne proteïque, *Bull. Soc. Chim. Biol.* **50:**31.
Raiha, N., 1959, On the placental transfer of vitamin C, *Acta Physiol. Scand.* **45**(suppl. 155):7.
Ramberg, C. F., Jr., Delivoria-Papadopoulos, M., Crandall, E. D., and Kronfeld, D. S., 1973, Kinetic analysis of calcium transport across the placenta, *J. Appl. Physiol.* **35:**682.
Rapoport, M. I., and Beisel, W. R., 1968, Circadian periodicity of tryptophan metabolism, *J. Clin. Invest.* **47:**934.
Ratanasopa, V., Schindler, A. E., Lee, T. Y., and Herrmann, W. C., 1967, Measurement of estriol in plasma by gas liquid chromatography, *Am. J. Obstet. Gynecol.* **99:**295.
Ravelli, G. P., Stein, Z. A., and Susser, M. W., 1976, Obesity in young men after famine exposure in utero and early infancy, *New Engl. J. Med.* **295:**349.
Reboud, P., Groslambert, P., Ollivier, C., and Groulade, J., 1967, Proteïnes et lipides plasmatiques au cours de la gestation normale et du post-partum, *Ann. Biol. Clin.* **25:**383.
Reinken, L., and Marigold, B., 1973, Pyridoxal phosphate values in premature infants, *Int. J. Vit. Nutr. Res.* **43:**472.
Reitz, R. E., Daane, T. A., Woods, J. D., and Weinstein, R. L., 1972, Human parathyroid hormone (HPTH) calcium interrelation in pregnancy and newborn infants, in: *Fourth International Congress of Endocrinology, Washington, D.C.*, Int. Congr. Ser. 256, p. 208, Excerpta Medica Foundation, Amsterdam.
Reynolds, M. L., and Young, M., 1971, The transfer of free alpha-amino nitrogen across the placental membrane in the guinea pig, *J. Physiol. (London)*, **214:**583.
Riopelle, A. J., 1975, Weight gain of non-pregnant and pregnant monkeys fed low protein diets, *Am. J. Clin. Nutr.* **28:**802.
Riopelle, A. J., Hill, C. W., and Li, S. C., 1975, Protein deprivation in primates, V. Fetal mortality and neonatal status of infant monkeys born of deprived mothers, *Am. J. Clin. Nutr.* **28:**989.
Roberts, R. A., Cohen, M. D., and Forfar, J. O., 1973, Antenatal factors neonatal nypocalcaemic convulsions, *Lancet* **2:**809.
Robertson, J. I. S., Weir, R. J., Dusterdieck, G. O., Fraser, R., and Tize, M., 1971, Renin, angiotensin and aldosterone in human pregnancy and the mentrual cycle, *Scot. Med. J.* **16:**183.
Robson, L. C., and Schwarz, M. R., 1975, Vitamin B_6 deficiency and the lymphoid systems, II. Effects of vitamin B_6 deficiency in utero on the immunological competence of the offspring, *Cell Immunol.* **16:**145.
Rodger, F. C., Saiduzzafar, H., Grover, A. D., and Fazal, A., 1963, A reappraisal of the ocular lesion known as Bitot's spot, *Br. J. Nutr.* **17:**475.
Romney, S. L., Reid, D. E., Metcalfe, J., and Burwell, C. S., 1955, Oxygen utilization by the human fetus in utero, *Am. J. Obstet. Gynecol.* **70:**791.
Rosahn, P. D., and Greene, H. S. N., 1936, The influence of uterine factors on the foetal weight of rabbits, *J. Exp. Med.* **63:**901.
Rose, D. P., Strong, R., Adams, P. W., and Harding, P. E., 1972, Experimental vitamin B_6 deficiency and the effect of oestrogen-containing oral contraceptives on tryptophan metabolism and vitamin B_6 requirements, *Clin. Sci.* **42:**465.
Rose, D. P., Strong, R., Folkand, J., and Adams, P. W., 1973, Erythrocyte amino transferase activities in women using oral contraceptives and the effect of vitamin B_6 supplementation, *Am. J. Clin. Nutr.* **26:**48.
Rosen, J. F., Roginsky, M., Nathenson, G., and Finberg, L., 1974, 25-Hydroxyvitamin D plasma levels in mothers and their premature infants with neonatal hypocalcemia, *Am. J. Dis. Child.* **127:**220.

Rosenfeld, C. R., Morris, F. H., Jr., Makowski, E. L., Meschia, G., and Battaglia, F., 1974, Circulatory changes in reproductive tissues of ewes during pregnancy, *Gynecol. Invest.* **5**:252.
Rosso, P., 1975, Changes in transfer of nutrients across the placenta during normal gestation in the rat, *Am. J. Obstet. Gynecol.* **122**:761.
Rosso, P., 1977a, Maternal nutrition, nutrient exchange and fetal growth, in: *Nutritional Disorders of American Women,* Vol. 5 (M. Winick, ed.), *Current Concepts in Nutrition,* pp. 3–25, John Wiley & Sons, Inc., New York.
Rosso, P., 1977b, Maternal–fetal exchange during malnutrition in the rat. Transfer of alpha-amino isobutyric acid, *J. Nutr.* **107**:2002.
Rosso, R., 1977c, Maternal–fetal exchange during malnutrition in the rat. Transfer of glucose and methyl (alpha-D-U-^{14}C glucose) pyranoside, *J. Nutr.* **107**:2006.
Rosso, P., 1978, Maternal nutritional status and plasma volume expansion in the pregnant rat, *Fed. Proc.* **37**:1465.
Rosso, P., and Luke, B., 1978, The influence of maternal weight gain on the incidence of fetal growth retardation (unpublished).
Rosso, P., and Norkus, E., 1976, Prenatal aspects of ascorbic acid metabolism in the albino rat, *J. Nutr.* **106**:767.
Rosso, P., and Winick, M., 1975, Effects of early undernutrition and subsequent refeeding on alkaline ribonuclease activity of rat cerebrum and liver, *J. Nutr.* **105**:1104.
Rosso, P., Wasserman, M., Rozovski, S. J., and Velasco, E., 1976, Effects of maternal undernutrition on placental metabolism and function, in: *The Neonate* (D. S. Young and J. M. Hicks, eds.), p. 59, John Wiley & Sons, Inc., New York.
Rush, D., Davis, H., and Susser, M. W., 1972, Antecedents of low birth weight in Harlem, New York City, *Int. J. Epidemiol.* **1**:393.
Russ, E. M., and Raymunt, J., 1956, Influence of estrogens on total serum, copper and ceruloplasmin, *Proc. Soc. Exp. Biol. Med.* **92**:465.
Russell, J. G. B., and Hill, L. F., 1974, True fetal rickets, *Br. J. Radiol.* **47**:732.
Samaan, N., Yen, S. C. C., Gonzales, D., and Pearson, O. H., 1968, Metabolic effects of placental lactogen (HPL) in man, *J. Clin. Endocrinol. Metab.* **28**:485.
Samaan, N. A., Hill, C. S., Jr., Beceiro, M., Jr., and Schultz, P. N., 1973, Immunoreactive calcitonin in medullary carcinoma of the thyroid and in maternal and cord serum, *J. Lab. Clin. Med.* **81**:611.
Sanstead, H. A., 1973, Zinc nutrition in the U.S., *Am. J. Clin. Nutr.* **26**:1251.
Sauberlich, H. E., Canham, J. E., Baker, E. M., Raica, N. J., and Herman, Y. F., 1970, Biochemical assessment of the nutritional status of vitamin B_6 in the human, *Am. J. Clin. Nutr.* **25**:629.
Sauberlich, H. E., Dowdy, R. P., and Skala, J. H., 1973, Laboratory tests for the assessment of nutritional status, "Critical Reviews," *Clin. Lab. Sci.* **4**(3):215–340.
Schaffer, A. J., 1971, *Diseases of the Newborn,* 3rd ed., W. B. Saunders Company, Philadelphia.
Scheinberg, I. H., Cook, C. D., and Murphy, J. A., 1954, The concentration of copper and ceruloplasmin in maternal and infant plasma at delivery, *J. Clin. Invest.* **33**:963.
Schlicker, S. A., and Cox, D. H., 1968, Maternal dietary zinc and development and zinc, iron, and copper content of the rat fetus, *J. Nutr.* **95**:287.
Schroder, H., Stolp, W., and Leichtweiss, H. P., 1972, Measurements of Na^+ transport in the isolated, artificially perfused guinea pig placenta, *Am. J. Obstet. Gynecol.* **114**:51.
Scott, D. E., and Pritchard, J. A., 1967, Iron deficiency in healthy young college women, *J. Am. Med. Assoc.* **199**:987.
Scott, J. M., 1962, Anaemia in pregnancy, *Postgrad. Med. J.* **38**:202.
Seitchik, J., Alper, C., and Szutka, A., 1963, Changes in body composition during pregnancy, *Ann. N.Y. Acad. Sci.* **110**:821.
Serrano, C. V., Talbert, L. M., and Welt, L. G., 1964, Potassium deficiency in the pregnant dog, *J. Clin. Invest.* **43**:27.
Sever, L. E., and Emanuel, I., 1973, Is there a connection between maternal zinc deficiency and congenital malformations of the central nervous system in man? *Teratology* **7**:117.

Shajania, A. M., Harnandy, G., and Barnes, P. H., 1969, Oral contraceptives and folate metabolism, *Lancet* **1**:886.

Shami, Y., Messer, H. H., and Copp, D. H., 1975, Calcium uptake by placental plasma membrane vesicles, *Biochim. Biophys. Acta* **401**:256.

Shoemaker, W. J., and Wurtman, R. J., 1971, Perinatal undernutrition: accumulation of catecholamines in rat brain, *Science* **171**:1017.

Shroder, R. E., and Zeman, F. J., 1973a, *In vitro* synthesis of anterior pituitary growth hormone as affected by maternal protein deprivation and postnatal food supply, *J. Nutr.* **103**:1012.

Shroder, R. E., and Zeman, F. J., 1973b, Skeletal development in rats as affected by maternal protein deprivation and postnatal food supply, *J. Nutr.* **103**:792.

Siassi, F., and Siassi, B., 1973, Differential effects of protein-calorie restriction and subsequent repletion on neuronal and nonneuronal components of cerebral cortex in newborn rats, *J. Nutr.* **103**:1625.

Siddal, P. C., and Mull, J. W., 1945, Thiamin status during pregnancy, *Am. J. Obstet. Gynecol.* **49**:672.

Singer, J. E., Westphal, M., and Niswander, K., 1968, Relationship of weight gain during pregnancy to birth weight and infant growth and development in the first year of life: A report from the collaborative study of cerebral palsy, *Obstet. Gynecol.* **31**:417.

Skinner, S. L., Lumbers, E. R., and Symends, E. M., 1972, Analysis of changes in the renin-angiotensin system during pregnancy, *Clin. Sci.* **42**:479.

Slemons, J. M., and Fagan, R. H., 1927, A study of the infant's birth weight and the mother's gain during pregnancy, *Am. J. Obstet. Gynecol.* **14**:159.

Slobody, L. B., Willner, M. M., and MesCern, J., 1949, Comparison of vitamin B_1 levels in mothers and their newborn infants, *Am. J. Dis. Child.* **77**:736.

Slonaker, J. R., 1924–1925, The effect of copulation, pregnancy, pseudopregnancy and lactation on the voluntary activity and food consumption of the albino rat, *Am. J. Physiol.* **71**:362.

Smith, C. H., and Depper, R., 1974, Placental amino acid uptake, II. Tissue pre-incubation, fluid distribution and mechanisms of regulation, *Pediatr. Res.* **8**:697.

Smith, C. H., Adcock, E. W., Teasdale, F., Meschia, G., and Battaglia, F. C., 1973, Placental amino acid uptake: tissue preparation, kinetics, and pre-incubation effect, *Am. J. Physiol.* **224**:55.

Smith, G. F. D., 1916, Effects of the state of nutrition of the mother during pregnancy and labour on the condition of the child at birth and for the first few days of life, *Lancet* **2**:54.

Snelling, C. E., and Jackson, S. H., 1939, Blood studies of vitamin C during pregnancy, birth, and early infancy, *J. Pediatr.* **14**:447.

Southgate, D. A. T., 1971, The accumulation of amino acids in the products of conception of the rat and in the young animal after birth, *Biol. Neonate* **19**:272.

Speert, H., Graff, S., and Graff, A. M., 1951, Nutrition and premature labor, *Am. J. Obstet. Gynecol.* **62**:1009.

Spellacy, W. N., 1971, Insulin and growth hormone measurements in normal and high-risk pregnancies, in: *Fetal Evaluation during Pregnancy and Labor* (P. G. Crosignani and G. Pardi, eds.), pp. 110–137, Academic Press, Inc., New York.

Spellacy, W. N., 1975, Maternal and fetal metabolic interrelationships, in: *Carbohydrate Metabolism in Pregnancy and the Newborn* (H. W. Sutherland and J. M. Stowers, eds.), pp. 42–57, Churchill Livingstone, Edinburgh.

Stegink, L. D., Pitkins, R. M., Reynolds, W. A., Filer, L. J., Jr., Boaz, C. P., and Brummel, D. C., 1975, Placental transfer of glutamate and its metabolites in the primate, *Am. J. Obstet. Gynecol.* **122**:70.

Stein, Z., Susser, M., Saenger, G., and Marolla, F., 1975, *Famine and Human Development, The Dutch Hunger Winter of 1944-45,* p. 284, Oxford Univeristy Press, Inc., New York.

Stephan, J. K., Chow, B., Frohman, L. A., and Chow, B. F., 1971, Relationship of growth hormone to the growth retardation associated with maternal dietary restriction, *J. Nutr.* **101**:1453.

Sternberg, J., and Pascoe-Dawson, E., 1959, Metabolic studies in atherosclerosis, I. Metabolic pathway of C^{14} labeled alpha-tocopherol, *Can. Med. Assoc. J.* **80**:266.

Stevenson, R. E., 1973, *The Fetus and Newly Born Infant*, The C. V. Mosby Company, St. Louis.

Stewart, E. L., and Welt, L. G., 1961, Protection of the fetus in experimental potassium depletion, *Am. J. Physiol.* **200**:824.

Straumfjord, J. V., and Quaife, M. L., 1946, Vitamin E levels in maternal and fetal blood, *Proc. Soc. Exp. Biol. Med.* **61**:369.

Streiff, R. R., and Little, A. B., 1967, Folic acid deficiency in pregnancy, *New Engl. J. Med.* **276**:776.

Strupeon, P., 1959, Studies of iron requirements of infants, III. Influences of supplemental iron during normal pregnancy on mother and infant, *Br. J. Haematol.* **5**:31.

Svanborg, A., and Vikrot, O., 1965, Plasma lipid fractions, including individual phospholipids at various stages of pregnancy, *Acta Med. Scand.* **178**:615.

Szabo, A. J., and Grimaldi, R. D., 1970, The effect of insulin on glucose metabolism of the incubated human placenta, *Am. J. Obstet. Gynecol.* **106**:75.

Szabo, A. J., Grimaldi, R. D., and Jung, W. F., 1969, Palmitate transport across perfused human placenta, *Metabolism* **18**:406.

Takahashi, Y. I., Smith, J. E., Winick, M., and Goodman, D. S., 1975, Vitamin A deficiency and fetal growth and development in the rat, *J. Nutr.* **105**:1299.

Tan, C. M., and Raman, A., 1972, Maternal-fetal calcium relationships in man, *Q.J. Exp. Physiol.* **57**:56.

Tan, C. M., Raman, A., and Sinnathyray, T. A., 1972, Serum ionic calcium levels during pregnancy, *J. Obstet. Gynaecol. Br. Commonw.* **79**:694.

Tao, T. W., and Hertig, A. T., 1965, Viability and differentiation of human trophoblast in organ culture, *Am. J. Anat.* **116**:315.

Tateno, M., and Ohshima, A., 1973, The relationship between serum vitamin E levels in the perinatal period and the birth weight of the neonate, *Acta Obstet. Gynaecol. Jpn.* **20**:177.

Temerley, I. J., Meehan, M. J. M., and Gatenby, P. B. B., 1968, Serum folic acid levels in pregnancy and their relationship to megaloblastic marrow change, *Br. J. Haematol.* **14**:13.

Thalme, B., 1966, Electrolyte and acid-base balance in fetal and maternal blood. An experimental and a clinical study, *Acta Obstet. Gynecol. Scand.* **45**(suppl. 8):1.

Thau, R. B., and Lanman, J. T., 1975, Endocrinological aspects of placental function, in: *The Placenta* (P. Gruenwald, ed.), p. 125, University Park Press, Baltimore.

Thomson, A. M., 1958, Diet in pregnancy, I. Dietary survey technique, *Br. J. Nutr.* **12**:446.

Thomson, A. M., 1959a, Diet in pregnancy, II. Assessment of the nutritive value of diets, especially in relation to differences between social classes, *Br. J. Nutr.* **13**:190.

Thomson, A. M., 1959b, Diet in relation to the course and outcome of pregnancy, *Br. J. Nutr.* **13**:509.

Thomson, A. M., 1963, Prematurity: Socio-economic and nutritional factors, *Mod. Prob. Pediatr.* **8**:197.

Todd, W. R., Chunard, E. G., and Wood, M. T., 1939, Blood calcium and phosphorous in the newborn, *Am. J. Dis. Child.* **57**:1278.

Tompkins, W. T., Mitchell, R. McN., and Wiehl, D. G., 1955, Maternal and newborn nutritional studies at Philadelphia Lying-in-Hospital. Maternal studies, II. Prematurity and maternal nutrition, The promotion of maternal and newborn health, Milbank Memorial Fund, New York.

Toombs, P. W., 1931, The relationship between mother's gain during pregnancy and infant's birth weight, *Am. J. Obstet. Gynecol.* **22**:851.

Toverud, K. U., 1939, The vitamin C requirements of pregnant and lactating women, *Acta Pediatr.* **24**:332.

Tulchinsky, D., and Kovenman, S. G., 1971, The plasma estradiol as an index of fetoplacental function, *J. Clin. Invest.* **50**:1490.

Tumbleson, M. E., Tinsley, O. W., Hicklin, K. W., and Mudler, J. B., 1972, Fetal neonatal development of Sinclair (S-1) miniature piglets affected by maternal dietary protein deprivation, *Growth* **36**:373.

Turk, D. E., Sunde, M. G., and Hoekstom, W. G., 1959, Zinc deficiency experiments with poultry, *Poultry Sci.* **38**:1256.

Twardock, A. R., and Austin, M. K., 1970, Calcium transfer in perfused guinea pig placenta, *Am. J. Physiol.* **219**:540.

Tyson, J. E., Austin, L., and Farinholt, J. W., 1971, Prolonged nutritional deprivation in pregnancy: changes in human chorionic somatomammotropin and growth hormone secretion, *Am. J. Obstet. Gynecol.* **109**:1080.

Ullberg, S., Kristofferson, H., Flodh, H., and Hanngren, A., 1967, Placental passage and fetal accumulation of labeled vitamin B_{12} in the mouse, *Arch. Int. Pharmacodyn. Ther.* **167**:431.

Unna, K. R., and Honig, C. R., 1968, Vitamin B_6 Group, XII, Pharmacology and toxicology, in: *The Vitamins: Chemistry, Physiology, Pathology, Methods,* Vol. II, 2nd ed. (W. H. Sebrell and R. S. Harris, eds.), pp. 104–108, Academic Press, Inc., New York.

Usher, R. H., and McLean, F. H., 1974, Normal fetal growth and the significance of fetal growth retardation, in: *Scientific Foundations of Pediatrics,* (J. A. Davis and J. Dobbing, eds.), p. 69, W. B. Saunders Company, Philadelphia.

Vande Wiele, R. L., Gurpide, E., Kelly, W. G., Laragh, J. H., and Lieberman, S., 1960, The secretory rates of progesterone and aldosterone in normal and abnormal late pregnancy, *Acta Endocrinol. (KBH) Suppl.* **51**:159.

Van Duyne, C. M., Havel, R. J., and Felts, J. M., 1962, Placental transfer of palmitic acid,1-C^{14} in rabbits, *Am. J. Obstet. Gynecol.* **84**:1069.

Vanier, T. M., and Tyas, J. F., 1967, Folic acid status in premature infants, *Arch. Dis. Child.* **42**:57.

Van Slyke, D. D., and Meyer, G. M., 1913, The fate of protein digestion products in the body III. The absorption of amino acids from the blood by the tissues, *J. Biol. Chem.* **16**:197.

Varadi, S., Abbott, D., and Elwis, A., 1966, Correlation of peripheral white cell and bone marrow changes with folate levels in pregnancy and their clinical significance, *J. Clin. Pathol.* **19**:33.

Varangot, J., 1942, Sur la teneur du sanguin en vitamin E au cours de la gestation humaine, *C.R. Acad. Sci.* **214**:691.

Vaz Pinto, A., Santos, F., Midlej, M. C., Almeida, A. M., and Gama, M. D., 1973, Vitamin B_{12} and folic acid in maternal and newborn sera, *Rev. Invest. Clin.* **25**:341.

Velasco, E., Rosso, P., Brasel, J. A., and Winick, M., 1975, Activity of alkaline ribonuclease in placentas of malnourished women, *Am. J. Obstet. Gynecol.* **123**:637.

Ventura, S., and Klopper, A., 1951a, Iron metabolism in pregnancy: the behavior of hemoglobin, serum iron, the iron-binding capacity of serum proteins, serum copper and free erthrocyte protoporphyrin in normal pregnancy, *J. Obstet. Gynaecol. Br. Emp.* **58**:173.

Ventura, S., and Klopper, A., 1951b, Iron metabolism in pregnancy, II. The behavior of serum iron in pregnancy after the administration of iron compounds by mouth and intravenously, *South Afr. Med. J.* **25**:969.

Vikbladh, I., 1951, Studies on zinc in blood, *Scand. J. Clin. Lab. Invest.* **3**(suppl. 2):1.

Vobecky, J. S., Vobecky, J., Shappott, D., and Munan, L., 1974, Letter: Vitamin C and outcome of pregnancy, *Lancet* **1**:630.

Von Studnitz, W., 1955, Studies on serum lipids and lipoprotein in pregnancy, *Scand. J. Clin. Lab. Invest.* **7**:329.

Von Studnitz, W., and Berezin, D., 1958, Studies on serum copper during pregnancy, during the menstrual cycle and after the administration of oestrogens, *Acta Endocrinol.* **27**:245.

Wachstein, M., and Gudiatis, A., 1953, Disturbance of vitamin B_6 metabolism in pregnancy, *Am. J. Obstet. Gynecol.* **66**:1207.

Wajcicka, J., and Zapalowski, Z., 1963, Serum copper levels in the blood of pregnant women in cases of normal and complicated pregnancies, *Ginekol. Pol.* **34**:693.

Walters, W. A. W., and Lim, Y. L., 1975, Blood volume and hemodynamics in pregnancy, *Clin. Obstet. Gynecol.* **2**:301.

Wardlaw, J. M., and Pike, R. L., 1963, Some effects of high and low sodium intake during pregnancy in the rat, IV. Granulation of renal juxtaglomerular cells and zona glomerulosa width, *J. Nutr.* **80**:355.

Warkany, J., and Schraffenberger, E., 1944, Congenital malformation induced in rats by maternal nutritional deficiency, *J. Nutr.* **27**:477.

Watanabe, M., Meeker, C. I., Gray, M. J., Sims, E. A. H., and Solomon, S., 1963, Secretion rate of aldosterone in abnormal pregnancy, *J. Clin. Endocrinol.* **25:**1619.

Watney, P. J. M., Chance, G. W., Scott, P., and Thompson, J. M., 1971, Maternal factors in neonatal hypocalcemia, a study in three ethnic groups, *Br. Med. J.* **2:**432.

Weir. R. J., Paintin, D. B., Robertson, J. I. S., Tree, M., Fraser, R., and Young, J., 1970, Renin, angiotensin and aldosterone relationships in normal pregnancy, *Proc. R. Soc. Med.* **63:**1101.

Weir, R. J., Paintin, D. B., Brown, J. J. Fraser, R., Lover, A. E., Robertson, J. I. S., and Young, J., 1971, A serial study in pregnancy of the plasma concentrations of renin, corticosteroids, electrolytes and protein and of hematocrit and plasma volume, *J. Obstet. Gynaecol. Br. Comonw.* **78:**590.

Whaley, W. H., Zuspan, F. P., and Nelson, G. H., 1968, Correlation between maternal and fetal plasma levels of glucose and free fatty acids, *Am. J. Obstet. Gynecol.* **94:**419.

Whalley, P. J., Scott, D. E., and Pritchard, J. A., 1969, Maternal folate deficiency and pregnancy wastage. Placental abruption, *Am. J. Obstet. Gynecol.* **105:**670.

Widdas, W. F., 1952, Inability of diffusion to account for placental glucose transfer in the sheep and consideration of the kinetics of a possible carrier transfer, *J. Physiol. (London)* **118:**23.

Widdowson, E. M., and Spray, C. M., 1951, Chemical development in utero, *Arch. Dis. Child.* **26:**205.

Widdowson, E. W., Crabb, D. E., and Milner, R. D. G., 1972, Cellular growth development of some human organs before birth, *Arch. Dis. Child.* **47:**652.

Widdowson, E. M., Dauncy, J., and Shaw, J. C. L., 1974, Trace elements in foetal and early postnatal development, *Proc. Nutr. Soc.* **33:**275.

Wideman, C. L., Baird, G. H., and Bolding, O. T., 1964, Ascorbic acid deficiency and premature rupture of fetal membranes, *Am. J. Obstet. Gynecol.* **88:**592.

Wiener, G., 1970, The relationship of birth weight and length of gestation to intellectual development at ages 8 to 10 years, *J. Pediatr.* **76:**694.

Wigglesworth, J. S., 1964, Experimental growth retardation in the fetal rat, *J. Pathol. Bacteriol.* **88:**1.

Wilken, H., 1960, Serum copper in late toxemia of pregnancy, *Arch. Gynaekol.* **194:**158.

Williams, P. F., and Fralin, F. G., 1942, Nutrition study in pregnancy: dietary analyses of 7-day food intake records of 514 pregnant women, comparison of actual food intakes with variously stated requirements and relationship of food intake to various obstetric factors, *Am. J. Obstet. Gynecol.* **43:**1.

Williamson, M. B., 1948, The vitamin A content of fetal rats from mothers on a high cholesterol diet, *J. Biol. Chem.* **174:**631.

Wills, L., Hill, G., Bingham, K., Miall, M., and Wrigley, J., 1947, Hemoglobin levels in pregnancy. The effects of the rationing scheme and routine administration of iron, *Br. J. Nutr.* **1:**126.

Winick, M., 1968, Changes in nucleic acid and protein content during growth of the human brain, *Pediatr. Res.* **2:**355.

Winick, M., 1969, Cellular growth of the placenta as an indicator of abnormal fetal growth, in: *Diagnosis and Treatment of Fetal Disorders* (K. Adamson, ed.), p. 83, Springer-Verlag New York Inc., New York.

Winick, M., and Noble, A., 1965, Quantitative changes in DNA, RNA, and protein during prenatal and postnatal growth in the rat, *Dev. Biol.* **12:**451.

Winick, M., and Noble, A., 1966a, Quantitative changes in DNA, RNA, and protein during normal growth of rat placenta, *Nature* **212:**34.

Winick, M., and Noble, A., 1966b, Cellular response in rats during malnutrition at various ages, *J. Nutr.* **89:**300.

Winick, M., and Rosso, P., 1969, Head circumference and cellular growth of the brain in normal and marasmic children, *J. Pediatr.* **74:**774.

Winick, M, and Rosso, P., 1975, Malnutrition and central nervous system development, in: *Brain Function and Malnutrition: Neuropsychological Methods of Assessment* (J. W. Prescott, M. S. Read, and D. B. Coursin, eds.), p. 41, John Wiley & Sons, Inc., New York.

Winick, M., Coscia, A., and Noble, A., 1967, Cellular growth of human placenta, I. Normal placental growth, *Pediatrics* **39**:248.

Winick, M., Brasel, J., and Rosso, P., 1972, Nutrition and cell growth, in: *Nutrition and Development*, Vol. 1, *Current Concepts in Nutrition* (Myron Winick, ed.), p. 49, John Wiley & Sons, Inc., New York.

Wong, C. T., and Morgan, E. H., 1953, Placental transfer of iron in the guinea pig, *J. Exp. Physiol.* **58**:47.

World Health Organization, 1965, Nutrition in Pregnancy and Lactation, p. 54, *WHO Tech. Rep. Ser. 302*, Geneva.

Wynn, R. M., 1975, Principles of placentation and early human placental development, in: *The Placenta* (P. Gruenwald, ed.), p. 18, University Park Press, Baltimore.

Yen, S. S., 1973, Endocrine regulation of metabolic homeostasis during pregnancy, *Clin. Obstet. Gynecol.* **16**:130.

Yendt, E. R., DeLuca, H. F., Garcia, D. A., and Colanim, M., 1970, Clinical aspects of vitamin D, in: *The Fat Soluble Vitamins* (H. F. DeLuca, and J. W. Suttie, eds.), pp. 87–104, The University of Wisconsin Press, Madison.

Young, M., and Prenton, M. A., 1969, Maternal and plasma amino acid concentrations during gestation and in retarded fetal growth, *J. Obstet. Gynaecol. Br. Commonw.* **76**:333.

Young, M., and Widdowson, E. M., 1975, The influence of diets deficient in energy, or in protein on conceptus weight, and the placental transfer of a non-metabolisable amino acid in the guinea pig, *Biol. Neonate* **27**:184.

Zamenhoff, S., Van Marthens, E., and Margolis, F. L., 1968, DNA (cell number) and protein in neonatal brain: alteration by maternal dietary protein restriction, *Science* **160**:322.

Zeman, F. J., 1967, Effect on the young of rat of maternal protein restriction, *J. Nutr.* **93**:167.

Zeman, F. J., 1968, Effects of maternal protein restriction on the kidney of the newborn young of rats, *J. Nutr.* **94**:111.

Zeman, F. J., 1970, Effect of protein deficiency during gestation on postnatal cellular development in the young rat, *J. Nutr.* **100**:530.

Zeman, F. J., and Widdowson, E. M., 1974, Lipid absorption in newborn young of guinea pigs fed a protein-deficient diet during gestation, *Biol. Neonate* **24**:344.

Zeman, F. J., Shrader, R. E., and Allen, L. H., 1973, Persistent effects of maternal protein deficiency in postnatal rats, *Nutr. Rep. Int.* **7**:421.

Zinneman, H. H., Johnson, J. J., and Seal, U. S., 1963, Effect of short-term therapy with cortisol on the urinary excretion of free amino acids, *J. Clin. Endocrinol.* **23**:996.

Zuspan, F. P. Whaley, W. H., Nelson, C. H., and Ahlquist, R. P., 1966, Placental transfer of epinephrine, *Am. J. Obstet. Gynecol.* **95**:284.

6

Early Infant Nutrition: Breast Feeding

Derrick B. Jelliffe and E. F. Patrice Jelliffe

1. Introduction

Functional and practical understanding of infant feeding can best be achieved by appreciating its dyadic nature—that is, as a nutritional, psychological, and biological interaction between mother and offspring, both in pregnancy and lactation, with *each* affecting the other, and, at the same time, by considering the early stages of the young human organism according to the biological classification of Bostock (1962)—that is, the fetus, the exterogestate fetus (up to 6–9 months postnatally), and the transitional (9 months to 2–3 years) (Jelliffe, 1967) rather than by the statistical calendar catagories of "infant" and "preschool child."

2. Fetus

All recent work endorses the truth of the old saying that "infant feeding begins in the uterus." Maternal nutrition affects both the newborn and the exterogestage fetus directly via the birth weight and levels of fetal stores of nutrients, and by the laying down of adequate lactation reserves in pregnancy in the form of some 4 kg of subcutaneous fat, needed as a major source of calories and fatty acids for subsequent breast milk production.

Additionally, the level of maternal nutritional needs to be viewed in relation to the cumulative impact of repeated reproductive cycles, compounded in some cultures by hard work and restrictive food customs. Such syndromes of "maternal depletion"—specific (e.g., iron, iodine, etc.) or general (protein–calorie)—are significant factors in unnecessary mortality in many developing countries in pregnancy and child birth (Jelliffe and Maddocks, 1964)—with, of

Derrick B. Jelliffe and E. F. Patrice Jelliffe • Division of Population, Family and International Health, School of Public Health, University of California at Los Angeles, Los Angeles, California.

course, severe consequences as regards infant feeding as a result of the lack of breast feeding and of the mother's close care and attention.

3. Exterogestate Fetus

For the first 6–9 months of independent life, the baby can best be considered as an exterogestate, or external, fetus—with the breast taking the place of the pacenta. During this period, the child's nutrient needs are obtained for the most part from fetal stores and from a diet of milk—human or cow's (or formula)—administered by the two quite different process of breast feeding or bottle feeding, respectively.

3.1. Adaptive Suckling

Lactation is a very ancient process of some 200 million years' duration. This long antedates placental gestation, as the first mammals were egg-laying, as are their present-day descendants, such as the duck-billed platypus and the echidna.

Man has existed for about 1 million years and in this vast period of time has kept cattle for only some 10,000 years (0.1%). The widespread use of cow's milk for feeding the exterogestate fetus has only been in vogue for about 50 years, or 5/100,000 of man's existence. This extremely recent change is, as Hambraeus *et al.* (1976) remark, "one of the world's biggest uncontrolled biological experiments." In fact, during millenia of evolution, species-specific patterns of "adaptive suckling" have developed, with the composition of the milk, the mechanism of administrating it and the lactatory apparatus specially modified for need. For example, in the whale, this process has to cater for singleton births, with high calorie requirements by the huge infant nursing rapidly while submerged, and with the mother having to conserve water in the saline ocean. Adaptive suckling is achieved in this species by a highly concentrated milk, containing up to 50% fat, and a very powerful let-down reflex, so that the baby whale is quickly and literally pumped with cream (Jelliffe and Jelliffe, 1978).

Similarly, human beings have evolved a pattern of "adaptive suckling" as an outcome of 1 million years of evolution. Basically, the female is equipped with two breasts, which can deal nutritionally with the usual range of offspring—that is, singletons and twins—and with the volume of milk secretion responsive to sucking stimulus. Mother–neonate bonding is achieved reflexly shortly after birth—an essential for the mobile life of early hunting man and the fragile immaturity of the newborn. Human milk appears to be adapted to the specific needs of the neonate, particularly the characteristically rapid growth of the brain and the protection against infection in the highly susceptible exterogestate phase of life. Breast feeding has an important child-spacing function—to permit maturity of one child before the arrival of the next. The introduction of other foods seems to be signaled biologically by the readiness of the infant to deal with them, as judged by the appearance of the "milk teeth," the decrease of the extrusion reflex, possibly a rise of intestinal enzymes, and by increasing manual dexterity.

3.2. Psychophysiology

Practical approaches to successful lactation (and explanations for many traditional practices) have been clarified by modern endocrinological research, particularly into the prolactin and letdown reflexes.

3.2.1. Prolactin Reflex

It is often not appreciated that human prolactin was only isolated as recently as 1971 by Hwang and colleagues, who devised a method of radioimmunoassay in the same year. This has lead to much new work, particularly confirming the clinical observation that the more sucking stimulus—that is, the number of feeds × the length of feedings × the vigor of the baby—the more the milk secreted. In other words, prolactin secretion is proportional to the sucking stimulus. Also, recent studies have shown the polyvalent hormonal role of prolactin—on the mammary alveoli with milk secretion, on the kidneys with a water-conserving antidiuretic effect, on the ovaries with anovulatory lactation amenorrhea (Van Ginneken, 1974), and possibly on the brain with increased maternal behaviour ("motherliness") (Newton *et al.,* 1968).

3.2.2. Letdown Reflex

Long known to traditional and modern dairy farmers, it has taken some time for it to be appreciated that the emotionally labile psychosomatic letdown reflex is the key to failure or success. Confidence enhances, while anxiety inhibits. The failure of the sophisticated well-to-do urbanite with no knowledge and much apprehension, and of the shantytown mother, with the environmental psychosocial stress of her grim and desperate life, are both based in part on emotional interference with the letdown reflex.

Conversely, it is now clear that the *doula,* or female assistant, occurring in pregnancy, child birth, and the puerperium in social mammals such as dolphins, and in traditional human cultures (Raphael, 1973), supplies physical and emotional support and information—and above all, generates confidence. A *doula* effect is the reason for success by such spontaneous modern women's groups concerned with breast feeding, such as the La Leche League International (LLLI) in the United States, the Nursing Mothers Association of Australia (NMAA), and Ammenhjelpen in Norway. They supply information and individual and group support through experienced mothers; they are, in fact, *doula* surrogates.

3.3. Recent Knowledge

Textbook accounts of the differences between human milk and cow's milk have usually given a ritual list of ten or so factors, based on long-outdated work. In the last two decades, a great deal of completely new nutritionally relevant information has been brought to light which has clarified the picture considerably. This includes information on antiinfective properties, child-spacing effects, and economic considerations, as well as biochemical and nutritional aspects.

3.4. Nutritional Aspects

Differences in the biochemical composition of breast milk occur with individual women, the stage of lactation, and, to some extent, the level of maternal nutrition, as well as the methods of sampling and analysis. Diurnal variations are well known, especially a higher fat content in the early morning.

General levels of "proximate principles" have been described for decades. A detailed account was given by Morrison (1952), and Fomon's (1974) figures give an approximate working summary of the main constituents compared with cow's milk (Table I). However, variations in results in different studies are difficult to interpret, especially those of minor degree, because of differences in methods of sampling and analysis, and of types and levels of maternal diet and nutrition.

3.4.1 Methods of Obtaining Information

Recent work has reemphasized the possibility of very considerable variations in results as a result of using different biochemical methods. This is particularly so with regard to protein (Section 3.5.1).

The actual obtaining of representative samples of human milk poses unique problems as the normal destination is the baby's stomach and not a test tube or laboratory container. Breast milk production is proportional to the

Table I. Composition of Mature Human Milk and Cow's Milk[a]

Composition	Human milk	Cow's milk
Water (ml/100 ml)	87.1	87.2
Energy (kcal/100 ml)	75	66
Total solids (g/100 ml)	12.9	12.8
Protein (g/100 ml)	1.1	3.5
Fat (g/100 ml)	4.5	3.7
Lactose (g/100 ml)	6.8	4.9
Ash (g/100 ml)	0.2	0.7
Proteins (% of total protein)		
Casein	40	82
Whey proteins	60	18
Nonprotein nitrogen (mg/100 ml)		
(% of total nitrogen)	15	6
Amino acids (mg/100 ml)		
Essential		
Histidine	22	95

(Continued)

Table I. (Continued)

Composition	Human milk	Cow's milk
Isoleucine	68	228
Leucine	100	350
Lysine	73	277
Methionine	25	88
Phenylalanine	48	172
Threonine	50	164
Tryptophan	18	49
Valine	70	245
Nonessential		
Arginine	45	129
Alanine	35	75
Aspartic acid	116	166
Cystine	22	32
Glutamic acid	230	680
Glycine	0	11
Proline	80	250
Serine	69	160
Tyrosine	61	179
Major Minerals per liter		
Calcium (mg)	340	1170
Phosphorus (mg)	140	920
Sodium (meq)	7	22
Potassium (meq)	13	35
Chloride (meq)	11	29
Magnesium (mg)	40	120
Sulfur (mg)	140	300
Trace minerals per liter		
Chromium (μg)	—	8–13
Manganese (μg)	7–15	20–40
Copper (μg)	400	300
Zinc (mg)	3–5	3–5
Iodine (μg)	30	47[b]
Selenium (μg)	13–50	5–50
Iron (mg)	0.5	0.5
Vitamins per liter		
Vitamin A (IU)	1898	1025[c]
Thiamin (μg)	160	440
Riboflavin (μg)	360	1750
Niacin (μg)	1470	940
Pyridoxine (μg)	100	640
Pantothenate (mg)	1.84	3.46
Folacin (μg)	52	55
B_{12} (μg)	0.3	4
Vitamin C (mg)	43	11[d]
Vitamin D (IU)	22	14[e]
Vitamin E (mg)	1.8	0.4
Vitamin K (μg)	15	60

[a] From Fomon (1974).
[b] Range 10–200 μg/liter.
[c] Average value for winter milk; value for summer milk, 1690 IU/liter.
[d] As marketed; value for fresh cow's milk, 21 mg/liter.
[e] Average value for winter milk; value for summer milk, 33 IU.

secretion of anterior pituitary hormone, prolactin, resulting from nipple stimulation ("prolactin reflex"), and to intraalveolar tension, related to emptying. The ejection of breast milk is mediated by the psychosomatic letdown reflex. Methods employed must, therefore, avoid interfering with these mechanisms as much as possible, but, in fact, all do considerably. As Hytten (1954a) noted, "the suckling of a baby has, for the mother, psychological overtones beyond the mere local stimulation of the areola and nipple; and the mechanical replacement of this local stimulus cannot hope to be an entirely effective substitute for the baby."

Also, diurnal variations in milk volume and composition have been noted (Hytten, 1954b). The invariable and considerable difference in fat content between fore- and hindmilk (1–2 g/100 ml) means that the time of sampling during a feeding can be significant.

In addition, in some areas of the world, seasonal variation has been noted—for example, lower levels of ascorbic acid in the "hungry season" in parts of Africa. The stage of lactation is another variable, as is frequency, intensity, and duration of nipple stimulation (and reflex prolactin secretion). Important reducers of such stimulation are complementary bottle feeds (*allaitement mixte*) and the introduction of semisolid foods.

Likewise, questions of total output are often based on daytime estimations. In fact, in traditional cultures, the baby sleeps by the mother's side and must obtain considerable quantities as "night feedings." For example, Omololu [1975 (personal communication)] in Nigeria found that about one-third of the number of daily feeds were given during the night (defined as from 8 P.M. to 6 A.M.).

Two methods can be employed to try to measure the volume produced: test feeding and expression.

3.4.1a. Test Feeding. In this venerable technique, the baby is weighed before and after feeding. Difficulties are numerous and obvious. Results depend on the vigor of the infant and the success of the mother–baby interaction. To ensure larger, more measurable samples, it may be considered preferable for mothers to nurse their babies at prescribed intervals rather than on demand, as under nonexperimental conditions. Expensive, accurate scales are required to measure relatively small weight increases. There is considerable likelihood of interference with the emotionally sensitive letdown reflex in the unnatural, anxiety-producing circumstances and embarrassment inevitably created by the investigations, which often have to be carried out in hospitals, sometimes when infants are too sick to take the breast normally.

Practically, there may be a need for continuous surveillance to ensure that no feeding takes place between testings, and, as noted earlier, ideally this should be on a 24-hr basis, which is almost never possible.

3.4.1b. Expression. Milk may be expressed from the breast manually and by some form of mechanical or electrical pump. These methods can be used to estimate total output and, of course, are required if samples of milk are to be obtained for analysis.

The same difficulties exist as with test feeding. Indeed, anxiety can be greater with expression, especially with an unfamiliar, uncomfortable apparatus, combined with the question of feeding the baby later with the expressed milk. Also, the influence of expression on prolactin secretion, compared with suckling, is unknown.

Comparative results have varied in different studies. In some, expression has given greater volumes; in others, test feeding. In either, it is apparent that results are approximations.

3.4.2. Maternal Nutrition

Comparisons are also made more difficult by variables in the nutritional status of mothers, both between mothers in the particular group and between various communities. The nutrients involved, the degree and duration of deprivation, methods of nutritional assessment, and the previous nutritional situation and stores can have many and varying combinations. There will, for example, be considerable differences between details of the nutritional past and present in poorer women in Sao Paolo, Brazil, in Ibadan, Nigeria, and in southern India.

In particular, the physiological weight gain in pregnancy, about one-third due to deposition of a subcutaneous fat energy bank ("lactation stores"), can vary considerably from the 12.5 kg suggested for western women to a total of only 5 kg (or even weight loss) in poorly fed communities. Dietary inadequacy in pregnancy may sometimes be complicated by associated hard physical work and by restrictive food customs. However, cultural and biological adaptation can occur, the latter in the form of a decreased activity.

Likewise, in all communities lactation itself leads to weight loss. In ill-fed mothers this can sometimes be as much as 7 kg after 1 year, even leading to the development of "nutritional edema" in very poorly nourished women. In well-fed women in Westernized societies, breast feeding has a "sliming" effect. Many accounts from different parts of the world suggest that malnourished women often lactate with unexpectedly little clinically obvious deterioration of their nutritional status. However, the cumulative effects of sequential reproductive cycles, including prolonged lactation, can lead to general "maternal depletion," as shown by progressive weight loss. More specific nutrient deficiencies may occur—for example, increasing goiter, osteomalacia, and nutritional edema.

3.5. Biochemical Composition

The biochemical composition of human milk has been examined in different parts of the world. Comparison will be made between various communities, and with well-nourished mothers, for protein, fat, lactose, vitamins, and calcium, keeping in mind previously mentioned variations with sampling (period in lactation cycle, season, one subject or pooled specimens, etc.) with laboratory techniques and with levels of maternal health and nutrition (Table II).

3.5.1. Protein

The protein content of human milk has been reported to vary between 1.0 and 1.6 g/100 ml in well-nourished women. In 1952, Morrison reported the mean of European analyses to be 1.6 g/100 ml, and American, 1.2 g/100 ml. He suggested that differences in methods of estimation might have been mainly responsible, and that 1.2 g/100 ml probably represented an overall mean. There seems little evidence for diurnal variation or fluctuation with age or parity.

However, very recent Swedish studies (Hambraeus *et al.,* 1976) have shown the protein content to be only 0.8–0.9 g/100 ml when determined by amino acid analysis. It has been pointed out that earlier investigations with breast milk were made with the same methods as those used with cow's milk. Thus, the protein content has usually been based on determination of the total nitrogen content using the Kjieldahl method and the conversion factors 6.25 or 6.38. However, this overestimates the protein content of breast milk, since about 20% of total nitrogen is present as nonprotein nitrogen.

In fact, it is now recognized that the relatively low protein/low solute characteristics of breast milk place the human as a constant contact, frequent suckling species in infancy, and also permits breast milk to supply water as well as nutrients. This contrasts, for example, with a high solute/high protein (10–13 g/100 ml) species like the rabbit, which suckles her litter once daily.

In poorer, technically developing countries, the average protein content of the milk of inadequately nourished mothers, based on total nitrogen assessment, is usually surprisingly high—in fact, a low normal (1.0–0.1 g/100 ml, although the range may be quite wide (Table II). However, in some places with probably poorer nutrition, the protein content can be lower (0.7–0.9 g/100 ml).

Varying results have been obtained with estimations of the protein content with what is termed by modern Western cultural definition "prolonged lactation"—that is, into the second year of life and later. Some have found a decline, some a rise, and others no material change. Likewise, one study has shown that breast feeding into pregnancy has been shown to be associated with milk with an unchanged protein content.

The effect of maternal dietary supplementation on protein content does not appear to have been investigated adequately. However, in the protein-supplementation study carried out by Gopalan and Belavady (1961), the increased output of milk was associated with a corresponding fall in protein concentration, with the result that the total protein output in 24 hr was not significantly altered.

Recent work indicated much greater complexity than previously appreciated and the need to reconsider some of the prior analyses. Detailed study shows breast milk to have a quite different *protein composition,* being richer in lactoferrin, lysozyme, and IgA, and with no beta-lactoglobulin, which is one of the main components of cow's milk protein (Hambraeus *et al.,* 1976). Human milk has a characteristic pattern of nucleotides, with no orotic acid (György, 1971) and a high level of polyamines, particularly spermidine and spermine (Sanguansensri *et al.,* 1974).

Table II. Fat, Lactose, Protein, and Calcium Content of Mature Human Milk from Some Well-Nourished and Poorly Nourished Communities[a]

	Fat (g/100 ml)	Lactose (g/100 ml)	Protein (g/100 ml)	Calcium (g/100 ml)
Well-nourished				
American (Macy)	4.5	6.8	1.1	34.0
British (Kon and Mawson)	4.78	6.95	1.16	29.9
Australian (Winikoff)	—	—	—	28.6–30.7
American and European (combined means) (Morrison)	3.33 (S.D. 0.57)	7.2 (S.D. 0.67)	1.32 (S.D. 0.32)	—
Poorly nourished	3.42	7.51	1.06	34.2
Indian (Belavady and Gopalan)				
Bantu, South Africa (Walker *et al.*)	3.90	7.10	1.35	28.7
Alexandria, Egypt (Hanafy *et al.*)				
Healthy	4.43	6.65	1.09	—
Malnourished	4.01	6.48	0.93	
New Guinea (Becroft)	2.3	6.48	0.93	—
Sri Lanka (De Silva)	2.8	6.8	1.5	—
Brazil (Carneiro and Dutra)				
High economic	3.9	6.8	1.3	20.8
Low economic	4.2	6.5	1.3	25.7
Pakistan (Underwood *et al.*)	—	1.3–2.9	1.2	—
Tanzania (Crawford *et al.*)	"Often below 2%"			
Nigeria (Ibadan) (Naismith)	4.05	7.67	1.22	—
Pakistan (Lindblad and Rahimtoola)	2.73	6.20	0.8–0.9	28.4
Chimbu, New Guinea Highlands (Venkatachalam)	2.36	7.34	1.01	
New Hebrides (Peters)	3.8	5.0	1.40	25.8
Wuppertal, Germany (Immediately Post World War II) (Gunther and Stanier)	3.59	—	1.20	
Nauru (Bray)	—	—	1.06	—
Ibadan, Nigeria (Jelliffe)	—	—	1.1	—
New Guinea (Biak) (Jansen *et al.*)	—	—	0.83–0.9	—

[a] Modified from Gopalan and Belavady (1961) with added data (Jelliffe and Jelliffe, 1978). References in the original.

Information available on variation in amino acid composition of human milk is inadequate, although major differences between the amino acid profile of cow's milk and human milk are clear. Breast milk is higher in cystine and lower in tyrosine, phenylalanine, and tryptophan than cow's milk. Physicochemically, it has been shown that there are "species-specific systems" (Ribadeau-Dumas *et al.,* 1975). However, there is a lack of investigations into the amino acid composition of human milk at different stages of lactation and maternal nutrition. The need for such work is suggested by the low levels of lysine and methionine in the breast milk of lower socioeconomic women in Pakistan (Lindblad and Rahimtoola, 1974).

3.5.2. Fats

Variations in total fat content have been reported from different parts of the world. In general, levels for well-nourished communities are above 4.5 g/100 ml although Morrison's 1952 combined mean of American and European mothers was only 3.33 g/100 ml. In poorly nourished mothers, the content has been reported as ranging from 2.3 to 3.9%. However, a further problem in comparing these figures is related to possible sampling differences between foremilk and the fattier hindmilk (after milk) (Hytten, 1954b), and to diurnal variations in fat content, which is higher in the early morning and decreased in the middle part of the day.

The significance of the fat content of milk has been underemphasized. It is the main source of calories; it contains fat-soluble vitamins, especially vitamin A. It is also the source of fatty acids needed for the growth and development of the central nervous system. Lastly, in breast feeding the higher fat in the after-milk may act as a physicochemical appetite control (Hall, 1975).

Human milk has a high level of polyunsaturated fatty acids (especially arachidonic and docosahexaenoic; Crawford *et al.,* 1973) and is relatively high in cholesterol, and it contains the enzyme lipase and palmitic acid in the 2-position, (Gyorgy, 1971)—both of which assist in the digestion of fat and absorption of calcium.

The fatty acid pattern of breast milk can be altered by variation in the types of dietary fat or by changes in the calorie intake. During energy equilibrium, milk fat resembles the fatty acid pattern of dietary fat, but when inadequate calories are eaten, the fat in human milk resembles the composition of human subcutaneous depot fat. Similarly, the fatty acid composition of the breast milk (formula based on cow's milk) is reflected in the fatty acids in the infant's subcutaneous fat (Widdowson *et al.,* 1975).

The relevance of fatty acid content of human milk (and cholesterol) is currently under consideration in relation to atheroma in adults in industrialized countries. In developing regions, the polyenoic fatty acids may be diminished in the breast milk of malnourished mothers, with possible ill consequences in relation to brain growth (Crawford *et al.,* 1973).

3.5.3. Lactose

This sugar—the third "proximate principle"—is unique to mammal milk and has its highest level in human milk (±7.0%). It is generally recognized as being most constant in concentration and shows no diurnal variation. Morrison's 1952 figures based on 1010 samples examined in various American and European studies showed a mean of 7.2 g/100 ml, while Kon and Mawson (1950) found 6.9% in 586 samples of mature milk. In poorly nourished mothers, lactose also does not seem to vary very much (Table II) (range 6.43–7.51 g/100 ml), except in one study in the New Hebrides, where 5.0% was reported.

Nitrogen-containing carbohydrates make up about 0.7% of milk solids, such as *N*-acetylglucosamine, *N*-acetylneuraminic acid (sialic acid), and the "bifidus factor" (Gyorgy, 1953).

3.5.4. Calories

The calorie intake from breast milk is a product of the volume produced (or taken by the baby) and its caloric content, which is primarily derived from fat, together with lactose. In well-fed communities, the caloric content varies. Macy *et al.* (1953) give a mean figure of 75 cal/100 ml (range 45–119). In very poorly nourished communities, both the volume secreted and the fat, the main calorie-containing constituent (and other nutrients), may be less than with well-fed mothers.

3.5.5. Vitamins

3.5.5a. Vitamin A. The concentration is influenced by the quantity and quality of the diet of the mother. The vitamin A content of breast milk is often much lower in parts of developing countries (India, Ceylon, Indonesia, Jordan), where this nutrient is marginal, than in Europe and North America. Maternal serum levels are also low. The intake is generally higher in the spring and summer months, owing to greater supplies of dark green leafy and yellow vegetables. Particularly high levels of retinol (vitamin A) (and linoleic acid) have been noted in western Nigeria, presumably because of the widespread use of palm oil in cooking.

Kon and Mawson (1950) noted that supplementation with vitamin A, before and after parturition or later during lactation, led to the production of milk richer in vitamin A than normally produced. In Central America, Arroyave *et al.* (1974) observed a rise in the breast milk vitamin A levels after the introduction of sugar fortified with vitamin A.

3.5.5b. Thiamine. The thiamine content of breast milk in areas with a high incidence of infantile beriberi has been found to be low, owing to insufficient maternal stores and intake in communities with diets largely based on polished rice. Under these circumstances, this specific form of malnutrition—infantile beriberi—occurs in exclusively breast-fed babies of normal weight and is due to a thiamine-deficient diet in the mother during pregnancy and lactation.

3.5.5c. Riboflavin. Human milk is a rich source of riboflavin, provided that the maternal diet is adequate. However, in south India, Gopalan and Belavady (1961) found an average of only 17.2 μg/100 ml of riboflavin in breast milk as compared with a value of about 25 μg/100 ml found by Kon and Mawson (1950) in Britain.

3.5.5d. Niacin. Although human milk is low in actual niacin, it has a high niacin value because this vitamin may be synthesized from the amino acid tryptophan.

3.5.5e. Vitamin B_{12}. Low levels of vitamin B_{12} have been found in the milk of poorer vegetarian women in Bombay. Also, in various parts of India, the "syndrome of tremors" has been described in solely breast-fed babies and has been ascribed to deficient vitamin B_{12} in the mother's milk (Jadhav *et al.*, 1962).

3.5.5f. Vitamin C. The level of ascorbic acid in breast milk is subject to variations in dietary intake, particularly the seasonal availability of fresh fruits and vegetables. In well-nourished mothers, human milk contains an average of 4 mg/liter of vitamin C. In Botswana, Squires (1952) found the content to be 1.7 mg/100 ml in the dry season and 2.7 mg/100 ml in the wet seasons. The subjects were poorly nourished Tswana women.

Since the ascorbic acid content of breast milk is greater than that of blood plasma, which is generally below 2.5 mg/100 ml, the secretory activity of mammary glands must play a part in determining the level of vitamin C in milk.

Apparent adaptation to low maternal intakes of vitamin C have been noted in Baroda, India (Rajalakshmi *et al.*, 1974), and elsewhere. The possibility of placental syntheses in pregnancy has been suggested. Certainly, both the placenta and the breasts appear to be able to secrete ascorbic acid from the mother to offspring.

3.5.5g. Vitamin Supplementation. Deodhar and Ramakrishnan (1960) carried out a dietary survey among women in south India with special reference to pantothenic acid, riboflavin, nicotinic acid, acorbic acid, and thiamine. Subsequently, their breast milk was analyzed for the concentrations of the same vitamins. A positive and significant correlation was found between dietary intake and vitamin content of the milk for the vitamins investigated and underlines the need for an adequate diet for the lactating women. The content of all the vitamins increased steadily with the dose of supplementation.

In a more recent investigation (Deodhar *et al.*, 1964), the supplementations of the vitamins, ascorbic acid, nicotinic acid, riboflavin, thiamine, pantothenic acid, cyanocobalamin, biotin, pyridoxine, and folic acid were carried out. As a result, the vitamin content of the milk increased steadily with the dose of supplementation.

3.5.6. Minerals

The concentration of minerals is over three times greater in cow's milk than in human milk. The CM/HM ratio is 5.9:1 for phosphorus, 3.7:1 for calcium, and 3.6:1 for sodium, with a greater risk of hypernatremia (Rosenbloom and Sills, 1975) and metabolic uremia (Davies and Saunders, 1973) in babies fed with cow's milk. Human milk is essentially a low-solute fluid, designed to supply nutrients and water. There is an obvious need for further studies on trace elements in human milk, especially in relation to possible alteration as a result of environmental contamination.

Levels of calcium reported in the milk of well-fed mothers vary considerably. The calcium content in poorly nourished mothers has been reported to range from "normal" levels to somewhat low concentrations (Table II). Again differences in sampling and technique may in part be responsible.

3.6. Volume Secreted

The volume secreted in human lactation is influenced by the mother's nutrition, but is even more related to the meshing of neonatal reflexes (rooting, sucking, swallowing) with maternal reflexes (prolactin, letdown). Recent endocrinological investigations have endorsed the relationships among sucking stimulus of the nipple and breast areola, prolactin production by the anterior pituitary, and milk secretion in the mammary alveoli (Kolodny *et al.*, 1974). The letdown reflex moves milk secreted to the terminal lacteals as a result of contraction of the myoepithelial cells surrounding the alveoli. It is psychosomatic reflex, easily inhibited by anxiety or uncertainty (Newton and Newton, 1967).

Assessment of the volume of milk secreted may be attempted directly by "test feeding," or by expression, or inferentially by "satisfactory" weight gain by the baby. Both direct methods have inherent difficulties, notably because lactation is a 24-hr phenomenon and because of the likelihood of interference with the letdown reflex, as a result of emotional upset from the procedure (Hytten, 1954a).

3.6.1. Well-Nourished Mothers

To interpret the adequacy of yields of mothers in communities with defective nutrition, it is plainly important to compare results with those from well-nourished communities. Unfortunately, such data are scanty, out of date, and difficult to compare because of differences in technique and sampling.

Typical outputs of mature breast milk in well-nourished mothers in the first 6 months of life may be between 600 and 700 ml/day, according to a recent (1976) interpretation by Thomson and Black of the data collected by Morrison up to 1952. However, much variation occurred between the results of different investigations, depending on the techniques used and the subjects. For example, a Detroit series of studies contained professional wet nurses.

A carefully conducted study of 363 babies of well-fed Swedish mothers (Table III) may be noted as being representative and yet as showing increased intakes by boys (Lönnerdal *et al.*, 1977).

Various studies have been undertaken which have sometimes shown minor variations between the volume produced by each breast, and on different days or times of nursing. Diurnal variation in the amount secreted has also been noted, often with maximal yields in the early morning and lowest yields toward evening. However, such studies were all undertaken prior to understanding the relationship of sucking stimulus to prolactin production, and certainly three decades or more prior to the isolation of human prolactin and the introduction of serum radioimmunoassay methods by Hwang *et al.* (1971).

Results concerning variation with age or parity of mother have been rightly termed "confused and inconclusive," but probably are of little material significance.

Variation in volume of milk secreted between individual women is recognized as being considerable, although difficulties in making comparison are great, including differences in the weight and sucking vigor of the baby. Breast size does not appear to be related to yield.

Variation in yield with stage of lactation is difficult to judge in well-nourished communities as, until the resurgence of interest in breast feeding of recent years, nursing into the second semester of life or longer has not been usual in Western industrialized countries. Also, there seems little doubt that the main stimulus responsible for the volume of milk produced is the amount of sucking at the breast, and hence of prolactin secreted. This is shown by "induced lactation" in some traditional societies by recently introduced "adoptive lactation" or relactation in nonlactating American women who wish to breast-feed their adopted babies. In both, frequent sucking at the breast is the main factor in initiating lactation.

Likewise, 24% of a series of twins have been shown to be solely breast-fed adequately for 3–6 months (Addy, 1976). Also, the "perpetual," or at least very prolonged, high output by traditional wet nurses is in part a reflection of continuing vigorous sucking stimulus by successive infants. In fact, the volume of milk secreted has to be viewed against the pattern of infant feeding in the particular family (or culture), and the consequent degree of sucking stimulus and its effect on the prolactin secretion. The common concept of the normal

Table III. Daily Volume of Breast Milk in Swedish Women[a]

Month postpartum	Number of mothers	Breast milk volume (ml)
0–½	15	558 ± 83
½–1½	11	724 ± 117
1½–3½	12	752 ± 177
3½–6½	12	756 ± 140

[a] From Lonnerdal *et al.* (1976).

pattern in Western-type cultures is of rise in output in the first month or so, followed by a decline thereafter to a plateau lasting approximately until the baby is about 6 months of age, followed by a slow decline thereafter. Consideration of lactation patterns elsewhere and the previously mentioned results in twins, in wet nurses, in induced lactation, and in relactation suggest that the Western pattern is not a biological inevitable but a response to a particular pattern of sucking stimulation in a certain cultural context.

Very few studies have been undertaken on dietary effects on volume of milk produced in well-nourished women. By contrast, numerous investigations have shown that variation in water intake between wide limits has no effect on the volume of milk secreted. This seems endocrinologically explicable in light of the renal antidiuretic, water-sparing effect of prolactin. The common belief that fluid intake affects milk yield probably operates more through a "sympathetic magic" effect on confidence and hence on the psychosomatic letdown reflex.

3.6.2. Poorly Nourished Mothers

Estimations of the volume of breast milk produced have also been undertaken in a variety of countries in Asia (Gopalan and Belavady, 1961) and Africa (Bassir, 1959) and in New Guinea (Bailey, 1965). The results have been extracted from published information and these *approximations* are presented in Table IV. Despite differences in methods of collection, sampling and analysis, and levels and forms of maternal undernutrition, it seems that the volumes produced were usually somewhat below those reported from well-nourished countries in Europe and North America (E. F. P. Jelliffe, 1976).

Volumes reported varied greatly. However, *working approximations in round figures can be suggested: between 500 and 700 ml/day in the first 6 months of life, 400–600 ml/day in the second 6 months, and 300–500 ml/day in the second year:* The few studies undertaken in the third year of lactation show very considerable differences, varying from 230 to 488 ml/day.

Also, it is well recognized in practice that the output of *very seriously* malnourished mothers in famines declines and ultimately ceases (Jelliffe and Jelliffe, 1971a), with fatal consequences for the nursing baby. The nutritional point at which human lactation becomes inhibited in famine circumstances is not known.

3.6.3. Supplementation

Some studies have been carried out on the effect of supplementation of the mother's diet on output. Gopalan and Belavady (1961) showed an increase in volume secreted, from 420 to 540 ml in poorly nourished Indian women following protein supplementation (from 61 to 99 g/day). Similar results were obtained in western Nigeria by Bassir (1959) using a vegetable protein supplement, 30 g of soya flour daily.

A study in England seemed to suggest that poor lactation could be related

Table IV. Approximate Quantities of Milk (ml) Produced Daily at Different Periods of Lactation in Some Poorly Nourished Communities[a,b]

Country (Author)	1–6 Months	6–12 Months	12–24 Months	24 Months and above
India (Belavady)	600	500	350	—
India (Baroda)				
Rajalakshmi	660 (3 mo) (350–1100)	—	—	—
Ramakrishnan	735 (3 mo) (540–1100)			
New Guinea (Chimbu	525	525	343	343–142
Biak Island (Jansen *et al.*)	427	390–430	127–338	243
New Guinea (Becroft)	720	660	705	488
Chimbu and Maprik	400	400	400	—
New Guinea (Bailey) Baiyer River	600	600	600	—
New Guinea (Oomen) Ajamaru	—	350-480	270-360 (12-18 mo) 210-210 (18-24 mo)	230-300
Nubuai	—	310-410	250-340 (12-18 mo) 150-210 (18-24 mo)	—
Egypt (Hanafy *et al.*)				
Healthy	922	—	—	—
Malnourished	733	—	—	—
Sri Lanka (De Silva)	475	495	506	—
Nigeria (Johnson *et al.*) Benin	555 (2–3 mo)	590 (6–9 mo)		
Mexico (Chavez *et al.*)	650	400–500	350	

[a] Results not strictly comparable, as varying collection techniques and methods of sampling employed.
[b] Modified from Gopalan and Belavady (1961) with added data. References in the original.

to a restricted or slimming diet and inadequate energy reserves—in the form of subcutaneous fat—laid down in pregnancy (Whichelow, 1975). Two recent studies in Nigeria by Edozien *et al.* (1976) and in Guatemala by Sosa *et al.* (1976) very clearly indicate prompt and substantial success in yield with maternal dietary supplementation. Calories seem likely to be the commonest nutrient lack, a fact known to dairy farmers for decades.

3.6.4. Adequacy

The nutritional adequacy of breast milk for the infant can be assessed by measuring 24-hr output and chemical composition, or by the recording of satisfactory growth, "good health," and absence of clinical malnutrition.

In this regard, it is often insufficiently appreciated that the RDAs given for infants are themselves derived from estimated intakes of breast-fed babies, with an additional safety factory added for the less certain situation of infants fed on cow's milk formulas. This is the case, for example, with the RDA of iron (10 mg/day) for the first 6 months of life. The derivations of RDAs for infants are often not understood by pediatricians or nutritionists, so the advertising of commercial baby foods can issue the following appeal to "logic" (Jelliffe and Jelliffe, 1978):

> A stimulating exercise for professionals would be sorting one's beliefs about breast feeding into those based on scientific fact and those stemming from hearsay and emotion. One could thus more objectively counsel that growing membership in lay organizations dedicated to breast feeding. Many sincerely believe that breast milk is all sufficient without any supplementation for at least six months if not the full duration of the breast feeding. The fallacy of this concept is obvious if one compares the nutrient content of breast milk with recommended RDA for infants.

The recommended daily allowance of protein and calories suggested for infants has been very largely derived from the important, but special studies by Fomon and May (1958) on babies bottle-fed *ad libitum* with pasteurized breast milk. Based on their findings, it has been suggested that infants in their first month need 836 ml/day (protein, 2.6 g/kg; calories, 143 kg) and infants in their sixth month need 990 ml/day (protein, 1.7 g/kg; calories 90 kg). However, as noted earlier, outputs in well-nourished groups are usually lower than this, and a common range is 600–700 ml/day for the first 6 months. Among poorly nourished tropical communities, volumes often range from 500 to 700 ml/day during the first 6 months (Table IV).

In addition, some of the constituents of breast milk are affected to some extent by inadequate maternal nutrition, depending upon its severity, length, and the mother's previous nutritional status. Water-soluble vitamins are particularly affected, including ascorbic acid and riboflavin. Likewise, levels of vitamin A and the pattern of fatty acids reflect maternal diet and stores. The protein content of human milk in poorly nourished women is usually within "normal" limits (1.0–1.1 g/100 ml) or sometimes somewhat below this (0.7–0.9 g/100 ml). Lactose levels are unaffected. (Table II). Levels of water-soluble vitamins, notably ascorbic acid (Section 3.5.5f), thiamine (Section 3.5.5b), and vitamin B_{12} (Section 3.5.5e), reflect the mother's diet and stores, as does the concentration of vitamin A (Section 3.5.5a).

By comparison with estimates of volume and composition in well-fed mothers, information derived from poorly nourished women suggests that their babies may sometimes receive lesser intakes of nutrients. However, five factors need to be taken into account before making too-sweeping conclusions.

1. The figures on volume obtained in general studies are often those obtained by expression or following test feeding with the errors implicit, often having a risk of underestimation because of interference with the psychosomatic letdown reflex and inadequate stimulation of prolactin production.

2. Such investigations are usually based on daytime output, when, in most traditional cultures, babies sleep by the mother's side, with frequent night feedings the normal practice.

3. There is a possibility of lower calorie intakes being needed by infants in truly tropical areas such as, for example, India.

4. Smaller babies have less need for nutrients, but also may have less sucking potential.

5. The data on which the RDA are based are slight and approximate for

infants. The babies in the Fomon and May (1958) investigation were receiving a pasteurized breast milk from a feeding bottle, with an easier rate of flow and without the probably appetite-controlling mechanism of the higher fat found in the hindmilk. In other words, the intake of milk noted, and hence of calories and protein recommended, are probably overestimates. As noted earlier, the reported consumption of milk in direct breast-feeding studies is often lower.

3.6.4a. Adequacy by Growth. It is apparent that the ultimate test of the adequacy of human milk output in breast feeding is not to be obtained by estimates of volume produced or by the composition, but rather by "external assessment"—that is, with adequate growth, usually judged by weight gain, and with the absence of nutritional deficiency, as the main practical yardsticks. Laboratory tests for specific nutrients may rarely be possible, but, in any case, are often difficult to evaluate. Clinical impressions concerning "good health," vigor, and well-being are helpful but impossible to measure objectively.

There is sufficient evidence from widely scattered areas, including such diverse sources as the United States (Jackson *et al.,* 1964) and China (Shanghai Child Health Coordination Group, 1975) that solely breast-fed infants do well and show satisfactory gains in weight during the first 6 months or so of life when the lactating mothers are well nourished. From about 6 months, flattening of the weight curve usually indicates that the intake of calories and protein is no longer adequate—in other words, the baby has outgrown the breast as a *sole* source of food.

The adequacy of unsupplemented human milk (and the prolactin-mediated "supply and demand" nature of feeding) is indicated by a recent investigation of 172 pairs of twins in southern California; 23.7% pairs were breast-fed and, of these, 59.5% received no food other than human milk for 3 months or more, 21.9% for 6 months or more (Addy, 1976). Their growth is being investigated in detail but in general was comparable to those bottle-fed.

With the babies of poorly nourished mothers in resource-poor, technically developing countries, difficulties in interpreting growth curves exist in relation to the prevalence and significance of low birth weights in the particular community, in the selection of reference standards thought to be the most appropriate for the particular groups, the significance of infections (including diarrhea with fluid loss) and the familiar problem of sorting out weight gain in relation to increases in fat, muscle, or water–electrolyte retention.

Earlier evidence from various parts of Africa in the 1950s indicated that breast feeding, with little or no supplementation, resulted in excellent growth for about the first 5–6 months of life (Welbourn, 1955; Jelliffe, 1954). However, as a generalization, more recent studies from such varied, but poorly nourished parts of the world as New Guinea (Bailey, 1965) sometimes show comparable weight curves to those of Western standards of reference only up to about 4 months of age. In other studies growth has continued to be similar up to 5 months of age, as in Malaysia (Dugdale, 1971), while in northeastern Tanzania (Poeplau and Schlage, 1971), in Mexico (Chavez *et al.,* 1975), and in Pakistan (Lindblad *et al.,* 1977), the weight increase in some babies was reported to have become inadequate after 3 months. These disquieting results probably reflect a combination of maternal malnutrition, anemia, and infections, together with environmental psychosocial stress, affecting the letdown reflex. The use of oral contraceptives containing estrogens may also have a role in some places, as they reduce the volume and composition of the milk secreted (Chopra, 1972).

3.7. Antiinfective Factors

It has long been recognized that human milk has a protective effect against various nutritional conditioning infections, especially diarrheal disease, particularly with low levels of environmental hygiene. Such differences were striking in Europe and North America only 40–50 years ago and are of critical significance in developing countries nowadays in relation to growth in young children.

Until recently, this antiinfective effect has been regarded as being due largely to the cleanliness and lack of opportunity for contamination of breast milk, which is delivered "direct from producer to consumer." However, in the past few years, it has become apparent that human milk contains positive, active host resistance factors, both cellular and humoral.

In fact, the cellular content of human milk is almost as great as in blood, with a hierarchy of cells, ranging from mobile, ameboid macrophages to interferon-producing lymphocytes (Murillo and Goldman, 1970). The humoral constituents are numerous and still being discovered. They include secretory IgA, lactoferrin, lysozyme (3000–4000 times the concentration found in cow's milk), and the bifidus factor (Winberg and Gothefors, 1975). The last, together with the differing buffering action of human milk and more acid stools, are responsible for the dominance of *Lactobacillus bifidus* in the intestinal flora, as opposed to the largely gram-negative flora in babies fed on cow's milk (Gyorgy, 1953).

3.8. Contraceptive Considerations

The child-spacing effect of breast feeding, well recognized in many cultures, has long been regarded as an old wive's tale. Recent endocrinological investigations have shown that the prolactin secreted by the anterior pituitary in response to the baby's suckling has an anovulatory effect, with lactation amenorrhoea.

That the human should have such a built-in biological method of child spacing is not surprising, as other mammals space their offspring by hormonally mediated breeding or estrous seasons. However, the effectiveness and length of lactation amenorrhea is related to the amount of sucking stimulus, and hence of prolactin secreted. It occurs, therefore, most effectively in those communities with biological breast feeding—that is, with frequent permissive feeds throughout the 24 hr, and without the use of the complementary bottle feeds or semisolids in the early months of life. In some ecologies it appears to be potentiated by accompanying maternal malnutrition and is, of course, extended in those communities where sexual abstinence is usual between husband and wife for a culturally defined period after birth, such as in parts of Africa. Conversely, with "nonnatural" or "dosed" breast feeding, the child-spacing effect is diminished. In general terms, the intervals between births are about 1 year in the non-breast-fed, 2 years with biological or natural lactation, and 3 years with natural lactation plus postpartum sexual abstinence (Van Ginneken, 1974; Van Balen and Ntabomvura, 1976).

At the moment, lactation amenorrhoea still affords more couple-years protection (CYC) annually than is achieved by technological contraceptives delivered through existing family planning services (Rosa, 1976). The continuing trend toward bottle feeding in periurban areas in some developing countries has an anticontraceptive effect with increasing population pressure.

3.9. Economics

The economics in the financial cost of breast feeding and bottle feeding are very large. In an individual family basis, the cost of formula can be compared with the extra nutrients needed by the lactating woman—that is, 500 cal and 20 g of protein. Plainly, this depends on the type and cost of the cow's milk formula used and of the foods recommended for the mother. For example, in Los Angeles recently, investigations showed that widely used ready-to-feed formulas were three times as expensive as the cost of supplying the additional nutrients to the mother in the form of everyday foods. In poorer circumstances, such considerations are theoretical. Comparison of the cow's milk formulas and basic wages in developing countries shows that the purchase of adequate quantities would take from 25 to 50% or more of the families' earnings. If artificial feeding is used, it is with homeopathic doses of milk and heavy loads of bacteria, with marasmus and diarrheal disease commonly resulting (Jelliffe, 1968).

On the macro scale, economists and agriculturists have only recently become aware of human milk as a national food resource which can be viewed in financial terms (Jelliffe and Jelliffe, 1975a). Apart from other consideration, declining lactation in a country represents loss of a specialized food, which has to be replaced either by locally produced cow's-milk-based formula, which is impossible in most developing countries, or by imported products, with loss of foreign currency. The magnitude of such losses can be very great—for example, the decline in breast feeding from the 1960s to the 1970s in Singapore would have needed the expenditure of the equivalent of U.S. $1.8 million annually to purchase formula as replacement for the breast milk lost (Berg, 1973).

Also, other financial costs need to be borne in mind, including the expenses incurred in developing and running the child health services to deal with the resulting undernutrition and dehydration, and for family planning services to make good the lost community contraceptive effect of breast feeding.

3.10. Mother–Infant Interaction

Much recent work has been undertaken in mother–infant interaction in the newborn period. Briefly, it seems increasingly clear that, as with other creatures, there is a critical or sensitive period after birth when species-specific reflex action occurs between mother and newborn, notably the eye-to-eye *en face* position and the head-to-toe stroking which appears to facilitate "bonding" between the dyad (Klaus *et al.*, 1970). Much remains to be learned concerning the consequences of mother–infant interaction, as regards subsequent personality development and behavior. This will be difficult to undertake in such a long-lived species as man, with so many variables in diverse cultures in child-rearing practices and later, which may also influence emotional development. Nevertheless, current evidence suggests that the present-day apparent increase in child abuse in some parts of the world is often a "disorder of mothering," related to various factors, including insufficient mother–neonate interaction. The concept that has developed in some Western countries that bottle feeding by an affectionate mother is the equivalent of breast feeding is plainly not so. The degree of somatosensory contact and the direct hormonal effects in the mother are obviously dissimilar. Also, bottle-feeding cultures are also more likely to include other practices limiting mother–baby contact, such as unphysiologically separating the newborn from the mother in nurseries and the use of scheduled feeds.

The breast-fed baby is at a considerably advantage in relation to the "stimulation," which has been shown to be a highly significant factor for growth and development, in addition to the intake of nutrients.

3.11. Nutrition of the Exterogestate Fetus

In traditional societies, those babies who survive the perils of birth trauma and neonatal infection usually do well for this phase of life on a diet of human milk. With recent declines in breast feeding in urbanized communities, new patterns of undernutrition are emerging. In poorer circumstances, infantile marasmus and diarrheal disease are on the increase as a result of overdilute, contaminated feeds from inadequate lactation in poorly nourished mothers and, in some areas, the side effects of estrogen-containing oral contraceptives. Conversely, in better-to-do circumstances, infantile obesity—or "protein-calorie malnutrition plus"—is becoming a public health problem, and in large measure is due to the practice of "double feeding" which has developed, in which the baby is bottle-fed, with the volume and concentration under the mother's control, and with the unnecessarily early introduction of semisolid foods in the first weeks of life (Jelliffe and Jelliffe, 1975b).

4. Transitional Period

The transitional period is characteristically difficult for all mammals, with the change from a diet of maternal milk to the full range of foods for the species. For traditional human communities, it is the time of maximal nutritional stress. In fact, the 2-year-old is so much at risk from the synergistic effects of an inadequate diet and infections that a special name has been suggested, the "secotrant," for the second-year transitional (Jelliffe, 1968). It is the classical period for kwashiorkor, for severe avitaminosis A, and for "late" marasmus, which, with increasing costs and scarcity of food supplies in many parts of the world, may be on the increase. In more affluent circumstances, little undernutrition is seen in this phase of life, except for the widespread problem of iron deficiency, and for obesity.

In poorer communities, the diet in this period should consist of as wide a range of mixtures of local foods as possible, with the small, but significant, volume of breast milk, which in domestic terms supplies at least one 8-oz glass of milk daily, with 3–5 g of protein (supplementary to the usual vegetable protein weaning diet), essential fatty acids, and 200–500 IU of vitamin A (Rohde, 1975). Such weaning foods should be present in the concept of complementary "multimixes" (Jelliffe, 1968), basically often as a cereal–legume "double mix," and with a realization that a major need is for calories, as well as for protein and other nutrients.

4.1. Consequences

Changes in patterns of feeding the exterogestate fetus in recent years may be summarized by saying that breast feeding has declined until recently in many parts of the world, excluding more rural and impoverished areas, but has begun to reemerge as the preferred method in educated families in some Western countries. Also, evidence of shorter periods of adequacy of breast feeding appears to be becoming evident in women in some poorly nourished, socially stressed communities, perhaps especially in urban slums.

As regards the feeding to the transitional, the frequently culturally defined difficulties in persuading mothers to use available foods in the second semester of life has often become more difficult with rising costs and more limited food supplies. By contrast, in more westernized, well-to-do communities, the very early introduction of semisolids in the first weeks of life has become the vogue—with increased possibilities of caloric overdosage.

The consequences of current patterns of infant feeding are universal (D. B. Jelliffe, 1976) but vary in importance in different parts of the world, notably with the prevalence and success of breast feeding and the availability and use of suitable foods for the transitional. A contrasting pattern of nutritional problems is also found in poorer and better-to-do communities—the former dominated by "early" marasmus and infantile diarrheal disease and/or by kwashiorkor and "late" marasmus; the latter by young child obesity and iron deficiency, cow's milk allergy, and increasingly relevant metabolic abnormalities in the early months of life, including, for example, hypernatraemia and amino acidemia (Jelliffe and Jelliffe, 1978).

5. Conclusions

An overview of the present-day scene suggests that, different as the problems of growth and nutrition in early childhood seem to be, in fact *the areas of emphasis needed in infant feeding are the same the world over,* and that there is no need to have two standards, one for the affluent and another for the poor (Jelliffe and Jelliffe, 1978).

Universally, the three main planks of scientifically guided biological infant feeding need to be (1) *to feed the pregnant and lactating mother with a mixed diet of locally available foods,* (2) *to breast-feed alone for 4–6 months, and* (3) *to introduce least-cost weaning foods,* based on the concept of "multi-mixes" from 4–6 months onward, preferably prepared from locally available foods, but with continuing lactation into the second year of life, particularly in less well-to-do-circumstances.

Such a regime would have preventive effect of very great dimensions. For example, a return to more widespread breast feeding can be estimated to protect some 10 million infants currently affected each year by early marasmus and diarrheal disease, and conversely, breast feeding (associated with the introduction of semisolids at 4–6 months) could prevent about 1 million cases of infantile obesity and 100,000 cases of cow's milk allergy annually (Jelliffe and Jelliffe, 1978).

As with all mammal milks, human milk is indeed unique (Jelliffe and Jelliffe, 1971b). Fortunately, increasing awareness of the unbridgeable biochemical differences between human and cow's milk, and of the much wider issues—of economics, child spacing, host resistance, and allergy—involved when comparing breast feeding and bottle feeding has recently become apparent in supportive statements from official governmental bodies in Sweden and Britain (Department of Health and Social Security, 1974) and the Committee on Nutrition of the American Academy of Pediatrics (1976) in the United States. Such a regime also obviously fits into the need to use local resources to best purpose with least waste, and to the aim of national self-sufficiency. However, such a regime may be difficult to achieve in practice, for various reasons.

It will require mental flexibility to appreciate that malnutrition in the transitional has part of its genesis in fetal life and maternal malnutrition, with a feeble, low-birth-weight baby starting a nutritionally hazardous life already at a disadvantage, with inadequate nutrient stores and probably deficient sucking vigor. The dyadic role of maternal nutrition in infant feeding is often difficult to appreciate, possibly because of conditioning in medical training to compartmentalize "obstetrics" and "pediatrics" as separate issues.

The unique value of human milk and of breast feeding is becoming increasingly clear for optimal growth and nutrition in the exterogestate in all parts of the world, for reasons mentioned earlier. The basic issue is to try to devise practical programs to improve the pattern of breast feeding or, at least, decelerate the decline, *and* to optimize the quantity and quality of milk production, particularly by maternal feeding.

It is often stated that attempts to reverse the trend to artificial feeding are impractical, idealistic, unrealistic, and, anyhow, have never succeeded. This is not the case—no adequate wide-spectrum program has ever been attempted. While field research is needed to make detailed community diagnoses of the factors responsible for success for failure in different areas or countries, modern knowledge of the psychophysiology of lactation, of the effectiveness of group support in engendering confidence, and of the major factors that are principally responsible indicate the main lines that such programs should include: reorientation of the education of health professionals, modification of maternity ward practices (Table V), monitoring of inappropriate advertising of infant-food companies, and consideration of introducing legislation and facilities for working mothers who wish to breast feed.

The investigations summarized in the present account of knowledge concerning the quantity and quality of human milk in both well and poorly nourished communities in different ecological circumstances are very far from being satisfactory. Understandably, studies have, on the whole, been piecemeal, with different emphases and techniques employed in various investigations. Until the recent studies in Sweden by Hambraeus *et al.* (1976), the most modern investigations in well-nourished communities were made about a quarter of a century ago.

Table V. Possible Modifications in Health Services Designed to Promote Breast Feeding in a Community[a]

Health service	Modifications
Prenatal care	Information on breast feeding (preferably from breast-feeding mothers). Breast preparation. Maternal diet. Emotional preparation for labor.
Puerperal care	Avoid maternal fatigue/anxiety/pain (e.g., allow to eat in early labor; avoid *unnecessary* episiotomy; relatives and visitors allowed; privacy and relaxed atmosphere; organization of day with breast feeding in mind). Stimulate lactation (e.g., no prelacteal feeds; first breast feeding as soon as possible; avoid *unnecessary* maternal anesthesia; permissive schedule; rooming in). Lactation "consultants" (advisers—preferably women who have breast-fed; adequate "lying-in" period. In hot weather, extra water to baby by dropper or spoon.
Premature unit	Use of expressed breast milk (preferably fresh). Contact between mother and baby with earliest return to direct breast feeding.
Children's wards	Accommodation in hospital (or nearby) for mothers of breast-fed babies. No "milk nurses" or free samples.
Home visiting	Encourage, motivate, support.
Health center	Supplementary food distribution (e.g., formula and weaning foods) according to defined, locally relevant policy.
General	Supportive atmosphere from all staff. Avoid promotion of unwanted commercial infant foods (e.g., samples, posters, calendars, brochures, etc). Adopt minimal bottle-feeding policy and practical health education concerning "biological breast feeding."

[a] Jelliffe and Jelliffe (1978).

Likewise, the technical difficulties of obtaining representative samples, especially on a 24-hr basis, can be almost insuperable, especially in the field. Also, the degrees of severity, chronicity, and specificity of maternal malnutrition (and of maternal depletion from numerous previous reproductive cycles) are rarely indicated, and indeed with the great number of variables are difficult to categorize. Finally, the question of genetic differences in lactation ability in different human groups has never been explored.

At another level altogether is the fact that most research undertaken has been concerned with gross analyses, with no recognition of the more newly recognized considerations, such as the pattern of amino acids, polyenoic fatty acids, polyamines, and nucleotides, and the presence of antiinfective substances. *There is, in fact, a need for coordinated modern studies into all aspects of human lactation in well-fed and in poorly nourished communities, using similar sampling methods and analytic techniques*. These would include investigations covering the volume and major nutrients, *as well as* more recently recognized biochemical components and physiochemical elements (such as the forms of casein present) in different communities with various levels of maternal over- and undernutrition.

Also, recent studies suggest very much that efficiency of utilization of human milk means that a "mathematical additive" approach to the nutrients available need not be valid. For example, the main source of calories, fat, is more rapidly and effectively used from human milk than from cow's milk, because of the presence of active lipase and because fat is absorbed less well from cow's milk, when insoluble calcium palmitate forms in the baby's stools. Also, the presence of an unique pattern of polyamines and nucleotides in human milk has been suggested as being responsible for superior utilization of protein. If this is so, it means that, quite apart from consideration of detailed differences in the fat and protein of cow's and human milks, it is not valid to equate them on a gram-for-gram basis.

5.1. Generalizations

Nevertheless, present-day incomplete knowledge appears to warrant the following generalizations for practical action.

1. Unsupplemented human milk is all that is required to sustain growth and good nutrition for the first 6 months of life in babies of well-nourished mothers who have produced fetuses with optimal stores and who have themselves laid down adequate nutritional reserves, including subcutaneous fat, in pregnancy.

2. The volume and composition of human milk from poorly nourished women is surprisingly good, possibly with some metabolic adaptations but probably usually to their cumulative nutritional detriment ("maternal depletion"), but is often suboptimal in quantity and in quality, with lower values of fat (calories), water-soluble vitamins, vitamin A, calcium, and protein than is the milk of well-nourished women.

3. Limited studies with supplementary feeding of poorly nourished lactat-

ing women (and commonsense probability) have shown improvement in the volume of output and in the nutritional quality of breast milk to be feasible.

4. The adequacy of breast milk as the sole food for the baby is related to the mother's diet in pregnancy; to maternal calorie reserves in the form of subcutaneous fat; to fetal stores, mainly hepatic; to birth weight; and to the iron obtained from the placental transfusion ("dyadic infant feeding").

5. Breast milk produced in "late lactation" (e.g., 6 months–2 years or more) is insufficient by itself for the rising nutrient needs (and declining stores) of the rapidly growing infant, but forms a small, but valuable, supplementary source of "complete" protein, and of fat, calcium, and vitamins.

5.2. *Practical Approaches*

If growth, as evidenced by the weight curve, becomes inadequate in solely breast-fed babies at 4 months or earlier in some poorly fed communities, an initial investigation needs to be made to ascertain the relative significance of the effects of interference with reflexes and of inadequate maternal nutrition.

5.2.1. *Interference with Reflexes*

Local infant-feeding practices should be scrutinized to identify possible practices which may be affecting the two main maternal reflexes responsible for the production and ejection of milk: the prolactin and letdown reflexes.

Various factors may be found in the traditional culture or, more often, in imported Westernized concepts in maternity units which reduce sucking stimulus (and hence diminish the prolactin secretion) and/or create uncertainty and anxiety (and thus inhibit the letdown reflex) (Table VI). There is little doubt that disturbed reflex behavior is more important to lactation failure than is maternal subnutrition, unless it is severe. In devising an appropriate program to improve breast feeding, *both* aspects need consideration.

5.2.2. *Inadequate Maternal Nutrition*

If the infant is judged to be receiving inadequate breast milk related to suboptimal maternal nutrition, three approaches need to be considered.

5.2.2a. Cow's Milk Supplementation. A "logical" possible solution may seem to be the introduction of bottle feeds with cow's-milk-based formulas from just prior to the usual age of flattening of the weight curve. In the circumstances of most resource-poor, less developed countries, this approach (*allaitement mixte*) usually needs to be avoided. It introduces the danger of weanling diarrhea at an early and vulnerable age. It decreases the secretion of breast milk as it interferes with sucking stimulation and proportional secretion of pituitary prolactin. It is an additional endorsement by the pediatric nutritionist of the unfortunate and unaffordable trend away from breast feeding.

Table VI. Puerperal Practices Interfering with Lactation via the Prolactin and Letdown Reflexes[a]

Practice	Effect on lactation (via maternal reflexes)
Delaying first breast feed Sedated newborn (excess maternal anesthesia) Supplying prelacteal and complementary feeds[b] Regular, limited feeds (4 hr) (with no or limited night feeds) Mother and infant separated ("nurseries")	Limitation of suckling and prolactin secretion
Uninformed, confused mother Tired mother (no food or drink in labor) Routine episiotomy (pain) Weighing before and after (test feeds) Restricting visitors Unsympathetic or ambivalent health staff ("antidoula effect") Lack of privacy Availability of free samples	Anxiety and interference with letdown reflex

[a] Jelliffe and Jelliffe (1978).
[b] Formula or glucose-water by bottle.

Finally, if viewed on a family or on a large-scale community basis, it has economic, agronomic, and food production consequences of very considerable dimensions.

5.2.2b. Early Introduction of Semisolids. Again, under the majority of circumstances in the world, the risks of weanling diarrhea are great if (necessarily) unclean semisolid foods are introduced before necessary. Also, in average kitchen circumstances in tropical countries and with the foods most usually available, it may be very difficult to prepare digestible, well-tolerated, and nutritionally adequate supplementary semisolid foods for a child of this early age. However, the introduction of semisolids by cup or by spoon would have less effort on lactation performance than would cow's milk by bottle and should be based on the use of mixtures of locally available foods, nutritionally blended as multimixes, particularly cereal–legume blends.

5.5.2c. Supplementation of Maternal Diet. Current knowledge suggests that the most economical, safe, physiological, and practicable method of approaching the situation is by laying maximum emphasis on feeding the mother during *both* pregnancy and lactation. Some nutrient stores may be made good rapidly (e.g., vitamin C); some will take months, particularly calories (subcutaneous fat) and protein (muscle).

Adequate feeding, again based on low-cost multimixes of locally available foods, or on maternal diet supplements, *during pregnancy* can assist in ensuring adequate maternal weight gain and sufficient nutrient stores, including calorie reserves in the form of subcutaneous fat, as well as a newborn of good weight and with optimal nutritional reserves. Similarly, an appropriate maternal diet, again based on locally available food mixtures, should be the emphasis during lactation, with semisolids slowly introduced to the baby, probably from about the age of 4–6 months onward, depending on local circumstances, particularly the usual weight curve in infancy in the community concerned.

Recent studies have shown that the main additional nutrients needed during lactation are less than previously thought, in part because of the high efficiency of conversion (90% for calories; 40–60% for protein). In the United States, the extra 500 kg of calories and 20 g of protein can be achieved with a peanut butter sandwich and a glass of milk; in Indonesia, by increased uptake of rice and *tempeh* (soya bean preparation) (Rohde, 1975).

The message seems clear. As with so much else concerning the health and nutrition of young children, the emphasis should be in large measure on the mother. By feeding her with locally available foods during *pregnancy and lactation,* it will become possible to optimize the volume and composition of breast milk, to avoid the economic and distributive complexities of introducing cow's milk and bottle feeding unnecessarily, and to avoid assisting still further a decline in breast feeding on a community basis.

In addition, it must be emphasized again that those concerned with infant feeding often do not give adequate appreciation to the associated effects of great nutritional significance of the antiinfective, child-spacing, mother–infant interaction, and economic significance of human milk and breast feeding.

Almost a quarter of a century ago, in 1952, Morrison concluded his painstakingly detailed analysis of available evidence concerning the yield, proximate principles, and inorganic constituents of human milk by commenting: "It is clear that there is plenty of room for work on every constituent. This review may serve to show where special care is needed in sampling technique and the spacing of samples." The need for further investigations is greater at the present day in view of the increasing realization of the significance of human milk on a world basis, but especially in resource-poor, less developed countries. Such studies need to take into account modern knowledge of the psychophysiology and endocrinology of lactation and, if feasible, should be undertaken on a collaborative and comparative basis in various representative ecologies in different parts of the world. Priority should be given to practical issues of immediate application, such as the interactions among lactation, conception, and contraception, and the effect of improved maternal diet on lactation. Such a practical priority is encapsulated by the declamatory title of a recent paper from INCAP by Sosa *et al.* (1976): "Feed the nursing mother; thereby the infant."

At the same time, within the labyrinth of available data, sufficient knowledge already exists to permit a rational practical approach to be suggested, based on such admittedly incomplete information, on probability, on an understanding of the biological background of man's mammalian needs, and on the wider significance of adaptive suckling to the nursing dyad and to the community.

6. References

Addy, H., 1976, The breast feeding of twins, *J. Trop. Pediatr. Environ. Child Health* **21**:231.

Arroyave, G., Beghin, I., Flores, M., de Guido, C. S., and Ticas, J. M., 1974, Efectos del consumo de azúcar fortificada con retino en las madres embarazadas y lactantes, *Arch. Soc. Latinoamer. Nutr.* **24**:485.

Bailey, K. V., 1965, Quantity and comparison of breast milk in some New Guinean populations, *J. Trop. Pediatr.* **11**:35.

Bassir, O., 1959, Nutritional studies on breast milk of Nigerian women, *Trans. Roz. Soc. Trop. Med. Hyg.* **53**:256.

Berg, A., 1973, *The Nutrition Factor,* Brookings Institution, Washington, D.C.

Bostock, J., 1962, Evolutionary approach to infant care, *Lancet* **1**:1033.

Chavez, A., Martinez, C., and Bourges, H., 1975, Role of lactation in the nutrition of low socio-economic groups, *Ecol. Food Nutr.* **4**:1.

Chopra, J., 1972, The effect of steroid contraceptives on lactation, *Amer. J. Clin. Nutr.* **25**:1202.

Committee on Nutrition, American Academy of Pediatrics, 1976, Commentary on breast feeding and infant formulas, including proposed standards of formulas, *Pediatrics* **57**:278.

Crawford, M. D., Sinclair, A. J., Msuya, P. M., and Munhango, A., 1973, in: *Dietary Lipids and Post-Natal Development,* p. 41, Raven Press, New York.

Davies, D. P., and Saunders, R., 1973, Blood urea: Normal values in early infancy related to feeding practices, *Arch. Dis. Child.* **48**:73.

Deodhar, A. D., and Ramakrishnan, C. V., 1960, Relationship between the dietary intake of lactating women and the chemical composition of milk with regard to vitamin content, *J. Trop. Pediatr.* **6**:44.

Deodhar, A. D., Rajalakshmi, R., and Ramakrishnan, C. V., 1964, Effect of dietary supplementation on vitamin content of breast milk, *Acta Paediatr.* **53**:42.

Department of Health and Social Security, 1974, Present-day Practice in Infant Feeding, Report on Health and Social Subjects 9, H.M. Stationery Office, London.

Dugdale, A. E., 1971, The effect of the type of feeding on weight gains and illness in infants, *Br. J. Nutr.* **26**:423.

Edozien, J. C., Rahim Khan, M. A., and Waslien, C. I., 1976, Protein deficiency in man: Results in a Nigerian village study, *J. Nutr.* **106**:312.

Fomon, S., 1974, *Infant Nutrition,* 2nd ed., W. B. Saunders Company, Philadelphia.

Fomon, S. J., and May, C. D., 1958, Metabolic studies of women in full-term infants fed pasteurized human milk, *Pediatrics* **22**:101.

Gopalan, C., and Belavady, B., 1961, Nutrition and lactation, *Fed. Proc.* **20**:3.

György, P. A., 1953, A hitherto unrecognized biochemical difference between human milk and cow's milk, *Pediatrics* **11**:98.

György, P. A., 1971, Biochemical aspects of human milk, *Am. J. Clin. Nutr.* **24**:976.

Hall, B., 1975, Changing composition of human milk and early development of appetite, *Lancet* **1**:779.

Hambraeus, L., Forsum, E., and Lönnerdal, B., 1976, in: Food and Immunology, Swedish Nutrition Foundation Symposium 12, Uppsala.

Horrobin, D. F., Burstyn, P. G., Lloyd, I. J., Durkin, N., Lipton, A., and Miururi, K. L., 1971, Action of prolactin on human renal function, *Lancet* **3**:352.

Hwang, P., Guyda, H., and Friesen, H., 1971, *Proc. Natl. Acad. Sci. USA* **168**:1902.

Hytten, F. E., 1954a, Collection of milk samples, *Br. Med. J.* **1**:175.

Hytten, F. E., 1954b, Diurnal variation in the major constituents of milk, *Br. Med. J.* **1**:179.

Jackson, R., Westerfeld, R., Flynn, M. A., Kimball, E. R., and Lewis, R. B., 1964, Growth of "well-born" American infants fed human and cow's milk, *J. Pediatr.* **33**:642.

Jadhav, M., Webb, J. K. G., Vaishuava, S., and Baker, S. J., 1962, Vitamin B12 deficiency in indian infants, *Lancet* **2**:903.

Jelliffe, D. B., 1954, Infant feeding among the Yoruba of Needan, *West Afr. Med. J.* **2**:114.

Jelliffe, D. B., 1967, The pre-school child as a biocultural transitional, *J. Trop. Pediatr.* **14**:217.

Jelliffe, D. B., 1968, Infant nutrition in the subtropics and tropics, 2nd ed., *WHO Monogr. 29,* Geneva.

Jelliffe, D. B., 1976, World trends in infant feeding, *Am. J. Clin. Nutr.* **29:**1227.

Jelliffe, D. B., and Jelliffe, E. F. P., 1971a, *The Effects of Starvation on the Functions of the Family and Society,* Swedish Nutrition Foundation Symposium on Nutrition and Relief Operations in Times of Disaster, Uppsala.

Jelliffe, D. B., and Jelliffe, E. F. P., 1971b, The uniqueness of human milk, *Am. J. Clin. Nutr.* **24:**968.

Jelliffe, D. B., and Jelliffe, E. F. P., 1975a, Human milk nutrition and the world resource crisis, *Science* **183:**557.

Jelliffe, D. B., and Jelliffe, E. F. P., 1975b, Fat babies: Perils, prevalence and prevention, *J. Trop. Pediatr. Environ. Child Health* **21:**123.

Jelliffe, D. B., and Jelliffe, E. F. P., 1978, *Human Milk in the Modern World,* Oxford University Press, Oxford.

Jelliffe, D. B., and Maddocks, L., 1964, Ecological malnutrition in the New Guinea Highlands, *Clin. Pediatr.* **3:**432.

Jelliffe, E. F. P., 1976, *Maternal Nutrition and Lactation, Ciba Foundation Symposium: Breast Feeding and the Mother,* Elsevier/North-Holland, Inc., New York.

Klaus, M. H., Kennell, J. H., Plumb, N., and Zuehlke, S., 1970, Human maternal behavior at the first contact with her young, *Pediatrics* **46:**187.

Kolodny, R. C., Jacobs, L. S., and Daughaday, W. H., 1974, Mammary stimulations causes prolactin secretion in non-lactating women, *Nature* **238:**284.

Kon, S. K., and Mawson, E. H., 1950, Human Milk, *Coun. Spec. Rept. Ser. 219, Med. Res.*

Lindblad, B. S., and Rahimtoola, R. J., 1974, A pilot study of the quality of human milk in a lower socio-economic group of karachi, Pakistan, *Acta Paediatr. Scand.* **63:**125.

Lindblad, B. S., Ljungquist, A., Gebre Mehdin, M., and Rahimtoola, A. R. J., 1977, *The Composition and Yield of Human Milk in Developing Countries,* Swedish Nutrition Foundation Symposium on Food and Immunology.

Lonnerdal, B., Forsum, E., and Hambraeus, L., 1977, *The Protein Content of Human Milk,* Swedish Nutrition Foundation Symposium on Food and Immunology.

Macy, I. G., Kelly, H. J., and Sloan, R. E., 1953, The composition of milks, *Nat. Acad. Sci., Nat. Res. Coun. Publ. 254.*

Morrison, S. D., 1952, *Human Milk: Yield, Proximate Principles and Inorganic Constituents,* Commonwealth Agriculture Bureau, Scotland.

Murillo, G. J., and Goldman, A. S., 1970, The cells of human colostrum, *Pediatr. Res.* **4:**71.

Newton, M., and Newton, N., 1967, Psychologic aspects of lactation, *N. Engl. J. Med.* **277:**1179.

Newton, N., Peeler, D., and Rawlins, C., 1968, The effect of lactation on maternal behavior in mice, with comparative data on humans, *J. Reproduct. Med.* **1:**257.

Poeplau, W., and Schlage, C., 1971, Nutrition and health—Usumbura, in: *Investigation into Health and Nutrition in East Africa,* Weltform Verlag, Munich.

Rajalakshmi, R., Subbulakshmi, G., and Kothari, B., 1974, Ascorbic acid metabolism during pregnancy and lactation, *Baroda J. Nutr.* **1:**117.

Raphael, D., 1973, *The Tender Gift: Breast Feeding,* Prentice-Hall, Inc., Englewood Cliffs, N.J.

Ribadeau-Dumas, B., Groschaude, F., and Mercier, J. C., 1975, Primary structure of the polymorphs of casein, *Mod. Prob. Pediatr.* **15:**46.

Rohde, J., 1975, Human milk in the second year, *Paediatr. Indones.* **14:**198.

Rosa, F. W., 1976, Breast feeding in family planning, *Protein–Calorie Advisory Group (PAG) Bull.* **5:**5.

Rosenbloom, L., and Sills, J. A., 1975, Hyperatraemic dehydration and infant mortality, *Arch. Dis. Child.* **50:**750.

Sanguansensri, J., Gyorgy, P., and Silluken, F., 1974, Polyamines in human and cow's milk, *Am. J. Clin. Nutr.* **27:**859.

Shanghai Child Health Coordination Group, 1975, Measurement of the growth and development of children up to 20 months in Shanghai, *J. Trop. Pediatr. Environ. Child Health* **21:**284.

Sosa, R., Klaus, M., and Urrutia, J. J., 1976, Feeding the nursing mother; thereby the infant, *J. Pediatr.* **88**:668.
Squires, B. T., 1952, Ascorbic acid content of Tswana women, *Trans. R. Soc. Trop. Med. Hyg.* **46**:95.
Thomson, A. M., and Black, A. E., 1976, *Nutritional Aspects of Human Lactation, Bull. WHO* **52**:168.
Van Balen, J., and Ntabomvura, A., 1976, Traditional methods of birth spacing, *J. Trop. Ped. Environ. Child Health,* **22**:50.
Van Ginneken, J. K., 1974, Prolonged breast feeding as a birth spacing method, *Stud. Fam. Plann.* **5**:117.
Welbourn, H. G., 1955, The danger period during weaning, *J. Trop. Pediatr.* **11**:34.
Whichelow, M. J., 1975, Caloric requirements for successful breast feeding, *Arch. Dis. Child.* **50**:669.
Widdowson, E. M., Dauncey, M. J., Cairdner, D. M. T., Janxis, J. H. P., and Pelika–Felipkoa, M., 1975, Body fat of British and Dutch infants, *Br. Med. J.* **1**:653.
Winberg, J., and Gothefors, L., 1975, Host resistance factors, *J. Trop. Ped. Environ. Child Health* **21**:260.

7

Early Infant Nutrition: Bottle Feeding

Lewis A. Barness

1. Introduction

The nutritional principles involved in satisfactory bottle feeding are similar to those of breast feeding. In both, the goals are well-nourished but not over-nourished infants, protected from diseases, imprinted with habits leading to a healthy and productive life, and pleasure for both parents and infants (Barness, 1961; Reina, 1975).

2. Nutritional Requirements

Nutritional requirements have been estimated from multiple empirical observations. Many of these requirements have been derived by observing, measuring, and evaluating the breast-fed infant. Requirements have been found to vary because infants vary in such factors as body stores, genetic potential, activity, rate of growth, and environment. Until goals of infant feeding are better specified, multiple and varied formulas remain acceptable and necessary.

The first goals of artificial feeding are the preservation of life and adequate growth. These can be met for the small infant if excesses are avoided while minimal requirements (Table I) are met. Recognized deficiencies are prevented by the addition of suitable minerals and vitamins.

2.1. Water

Water allowance for term infants is about 130–150 ml/kg. This allowance is obtained by the breast-fed infant consuming sufficient calories to grow, since human milk contains approximately 67 cal/100 ml.

Lewis A. Barness • Department of Pediatrics, University of South Florida, Tampa, Florida.

Table I. Estimated Water Expenditure of a 1-Month-Old 4.2-kg Infant in a Thermoneutral Environment[a]

Sources of water loss	ml/kg	Percent
Growth	5	6
Skin and lungs	50	64
Feces	10	13
Urine	15	17
	80	100

[a] Modified from Bergman *et al.* (1974).

Requirements for water can be much less than 150 ml/kg. Bergmann *et al.* (1974) have estimated that a month-old infant requires approximately 80 ml/kg/day, if the infant is in a thermoneutral environment and is able to concentrate urine to 100 mosm/liter (Table I).

While infants of nursing age may be able to concentrate to 1000 mosm (Bergmann *et al.*, 1974; Edelman and Spitzer, 1969; American Academy of Pediatrics, 1957), acceptable maximal urinary concentration for this age infant is approximately 600–700 mosm/liter. An isoosmolal urine of approximately 300 mosm/liter is probably associated with the lowest energy requirement for renal processes. Urinary concentration of 1000 mosm/liter requires a volume excretion of about 15 ml/kg; 700 mosm/liter, approximately 25 ml/kg; and 300 mosm/liter, approximately 50 ml/kg.

Much of the osmolality of the urine is due to the electrolyte and protein content of the formula fed. Each milliequivalent of electrolyte contributes one milliosmole, and each gram of dietary protein probably contributes about four milliosmoles of renal solute load.

Thermoneutral environmental temperature causes the least amount of sweating, and at thermoneutral temperatures, approximately 50 ml/kg of water is required for sweat and insensible water. As the environmental temperature increases, insensible water loss may double at moderate temperatures and increase even more in warm temperatures. Similarly, fever in the patient will require more water excretion. A figure of 10 ml of increment for each degree of fever (centigrade) is useful.

Furthermore, since water is usually calculated as that which is contained in the formula, a further allowance of 5–10% of formula volume is necessary in the calculations, since most formulas contain 90–95% free water.

Roy and Sinclair (1975) have estimated similar water requirements, 85–170 ml/kg, for low-birth-weight infants. Those who receive phototherapy for hyperbilirubinemia should receive a 20% increase in water allowance.

2.2. Calories

Requirements for calories approximate 100–110/kg/day for the first 6 months of age. After 6 months, calories required decrease to about 90–100/kg/

day to the end of the first year. Calories are utilized as follows: basal metabolism, 50%; specific dynamic action, the increase in metabolism over the basal rate caused by the ingestion and assimilation of food, 7-8%; growth, 12-15%; physical activity, 22-25%; and fecal losses, 3-5%. Most simulated human milk formulas, as well as cow's milk, provide 65-67 cal/100 ml. Therefore, a caloric intake of 100-110 cal/kg/day provide 150-165 ml of fluid/kg/day. Thus, standard formulas provide sufficient water intake when adequate calories are consumed; conversely, sufficient calories are consumed when fluid intake is adequate if the caloric density of the formula is suitable.

The infant of low birth weight may have lower requirements in the first week of life. An intake of 50–80 cal/kg/day in the first week of life may suffice for such a baby; however, he, too, should be consuming approximately 100–130 cal/kg/day by the second week. The optimal rate of weight gain for the low-birth-weight infant is as yet uncertain; however, a suggestion has been made that too slow a rate of gain in the first few weeks may be a cause of inadequate growth of the brain.

The distribution of calories in most formulas is similar to that in human milk. Approximately 9–15% of the calories are protein, 45–55% are carbohydrate, and 35–45% are fat. Approximately 10% of the fat content should be supplied as polyunsaturated fatty acids, including essential fatty acids.

Overfeeding the infant may result in obesity, which may be resistant to treatment later. One limitation of bottle feeding is that the mother tends to put a certain amount of fluid in the bottle and expects the child to drain the bottle completely. This does not allow the child to adjust his fluid intake to his needs, but rather to the expectations of another person. Unfortunately, some mothers tend to put a bottle of formula in the baby's mouth whenever he cries, since she may not recognize that the crying may be expressing needs other than that for food. In warm weather, particularly if the formula is relatively high in osmolarity, the baby may cry because he is thirsty for a diluent, particularly water. Instead of water, the mother may give such children the same formula to satisfy the infant's thirst. These practices may lead to obesity of the infant.

It has been shown that infants take food and fluid according to both their volume and caloric density. Thus, if the caloric density is high, the child attempts to satisfy his fluid requirements, and therefore will consume more calories. Conversely, if the caloric density is low, the child will tend to satisfy his caloric needs by increasing the volume of food ingested. This can be useful in that obese babies can be treated by diluting the formula with water, and thin babies can be given a calorically more dense formula than the usual one containing approximately 67 cal/100 ml.

2.3. Protein

Dietary protein is necessary for infants for growth, and for the formation of serum proteins, hemoglobin, enzymes, hormones, and antibodies. Low dietary protein is associated with decreased growth, decreased resistance to infection, and, if extended, may be associated with mental retardation. Excessive protein not only may be not of advantage to the infant, but may be disadvantageous (Barness *et al.*, 1963).

Proteins fed to infants must contain all the essential amino acids and may contain nonessential amino acids as well. Histidine, tyrosine, and cystine are amino acids which may not be essential for older infants but are essential for small infants (Raiha *et al.*, 1976). Protein values may be determined by their content of essential amino acids compared with a standard protein mixture. Other methods include determining the amount of protein necessary to cause a measured weight gain, the basis of the term protein efficiency ratio (PER). Excess individual amino acids in a protein may cause an imbalance in the metabolism of protein and prevent efficient utilization of the other amino acids present.

In almost all infants, protein is hydrolyzed by peptidases and trypsin to amino acids in the intestinal tract and efficiently absorbed. Although 1 g of protein contains 4 cal, approximately 15% of these calories are consumed in the digestion and assimilation of the protein, specific dynamic action. Excess amino acids are excreted in the urine. Some are deaminated, and the ammonia is excreted or detoxified to urea and excreted.

The protein content of the formula should be such that it supplies 9–16% of the calories. The Committee on Nutrition of the American Academy of Pediatrics (1976) has recommended that formulas must provide a minimum of 1.8 g of protein/100 cal, provided that the protein has a PER at least 100% that of casein. Human milk contains approximately 1.6 g of protein/100 cal, with a PER much higher than that of casein. Most of the simulated human milk formulas contain approximately 2.3 g of protein/100 cal. The Committee on Nutrition further proposed a maximum protein level of 4.5 g/100 cal. Higher protein concentrations than these provide increasing solute load and have not been shown to be beneficial to the infant. Some special-purpose formulas do contain more protein, but these should be used only for specific indications.

For low-birth-weight infants, the optimal protein intake has not been well defined. Most low-birth-weight infants will grow satisfactorily when fed formulas containing good-quality protein which supplies 2.5–5 g/kg/day.

Since included in the protein efficiency ratio is an evaluation of the amino acid content of the protein, the recommendation concerning casein equivalent becomes more important as more vegetable proteins are used in the construction of infant formulas.

2.4. Fat

Fat is a convenient and efficient method of energy storage. Fat has a high fuel value, and thus spares essential dietary ingredients so that they are not used for calories. Fat in the diet also is a carrier of the fat-soluble vitamins and essential fatty acids.

Triglycerides, esters of fatty acids with glycerol, usually contain two and sometimes three different fatty acids. These fatty acids may be saturated or unsaturated. Depending on the position and type of fatty acid present on the glycerol molecule, absorption of the fat can be increased or decreased. Phospholipids contain phosphoric acid esterified with glycerol or sphingosine. Fatty acids with 4–6 carbon atoms per molecule are classified as short chain, with 8–10 carbon atoms as medium chain, and with 12 or more carbon atoms as long chain. Short-chain fatty acids provide approximately 5.3 cal/g, medium chain 8.3 cal/g, and long chain 9 cal/g.

Fat is ingested and converted to a coarse emulsion in the stomach by churning and mixing with the phospholipids in foods. Emulsified foods are then mixed in the duodenum with bile, pancreatic juice, and chyme. The triglycerides are subject to hydrolysis by pancreatic and intestinal lipases. Absorption of triglycerides containing long-chain fatty acids is dependent upon the presence of intestinal lipase. Absorption of triglycerides containing short- or medium-chain fatty acids is not lipase-dependent. Short- and medium-chain fatty acids complex with albumin, and this fatty acid–albumin complex is transported to the liver via the portal vein without going through the lymphatics.

The American Academy of Pediatrics (1976) recommends that a minimum of 30% of the calories or 3.3 g of fat/100 cal and 300 mg of linoleic acid or approximately 10% of the fat calories be provided in the daily formula. Human milk contains about 50% of the calories as fat, and approximately 10% of the fat calories as linoleic acid.

Essential fatty acids are those which cannot be synthesized by the human body, and include linoleic (18:2), linolenic (18:3), and arachidonic (20:4) acids. In humans they are necessary for growth, skin and hair growth and integrity, regulation of cholesterol metabolism and lipotropic activity, synthesis of prostaglandins, decreased platelet adhesiveness and coagulation, and reproduction (Schlenk, 1972). They help maintain cell membranes.

Excessive linoleic acid in the diet may increase peroxidation and apparently increases vitamin E requirements. The source of polyunsaturated fatty acids for formulas is obtained chiefly from plants. Animal fats are largely saturated. One consequence of increasing the quantity of plant sources of fat in infant formulas is the increase in plasma phytosterols in children fed these diets (Mellies *et al.*, 1976). Effect of phytosterols on the human is not known.

The maximum amount of fat recommended for infant formulas is 6 g/100 cal. Larger amounts of fat may induce ketosis or acidosis. Lower-fat formulas than those specified should be considered special diets for medical problems. Similarly, higher-fat diets, such as ketogenic diets, should be used only for special medical indications.

While high-fat or high-cholesterol-containing diets may be risk factors in the later development of atherosclerosis, no convincing evidence has been produced to indicate that feeding infants fats during the time that bottle feeding is indicated would affect the development of atherosclerosis. Some have recommended low-fat diets for those children with Type II hyperbetalipoproteinemia. The American Academy of Pediatrics (1972) has as yet not felt that there is any convincing evidence that this should be started in infancy. If low-fat diets are used for infant feeding, especially for reducing diets, such diets will have an increased amount of protein and carbohydrate and an increased electrolyte and amino acid content which can be detrimental to the infant. Infants fed such diets must be supplied with adequate water.

In the low-birth-weight infant, fat absorption is decreased compared to that of the term infant. Presumably this is largely due to the inadequacy of biliary secretion (Katz and Hamilton, 1974). Medium-chain triglycerides and polyunsaturated fatty acids are better absorbed by the low-birth-weight infant than other types of triglycerides. Some recommend that formulas for the low-birth-weight infant contain increased amounts of medium-chain or polyunsaturated fatty acids. However, there are few data on the long-term effects of medium-chain triglycerides, and vitamin E deficiency is more likely to develop with formulas containing increased amounts of linoleic acid.

2.5. *Carbohydrate*

In most formulas, as well as in human milk, approximately 40–50% of the calories are provided as carbohydrate. Lactose is the main carbohydrate in these formulas. Lactose is the milk sugar of most mammals. Some of the advantages of lactose in infant feeding include the production of an acid flora in the stool, which increases the absorption of calcium and decreases the absorption of phenols.

Carbohydrate is absorbed after starches or complex sugars are broken down into simple sugars. There is some evidence that amylase is decreased in the newborn. Apparently the quantity of amylase is sufficient for adequate starch absorption by 4 months of life. The infant is born with sufficient lactase to split lactose; however, in certain children and certain populations lactase deficiency may develop as early as 1 year of life. Lactase deficiency also may be induced in those children with diarrhea, and lactose intolerance then develops. In those children with lactase deficiency, lactose should be avoided.

Most infants are able to metabolize other simple sugars easily, so that lactose-free formulas are available with sucrose, dextrins, maltose, and other simple sugars. Rarely, infants with sucrase–isomaltase deficiency or fructose intolerance will be found. Such infants usually tolerate glucose or lactose as the sole carbohydrate.

Intolerance to carbohydrate usually is noted when the infant is introduced to a food and he rapidly develops diarrhea. Stools become very acid, with a pH below 5.5, and usually contain detectable quantities of the offending sugar. Blood glucose is usually low or normal until the infant is given the offending sugar, after which the blood glucose descends to truly hypoglycemic levels, and he may develop signs of hypoglycemia. An occasional infant has starch intolerance and develops similar symptomatology on receiving complex but not simple carbohydrates.

Infants with galactosemia usually appear normal at birth and may develop vomiting on first feedings. They become jaundiced and develop hepatomegaly, cataracts, dehydration, and signs of hypoglycemia. However, some of these infants have few symptoms, except for jaundice. Because of the potential dire consequences of this treatable condition, all infants with otherwise unexplained jaundice should be studied promptly for this disorder.

With proper carbohydrate metabolism, even small infants rapidly convert absorbed carbohydrate to immediate use via glycolytic pathways or store carbohydrate as glycogen or fat. Blood glucose in infants is narrowly controlled, as in older children. Concern has been expressed that high carbohydrate intake, particularly of refined sugars, has been associated with later atherosclerosis. Sugars are consumed mainly for taste. Later sugar consumption may be decreased if early infant feedings are not overly sweetened.

2.6. Vitamins

Vitamin requirements for term infants have recently been summarized by the American Academy of Pediatrics (1976). The following is a brief review of their recommendations.

Vitamin A. The recommended daily allowance for vitamin A is 1400 IU, or approximately 200 units/100 cal. Human milk, cow's milk, and commercially prepared formulas are excellent sources of vitamin A. Vitamin A is stored in the liver.

Vitamin D. Most formulas contain 62 IU/100 cal, or 400 IU/liter. This is double the estimated requirement of term infants.

Vitamin E. Full-term infants require approximately 0.3 IU of vitamin E/100 cal. The requirement for premature infants is not yet ascertained, although it is recognized that iron-containing formulas and formulas with higher linoleic acid content may necessitate additional vitamin E.

Vitamin K. Most infants born to well-nourished mothers will have adequate vitamin stores at birth except for vitamin K. The newborn infant or his mother is given vitamin K parenterally close to the time of birth. Vitamin K is present in most formulas, and the bottle-fed infant ordinarily does not need added vitamin K.

Water-Soluble Vitamins. In general, the Committee on Nutrition agrees with the recommended daily allowances of the NAS–NRC. Deficiencies of the water-soluble vitamins of formula-fed infants is rare.

Some children have a vitamin B_6 dependency state, for which extra vitamin B_6 may be necessary. Folic acid requirements are not met when goat's milk is fed. Some infants fed goat's milk also develop vitamin B_{12} deficiency. Other vitamin-dependency states are known and present as inborn errors of metabolism.

2.7. Minerals

The mineral requirements are similar to those allowances proposed by the FDA in 1974. Recommendations for sodium, potassium, and chloride include the warning that the ratio of sodium to potassium expressed in milliequivalents should not exceed 1.0, and the ratio of sodium plus potassium to chloride should be at least 1.5. In human milk, the ratio of sodium to potassium is 0.5

and sodium plus potassium to chloride is 2.0. Increased sodium in relation to potassium may imprint the child later to higher sodium intake, with the possible increased risk of developing hypertension.

2.7.1. *Iron*

While iron deficiency is not common in term infants fed human milk, iron availability is less in formulas with higher casein content. If a baby is fed a formula, the formula should contain 8–15 mg of iron/liter to decrease or obviate the development of iron deficiency and iron-deficiency anemia. Iron is transported via iron-carrying proteins, including transferrin. Saturation of transferrin of more than 50% may make iron available to bacteria in the intestinal tract. Thus, both iron deficiency and iron excess may be responsible for increased numbers of infections in infants.

2.7.2. *Calcium and Phosphorus*

Calcium absorption parallels fat absorption (Barness *et al.*, 1974). Uncomplicated nutritional neonatal hypocalcemia with tetany has almost disappeared with the use of mother's milk or many of the propietary milk formulas which have 2:1 calcium/phosphorus ratio. The relative phosphorus load in the newborn with possibly limited parathyroid and renal function can cause increased serum phosphate and decreased serum calcium, resulting in neonatal seizures or tetany. In such infants serum phosphorus is usually elevated about 7 mg/100 ml. Formulas with supplementary calcium are used as treatment. This hypocalcemia is usually transient with a good prognosis.

Complex hypocalcemic seizures are associated with complications of pregnancy, labor or delivery, including low birth weight, prematurity, maternal diabetes or toxemia, and chilling of the infant. Infants with severe renal anomalies, exchange transfusion, or those born to mothers with latent hypoparathyroidism may develop hypocalcemia or hypomagnesemic seizures within 24–48 hr (Rose and Lombroso, 1970; Mizrachi *et al.*, 1968).

2.7.3. *Other Minerals*

Zinc deficiency has been reported in infants, particularly in those fed formulas containing high phytate. Zinc requirement is approximately 3 mg/day. Zinc deficiency has been associated with the development of acrodermatitis enteropathica, other skin lesions, and growth failure.

Fluoride as drops or in the formula at 0.25 mg/day is effective in decreasing the incidence of caries. If the formula is diluted with fluoride-containing water, other supplements are not needed.

Copper deficiency has been associated with the development of edema and anemia. Small amounts of copper as well as other trace metals should be included in the formula.

3. Special Formulas

Certain infants are intolerant to certain aspects of formulas. Others may require special additions to the formula. Lactose-free formulas are available, such as soybean, hydrolyzed casein, or meat-based formulas. A carbohydrate-free formula is also available as a base with carbohydrate to be added as indicated. Formulas are available with medium-chain triglycerides as the main source of fat. These formulas are better absorbed by those children with fat intolerance. Evaporated milk formulas can be made by diluting one 13-oz can of evaporated milk with 17 oz of water, and 3 tablespoons of cane sugar. The incidence of diarrhea appears to be greater with infants fed evaporated milk mixtures than with those fed other milk formulas. Skim milk may be used when a low-fat formula is indicated; however, the essential fatty acids, particularly linoleic acid, are not contained in skim milk or in evaporated milk formulas.

4. Feeding Regimens

The first feeding may be given as soon as the infant recovers from the process of delivery. The first feeding should consist of sterile water, in the event that the infant vomits. After the first feeding, if the baby is to be fed from a bottle, any of the formulas that simulate human milk may be fed. Usually the feedings are given every 4 hr. Both the mother and infant should be kept as comfortable as possible and in as happy surroundings as possible. As more is learned of the psychology of infant feeding, it becomes apparent that it is difficult to separate the psychological from the nutritional factors. Both overfeeding or underfeeding as well as overindulgence or neglect can result in failure to thrive, and all should be avoided.

At present, commercial efforts are directed toward formulating cow's milk formulas more nearly like those of human milk. At present, human milk still appears to be the better feeding for infants. As more technological advances are made, it may be found that artificial formulas can more closely simulate human milk.

5. Solid Feedings

Solid feedings are necessary for infants since they cannot live on bottles or breast alone indefinitely. The best time to introduce solid feedings depends on factors other than nutritional needs.

From a nutritional standpoint, solid feedings offer a safety factor that may be missing from the diet which contains only a single source of calories and trace elements. It provides further a flexibility beyond that available from a single source. For example, if the protein source is deficient in certain amino acids, the total protein quality may be improved by supplementing with a different protein, which may be deficient in other amino acids.

Age should not be a rigid standard for introducing solid foods. Early introduction of solid foods may have certain disadvantages. Food allergy is more likely to develop in sensitive infants. The extrusion reflex does not disappear until 2–4 months, and so makes spoon-feeding difficult. Excessive solid feedings in addition to bottle or breast may increase total caloric intake and enhance the development of obesity. The Committee on Nutrition of the American Academy of Pediatrics has stated that "no nutritional advantage or disadvantage has as yet been proven for supplementing adequate milk diets with solid foods in the first 3 or 4 months of life . . ." (1958).

When solid foods are given, foods should be introduced singly, with a full week interval before introducing a second new food. Suitable starting foods include single cereals, apples, pears, peaches, beef, lamb, liver, and peas, beans, and carrots. Later, egg yolk can be offered but egg white probably should not be given before 1 year.

If a long delay occurs before solid feedings are offered, the attendant may find it easier to continue bottle feedings, delaying weaning. If the contents of the bottle are iron-deficient, anemia commonly occurs.

6. Summary

Formulas for early infant feeding have been developed as substitutes for breast feeding. Historically, formulas were used as a convenience to mothers or as the only nourishment of those infants who could not nurse. Goals at that time were to make the infant live and grow.

As investigative techniques have improved, unique properties of human milk for the human infant have been learned and relearned. As chemical properties of human milk have been identified, attempts have been made to make formulas similar chemically to human milk. Some commercially available formulas simulate human milk more closely than others; some manufacturers have stressed one chemical property and ignored others.

No presently available formula is a true substitute for human milk. The immunoglobulins, the polysaccharides, and some of the fats of human milk, for example, remain unique. Some properties of human milk, the effects of which are beginning to be noted, for example macrophages in milk as obtained from the human breast, are presently beyond technological duplication. Yet, not all mothers can nurse their babies; and obviously, mothers should not be made to feel guilty if they do not nurse their babies.

If babies are bottle-fed, those principles of nutrition which have been learned from observations of breast feeding should be followed. For the apparently normal baby, a formula which most closely simulates human milk should be chosen. Caloric density should ensure adequate water and caloric intake. Protein content should be sufficient to avoid protein deficiency, and the quality should be such as to provide all essential amino acids. The quantity of protein, however, should be restricted so as not to tax detoxifying or electrolyte-excreting mechanisms.

For most infants, carbohydrate probably should be restricted to lactose, although other sugars may have some advantages in certain situations. Fat should include the known essential fatty acids, as well as other unsaturated fatty acids to improve absorption of fat and calcium.

Since iron is not present in many human-milk substitutes, formulas should be supplemented with easily absorbable iron. Other inorganic substances, particularly fluoride to prevent dental caries, should be added. Sodium, potassium, and chloride should not be excessive. Other minerals, particularly zinc, are certainly necessary; optimal levels of these await better elucidation.

Vitamin deficiencies are uncommon in most artifically fed infants; formulas should be used which provide minimal requirements for the growing infant.

Eventually, other requirements of human infants may be recognized, and indeed it may be proven that some improvements can be made on human milk for the human infant. Any change away from human milk, however, should be accompanied by caution and proof of ultimate benefit.

7. References

American Academy of Pediatrics, Committee on Nutrition, 1957, Water requirement in re: Osmolar load as it applies to infant feeding, *Pediatrics* **19:**33.

American Academy of Pediatrics, Committee on Nutrition, 1958, On the feeding of solid foods to infants, *Pediatrics* **21:**685–692.

American Academy of Pediatrics, Committee on Nutrition, 1972, Childhood diet and coronary heart disease, *Pediatrics* **49:**305–307.

American Academy of Pediatrics, Committee on Nutrition, 1976, Commentary on breast feeding and infant formulas, including proposed standards for formulas, *Pediatrics* **57:**278–285.

Barness, L., 1961, Infant feeding (formula feeding), *Pediatr. Clin. North Am.* **8:**639–649.

Barness, L. A., Omans, W. B., Rose, C. S., and Gyorgy, P., 1963, Progress of premature infants fed a formula containing demineralized whey, *Pediatrics* **32:**52.

Barness, L. A., Morrow, G., Silverio, J., Finnegan, L. P., and Heitman, S. E., 1974, Calcium and fat absorption from infant formulas with different fat blends, *Pediatrics* **54:**217–221.

Bergmann, K. E., Ziegler, E. E., and Fomon, S. J., 1974, Water and renal solute load, in: *Infant Nutrition* (S. J. Fomon, ed.), Chap. 10, pp. 245–266, W. B. Saunders Company, Philadelphia.

Edelman, C. M., Jr., and Spitzer, A., 1969, The maturing kidney, *J. Pediatr.* **75:**509.

Kanaanch, H., 1972, Detrimental effects of bottle feeding, *New Engl. J. Med.* **286:**791–892.

Katz, L., and Hamilton, J. R., 1974, Fat absorption in infants of birth weight less than 1300 gm., *J. Pediatr.* **65:**608.

Mellies, M., Glueck, C. J., Sweeney, C., Fallat, R. W., Tsang, R. C., and Ishikawa, T. T., 1976, Plasma and dietary phytosterols in children, *Pediatrics* **57:**60–68.

Mizrachi, A., London, D., and Gribetz, D., 1968, Neonatal hypocalcemia, its causes and treatment, *New Engl. J. Med.* **279:**1163–1165.

Raiha, N. C. R., Heinonen, K., Rassin, D. K., and Gaull, G. E., 1976, Milk protein quantity and quality in low birth weight infants, *Pediatrics* **57:**659.

Reina, D., 1975, Infant nutrition, *Clin. Perinat.* **2:**373–391.

Rose, A. L., and Lombroso, C. T., 1970, Neonatal seizures states, *Pediatrics* **45:**404–407.

Roy, R. N., and Sinclair, J. C., 1975, Hydration of the low birth weight infant, *Clin. Perinat.* **2:**393–417.

Schlenk, H., 1972, Odd numbered and new essential fatty acids, *Fed. Proc.* **31:**1430–1435.

8

Malnutrition, Learning, and Animal Models of Cognition

David A. Levitsky, Larry Goldberger, and Thomas F. Massaro

1. Introduction

The possibility that severe malnutrition experienced early in the life of a child may permanently lower ultimate cognitive potential has stimulated a great deal of multidisciplinary research. The major objectives were to determine (1) whether malnutrition does alter the cognitive functioning of the brain, (2) whether the effects are permanent, and (3) the mechanism(s) through which malnutrition affects cognition. Although the major concern of these studies is the consequences of malnutrition on human cognition, the strategy of many researchers is to investigate the effects of early malnutrition on animal learning and behavior.

Several things necessitated the use of animal behavior as a model for the study of malnutrition and cognition in man. First, it is very difficult to scientifically answer the question of whether malnutriton, per se, impedes cognition in humans. Typically, in a population of children who are suffering from malnutrition, there coexist poor health care, poor sanitary conditions, poor educational facilities, and all the other concomitants of poverty which are strongly suspected to affect cognitive functioning.

Second, ethical considerations prohibit the use of strict experimental design in human studies. For example, it is ethically impossible to study the behavior of a malnourished and sick child for an extended period of time without medically treating the child. Moreover, it is very difficult, if not impossible in many cases, to manipulate environmental variables in human studies which may be important in understanding the etiology of malnutrition-

David A. Levitsky, Larry Goldberger, and Thomas F. Massaro • Division of Nutritional Sciences and Department of Psychology, Cornell University, Ithaca, New York.

induced behavioral changes. The use of animal models allows more precise manipulation of both nutritional and environmental variables.

Finally, the use of animal models enables one to carry out extensive behavioral testing and biochemical measurements, particularly measurement of brain chemistry under various dietary and environmental conditions. This kind of research design is extremely important if we are ever to understand the relationships among brain structure, chemistry, and behavioral function. These measurements are extremely difficult to obtain in man.

However, the use of animal models is not without serious drawbacks. It is sometimes very tempting to extrapolate the determinants of seemingly similar behaviors such as aggression or maternal behavior from animal to man (e.g., Lorenz, 1966; Ardrey, 1966) and is often quite hazardous (Montagu, 1976). In this chapter we have attempted to review the literature on the effects of malnutrition on learning and behavior of animals. Particular attention is devoted to the experimental confounds many researchers have neglected to take into account when interpreting learning studies and to a discussion of an appropriate animal learning model for the study of malnutrition and cognitive functioning.

2. *Techniques for Producing Early Malnutrition in Animals*

Many different techniques have been used to produce malnutrition during gestation, lactation, and early weaning in the experimental animal. Each has its advantages and disadvantages (Altman *et al.*, 1971a; Plaut, 1970). There has been little work directed at comparing the different techniques as to behavioral outcome. Nevertheless, we must be cautious in assuming that the different nutritional techniques for producing early malnutrition have the same effect on behavior and behavioral development.

3. *Behavior Effects of Malnutrition in the Young Animal*

There is no question that dietary restriction has a very profound effect on animal behavior. During the period of lactation, malnutrition imposed on the young mammal by either food restriction or the administration of a low-protein diet causes significant delays in the rate of development of various psychomotor reflexes (Altman *et al.*, 1971b; Salas, 1974; Simonson *et al.*, 1968). In the rat, developmental indices such as eye opening, ear flap opening, grasping reflex, and rearing and righting reflexes are all significantly delayed. Moreover, the emergence of more highly integrated behaviors, such as food and water ingestion, climbing (Massaro *et al.*, 1974), and development of sexual behavior (Larsson *et al.*, 1974), are delayed as a consequence of undernutrition. In the monkey, the normal development of social behavior has been shown to be delayed during the period of malnutrition (Zimmermann and Strobel, 1969).

It is important to recognize that malnutrition imposed during lactation affects not only the behavior of the pups but also the behavior of the dam. When the pups are removed from the litter by the experimenter, the time required to retrieve them is increased in dams malnourished with a low-protein diet (Frankova, 1971), restricted food intake (Smart, 1976; Smart and Preece, 1973), or for dams nursing an abnormally large litter (Seitz, 1954). Observations of the undisturbed dam in the home cage indicate that malnutrition imposed on the pups through a low-protein diet (Massaro *et al.*, 1974), nipple ligation (Lynch, 1976b), or maternal adrenalectomy (Levine and Wiener, 1976) increases the time spent with the litter. Other aspects of the dam's behavior are also affected by malnutrition during this period (Massaro *et al.*, 1974).

Undernutrition imposed on the pregnant mouse by feeding a low-protein diet (5% casein) has been shown to produce a significant depression in the ontogeny of various psychomotor reflexes in the offspring (Bush and Leathwood, 1975). The data in the rat, however, are more equivocal, although a tendency toward delayed psychomotor development has been observed (Simonson *et al.*, 1968; Smart and Dobbing, 1971; Massaro, 1977). These effects are evident despite the fact that the pups are cross-fostered at birth to well-nourished dams.

Moreover, disturbances in normal maternal behavior are also evident in well-nourished dams suckling gestationally malnourished pups. These dams spend more time in the nesting area than dams nursing pups well nourished *in utero*, particularly at the end of lactation (Massaro, 1977).

These alterations in the interaction of the young pup with its immediate environment make the interpretation of long-term effects of early nutritional insult very difficult, if not impossible. As we have seen, malnutrition imposed during gestation or lactation alters the behavior of the dam and the pups during this preweaning period. Thus, from a strict experimental viewpoint, any long-term effect of early malnutrition may result from an altered early environment rather than from any effects of nutrition on brain structure, per se.

4. Learning and Motivation

From the previous discussion it is clear that malnutrition imposed either during gestation or lactation profoundly affects the development of those behaviors and reflexes which are normally expressed prior to weaning. These behaviors, although reflective of the development of the brain, do not reflect its ability to process the kind of information we generally refer to as learning or cognition. One of the most important questions regarding man is whether malnutrition severe enough to affect brain growth affects the cognitive functioning of the brain.

Cognition generally refers to the ability of the brain to acquire, process, and retrieve information about its environment. Traditionally, psychologists have used the rate of learning a solution to a problem as the model to study cognition in animals.

One problem with this model of cognition is that for the animal as well as the human, several different kinds of learning may exist. It is fairly clear, for example, that learning to avoid a food which produces illness is quite different from learning a response to procure food or avoid an electric shock (Garcia *et al.*, 1966). Most of the work in the area of malnutrition and learning in animals has used almost exclusively a model of learning based on the acquisition of a response to procure food or avoid aversive (non-food-related) events.

Another problem in the area of malnutrition and learning in animals is that the learning process cannot be measured directly, but can only be inferred from the performance of an animal in a learning situation. Many factors are known to affect the performance of an animal in a learning situation but do not affect the learning process. For example, if one observed the rate of learning to bar-press for water following a condition of either mild or severe water deprivation, one would observe a greater rate of performance under the more severe deprivation (Collier and Levitsky, 1967). This does not mean that learning is enhanced under severe deprivation conditions; it simply means that performance is enhanced. To test this, a rather complicated factorial design is required (Kimble, 1961). Similarly, performance in a learning situation is sensitive to the kind of reinforcement used, the magnitude of reinforcement, the interval between trials, the novelty of the learning situation, and many other factors. Therefore, one must use much caution in interpreting any alteration in performance of an animal in a learning situation as demonstrating an effect on the learning process (Levitsky and Barnes, 1973).

5. *Concurrent Effects of Protein and Calorie Malnutrition on Stimulus–Response Learning*

This difficulty in interpreting learning studies is clearly demonstrated in the early attempts to establish an effect of concurrent protein or calorie malnutrition on simple stimulus–response learning. Stimulus–response learning involves the presentation of a stimulus to the animal to which the animal must respond in order to obtain a nutrient or escape an aversive stimulus.

The early literature on the relation between dietary restriction and S-R learning seems to indicate that undernutrition "enhances" learning a solution for food reinforcement. In rats, malnutrition begun at weaning and maintained throughout testing results in *faster* learning of a simple S-R problem as measured by a decrease in the number of errors made in learning the problem or an increase in the number of trials necessary to learn a problem (Anderson and Smith, 1926; Riess and Block, 1942; Bevan and Freeman, 1952). Koch and Warden (1936) found similar effects in mice. Biel (1938) found no better performance as judged by the number of errors made in learning a maze, but did observe better performance in the speed of running a maze in animals maintained on low-protein diets since weaning.

Others, however, using the same basic kinds of dietary restrictions, found evidence of *poorer* learning of a problem for food reinforcement (Rajalakshmi *et al.*, 1965). One reason is that the poor protein diet on which the rats were raised was used as the reinforcement. Griffiths and Senter (1954) have demonstrated that rats maintained on a low-protein diet make significantly more errors than controls when the reinforcement consists of the low-protein diet. When the reinforcement is a normal control chow, no significant difference in errors is observed.

Zimmerman *et al.* (1974) reported a series of studies in the monkey in which great care was taken to equate the reinforcement value for protein malnourished and control groups. They observed no effect of concurrent protein restriction on simple discrimination learning, simple reversal learning, learning set formation, delayed responding, or short- or long-term memory. Thus, concurrent protein or calorie malnutrition does produce behavioral differences, as evidenced by the performance differences in learning situations. However, there exists no evidence that concurrent malnutrition, even when started early in life, detrimentally affects learning a solution of a problem to obtain food.

One way to circumvent the food reinforcement problem in learning studies is to use other types of reinforcement. Typically, rate of learning an escape response from water or electric shock is used. However, these kinds of studies are not immune to the criticism that differences in learning performance may be due to factors other than alterations in learning mechanisms. For example, Cowley and Griesel (1962) tested young rats protein-malnourished from conception in a Hebb-Williams water maze. The maze was filled with water and the animals were forced to learn a path that removed them from the water. When the water temperature was approximately at room temperature, the rats maintained on a low-protein diet learned the maze more rapidly than the well-nourished animals. However, when the water temperature was lowered, the protein-malnourished rats made more errors and displayed slower learning rates than their well-nourished controls. Instead of showing learning differences, these studies suggest that the protein-malnourished rats are more affected by the low temperature than control animals. Others have also found poor learning performance in water escape mazes when the water temperature was low (Barnes *et al.*, 1966; Bernhardt, 1936). Pilgrim *et al.* (1951) did not find any effect of low-protein diets in water-maze escape learning. One reason for this was that the "low"-protein diet of Pilgrim *et al.* (1951) contained 12% protein, whereas the Bernhardt (1936) and Barnes *et al.* (1966) diets contained only 5% protein. Ruch (1932) also found no difference in learning a water maze at low temperatures in calorically deprived rats maintained at 80% of the body weight of controls from weaning, yet he found better learning performance in animals limited to no increase in growth until the time of testing at approximately 60 days.

Thus, there exists no evidence that concurrent calorie or protein malnutrition introduced early in life can clearly affect S-R learning independent of the potent effects calorie and protein malnutrition have on performance in simple learning situations.

6. Concurrent Effects of Protein Malnutrition on Complex Stimulus–Response Learning

As seen above, concurrent protein or calorie malnutriton has not been clearly shown to affect the rate of learning a response to a stimulus either to obtain food reinforcement or to avoid an aversive reinforcement. Therefore, either concurrent malnutriton does not affect the learning process, or the learning model based on simple S-R associations is insensitive and therefore invalid as a model for human intelligence. To test which conclusion is correct, it is necessary to review the effects of malnutrition on tests of learning other than simple S-R learning. One such alternative is the Hebb-Williams test of animal "reasoning." The test procedure was developed by Hebb and Williams (1946) and modified by Rabinovitch and Rosvald (1951). The animals are first adapted to a large box in which they learn the location of the start box (site of maze entrance) and goal box (site of food placement). After a period of adaptation a pattern of barriers is constructed in the maze between the start and goal boxes. Each pattern of barriers makes up a different problem. The animals are then given a series of trials on each of several problems. The rate at which the animals learn the shortest path between the start and goal box measured by the number of errors committed is used as the measure of performance. The ability of this test to discriminate between brain-damaged and normal rats (Rabinovitch and Rosvald, 1951; Hebb and Williams, 1946) as well as environmentally enriched and deprived rats (Hebb and Williams, 1946; Rabinovitch and Rosvald, 1951; Hymovitch, 1952; Forgays and Reid, 1962; and Denenberg *et al.*, 1968), has been clearly demonstrated. Moreover, performance in the Hebb-Williams maze is supposed to be relatively insensitive to food motivation and emotionality (Das and Broadhurst, 1959).

Young or adult rats maintained on low-protein diets since early life display poorer performance in the Hebb-Williams test than do well-nourished controls. Significant differences have been found between controls and experimentals in which malnutrtion was initiated postweaning (Baird *et al.*, 1971), at parturition (Zimmermann and Wells, 1971; Baird *et al.*, 1971), or in dams malnourished at conception (Cowley and Griesel, 1959, 1963, 1966) and continued through the testing period. One exception, however, is a study reported by Baird *et al.* (1971) in which no significant effect of protein or calorie restriction was observed in Hebb-Williams performance of young rats malnourished since conception.

Overall, protein or calorie malnutriton induced early in the rat and continued throughout life does seem to affect performance in a Hebb-Williams test of learning, whereas simple S-R learning seems to be unaffected. Although all the studies cited above introduce dietary manipulation fairly early in life, they do not test whether or not malnutrition during sensitive phases of early life causes permanent alterations in learning ability. This question can only be answered by studying the animal malnourished early in life but later given a nutritionally rehabilitative diet.

7. Long-Term Effects of Early Malnutrition on Simple Stimulus–Response Learning Following Rehabilitation

Malnutrition experienced during gestation or lactation has been reported to produce behavioral effects which persist after nutritional rehabilitation. Malnutrition during gestation has been observed in adult rats to enhance passive avoidance behavior (Hanson and Simons, 1971; Rider and Simonson, 1973, 1974) and increase errors in learning a water maze (Caldwell and Churchill, 1967), a shock–escape response (Vore and Ottinger, 1970), or an elevated T maze for water reinforcement (Simonson and Chow, 1969; Hsueh *et al.*, 1974). In mice malnourished only during gestation, Leathwood (1973) has observed poorer active avoidance behavior in adult mice. Malnutrition imposed during both gestation and lactation or lactation alone improves passive avoidance (Levitsky and Barnes, 1970; Smart *et al.*, 1973; Sobotka *et al.*, 1974), and active avoidance behavior (Levitsky and Barnes, 1970; Morris, 1974) in the adult rat. Others have found postnatal malnutrition to result in poorer active avoidance in rehabilitated mice (Leathwood, 1973; Leathwood *et al.*, 1974), pigs (Barnes, *et al.*, 1970), and rats (Sobotka *et al.*, 1974). Frankova and Barnes (1968) found no differences in rate of acquisition of an avoidance response in adult rats malnourished during lactation alone, but found significantly more intertrial responses in the previously malnourished rats. Barnes *et al.* (1966) observed poor performance of previously malnourished male rats in learning a water maze discrimination, but did not observe poor performance in females. Guthrie (1968) and Randt and Derby (1973) observed no significant effect in active avoidance.

One of the major problems with these studies is not the question of whether early malnutrition produces an effect on behavior, but whether the interpretation of these differences in performance reflects long-term alterations in the ability to learn. Although it would seem that long-term studies involving early malnutrition should be relatively immune to the motivational and reinforcement confounds prominent in studies of concurrent malnutrition, studies have indicated that malnutrition experienced early in life does affect appetitive and aversive motivation.

It is known that food restriction in the weanling rat can be shown to have a long-term effect on food-related activity. Early deprivation has been observed to increase hoarding of food (Wolfe, 1939; Hunt, 1941; Hunt *et al.*, 1947), increase rate of food consumption (Marx, 1952; McKelvey and Marx, 1951), and increase rate of bar pressing for food (Mandler, 1958). Malnutrtion induced during lactation and early postweaning also increases motivation for food in the food-deprived adult rat (Levitsky, 1975; Smart *et al.*, 1973) and results in increased food intake when expressed per unit surface area (Barnes *et al.*, 1968; Levitsky, 1975). This increased food intake in the adult has also been observed in rats malnourished during gestation and lactation (Chow and Lee, 1964; Blackwell *et al.*, 1969). Thus, if food is used as a reinforcement in learning situations for previously malnourished but rehabilitated animals, then the question as to whether the reinforcement may have the same "incentive value" when compared to the well-fed controls is important. Differences in incentive value of the reinforcement may affect the performance of the animal in a learning situation and thus confound an interpretation of a "learning deficit." Whether motivational or incentive differences result in better or worse performance in a learning situation depends upon the complexity of the learning task. The more difficult the task, the more detrimental high motivation or incentive is to the learning performance (Kimble, 1961).

The motivational confound observed with food also may hold for studies in which water is used as a reinforcement. Malnutrition induced during lactation and early postweaning period followed by nutritional rehabilitation also produces an increase in *ad libitum* water intake (Levitsky, 1975).

In an attempt to find a neutral reinforcement for the previously malnourished animal, many researchers have used aversive stimulation. Unfortunately, early malnutrition has been demonstrated to increase the adult rat's reaction to a whole host of aversive stimuli (Levitsky and Barnes, 1970), most notably demonstrated by a decrease in foot shock threshold (Levitsky, 1975; Smart *et al.*, 1975; Lynch, 1976a). Thus, as with concurrent malnutrition, since differences in performance may result from different reactions to the reinforcement conditions, it is almost impossible to conclude from the studies cited above the malnutrition imposed early in life can alter the ability of the rat to learn.

To try to parcel out these confounding performance factors from learning studies, Levitsky (1975) developed a food titration technique. The level of food motivation of adult animals malnourished during lactation and the first 4 weeks of postweaning was equated to their well-nourished controls. When motivation was equated, the animals learned a visual discrimination task, then a discrimination reversal. No difference was observed in either the rate-of-learning task or the asymptotic level of performance between the two groups. The animals were further tested on a response alternation task where differences in performance emerged, but there was no indication of any learning deficit.

8. Long-Term Effects of Early Malnutrition on Hebb-Williams Maze Learning

It may be argued that just as with concurrent malnutrition, these tests involved simple S-R learning, and this kind of learning may not be sensitive to the effects of malnutrition. Moreover, we have seen that Hebb-Williams performance was more sensitive to concurrent malnutrition than S-R learning. The results of studies of the long-term effects of early malnutrition on Hebb-Williams tests, however, are contradicting and confusing.

Adult rats which have been only gestationally malnourished score significantly poorer on the Hebb-Williams test than do well-nourished controls (Vore and Ottinger, 1970; Cowley and Griesel, 1966). On the other hand, others have found little or no effect of malnutrition during gestation alone on adult Hebb-

Williams tests (Ottinger and Tanabe, 1969; Smart *et al.*, 1973). When additional protein–calorie malnutrition is carried from gestation through lactation, poor adult performance is observed (Randt and Derby, 1973; Vore and Ottinger, 1970; Ottinger and Tanabe, 1969; Tanabe, 1972; Cowley and Griesel, 1966).

However, when malnutrition is induced only during the lactational period, a number of studies have found *superior* performance of rehabilitated adult rats in a Hebb-Williams maze. This result has been observed when malnutrition was precipitated by limiting the dam's food to 50% of control intake (Smart *et al.*, 1973), or by placing the pups with a nonlactating foster mother for half of each day (Slob *et al.*, 1973). Two studies have been reported with the opposite result; that is, lactationally malnourished rats performed more poorly in a Hebb-Williams maze (Vore and Ottinger, 1970; Ottinger and Tanabe, 1969). When malnutrition is continued from birth until 7 or 27 weeks of life, rehabilitated, poorer performance results (Zimmermann and Wells, 1971).

Baird *et al.* (1971, Exp. I) observed highly significant differences in rehabilitated rats that had suffered only postweaning malnutrition in one group of animals, yet no differences in another group. It is interesting to note that in the same study Baird *et al.* (1971) also found that a combination of gestational, lactational, and postweaning malnutrition had no effect on Hebb-Williams performance.

If the Hebb-Williams test measures some aspects of ability to learn or process information and is independent of factors which affect traditional measures of learning, it is quite difficult to understand how malnutrition introduced during gestation produces poorer performance while malnutrition during lactation can produce superior performance in the Hebb-Williams maze, and malnutrition carried throughout gestation, lactation, and postweaning has no effect. It is also difficult to understand why malnutrition induced during lactation was the only period when increased performance was observed, despite the fact that this is the period where the growth of the brain is most severely affected by malnutrition. It seems more likely that contrary to earlier beliefs, the Hebb-Williams maze may be sensitive to incentive, motivational, and emotional factors which may affect performance, as suggested by Stein (1972, 1974).

Thus, these spurious and inconsistent results of experiments on the long-term effects of early malnutrition on Hebb-Williams performance raise serious doubts as to whether any permanent disability in learning process exists. This is not to say that long-term effects of early malnutrition do not exist. It may only mean that traditional techniques of analyzing the learning processes may be insensitive. It is fair to conclude that little evidence exists that malnutrition induced at any stage of development detrimentally affects the ability of an animal to learn an association leading to the solution of a problem. Does this mean that malnutrition does not affect cognition? It may not. Malnutrition induced early in the life of an animal does produce consistent long-term effects on many types of behavior (e.g., emotional reactivity). It is through the study of the long-term effects of early malnutrition that another model of learning for human cognitive development has emerged.

9. Long-Term Effects of Early Malnutrition on Emotional Reactivity

One of the most consistently reported effects of early malnutrition in adult, nutritionally rehabilitated animals is an increase in "emotional reactivity," which is defined as an exaggeration of any behavioral response to a stressful stuation. Barnes *et al.* (1968) observed that adult rats malnourished early in life become very excited when feeding if access to food is limited to 1 hr/day. This restriction resulted in a significantly greater amount of food spillage in malnourished animals than in well-nourished controls. Moreover, previously malnourished animals display more sensitive motivated behavior for food in response to body weight loss than do controls (Levitsky, 1975; Smart *et al.*, 1973).

Differences in emotional reactivity of the early malnourished, yet rehabilitated rat have also been observed by several researchers in "open-field" behavior tests (see Levine—Review). An open field is a large observation chamber. Differences in locomotor activity in the open field has been observed in nutritionally rehabilitated rats which were malnourished during the combined pre- and postweaning period (Frankova and Barnes, 1968; Levitsky and Barnes, 1972; Smart, 1974) or during gestation and lactation (Simonson *et al.*, 1968). Similar findings have been observed in pigs which were either protein- or calorie-restricted from 3 to 11 weeks of age (Barnes *et al.*, 1976).

The behavioral differences observed in the open field usually fade fairly rapidly with repeated exposure. These differences will reappear if the animals are disturbed by a loud noise (Levitsky and Barnes, 1970) or an unfamiliar object (Zimmermann and Zimmermann, 1972), or if another rat is placed in the testing apparatus (Frankova, 1973). All animals are disturbed by these alterations, but it is the previously malnourished animals that display the greater disruption in behavior.

One of the most consistent findings in this area is the long-term effect of early malnutrition on reaction to an aversive stimulus. Increased reactivity to electric shock has been observed in the rat (Levitsky and Barnes, 1970; Sobotka *et al.*, 1974; Smart and Dobbing, 1972) mouse (Leathwood *et al.*, 1974), pig (Barnes *et al.*, 1976), and monkey (Wise and Zimmermann, 1973). More recently, disturbances in social behavior, particularly increased aggressive behavior, have been observed in the early malnourished rat (Frankova, 1973; Levitsky and Barnes, 1972; Whatson *et al.*, 1974; Randt *et al.*, 1975). Thus, early malnutrition does consistently result in an increase in emotional reactivity.

10. Malnutrition and Functional Isolation

Malnutrition is not the only variable which leads to long-term changes in emotionality in animals. Early environmental isolation causes an increase in the emotional reactivity of adult dogs (Melzack, 1954; Thompson *et al.*, 1956), cats (Riesen, 1961), rats (Ader and Plaut, 1968; Stern *et al.*, 1960), mice

(Weltman *et al.*, 1966), and monkeys (Mason *et al.*, 1968). To explain these effects, Melzack (1965) proposed the following model.

He suggested that the emotional reactivity of an organism to a stimulus was directly related to its familiarity. The more an organism "knows" about a particular stimulus or situation, the less reactive is its behavior in that situation. By isolating the animal early in life, the experimenter restricts the animal's access to all kinds of environmental information. When the animal is later tested, it responds to almost any situation as if it were new, and therefore reacts in an exaggerated emotional manner compared to the more experienced animal.

It is important to understand the kind of learning which is implicit in Melzack's theory. For Melzack (1965) and Hebb (1949), as well as human developmental theorists such as Piaget (1952), the young organism is continuously assimilating environmental information. This learning is internally motivated, and no overt behavioral response is required by an experimenter, parent, or teacher in order for the learning to occur. Psychologists generally refer to this kind of learning as intrinsically motivated learning.

In 1969 Levitsky and Barnes suggested the hypothesis that malnutrition may affect cognitive development not by altering the ability of the brain to make associations (i.e., the ability to learn) but by altering the early experience of the animal (i.e., what is learned). They suggested that malnutrition may "functionally isolate" the young animals from those stimuli in their environment, which may be important for normal behavioral adjustment of the adult. Support for this concept comes from experiments in which both nutrition and environment are covaried.

Levitsky and Barnes (1972) observed in the adult rat that the behavioral effects of early isolation were exacerbated by early malnutrition. Moreover, raising rats in an "enriched" environment almost totally ameliorated the behavioral effects of early malnutrition. Tanabe (1972) has demonstrated similar effects. Frankova (1973) has also shown that the long-term behavioral effects of early malnutrition can be dramatically reduced in rats by social stimulation during the period of malnutrition. Elias and Sammonds (1974) have shown the ameliorating effect of environmental enrichment on malnutrition-induced behavioral abnormalities in the monkey. These findings raise the possibility that malnutrition may produce a set of behaviors which are inconsistent with environmental learning, but that forcing in the environmental information with additional stimulation may eliminate the effect of malnutrition.

11. Mechnisms Through Which Malnutrition May Produce "Functional Isolation"

The functional isolation model was proposed as an alternative to the traditional learning model to explain how nutrition may affect cognitive development in man. Since the essence of the model rests upon the concept that poor nutrition removes the animal from important sources of environmental information, it is important to demonstrate how malnutrition functionally isolates the animal. Recent research has suggested at least four ways in which this is accomplished.

11.1. Slower Psychomotor Development

The effect of concurrent malnutrition in delaying psychomotor development is well documented both in animals (see above) and man. This decreased rate of psychomotor development impedes the independent exploration of the young mammal's environment, thus depriving it of sources of environmental information. It is possible, however, that the malnourished animal is conserving energy and thus optimizing its limited nutrient intake for growth and maintenance.

11.2. Increased Maternal Care

The decreased size and depressed rate of psychomotor development seems to stimulate maternal behavior. Both in the well-nourished adult nursing gestationally malnourished rats and the lactationally malnourished adult, the nursing dam spends considerably more of her time with the malnourished pups than do dams nursing well-fed pups. This may also be an important strategy for survival. The dams may be attempting to supply both nutrients and thermal comfort to the undernourished pups. The consequence, however, is a decrease in the exploration of the environment by the young pup.

It is interesting to note that Chavez *et al.* (1975), in working with humans under marginal nutritional conditions, finds the same kinds of effects as observed in the rat. The mother of a malnourished child spends more time holding and carrying the child than does the mother of a well-nourished child. The malnourished child spends more time sleeping and staying inside the home, whereas the well-nourished child spends more time being active and playing outside the home. The net effect may be greater exposure of the well-nourished child to environmental information.

11.3. Exploratory Behavior

In a study of learning in protein malnourished monkeys, Zimmermann and Strobel (1969) noticed that whenever an unfamiliar stimulus was presented, the malnourished monkeys tended to avoid it. In a later study, Zimmermann *et al.* (1970) further documented this "neophobic" response in concurrently malnourished monkeys.

Atinmo *et al.* (1978) showed that both protein- and calorie-restricted pigs were not at all attracted to a novel object introduced into their testing chamber, whereas the novel object dominated the attention of the well-nourished controls.

It is clear from these studies that the approach behavior of a young mammal to novel objects in its environment, a behavior so characteristic of the young mammal, is inhibited by malnutrition. Indeed, the apathy and lack of play behavior is well recognized as a major sign of kwashiorkor in humans (Latham, 1974). This approach behavior of the young organisms to novel stimuli, which is particularly sensitive to nutritional status, may be a very basic behavior mechanism involved in acquiring, processing, and storing information of the environment. The biochemical mechanism involved in this process needs further exploration.

11.4. Nondirected Learning

The fourth and most subtle process through which malnutrition may functionally isolate the animal from its environment is by limiting the amount of "nondirected" learning that occurs. Nondirected learning or incidental learning is learning which occurs in a learning situation but is not required to solve a particular problem for reinforcement. Several experimental paradigms have been developed which can demonstrate nondirected learning: these include latent learning, redundant learning, incidental learning, and latent inhibition. In all cases, the animal is merely exposed to a set of stimuli and then tested to determine what the animal learned during the exposure. Concurrent protein and calorie malnutrition depresses latent learning (Levitsky, 1975) and redundant learning (Levitsky, 1975), although as mentioned above, malnutrition does not affect directed learning. It appears that even under severe malnutrition, the animal can learn just as rapidly as the well-nourished animal. However, malnourished animals seem to learn the minimum amount of information needed to solve the problem, whereas the well-nourished animals seem to learn and process much more about a situation than only that information necessary for the solution of the problem. This kind of endogenously motivated environmental learning may be of considerably more importance to the accumulation of environmental information than is learning motivated only by reinforcement.

12. Implications of an Animal Model for Human Cognitive Development

It is fairly clear from this review that the traditional directed learning model is not sensitive to the effects of either concurrent or previous malnutrition. A more appropriate model may be based on a nondirected or intrinsically motivated learning paradigm. For this model the young organism may be viewed as continuously exploring and processing information: in essence, programming the knowledge of the world which it may need as an adult to survive and raise its offspring. This kind of learning is endogenously motivated. It may express itself in behavioral responses, such as exploratory behavior, or may be assessed by more sophisticated tests of learning, such as tests of latent learning, redundant learning, or latent inhibition, the behaviors which seem to be most sensitive to malnutrition.

We must not think of malnutrition as incapacitating the brain in such a way as to make it unable to learn or make association. Rather, malnutrition may be thought of as producing another behavioral strategy for survival, one in which the organism may respond only to those stimuli which are biologically meaningful for its survival. The consequence, however, is that organisms may process less of the other kinds of environmental information which may be important for development of maximum cognitive potential.

13. References

Ader, R., and Plaut, S. M., 1968, Effects of prenatal maternal handling and differential housing on offspring emotionality, plasma corticosteroid levels, and susceptibility to gastric erosions, *Psychosom. Med.* **30**:277–286.

Altman, J., Das, G. D., Sudarshan, K., and Anderson, J. B., 1971a, The influence of nutrition on neural and behavioral development. II. Growth of body and brain in infant rats using different techniques of undernutrition, *Dev. Psychobiol.* **4**(1):55–70.

Altman, J. S., Sudarshan, K., Das, G. D., McCormick, N., and Barnes, D., 1971b, The influence of nutrition on neural and behavioral development. III. Development of some motor, particularly locomotor patterns during infancy, *Dev. Psychobiol.* **4**(2):97–113.

Anderson, J. E., and Smith, A. H., 1926, The effect of quantitative stunting upon maze learning in the white rat, *J. Comp. Psychol.* **6**337–359.

Ardrey, R. 1966, *Territorial Imperative*, Atheneum Publishers, New York.

Atinmo, T., Levitsky, D. A., Pond, W. G., and Barnes, R. H., 1978, Behavioral abnormalities in offsrping caused by prenatal malnutrition in swine, in preparation.

Baird, A., Widdowson, E. M., and Cowley, J. J., 1971, Effects of calorie and protein deficiencies early in life on subsequent learning abilities of rats, *Br. J. Nutr.* **25**:391–403.

Barnes, R. H., Cunnold, S. R., Zimmermann, R. R., Simons, H., McLeod, R. B., and Krook, L., 1966, Influence of nutritional deprivations in early life on learning behavior of rats as measured by performance in a water maze, *J. Nutr.* **89**:399–410.

Barnes, R. H., Neely, C. S., Kwong, E., Labadan, B. A., and Frankova S., 1968, Postnatal nutritional deprivations as determinants of adult rat behavior toward food, its consumption, and utilization, *J. Nutr.* **96**:467–476.

Barnes, R. H., Moore, A. U., and Pond, W. G., 1970, Behavioral abnormalities in young adult pigs caused by malnutrition in early life, *J. Nutr.* **100**:149–155.

Barnes, R. H., Levitsky, D. A., Pond, W. G., and Moore, A. U., 1976, Effect of dietary protein and energy restriction on exploratory behavior in young pigs, *Dev. Psychobiol.*, **9** (5):425–435.

Bernhardt, K. S., 1936, Protein deficiency and learning in rats, *J. Comp. Psychol.* **22**:269-272.

Bevan, W., and Freeman, O. I., 1952, Some effects of an amino acid deficiency upon the performance of albino rats in a simple maze, *J. Genet. Psychol.* **80**:75-82.

Biel, W. C., 1938, The effect of early inanition upon maze learning in the albino rat, *Comp. Psychol. Monogr.* **15**:1-33.

Blackwell, B., Blackwell, R. Q., Yu, T. T. S., Weng, Y., and Chow, B. F., 1969, Further studies on growth and food utilization in progeny of underfed mother rats, *J. Nutr.* **97**:79-84.

Bush, M., and Leathwood, P. D., 1975, Effect of different regimes of early malnutrition on behavioral development and adult avoidance learning in Swiss white mice, *Br. J. Nutr.* **33**:373-385.

Caldwell, D. F., and Churchill, J. A., 1967, Learning ability in the progeny of rats administered a protein deficient diet during the second half of gestation, *Neurology* **17**:95-99.

Chavez, A., Martinez, C., and Yaschine, T., 1975, Nutrition, behavioral development and mother-child interaction in young rural children, *Fed. Proc.* **34**(7):1574-1582.

Chow, B. F., and Lee, C., 1964, Effect of dietary restriction of pregnant rats on body weight gain of the offsrping, *J. Nutr.* **82**:10-18.

Collier, G., and Levitsky, D. A., 1967, Defense of water balance in rats: Behavioral and physiological response to depletion, *J. Comp. Physiol. Psychol.* **64:**59–67.

Cowley, J. J., and Griesel, R. D., 1959, Some effects of a low protein diet on first filial generation of white rats. *J. Genet. Psychol.* **95:**187–201.

Cowley, J. J., and Griesel, R. D., 1962, Pre- and postnatal effects of a low protein diet on the behavior of the white rat, *Psychol. Africana* **9:**216–225.

Cowley, J. J., and Griesel, R. D., 1963, The development of second generation low protein rats, *J. Genet, Psychol.* **103:**223–242.

Cowley, J. J., and Griesel, R. D., 1966, The effect of growth and behavior on rehabilitating first and second generation low protein rats, *Anim. Behav.* **14:**506–517.

Das, G., and Broadhurst, P. L., 1959, The effect of inherited differences in emotional reactivity on a measure of intelligence in the rat. *J. Comp. Physiol. Psychol.* **52:**300–303.

Denenberg, V. H., Woodcock, J. M., and Rosenbergh, K. M., 1968, Long term effects of preweaning and postweaning free environment experience on rats problem solving behavior, *J. Comp. Physiol. Psychol.* **66:**533–535.

Elias, M. G., and Sammonds, K. W., 1974, Exploratory behavior and activity of infant monkeys during nutritional and rearing restriction, *Am. J. Clin. Nutr.* **27:**458–463.

Forgays, D. G., and Reid, J. M., 1962, Crucial periods for free environment experience in the rat, *J. Comp. Physiol. Psychol.* **55:**816–818.

Frankova, S., 1971, Relationship between nutrition during lactation and natural behavior in rat, *Dev. Psychobiol.* **6:**33–43.

Frankova, S., 1973, Effect of protein-calorie malnutrition on the development of social behavior in rats, *Dev. Psychobiol.* **6:**33–43.

Frankova, S., and Barnes, R. H., 1968, Influence of malnutrition in early life on exploratory behavior in rats, *J. Nutr.* **96:**477–484.

Garcia, J., Ervin, F. R., and Koelling, R. A., 1966, Learning with prolonged delay of reinforcement, *Psychon. Sci.* **5:**121–122.

Griffiths, W. J., and Senter, R. S., 1954, The effect of protein deficiency on maze performance of domestic Norway rats, *J. Comp. Physiol. Psychol.* **47:**41–43.

Guthrie, H. A., 1968, Severe undernutrition in early infancy and behavior in rehabilitated albino rats, *Physiol Behav.* **3:**619–623.

Hanson, H. M., and Simonson, M., 1971, Effects of fetal undernourishment on experimental anxiety, *Nutr. Rep. Int.* **4**(5):307.

Hebb, D. O., 1949, *The Organization of Behavior: A Neuropsychological Theory,* John Wiley & Sons, Inc., New York.

Hebb, D. O., and Williams, L., 1946, A method of rating animal intelligence, *J. Genet. Psychol.* **34:**59–65.

Hsueh, A. M., Simonson, M., Chow, B. F., and Hanson, H. M., 1974, The importance of the period of dietary restriction of the dam on behavior and growth in the rat, *J. Nutr.* **104:**37–46.

Hunt, J. McV., 1941, The effects of infant feeding frustration upon hoarding in the albino rat, *J. Abnorm. Soc. Psychol.* **36:**338–360.

Hunt, J. McV., Scholsberg, H., Solomon, R. L., and Stellar, E., 1947, Studies of the effects of infantile experience on adult behavior in rats, *J. Comp. Physiol. Psychol.* **40:**291–304.

Hymovitch, B., 1952, The effects of experimental variations on problem solving in the rat, *J. Comp. Physiol. Psychol.* **45:**313–321.

Kimble, G. A., 1961, *Conditioning and Learning* (E. R. Hilgard, D. G. Marquis, and R. M. Elliott, eds.), Appleton-Century-Crofts, New York.

Koch, A. M., and Warden, C. J., 1936. The influence of quantitative stunting on learning ability in mice, *J. Genet. Psychol.* **48:**215–217.

Larsson, K., Carlsson, S. G., Sourander, P., Fornstrom, B., Hansen, S., Henriksson, B., and Lindquist, A., 1974, Delayed onset of sexual activity of male rats subjected to pre- and postnatal undernutrition, *Physiol Behav.* **13:**307–311.

Latham, M. C., 1974, Protein-calorie malnutrition in children and its relation to psychological development and behavior, *Physiol Rev.* **54:**541–565.

Leathwood, P. D., 1973, Early undernutrition and behavior, in *Nestle Research News*, Nestle Products Technical Assistance Co. Ltd., Lausanne, Switzerland.
Leathwood, P., Bush, M., Berent, C., and Mauron, J., 1974, Effects of early malnutrition on Swiss white mice: avoidance learning after rearing in large litters, *Life Sci.* **14:**157-162.
Levine, S., and Wiener, S., 1976, A critical analysis of data on malnutrition and behavioral deficits, *Adv. Pediatr.*, **22:**113-136.
Levitsky, D. A., 1975, *Malnutrition and Animal Models of Cognitive Development, Proceedings* Kittay Scientific Foundation (G. Serban, ed.), Plenum Press, New York.
Levitsky, D. A., and Barnes, R. H., 1969, Effects of early protein calorie malnutrition on animal behavior, Paper read at Meeting of American Association for Advancement of Science, December.
Levitsky, D. A., and Barnes, R. H., 1970, Effects of early malnutrition on the reaction of adult rats to aversive stimuli, *Nature* **225:**468–469.
Levitsky, D. A., and Barnes, R. H., 1972, Nutritional and environmental interactions in the behavioral development of the rat. Long-term effects, *Science* **176:**68–71.
Levitsky, D. A., and Barnes, R. H., 1973, Malnutrition and animal behavior, in: *Nutrition, Development and Social Behavior* (D. J. Kallen, ed.), DHEW Publ. (NIH) 73–242.
Lorenz, Konrad, 1966, *On Agression,* Harcourt Brace and World, New York.
Lynch, A., 1976a, Passive avoidance behavior and response thresholds in adult male rats after early postnatal undernutrition, *Physiol. Behav.* **16:**27–32.
Lynch, A., 1976b, Postnatal undernutrition: An alternative method, *Dev. Psychobiol.* **9**(1):39–48.
McKelvey, R. K., and Marx, M. H., 1951, Effects of infantile food and water deprivation on adult hoarding in the rat, *J. Comp. Physiol. Psychol.* **44:**423–430.
Mandler, J. M., 1958, Effects of early food deprivation on adult behavior in the rat, *J. Comp. Physiol. Psychol.* **51:**513–517.
Marx, M. H., 1952, Infantile deprivation and adult behavior in the rat. Retention of increased rate of eating, *J. Comp. Physiol. Psychol.* **45:**43–49.
Mason, W. A., Davenport, R. K., and Menzel, E. M., 1968, Early experience and the social development of rhesus monkeys and chimpanzees, in: *Early Experience and Behavior* (G. Newton and S. Levine, eds.), Charles C Thomas, Publisher, Springfield, Ill.
Massaro, T. F., Levitsky, D. A., and Barnes, R. H ., 1974, Protein malnutrition in the rat. Its effects on maternal behavior and pup development, *Dev. Psychobiol.* 7(6):551–561.
Massaro, T. F., Levitsky, D. A., and Barnes, R. H., 1977, Protein malnutrition induced during gestation: Its effect on pup development and maternal behavior, *Dev. Psychobiol.* **10** (4):339–345.
Melzack, R., 1954, The genesis of emotional behavior: an experimental study of the dog, *J. Comp. Psychol.* **47:**166–168.
Melzack, R., 1965, Effects of early experience on behavior: experimental and conceptual considerations, in: *Disorders of Perception* (P. H. Hock and J. Zubin, eds.), Grune & Stratton, Inc. New York.
Montagu, A., 1976, *The Nature of Human Aggression,* Oxford University Press, Inc., New York.
Morris, C. J., 1974, The effects of early malnutrition on one-way and two-way avoidance behavior, *Physiol Psychol.* **2:**148-150.
Ottinger, D. R., and Tanabe, G., 1969, Maternal food restriction: effects on offspring behavior and development, *Dev. Psychobiol.* **2:**7-9.
Piaget, J., 1952, The origins of intelligence in children (M. Cook, trans.), International Universities Press, Inc. New York.
Pilgrim, F. J. Zabarenko, L. M., and Patten, P. A., 1951, The role of amino acid supplementation and dietary protein level in serial learning performance of rats, *J. Comp. Physiol. Psychol.* **44:**26-36.
Plaut, S. M., 1970, Studies of undernutrition in the young rat: methodological considerations, *Develop. Psychobiol.* **3:**157-167.
Rabinovitch, M. S., and Rosvald, H. E., 1951, A closed field intelligence test for rats, *Can. J. Psychol.* **5:**122-128.
Rajalakshmi, R., Govindarajan, K. R., and Ramakrishnan, C. U., 1965, Effect of dietary protein

content on visual discrimination learning and brain biochemistry in the albino rat, *J. Neurochem.* **12:**261–271.

Randt, C. T., and Derby, B. M., 1973, Behavioral and brain correlates in early life nutritional deprivation, *Arch. Neurol.* **28:**167–172.

Randt, C. T., Blizzard, D. A., and Friedman, E., 1975, Early life undernutrition and aggression in two mouse strains, *Dev. Psychobiol.* **8**(3):275–279.

Rider, A. A., and Simonson, M., 1973, Effect on rat offspring of maternal diet deficient in calories but not in protein, *Nutr. Rep. Int.* **7**(5):36–37.

Rider, A. A., and Simonson, M., 1974, The relationship between maternal diet, birth weight and behavior of the offspring in the rat, *Nutr. Rept. Int.* **10**(1):19–24.

Riesen, A. H., 1961, Excessive arousal effects of stimulation after early sensory deprivation, in: *Sensory Deprivation* (P. Solomon, P. E. Kubzansky, P. H. Liederman, J. H. Mendelson R. T. Trumbull, and D. Wexler, eds.), Harvard University Press, Cambridge, Mass.

Riess, B. F., and Block, R. J., 1942, The effect of amino acid deficiency on the behavior of the white rat: I. Lysine and cystine deficiency, *J. Psychol.* **14:**101–113.

Ruch, F. L., 1932, The effect of inanition upon maze learning in the white rat, *J. Comp. Psychol.* **14:**321–329.

Salas, M., 1974, Effects of early malnutrition on the development of swimming ability in the rat, *Physiol. Behav.* **8:**119–122.

Seitz, P. F. D., 1954, The effects of infantile experiences upon adult behavior in animal subjects. I. Effects of litter size during infancy upon adult behavior in the rat, *Am. J. Psychiat.* **110:**916–927.

Simonson, M., and Chow, B. F., 1969. Maze studies on progeny of underfed mother rats. *J. Nutr.* **100:**685–690.

Simonson, M., Sherwin, R. W., Anilane, J. K., Yu, W. Y., and Chow, B. F., 1968, Neuromotor development in progeny of underfed mother rats, *J. Nutr.* **98:**18–24.

Slob, A. K., Snow, C. E., and Mathot, E., 1973, Absence of behavioral deficits following neonatal undernutrition in the rat, *Dev. Psychobiol.* **6**(6):177–186.

Smart, J. L., 1974, Activity and exploratory behavior of adult offspring of undernourished mother rats, *Dev. Psychobiol.* **7:**315–321.

Smart, J. L., 1976, Maternal behavior of undernourished mother rats towards well fed and underfed young, *Physiol. Behav.* 47–49.

Smart, J. L., and Dobbing, J., 1971, Vulnerability of developing brain. VI. Relative effects of foetal and early postnatal undernutrition on reflex ontogeny and development of behavior in the rat, *Brain Res.* **33:**303–314.

Smart, J. L., and Dobbing, J., 1972, Vulnerability of developing brain. IV. Passive avoidance behavior in young rats following maternal undernutrition, *Dev. Psychobiol.* **5:**129–136.

Smart, J. L., and Preece, J., 1973, Maternal behavior of undernourished mother rats, *Anim. Behav.* **21:**613–619.

Smart, J. L., Dobbing, J., Adlard, B. P. F., Lynch, A., and Sands, J., 1973, Vulnerability of developing brain. Relative effects of growth restriction during the fetal and suckling period on behavior and brain composition of adult rats, *J. Nutr.* **103:**1327–1338.

Smart, J. L., Whatson, T. S., and Dobbing, J., 1975, Thresholds of response to electric shock in previously undernourished rats, *Br. J. Nutr.* **34:**511–516.

Sobotka, T. J., Cook, M. P., and Brodie, R. E., 1974, Neonatal malnutrition: Neurochemical, hormonal, and behavioral manifestations, *Brain Res.* **65:**443–457.

Stein, D. G., 1972, The effects of saline or blank injections during development on maze learning at maturity, *Dev. Psychobiol.* **5:**319–322.

Stein, D. G., 1974, The effects of early saline injections and pentylenetetrazol on Hebb-Williams maze performance in the adult rat, *Behav. Biol.* **11:**415–422.

Stern, J. A., Winokur, G., Eisenstein, A., Taylor, R., and Sly, M., 1960, The effect of group vs. individual housing on behavior and physiological responses to stress in the albino rat, *J. Psychosom. Res.* **4:**185–190.

Tanabe, G., 1972, Remediating maze deficiencies by the use of environmental enrichment, *Dev. Psychobiol.* **7**(2):224.

Thompson, W. R., Melzack, R., and Scott, T. H., 1956, Whirling behavior in dogs as related to early experience, *Science* **123**:939.

Vore, D. A., and Ottinger, D. R., 1970, Maternal food restriction: Effect on offspring development, learning, and a program of therapy, *Dev. Psychol.* **3**(3):337–342.

Weltman, A. S., Sackler, A. M., and Sparber, S. B., 1966, Endocrine metabolic and behavioral aspects of isolation stress on female albino mice, *Aerosp. Med.* **37**:804–810.

Whatson, T. S., Smart, J. L., and Dobbing, J., 1974, Social interactions among adult male rats after early undernutrition, *Br. J. Nutr.* **32**:413–419.

Wise, L. A., and Zimmermann, R. R., 1973, Shock thresholds and high protein reared rhesus monkeys, *Percept. Mot. Skills* **36**:674.

Wolfe, J. B., 1939, An exploratory study of food storing in rats, *J. Comp. Psychol.* **28**:97–108.

Zimmermann, R. R., and Strobel, D. A., 1969, Effect of protein malnutrition on visual curiosity, manipulation, and social behavior in the infant rhesus monkey, *Proc. 77th Ann. Conv. APA* **5**:241–242.

Zimmerman, R. R., and Wells, A. M., 1971, Performance of malnourished rats on the Hebb-Williams closed field maze learning task, *Percept. Mot. Skills* **33**:1043–1050.

Zimmermann, R. R., and Zimmermann, S. J., 1972, Responses of protein malnourished rats to novel objects, *Percept. Mot. Skills* **35**:319–321.

Zimmermann, R. R., Strobel, D. A., and Maguire, D., 1970, Neophobic reactions in protein malnourished infant monkeys, *Proc. 78th Ann. Conv. APA* **6**:197–198.

Zimmermann, R. R., Geist, C. R., and Wise, L. A., 1974, Behavioral development, environmental deprivations, and malnutrition, in: *Advances in Psychobiology*, Vol. 2 (C. Newton and A. H. Riesen, eds.), John Wiley & Sons, Inc., New York.

9

Nutrition and Mental Development in Children

Patricia L. Engle, Marc Irwin, Robert E. Klein, Charles Yarbrough, and John W. Townsend

1. Introduction

Over the past two decades a number of investigators have addressed the question of whether childhood malnutrition causes retardation in mental development. To date, an unequivocal answer has not been forthcoming. This is understandable because of the many problems involved in designing an appropriate study to demonstrate causality.

Among these problems is the fact that the definition and measurement of nutritional status are still debated. Furthermore, the usual research strategies are difficult to apply to the study of the effects of malnutrition, since one can neither cause malnutrition in order to observe its effects, nor erase the poverty, illness, and social deprivation which coexist with malnutrition and constitute plausible alternative explanations of deficient mental development.

Malnutrition is not an easy concept to operationalize because, while it refers in part to input, or the food a person eats, the only measures usually available are output measures (i.e., growth or health status). Thus, the criteria for classification (summarized in Jelliffe, 1966) are all inferential, with past malnutrition inferred from present physical state. In particular, protein–calorie malnutrition (PCM) has been defined using indicators of physical growth, various clinical signs, and biochemical indicators.

The primary indicator of malnutrition is low weight for age, and the severely malnourished child has less than 70% of the expected weight for age (below the 30th percentile). The moderately malnourished child is usually

Patricia L. Engle, Marc Irwin, Robert E. Klein, and John W. Townsend • Division of Human Development, Institute of Nutrition of Central America and Panama (INCAP), Guatemala, Central America. *Charles Yarbrough* • Computers for Marketing Corporation, Kenwood, California.

identified by a growth retardation of less than 90% of the normal weight for age. The low weight for age of the moderately malnourished child may be accompanied by biochemical indications and/or evidence of reduced food intake, but almost never by clinical signs.

The majority of studies on human malnutrition and mental development have dealt exclusively with the effects of severe malnutrition (Birch *et al.*, 1971; Botha-Antoun *et al.*, 1968; Brockman and Ricciuti, 1971; Cabak and Najdanvic, 1965; Champankan *et al.*, 1968; Chase and Martin, 1970; Cravioto and DeLicardie, 1968; Cravioto and Robles, 1965; Cravioto *et al.*, 1966; DeLicardie and Cravioto, 1974; Edwards and Craddock, 1973; Evans *et al.*, 1971; Hertzig *et al.*, 1972; Liang *et al.*, 1967; Mönckeberg, 1968; Montelli *et al.*, 1974; Pollitt and Granoff, 1967; Stein *et al.*, 1972a). Typically, subjects have been children hospitalized for one of two forms of acute malnutrition, marasmus or kwashiorkor (Birch *et al.*, 1971; Brockman and Ricciuti, 1971; Cabak and Najdanvic, 1965; Champankan *et al.*, 1968; Chase and Martin, 1970; Cravioto and Robles, 1965; DeLicardie and Cravioto, 1974; Evans *et al.*, 1971; Hertzig *et al.*, 1972; Mönckeberg, 1968; Montelli *et al.*, 1974; Pollitt and Granoff, 1967). In these studies of severe malnutrition an association between early PCM and later poor cognitive performance has usually been found. However it is likely that the trauma associated with having a serious illness, being hospitalized, and being separated from the home environment in itself has negative effects on mental development.

Fewer investigations have focused on the consequences of mild-to-moderate malnutrition, although the number of children affected is far greater. In Guatemala, for instance, a recent report estimated that 80% of the country's children were undernourished, compared to a prevalence of severe malnutrition of 3% (INCAP/HEW, 1972). The effects of mild-to-moderate PCM have been studied by Boutourline *et al.* (1973), Chávez *et al.* (1974), McKay *et al.* (1969), Montelli *et al.* (1974), Mora *et al.* (1974), and Patel (1974). Like the research work on severe PCM cited, these studies have typically reported an association between malnutrition and poor cognitive performance, although causality has not been clearly demonstrated.

Up to the present, it has not been possible to establish a causal link between malnutrition and deficient mental development, largely because most studies have employed observational designs in which naturally occurring malnutrition has been identified and correlated mental development has been measured rather than experimental designs. Although most of these studies have found associations between early malnutrition and deficient mental development, an alternative causal explanation for these findings has always existed. Poverty and accompanying social deprivation almost invariably coexist with malnutrition (e.g., Chase and Martin, 1970; Chávez *et al.*, 1974; Cravioto and Robles, 1965; Evans *et al.*, 1971; Mönckeberg *et al.*, 1972; Mora *et al.*, 1974; Patel, 1974; Schlenker *et al.*, 1968; Stoch and Smythe, 1963, 1967). Since social deprivation may well play a causal role in deficient mental development (e.g., Hess *et al.*, 1968; Whiteman and Deutsch, 1968), the effects of nutritional and social variables on mental development are confounded.

One useful strategy employed in malnutrition/mental development research has been to identify situations in which the usual association between malnutrition and poor social and economic background does not hold. Winick *et al.* (1975) compared Korean infants with varying histories of malnutrition adopted by U.S. middle-class families and found that a good environment could substantially increase a previously malnourished child's chances for adequate development. Lloyd-Still *et al.* (1974) identified 41 middle-class children who had suffered severe malnutrition in the first 6 months of life as a result of cystic fibrosis or various congenital defects. These children showed mental deficits compared to siblings only at age 5. For older children differences were not present. Richardson (1976) studied 71 Jamaican schoolboys with a history of early severe malnutrition. Only the children who had inadequate growth and a poor social background after their episode of malnutrition continued to be retarded in comparison with siblings, whereas those with either adequate growth or a good home environment did not continue to show retarded development. These three studies suggest an independent causal role of malnutrition, and underline the important role of social environment for mental development, particularly in later childhood.

A second design solution to the problem of establishing causality consists of introducing experimental manipulation in the form of improvement of baseline nutritional status of a study population, along with prospective measurement of both nutritional status and social deprivation data. Because it involves prospective measurement of nutritional status, and because of the greater precision of measurement made possible by experimental manipulation, such a design is probably optimal. This design has been employed in the INCAP longitudinal study described in the present chapter.

2. *The INCAP Study*

Since 1969 INCAP has been collecting data on pregnant and lactating mothers and children from birth to 7 years of age in a longitudinal quasi-experimental intervention study. Four villages in the eastern, Spanish-speaking section of Guatemala, where malnutrition is endemic, were matched according to a number of demographic, social, and economic characteristics. The experimental intervention was differential supplemental feeding of two matched groups; two "experimental" villages were selected at random, where a high-protein/calorie drink similar to a popular corn-based gruel (atole) was made available twice daily for all residents at a central dispensary; and two were selected as "control" villages, where a drink was made available to all who wished to partake; this drink (fresco), similar to Kool-Aid, contains about one-third of the calories of the atole. Both drinks provide enough iron, vitamins, and minerals to ensure that none of these substances should be limiting in the recipients' diets. In both experimental and control villages, free outpatient medical care has also been provided since the inception of the study.

The study design is longitudinal, and as of 1969 all children in the villages from birth through 7 years of age have been measured in terms of the independent and dependent variables. Our primary concern in this chapter is the preschool children for whom we have information on supplemental food ingested by mother or child since the child's conception. The design is prospective in that data about a child's health and feeding are collected along with mental testing.

This design permits the examination of change over time from one testing to another, and thus makes possible investigation of whether supplemental feeding during a certain period is related to increases in test performance during the same period, or at what stage of mental development the largest effects of the supplemental feeding are likely to appear. It also permits determination of whether there are periods in development during which nutritional supplementation most affects subsequent mental development. Finally, the longitudinal design allows one to ask whether effects seen at one age persist to later ages or are erased by intervening circumstances.

2.1. *Definition of the Independent Variable*

The optimal measure of nutritional status is total nutrient ingestion by the mother during pregnancy and lactation and by the child up to the point of his mental testing. This measure would be the sum of supplemental feeding and level of home diet of the mother and child during these periods. Individually reliable home diet information is extremely difficult to obtain because of vagaries and biases of individual reporting, variability in what a child eats from day to day, and changes in the biochemical and nutrient composition of various foods from year to year. In fact, in the INCAP study, estimates of the total home diet based on dietary surveys are less precise than supplemental food ingested, which can be measured very accurately. Supplement ingestion has consistently been related to growth in the study population, and does not significantly replace home dietary consumption (Martorell *et al.*, 1978). We have, however, continued to measure home diet, employing it as a family background variable that assures us that high- and low-supplement ingestion groups do not differ in their family home diet.

2.2. *Measurement of Mental Development*

Measurement of preschool mental development in the longitudinal study has been purposely broad and eclectic. At 3, 4, 5, 6, and 7 years of age, the INCAP Preschool Battery (DDH/INCAP, 1975) is administered. It consists of 10 tests at ages 3 and 4, and 22 tests from age 5 on. The tests in the battery were selected to represent diverse theoretical orientations (e.g., learning, psychometric, and Piagetian), and to broadly sample attentional, perceptual analytic, learning, memory, and reasoning processes.

Reliabilities of Preschool Battery tests have been assessed, and are generally high. Interobserver reliability is at least 0.99 for each test. Test–retest reliability (1-week intervals) coefficients are also acceptable high (DDH Progress Report, 1975).

A major concern of the longitudinal study has been to assure that mental tests employed possessed emic validity (Berry, 1969; Irwin *et al.*, 1977), or were really measuring significant aspects of intellectual competence in the rural Guatemalan context. From the inception of the study, considerable effort has gone into achieving this objective. The tests in the Preschool Battery were adapted to the research setting by a team consisting of American and Guatemalan psychologists, a Guatemalan cultural anthropologist, and Guatemalan testers and cultural informants. Two years of pretesting, during which some tests went through as many as 10 revisions, were devoted to developing test materials and instructions which both the intuitions of the testers and the performances of local (pilot sample) children of various ages suggested were appropriate and meaningful.

Attempts to establish the emic validity of these measures have been made with subsamples of the longitudinal study population. In one such study, adults' ratings of smartness (Klein *et al.*, 1976; Nerlove *et al.*, 1974), which is translated as "listura" in Spanish and is associated with the concept of alertness, verbal facility, good memory, and a high level of physical activity, were correlated with a representative set of Preschool Battery tests (Vocabulary Recognition, Verbal Inferences, Discrimination Learning, Memory for Digits, and Embedded Figures) and found to be strongly related (r's up to 0.75) to these Preschool Battery test scores.

A second study has compared the school attendance and performance of children who received high and low scores on our Preschool Battery, and found consistent relationships between Preschool Battery scores and the age at which a child was first sent to school (higher-scoring children tending to be sent at younger ages), how long they remained (higher-scoring children tending to attend longer), number of years of school passed, and scores on nationally standardized achievement tests (Irwin *et al.*, 1978).

3. Results

3.1. Nutritional Status and Mental Development

The central question in the longitudinal study is whether mild-to-moderate malnutrition, a condition affecting the majority of children growing up today in developing countries, affects the mental development of these children. Our data argue strongly that the answer is yes.

Children in the sample were categorized as adequately, inadequately, and moderately supplemented from conception (DDH/INCAP, 1975). Mean psychological test scores at ages 3, 4, and 5 years for each supplementation category, and F values from a one-way analysis of variance by supplement category are presented in Tables I, II, and III. Since sex differences in test

Table I. Means, Pooled Standard Deviations, and Analysis-of-Variance F Values of Psychological Test Scores at 36 Months for Children Falling into Various Cumulative Supplementation Categories[a]

	1	2	3	S.D.	F^b
Composite Score	−0.41 (452)	0.60 (119)	0.78 (147)	3.64	8.09**
Embedded Figures	9.78 (440)	9.94 (119)	10.12 (144)	3.32	0.60
Digit Memory	10.19 (367)	12.42 (107)	11.96 (129)	8.23	4.28**
Sentence Memory	12.67 (385)	15.05 (108)	15.87 (134)	12.85	3.78*
Vocabulary Naming	6.66 (426)	8.28 (114)	8.04 (143)	4.18	10.52**
Vocabulary Recognition	19.90 (426)	20.98 (114)	20.66 (143)	5.72	2.11
Verbal Inferences	1.38 (191)	1.87 (53)	2.06 (50)	1.27	7.63**
Memory for Objects	1.94 (199)	1.95 (83)	2.19 (114)	1.39	1.32
Reversal Discrimination Learning	23.79 (393)	25.70 (115)	24.18 (137)	20.61	0.38
Knox Cubes	0.62 (88)	0.41 (46)	0.69 (62)	1.30	0.65
Draw-a-Line Slowly	10.22 (412)	9.57 (113)	8.90 (147)	4.44	5.09**
Persistence on an Impossible Puzzle	5.91 (364)	7.03 (116)	6.51 (148)	5.34	2.15

[a] Caloric supplementation per 3-month period: category 1 (≤ 5000 cal); category 2 (5000–9999 cal); category 3 (≥ 10,000 cal).
[b] Symbol: $*p < 0.05$; $**p < 0.01$.

performance were few, data for boys and girls have been combined. A composite score constructed by standardizing and then summing test scores, as well as several individual tests, show significant effects of nutritional status at each age. Tests which show the effects of food supplementation cover a wide

Table II. Means, Pooled Standard Deviations, and Analysis-of-Variance F Values of Psychological Test Scores at 48 Months for Children Falling into Various Cumulative Supplementation Categories[a]

	1	2	3	S.D.	F^b
Composite Score	−0.41 (473)	0.59 (132)	0.98 (120)	3.87	8.21**
Embedded Figures	2.71 (451)	2.74 (129)	3.24 (119)	1.80	4.31**
Digit Memory	21.56 (448)	22.18 (128)	21.65 (115)	11.90	0.14
Sentence Memory	31.94 (442)	33.91 (128)	37.33 (113)	20.04	3.37*
Vocabulary Naming	12.11 (461)	13.95 (130)	14.49 (118)	5.13	14.36**
Vocabulary Recognition	25.70 (461)	27.79 (130)	27.39 (118)	5.34	10.65**
Verbal Inferences	2.92 (364)	3.19 (110)	3.22 (91)	1.49	2.41
Memory for Objects	3.63 (221)	3.80 (91)	3.68 (105)	1.68	0.32
Reversal Discrimination Learning	35.52 (458)	39.00 (131)	40.80 (118)	19.62	4.24**
Knox Cubes	1.36 (163)	1.75 (83)	1.22 (82)	2.02	1.57
Draw-a-Line Slowly	6.88 (449)	6.36 (129)	4.69 (120)	4.07	14.18**
Persistence on an Impossible Puzzle	8.88 (127)	9.78 (127)	10.01 (120)	6.46	1.89

[a] Caloric supplementation per 3-month period: category 1 (≤ 5000 cal); category 2 (5000–9999 cal); category 3 (≥ 10,000 cal).
[b] Symbols: $*p < 0.05$; $**p < 0.01$.

Table III. Means, Pooled Standard Deviations, and Analysis-of-Variance F Values of Psychological Test Scores at 60 Months for Children Falling into Various Cumulative Supplementation Categories[a]

	1	2	3	S.D.	F^b
Composite Score	−0.70 (452)	0.87 (135)	1.90 (104)	7.64	6.05**
Embedded Figures	4.12 (450)	4.40 (135)	4.84 (103)	2.21	4.72**
Digit Memory	35.82 (441)	35.20 (135)	34.09 (103)	13.10	0.75
Sentence Memory	51.27 (442)	51.56 (134)	52.17 (104)	21.16	0.08
Vocabulary Naming	17.08 (449)	18.78 (135)	18.40 (104)	5.06	7.53**
Vocabulary Recognition	30.08 (449)	31.13 (135)	31.35 (104)	4.39	5.40**
Verbal Inferences	4.07 (422)	4.44 (128)	4.23 (99)	1.89	1.98
Memory for Objects	5.51 (197)	5.89 (88)	5.81 (88)	1.85	1.60
Reversal Discrimination Learning	23.73 (438)	25.56 (133)	29.68 (103)	21.20	3.37*
Knox Cubes	3.44 (202)	4.74 (88)	4.91 (89)	3.53	7.57**
Conservation of Material	0.29 (197)	0.34 (86)	0.26 (86)	0.60	0.40
Conservation of Area	0.35 (193)	0.40 (85)	0.38 (87)	0.67	0.16
Conservation of Continuous Quantity	0.20 (202)	0.21 (86)	0.17 (87)	0.43	0.17
Incomplete Figures	5.59 (203)	5.73 (86)	5.15 (84)	2.64	1.15
Elimination	5.10 (199)	5.24 (88)	5.33 (87)	2.16	0.40
Block Design	39.58 (202)	42.84 (88)	47.75 (88)	20.46	5.04**
Incidental Learning	1.34 (448)	1.77 (135)	1.79 (103)	1.27	9.36**
Intentional Learning	2.18 (448)	2.44 (135)	2.87 (103)	1.40	10.96**
Haptic-Visual	3.86 (351)	4.22 (126)	4.45 (103)	2.87	2.01
Face-Hands	2.44 (200)	2.37 (87)	2.42 (87)	1.39	0.09
Matching Familiar Figures	2.75 (443)	2.85 (134)	2.86 (103)	1.35	0.51
Memory for Designs	11.66 (444)	13.04 (131)	14.76 (102)	11.37	3.37*
Animal House	118.64 (196)	114.76 (88)	120.16 (87)	54.08	0.15
Draw-a-Line Slowly	4.42 (430)	3.39 (131)	3.18 (104)	3.13	10.27**
Persistence on an Impossible Puzzle	12.01 (365)	13.68 (130)	13.74 (102)	6.20	5.36**

[a] Caloric supplementation per 3-month period: category 1 (≤ 5000 cal); category 2 (5000–9999 cal); category 3 (≥ 10,000 cal).
[b] Symbols: *$p < 0.05$; **$p < 0.01$.

range of skills, including verbal reasoning, learning and memory, visual analysis, and motor control.

We have also attempted to determine if crucial periods exist in which nutritional status has a particularly strong impact upon subsequent mental test performance. To examine this question, we have employed regression analyses wherein supplementation during the distinct developmental epochs of pregnancy, 0 to 24 months, 24 to 36 months, and 36 to 48 months, was regressed on mental test performances at ages 3 and 4. These analyses could not be interpreted at age 5, since relatively few children tested as yet at age 5 have been well supplemented during pregnancy. However, at ages 3 and 4 these analyses suggest that supplementation during the periods of gestation and birth to 24 months is most important in determining mental development. Table IV presents simple slopes of supplementation on test performance during these

Table IV. *Regressions on Test Performance Using a Caloric Supplement Varying by Time Periods*

Test	*N*	Supplementation variable	Simple slope[a]	Partial slope[a,b]	S.D. of test score
		At 36 months			
Cognitive	718	*In utero*	0.218**	0.112	3.64
Composite		To child 0–24	0.126***	0.108***	
Vocabulary	683	*In utero*	0.212*	0.076	4.18
Naming		To child 0–24	0.150***	0;138***	
Verbal	294	*In utero*	0.223***	0.203***	1.26
Inferences		To child 0–24	0.051**	0.024	
		At 48 months			
Cognitive	725	*In utero*	0.281**	0.142***	3.87
Composite		To child 0–24	0.142***	0.122***	
Vocabulary	709	*In utero*	0.343**	0.093	5.12
Naming		To child 0–24	0.233***	0.220***	
Verbal	565	*In utero*	0.120**	0.096*	2.12
Inferences		To child 0–24	0.034*	0.020	

[a] Symbols: *$p < 0.05$; **$p < 0.01$; ***$p < 0.001$.
[b] Partial slope of the supplement period indicated controlling for the other period.

two periods, and partial slopes with each, after variance from the other period of supplementation was removed. These partial slopes indicate how much variance in test performance can be attributed to supplement during one period independent of the other. Data are presented for a reduced set of tests: the Composite Score, Vocabulary Naming, and Verbal Inferences. These last two tests were selected in addition to the Composite Score because data for Vocabulary typify effects seen at ages 3 and 4, while Verbal Inferences behaves uniquely, showing strong effects of early supplementation. All simple slopes for supplement both during pregnancy and from 0 to 24 months are significantly different from zero. In addition, the partial slopes of gestational supplementation controlling for supplement ingested from 0 to 24 months are significant for the Composite Score at 48 months and for Verbal Inferences at 36 and 48 months. In contrast, significant partial slopes for supplement ingestion from birth to 24 months controlling for gestational supplement are seen for the Composite Score and for Vocabulary Naming at 36 and 48 months. Although the partial effects vary somewhat by age and test, both periods of supplement appear to play an independent role in the mental development of these children at ages 3 and 4.

3.2. Alternative Explanations for Findings, and the Roles of Social, Economic, and Biomedical Factors in Mental Development

We have considered the possibility that nonnutritional variables may be confounded with and responsible for the apparent effects of supplementation on mental development observed in our analyses. Among the variables considered have been: effect of repeated testing, morbidity of *S*'s, parental cooperation with the project, village differences, and attendance to the supplementation centers. Testing effects have been examined, and they do not appear to constitute a vaiable alternative explanation for our findings among 3- to 5-year-olds. The same may be said for child morbidity history and for parental cooperation with the project: correlations between morbidity measures and psychology scores are miniscule. So are correlations between parental cooperation ratings (made by staff personnel) and psychology scores.

We have also examined the possibility that thc frcsco and atole villages differ in characteristics other than treatment that might be related to mental development. These differences would confound the results, since well-supplemented children are more common in the atole villages, while nonsupplemented children are more common in the fresco villages. To explore village differences, an analysis of variance of year of project by village type (atole or fresco) was run on all 36-month psychology variables using only nonsupplemented children from the fresco and atole villages. Since none of these children have received much supplementation, any differences by type of village (atole or fresco) between them would support the existence of artifactual village differences. However, no main effect of village type above the 0.05 level was found for any test. Thus, village differences do not appear to constitute an alternative explanation for the observed supplement effects.

It is also possible that the social stimulation inherent in attending the supplementation center per se may increase test scores. Days of attendance at the supplementation center from 36 months on does appear to affect test performance byond that age. However, the magnitude of this attendance effect is quite small. Regression analyses examining the effect of number of days of attendance to our supplementation centers on Preschool Battery performance reveal that for every 100 days of attendance, an increment of 0.007 standard deviation unit on the Composite Score is produced. Thus, the size of this effect renders it of little importance as an alternative explanation of our previously observed effects of nutritional supplementation on test performance.

Another possible source of confounding secular change, or change over historial time, has been investigated. To address the problem of secular change, mean composite scores of the least supplemented children, who would have been relatively unaffected by the treatment, were calculated by year of the program. Inspection of the means has demonstrated that there is no linear trend over the duration of the program.

Probably the best candidate for an alternative explanation of our finding of an association between food supplementation and mental development is family socioeconomic status (SES). There are two important reasons for examining economic and social variables in our study. First, it is generally known that family characteristics are frequently associated with mental test performance (e.g., Whiteman and Deutsch, 1968; Hess, 1970), as was found in these villages. Second, family SES level usually covaries with children's nutritional status. Thus, family SES must be carefully measured in tested populations in studies of malnutrition and mental development to control for possible confounding effects.

The second reason for investigating the role of economic and social factors is that mental development almost certainly has multiple causes, and to understand the effects of nutrition on mental development we must first investigate interactions of nutritional status with other possibly causal variables. Thus, by employing measures of family SES level and intellectual stimulation in the home, we may be able to identify the conditions under which nutrition has its largest effects.

In the present study, family SES level is unrelated to supplement ingestion. This is not surprising, because the supplement is provided to all, regardless of their SES level. Furthermore, we know that the associations we have found between food supplementation and mental test performance are not due merely to SES differences between good and poor test performers; regression analyses in which successive offspring of the same mother are compared have been performed. Because these analyses make within-family comparisons, all family-level variables, including SES, are held constant. They indicate that mental test scores are significantly related to differences in nutritional status between successive offspring of the same mother, and that this within-family association is as strong as the between-family association of nutritional status and mental development described earlier. Although it does not explain the association between nutritional status and mental development, family SES level does appear to interact with nutritional status in affecting mental development. Although effects of supplement ingestion exist within both the upper and lower halves of the family SES distribution, these effects are somewhat different for children of low and high SES families.

A child's relative risk of falling into the lowest pentile of the Composite Score distribution if he is both low SES and inadequately supplemented is shown in Table V. The table presents the number of 3-, 4- and 5-year-old

Table V. Numbers of Children in Lowest Supplementation Category[a] during the Period from Birth to 24 Months Falling into the Lowest Pentile of Composite Score Performance or Not, by SES Level

	Lowest pentile	Higher pentile	Relative risk[b]
3 years			
Low SES	61	166	1.41*
High SES	47	200	
4 years			
Low SES	82	169	2.10**
High SES	43	233	
5 years			
Low SES	69	183	1.68**
High SES	47	241	

[a] Caloric supplementation per 3-month period: category 1 (≤ 5000 cal); category 2 (5000–9999 cal); category 3 (≥ 10,000 cal).
[b] Symbols: $*p < 0.05$; $**p < 0.01$.

children in high and low pentiles of the Cognitive Composite distribution who fall into the lowest supplement category for the period of birth through 24 months as a function of family SES level. Low-SES children who were in the lowest supplementation category were at greater risk of being in the lowest pentile of test performance than were poorly supplemented children from higher SES families. This finding is more impressive because the range of family SES variation in the study communities is very small.

Family SES not only interacted with nutritional status to affect mental development, but also appears to have exerted an effect directly upon mental development. Table VI presents correlations between mental test performance and family economic indicators of House Quality and quality of Parents' Clothing, as well as with a number of family measures constructed to index aspects of intellectual stimulation available in the home, following to some degree Caldwell's Home Environment instrument (Bradley and Caldwell, 1976). These measures are: Mother Composite, which combines the mother's vocabulary test score, literacy, years of school passed, and modernity into a single index of intellectual characteristics; Father's Schooling, or his years of school passed; Sibling Schooling, or average years passed by all older siblings living at home; Toys, or the quantity and quality (e.g., presence of moving parts, etc.) of toys in the home; and Books and Objects, or the number of books, magazines, and visually stimulating objects such as drawings, photographs, pictorial calendars, diplomas, and decorated furniture found in the home.

As Table VI indicates, the Mother Composite of Intellectual Characteristics and Sibling Schooling both correlated consistently with test performances for boys, although Mother Composite was unrelated to test performance for girls and Sibling Schooling only related significantly to performance at age 5. Father's Schooling was unrelated to test performances. Of the two measures of material stimulation, Books and Objects, but not Toys, also correlated with test performance, but only at age 5. The economic indicators of House Quality and Parents' Clothing related significantly to test performances of both boys and girls at 5 years of age, and Parents' Clothing also related to boys' test performances at ages 3 and 4, and to girls' performances at 5.

In summary, family economic level, as well as human and material sources of intellectual stimulation, appears to influence mental development in the study communities. Work is presently proceeding on the development of various additional economic and social variables measured at the family level.

4. Discussion

Findings of the INCAP study to date suggest that there is an effect of malnutrition of mental development at least up to 60 months of age (the oldest children we have been able to study fully longitudinally to date), that nutritional input is of greatest importance during gestation and the first 2 years of life, and that the effects of a poor socioeconomic environment exacerbate the

Table VI. Correlations between Various Family Variables Related to Intellectual Stimulation and of Family Wealth Indicators, and Composite Score on the Preschool Battery at Ages 3, 4, and 5

Age (years)	Mother Composite[a,b]		Father Schooling[b]		Sibling Schooling[b]		Toys		Books and Objects[b]		House Quality[b,c]		Parents' Clothing[b,c]	
	Boys	Girls	Boys	Girls	Boys	Girls	Boys	Girls	Boys	Girls	Boys	Girls	Boys	Girls
3	0.13*	0.02	0.02	0.10	0.14*	0.01	0.05	0.00	0.02	−0.02	0.05	0.03	0.16**	0.02
4	0.17**	0.06	0.09	−0.01	0.26**	0.00	0.00	0.01	0.04	−0.05	0.05	0.07	0.16**	0.10*
5	0.20**	0.12*	0.13*	0.12*	0.31**	0.22**	0.06	0.10	0.20**	0.23**	0.13*	0.20**	0.12**	0.15**

[a] Cell n's for Mother Composite, Father Schooling, Sibling Schooling, Toys, Books and Objects, all ≈ 300.
[b] Symbols: $*p < 0.05$; $**p < 0.01$.
[c] Cell n's for House Quality and Parents' Clothing ≈ 600.

effects of early malnutrition. Furthermore, these data suggest, more strongly than have those of any previous study, that the effects of malnutrition on mental development are causal.

Two major questions regarding the effects of malnutrition on mental development remain to be answered. The first question concerns the nature of the mechanism through which malnutrition affects mental development. Two main possibilities seem consistent with the INCAP findings. The first of these possible mechanisms is that of structural damage to the developing nervous system. The animal research work of Winick *et al.* (1972) and of others, which has identified various anatomical differences in the brains of lower animals subjected to severe malnutrition during periods of rapid CNS development, certainly supports such a possibility.

The child's relationship to his environment in the first 2 years may also play a role. As Piaget and Inhelder (1969) have argued, intellectual development is an interactive process in which the growing child must both act and be acted upon by the world around him in order to mature. One type of interaction that has been demonstrated to be particularly important in subsequent intellectual development is that between mother and child (e.g., Hess *et al.*, 1968). Chávez *et al.* (1975) and Graves (1976) have recently presented evidence that mother–child interaction patterns for malnourished children differ from those for well-nourished children. Thus, it is possible that deleterious effects of insufficient energy on the child's interactions with his mother or other sources of environmental stimulation may underlie the increasingly well documented effects of malnutrition on mental development. This issue is currently being explored at INCAP.

The second question concerns whether the effects seen in childhood normally last into adolescence and beyond, condemning victims of early malnutrition to poor school achievement and less than competent adulthood. The studies of Winick *et al.* (1975) with Korean orphans adopted by middle-class families, of Stein *et al.* (1972a,b, 1975) with offspring of pregnant World War II famine victims in Holland, and of Lloyd-Still *et al.* (1974) with middle-class children who suffered severe infant malnutrition due to cystic fibrosis or congenital defects of the gastrointestinal tract suggest that effects of early malnutrition on subsequent intellectual development may be ameliorated in later childhood by a comfortable and stimulating home environment. However, the vast majority of victims of childhood malnutrition have little hope of substantial improvement in their deprived environments. Whether our well-nourished children will continue to show the effects of early good nutrition on their intellectual competence can best be determined in a longitudinal prospective study. This question is currently under investigation by the INCAP research team in Guatemala.

ACKNOWLEDGMENT. The research described was financed by Contract No. 1-HD-5-0640 from the National Institute of Child Health and Human Development, National Institutes of Health (NIH), Bethesda, Maryland.

5. References

Berry, J. W., 1969, On cross-cultural comparability, *Int. J. Psychol.* **23**:1.

Birch, H. G., Piñero, C., Alcalde, E., Toca, T., and Cravioto, J., 1971, Relation of kwashiorkor in early childhood and intelligence at school age, *Pediatr. Res.* **5**:579.

Botha-Antoun, E., Babayan, S., and Harfouche, J. K., 1968, Intellectual development related to nutritional status, *J. Trop. Pediatr.* **14**:112.

Boutourline, H., Manza, B., Louyot, P., El Amour, T., Redje, H., Boutourline, E., and Tesi, G., 1973, Social and environmental factors accompanying nutrition, Presented at Biannual Meeting of the International Association of Behavioral Sciences, Symposium on the Effects of Nutrition and Development, Ann Arbor, Michigan, August 21–24, 1973.

Bradley, R. H., and Caldwell, B. M., 1976, The relation of infants' home environment to mental test performance at 54 months: a follow-up study, *Child Dev.* **47**:1172.

Brockman, L. M., and Ricciuti, H. N., 1971, Sevee protein-calorie malnutrition and cognitive development in infancy and early childhood, *Dev. Psychol.* **4**:312.

Cabak, V., and Najdanvic, R., 1965, Effect of undernutrition in early life on physical and mental development, *Arch. Dis. Child.* **40**:532.

Champankan, S., Srikantia, S. G., and Gopalan, C., 1968, Kwashiorkor and mental development, *Am. J. Clin. Nutr.* **21**:844.

Chase, H. P., and Martin H. P., 1970, Underntrition and child development, *New Engl. J. Med.* **282**:933.

Chávez, A., Martínez, C., and Yoschine, T., 1974, The importance of nutrition and stimuli on child mental and social development, in: *Early Malnutrition and Mental Development* (J. Cravioto, L. Hambreus, and B. Vahlquist, eds.), pp. 211–214, Almqvist & Wiksell, Uppsala.

Chávez, A., Martínez, C., and Yoschine, T., 1975, Nutrition, behavioral development and M. C. interaction in young rural children, *Fed. Proc.* **34**:1574.

Cravioto, J., and DeLicardie, E. R., 1968, Intersensory development of school age children, in: *Malnutrition, Learning and Behavior* (N. S. Scrimshaw and J. E. Gordon, eds.), pp. 252–269, The MIT Press, Cambridge, Mass.

Cravioto, J., and Robles, B., 1965, Evolution of adaptive and motor behavior during rehabilitation from kwashiorkor, *Am. J. Orthopsychiat.* **35**:449.

Cravioto, J., DeLicardie, E. R., and Birch, H. G., 1966, Nutrition, growth and neurointegrative development: An experimental and ecologic study, *Pediatrics* **38**:319.

DDH/INCAP, 1975, Progress Report of the Division of Human Development, Institute of Nutrition of Central America and Panama, Guatemala.

DeLicardie, E. R., and Cravioto, J., 1974, Behavioral responsiveness of survivors of clinical severe malnutrition to cognitive demands, in: *Early Malnutrition and Mental Development* (J. Cravioto, L. Hambreus, and B. Vahlquist, eds.), pp. 134–153, Almqvist & Wiksell, Uppsala.

Edwards, L. D., and Craddock, L. J., 1973, Malnutrition and intellectual development. A study in school-age aboriginal children at Walgett, New South Wales, *Med. J. Aust.* **1**:880.

Evans, D. E., Moodie, A. D., and Hansen, J. D. L., 1971, Kwashiorkor and intellectual development, *S. African Med. J.* **45**:1413.

Graves, P. L., 1976, Nutrition, infant behavior, and maternal characteristics: a pilot study in West Bengal, India, *Am. J. Clin. Nutr.* **29**:305.

Hertzig, M. E., Birch, H. G., Richardson, S. A., and Tizard, J., 1972, Intellectual levels of school children severely malnourished during the first two years of life, *Pediatrics* **49**:814.

Hess, R. D., 1970, Social class and ethnic influences on socialization, in: *Charmichael's Manual of Child Psychology*, Vol. 2 (P. Mussen, ed.), pp. 457–558, John Wiley & Sons Inc., New York.

Hess, R. D., Shipman, V. C., Brophy, J. E., and Bear, R. M., 1968, The Cognitive Environments of Urban Preschool Children, Report to the Graduate School of Education, University of Chicago, Chicago.

INCAP/HEW, 1972, Nutritional evaluation of the population of Central America and Panama, regional summary, Institute of Nutrition of Central America and Panama (INCAP) and Nutrition Program, Center for Disease Control (formerly Interdepartmental Committee on Nutrition for National Development), *DHEW Publ.* (*HSM*) 72-8120, Washington, D.C., 165 pp.

Irwin, M., Klein, R. E., Engle, P. L., Yarbrough, C., and Nerlove, S., 1977, The problem of establishing validity in cross-cultural measurements, *Ann. N.Y. Acad. Sci.* **285**:308.

Irwin, M., Engle, P., Klein, R. E., and Yarbrough, C., 1978, The relationship of prior ability and family characteristics to school attendance and performance in rural Guatemala, *Child Dev.* **49**:415.

Jelliffe, D. G., 1966, The assessment of nutritional status of the community (with special reference to field surveys in developing regions of the world), *WHO Monogr. Ser. 53,* Geneva, 271 pp.

Klein, R. E., Freeman, H. W., Spring, B., Nerlove, S. B., and Yarbrough, C., 1976, Cognitive test performance and indigenous conceptions of intelligence, *J. Psychol.* **93**:273.

Liang, P. H., Hie, T. T., Jan, O. H., and Giok, L. T., 1967, Evaluation of mental development in relation to early malnutrition, *Am. J. Clin. Nutr.* **20**:1290.

Lloyd-Still, J. D., Hurwitz, I., Wolff, P. H., and Scwachman, H., 1974, Intellectual development after severe malnutrition in infancy, *Pediatrics* **54**:306.

McKay, H. E., McKay, A. C., and Sinisterra, L., 1969, Behavioral effects of nutritional recuperation and programmed stimulation of moderately malnourished preschool age children, Presented at the Meeting of the American Association for the Advancement of Science Symposium, Boston, Mass.

Martorell, R., Lechtig, A., Yarbrough, C., Delgado, H., and Klein, R. E., 1978, Relationships between food supplementation, home dietary intake and physical growth, Manuscript, Institute of Nutrition of Central America and Panama, Guatemala.

Mönckeberg, F., 1968, Effects of early marasmic malnutrition on subsequent physical and psychological development, in: *Malnutrition, Learning and Behavior* (N. S. Scrimshaw and J. E. Gordon, eds.), pp. 269–278, MIT Press, Cambridge, Mass.

Mönckeberg, F., Tisler, S., Toro, S., Gattás, V., and Vega, L., 1972, Malnutrition and mental development, *Am. J. Clin. Nutr.* **25**:766.

Montelli, T. B., Moura Ribeiro, V., Moura Ribeiro, R., and Ribeiro, M. A. C., 1974, Electroencephalographic changes and mental development in malnourished children, *J. Trop. Pediatr.* **20**:201.

Mora, J. O., Amézquita, A., Castro, L., Christiansen, J., Clement-Murphy, J., Cobos, L. F., Cremer, H. D., Dragastin, S., Elías, M. F., Franklin, D., Herrera, M. G., Ortíz, N., Pardo, F., Paredes, B. de, Ramos, C., Riley, R., Rodríguez, H., Vuori–Christiansen, L., Wagner, M., and Stare, F. J., 1974, Nutrition and social factors related to intellectual performance, *World Rev. Nutr. Dietet.* **19**:205.

Nerlove, S. B., Roberts, J. M., Klein. R. E., Yarbrough, C., and Habicht, J.-P., 1974, Natural indicators of cognitive development: An observational study of rural Guatemalan children, *Ethos* **2**:265.

Patel, B. D., 1974, Influence of malnutrition and environmental deprivation on the development of the child. Observations on a tribal population in Darg (Western India), in: *Early Malnutrition and Mental Development* (J. Cravioto, L. Hambreus, and B. Vahlquist, ed.), p. 155, Almqvist & Wiksell, Uppsala.

Piaget, J., and Inhelder, B., 1969, *The Psychology of the Child,* Basic Books, Inc., Publishers, New York.

Pollitt, E., and Granoff, D., 1967, Mental and motor development of Peruvian children treated for severe malnutrition, *Rev. Interam. Psicol.* **1**:93.

Richardson, S. A., 1976, The relation of severe malnutrition in infancy to the intelligence of school children with differing life histories, *Pediatr. Res.* **10**:57.

Schlenker, J. D., Bossio, V., and Romero, E. R., 1968, Desarrollo social de niños preescolares con kwashiorkor y marasmo, *Arch. Latinoam. Nutr.* **18**:173.

Stein, Z., Susser, M., Saenger, G., and Marolla, F., 1972a, Nutrition and mental performance, *Science* **178**:708.

Stein, Z., Susser, M., Saenger, G., and Marolla, F., 1972b, Intelligence test results of individuals exposed during gestation to the World War II famine in The Netherlands, *T. Soc. Geneesk.* **50:**766.
Stein, Z., Susser, M., Saenger, G., and Marolla, F., 1975, *Famine and Human Development: The Dutch Hunger Winter of 1944–45,* Oxford University Press, Inc., New York.
Stoch, M. B., and Smythe, P. M., 1963, Does undernutrition during infancy inhibit brain growth and subsequent intellectual development? *Arch. Dis. Child.* **38:**546.
Stoch, M. B., and Smythe, P. M., 1967, The effect of undernutrition during infancy on subsequent brain growth and intellectual development, *S. African Med. J.* **41:**1027.
Whiteman, M., and Deutsch, M., 1968, Social disadvantage as related to intellective and language development, in: *Social Class, Race and Psychological Development* (M. Deutsch, I. Katz, and A. R. Jensen, eds.), pp. 86–114, Holt, Rinehart and Winston, Inc., New York.
Winick, M., Rosso, P., and Brasel, J. A., 1972, Malnutrition and cellular growth in the brain: Existence of critical periods, in: *Lipids, Malnutrition and the Developing Brain,* pp. 199–213, Ciba Foundation, Amsterdam.
Winick, M., Meyer K., and Harris, R. C., 1975, Malnutrition and environmental enrichment by early adoption, *Science* **190:**1173.

10

Malnutrition and Infection

Gerald T. Keusch and Michael Katz

1. Introduction

Severe malnutrition, such as is seen in the developing countries, results from many causes. Insufficiency of food and its consequence, marasmus, and inadequacy of specific nutrients in the diet resulting in a variety of syndromes of malnutrition, such as kwashiorkor, xerophthalmia, and anemia, are the most obvious and dramatic examples of malnutrition. Many other factors, however, contribute primarily or secondarily to the development of malnutrition. These include unfamiliarity with good nutritional practices, cultural food taboos, and faulty management of food distribution. The ultimate consequence of malnutrition is inadequate growth and probably also failure to achieve optimal intellectual potential.

Most important in the understanding of the pathogenesis of malnutrition is the concept that it results not from a single event, or even a number of events operating at a particular time, but rather from cyclic, repetitive phenomena. Recurrent illness may be difficult to appreciate when one examines a community cross-sectionally—as had been the case in past studies—but the concept of periodicity becomes obvious on a longitudinal analysis, which reveals periodic faltering of growth in individuals. Recently, attention has been directed toward the forces responsible for this periodicity and at least two important principles have emerged. Contrary to the previously espoused concept, intrauterine malnutrition has been shown to have a long-term influence on postnatal growth. Second, multiple and frequent infections strongly interfere with normal growth patterns.

Gerald T. Keusch • Department of Medicine, Tufts University, School of Medicine, Boston, Massachusetts. *Michael Katz* • Departments of Pediatrics and Public Health, Columbia University College of Physicians and Surgeons, New York, New York.

2. Perspective from the Developing Nations

For many reasons malnutrition is prevalent in both the rural and urban poor of developing nations. Overcrowding and bad sanitation contribute to the ready transmission of the ubiquitous infectious agents and to a high incidence of disease. The magnitude and importance of each of these forces is difficult to assess. Maintenance of surveillance of both nutritional status and clinical as well as subclinical infections presents enormous technical and logistic problems. Moreover many of the poor in developing countries tend to be highly mobile; some are quite suspicious of and resistant to the social intrusion of a scientific study. It is understandable why there are few adequately documented studies of a population sufficiently large for analysis from which one can evaluate the importance of infection in the malnourished.

One notable exception has been the study in a Mayan Indian village, Santa Maria Cauque, in the highlands of Guatemala (Mata, 1978). This is attributable in large part to the stable nature of the village population and to several features of the study, including long-term prospective design, rapport between the investigators and the villagers, constant contact and sensitive social interaction between them, the presence of a continuously attended health center in the village square, and intensive surveillance and laboratory investigation. Although geography, climate, diet, genetic and cultural heritage, and living conditions are not necessarily representative of those of villages elsewhere in the world, or of urban centers of Guatemala, there are many basic principles that can be learned from the Santa Maria experience.

The study showed that pregnant women have a high rate of infection (Urrutia *et al.*, 1975). These infectious illnesses affect not only the urinary tract, commonly infected among pregnant women in all societies, but also the upper respiratory tract, including the middle ear, gastrointestinal tract, and skin. Among 82 pregnant women followed in Santa Maria Cauque, Urrutia *et al.* (1975) recorded 104 episodes of upper respiratory and 25 of lower respiratory infections, 29 of diarrhea or dysentery, and 22 of urinary tract infections. The incidence of infection was 2.45 per pregnancy, with 21 days of illness per 280-day pregnancy.

The women under study showed evidence of a background of nutritional deprivation, expressed as short stature, low body weight, and sparse subcutaneous fat (Mata *et al.*, 1972b). Dietary adequacy for most nutrients in the majority was less than 75% of recommended allowances based on age, weight, and height. Even this is an underestimate of the degree of deficiency, because standards for dietary adequacy become higher during pregnancy (Arroyave, 1975). The average pregnancy weight gain in Santa Maria Cauque was only 7 kg (Mata, 1978), in contrast to approximately 13 kg in New York, where many women monitor and deliberately limit their weight gain (Hellman and Pritchard, 1971). The observations in Santa Maria Cauque are similar to those in non-Indian (Ladino) villages in Guatemala and elsewhere in the world (Lechtig *et al.*, 1975).

One consequence of these factors in the pregnant woman is low birth weight of the newborn. Over a period of 8 years in Santa Maria Cauque, 424 singleton births were observed whose mean birth weight was 2555 g (Mata *et al.*, 1975b). One of every three newborns was full term (≥37 weeks gestation) and of low birth weight (<2501 g); only 7% of the infants were both of low birth weight and premature. Whereas mean birth weight of infants in rural Ladino villages was 3050 g and the highest recorded birth weight was 4800 g (Lasky *et al.*, 1975), only 10% of Santa Maria babies were above 3000 g and the highest weight recorded was only 3903 g. In contrast, mean birth weight in the United States is close to 3500 g and the total incidence of low birth weight is about 7%, the majority being preterm rather than full-term infants (Niswander and Gordon, 1972).

Like their mothers, these infants have been exposed to many infectious agents during pregnancy. The fetus can develop immune responses to intrauterine infections, symptomatic, asymptomatic, or latent (Stiehm, 1975). Because IgM is not transported across the placenta, the infant serum IgM level in the neonatal period serves as an index of intrauterine antigenic stimulation (Alford *et al.*, 1975). When specific antibody activity of these macroglobulins can be shown, intrauterine infection can be presumed to be the cause, and the etiology can be determined. The proportion of newborns in Santa Maria Cauque with excessive IgM serum levels at birth is extraordinarily high, reaching a figure of 42% in one survey (Urrutia *et al.*, 1975). Similar high incidence was found in another nearby Indian village, whereas among Ladino infants with higher mean birth weight, the incidence is less (14%). Among infants from the United States with a still higher average birth weight, abnormally elevated cord serum IgM levels are noted in fewer than 5% (Alford *et al.*, 1975). Average birth weight also tends to be higher in those with higher socioeconomic status, and the incidence of elevated IgM levels is correspondingly less.

The picture, then, is one in which both pregnant women and newborns suffer nutritional deprivation and excess of infections. Following birth the infant continues to consume an inadequate diet and experiences frequent infections, particularly after the period of exclusive breast feeding (Mata *et al.*, 1976). Weaning foods are often inadequate in nutrients and quantity, and sometimes are contaminated by microorganisms (Capparelli and Mata, 1975). Incidence of infection rises steadily, reaching a peak during the latter part of the second year of life (Mata *et al.*, 1977a). At this age children in Santa Maria Cauque experience a new infection every 3 weeks. These infections are associated with anorexia, decreased food intake, weight loss, and cessation of linear growth (Mata *et al.*, 1972a, 1977a). The most severe effects are noted in children with greatest preexisting deficits (Mata *et al.*, 1977a). There follows in many such children a cyclic deterioration in health which finally leads to overt, severe, and often fatal malnutrition (Gordon *et al.*, 1967). Thus, nutrition, growth, infection morbidity, and mortality are associated with one another throughout childhood and they interact in a manner detrimental to the growing child. Mortality of 96 per 1000 live births during the first year of life in Santa Maria Cauque is followed by a death rate of 52 per 1000 children during the second year (Mata, 1978; Mata *et al.*, 1975a).

The fate of children in the village improves thereafter. Mata considers this to be related in part to cumulative immunization against the prevailing infectious diseases. As a consequence of this, "general health is better and deaths become significantly fewer. These changes occur without obvious alterations in the composition of the diet, although the quantity consumed is greater, coincidental with fewer episodes of infectious diseases" (Mata *et al.*, 1977a). Other factors that may account for these findings are a degree of natural selection of the heartiest children, and the ability of the older child to compete more successfully for available food.

3. Perspective from the Industrialized Nations

Serious infection and profound malnutrition are uncommon in the very young children of industrialized nations. It is among the adult population of these countries that severe malnutrition is found, generally secondary to debilitating chronic diseases, including neoplasms, alcoholism, diffuse gastrointestinal processes, or renal failure (Bistrian *et al.*, 1976). Anorexia due to the primary disease and catabolic or malabsorptive losses are the main causes of this recently recognized adult protein–calorie malnutrition syndrome (Bistrian, 1977; Bistrian *et al.*, 1974).

It is now obvious that malnutrition can also develop as a consequence of hospitalization itself. This is in large part caused by the routine semistarvation regimens used for maintenance fluid and electrolyte therapy in many acute medical or surgical illnesses (Bistrian, 1977). Negative nutrient balances are amplified by the catabolic response to many acute disease processes in which the host uses stored protein in muscle for gluconeogenesis and in addition, is obligated to divert new protein synthesis to cellular and humoral factors involved with defense, repair, and survival (Beisel, 1975). In fact, in these patients the usual fluid therapy with small quantities of carbohydrate (5% dextrose in water) may adversely alter the host response to inflammation (Bistrian *et al.*, 1975). The resulting hyperinsulinemia may impair fat mobilization and ketogenesis for energy to a greater extent than the calories infused will supply. As a result of this "semistarvation" there may occur a rapid fall in serum albumin (Bistrian *et al.*, 1975). Surveys of adult patients in hospitals in the United States disclose a 50% prevalence rate of adult protein–calorie malnutrition as assessed by weight: height ratio, arm muscle circumference, triceps skinfold measurement, and determination of serum albumin levels (Bistrian, 1977; Bistrian *et al.*, 1974).

Malnutrition adversely affects immunological mechanisms, particularly cell-mediated immunity (Bistrian *et al.*, 1975; Law *et al.*, 1974). Altered immune functions render the patient more susceptible to infection, which may lead to the cycle of malnutrition and infections. Progressive deterioration of the health status of the patient is often the result; when superimposed on the underlying disease, the added insult may tip the balance from survival to death.

When total parenteral nutrition was first introduced, the recipients often developed systemic *Candida albicans* infections (Curry and Quie, 1971). Originally, this was presumed to be due to the rich nutrient solution infused, which represented a good culture medium for the yeast and to the direct route to the bloodstream provided by the indwelling intravenous catheter (Goldmann *et al.*, 1973). However, patients receiving parenteral alimentation are almost invariably suffering from adult protein-calorie malnutrition and precisely for this reason are at risk of *Candida* infection because of associated acquired immunodeficiency (Copeland *et al.*, 1975).

The population affected and the pathogenesis of malnutrition in industralized countries are quite different from those in developing nations. Therefore, different forms of intervention are required. It would seem that a major effort could be expended toward recognition of adult protein-calorie malnutrition, identification of patients at high risk, and nutritional rehabilitation or therapy as preventive measures. Appropriate dietary management during periods of high risk of catabolic and anorectic stress can successfully circumvent later problems due to hospital-acquired malnutrition (Bistrian, 1977; Blackburn *et al.*, 1976).

4. *Effects of Infection on Nutritional Status*

It has long been known that infections result in metabolic changes that affect the nutritional state of the patient (Beisel, 1975). These nutritional consequences are determined in part by the severity and duration of the infection, nutritional status of the subject at the time the infection occurs, and cultural responses to the illness [e.g., alterations in dietary intake and the administration of improper or even harmful remedies (Beisel, 1975; Latham, 1975; Mata *et al.*, 1977a)]. Typically, the patients are prevented from eating and some are even subjected to purgatives, as if these measures would help the body rid itself of the disease. Unfortunately, they result in severe nutritional-deficiency states.

Careful field observation of patients in developing countries and of experimentally infected volunteers in the United States demonstrate that one of the earliest manifestations of infection is anorexia (Beisel, 1975, 1977a; Latham, 1975; Mata *et al.*, 1972a, 1977a). It occurs with illness caused by all classes of infectious agents, whatever the organ system involved. Loss of appetite is apparently a basic response to infection, although its value for function and survival of the patient during infection is unclear. In addition, the mechanism by which anorexia is triggered is unknown.

When rapidly growing young children are allowed free access to food during acute infections and in convalescence, a remarkable pattern of dietary response is observed. Early anorexia certainly contributes to the wasting effect of the infection; the child rapidly begins to lose weight and his linear growth ceases (Latham, 1975; Mata *et al.*, 1977a). Wasting persists during the period of active infection and fever. Once convalescence begins, however, a marked hyperphagia is often noticed, with consumption of energy exceeding usual intakes by as much as 100%. Even more remarkable is the observation by Whitehead (1977; cited in Beisel, 1977b), that the hyperphagia continues until weight-for-height indices return to normal standards. At this time intake is voluntarily reduced to usual levels. The hyperphagic period has been referred to as one of "catch-up growth"; it appears to be as characteristic a response to infection as is anorexia. Catch-up growth begins at about the time that cumulative losses of various nutrients during the acute infection have reached their maxima (Beisel, 1977b). It is at this stage that the patient is most vulnerable to nutritional-depletion effects on tissue repair and host defenses. A second infection occurring at this time in a partially depleted host would have a greater impact than in the already repleted patient. Thus, there is an apparent physiologic purpose to a rapid repair of all incurred deficiences in early convalescence. However, limitations of food availability or the bulk quantity required to supply needs can preclude catch-up growth from ever happening.

Fever is usually a cardinal manifestation of infection, and although much has been learned about the mechanism of fever, little is known about its purpose (Keusch, 1977). It is difficult in clinical situations to demonstrate a clear-cut benefit of fever on morbidity or mortality (Atkins and Bodel, 1972). Occasional experimental manipulations in animals appear to favor the infected host; others favor the infection. A recent experiment with the cold-blooded lizard *Dipsosaurus dorsalis* demonstrates that even poikilotherms develop fever during infection if they are permitted to choose environmental conditions that raise core temperature (Kluger *et al.*, 1975). If this fever response is prevented during a bacterial infection, mortality increases. Thus, the presence of fever correlates positively with survival. It is unlikely that fever has withstood the forces of natural selection without being beneficial in some manner (Atkins and Bodel, 1972; Keusch, 1977).

Whatever its survival value, however, the fever response is not without metabolic cost. Body temperature is the resultant of heat production and heat loss, regulated by a finely tuned thermostat in the hypothalamus (Atkins and Bodel, 1972). Heat production is in part a by-product of metabolic activity, but principally it is the result of involuntary work by muscles—the shivering response. It requires a 25–40% increase in resting metabolic expenditures, as was calculated in recent energy-balance studies in septic patients (Long, 1977). A portion of this is used for heat produciton alone. An average figure of 13% increase in metabolism per centrigrade degree of fever has been determined in a number of different circumstances of fever; acute febrile infections tend to cost more than this and chronic processes less (DuBois, 1937; Keusch, 1977). This is in marked contrast to the usual caloric cost of daily activity above resting requirements, an equivalent of only about 2°C of temperature. The sick, lethargic patient is not saving calories in his inactivity because of the physiologic cost of fever.

In addition to the calories required as fuel to produce fever, elevated temperature per se causes a number of important metabolic alterations. Whether the result of infection or artificially induced, fever results in negative balance of nitrogen, magnesium, potassium, and phosphate, and through the sweating mechanism contributes to losses of electrolytes and other substances (Beisel, 1977a; Powanda, 1977; Rowell, 1974). Fever also induces tachypnea, which leads to alkalosis as CO_2 is blown off along with water. Moreover, fever promotes an increased secretion of adrenal steroids, growth hormone, glucagon, and insulin and thus has secondary effects on metabolism (Powanda, 1977; Rayfield *et al.*, 1973).

Absolute losses of nutrients have been recognized for many years to be a consequence of infections with fever; nutrient loss has been considered to be a major cause of the weight loss associated with infection (Beisel, 1975). However, wasting in infected patients is not simply the result of absolute losses of body nutrients, such as those occurring through fecal, urinary, or dermal excretion, compounded by the effects of diminished dietary intake caused by anorexia. Rather, wasting is also due in part to "functional wastage," by which is meant the removal of nutrients from normal biosynthetic pathways to other pathways, the purpose or function of which has not been identified (Beisel, 1975). This process is exemplified by the sequestration of Fe, Zn, and other trace metals within the mononuclear phagocyte system, particularly macrophages in liver and spleen. Furthermore, the liver begins to make a large number of proteins in response to infection and in so doing, turns off synthesis of albumin.

Recent work has indicated that a single mechanism may underlie the majority of these diverse responses, and that the diversion of nutrients may have a function. The common mechanism appears to be production of one or group of polypeptide mediators collectively called leukocyte endogenous mediator (LEM), released by phagocytic leukocytes under appropriate stimulation, for example in the process of phagocytosis (Beisel, 1975; Kampschmidt *et al.*, 1973; Powanda, 1977). In essence, this is analogous to the release of endogenous pyrogen (EP) as a mediator of fever (Atkins and Bodel, 1972). Current knowledge suggests that LEM and EP are not the same peptides, although they have much in common (Mapes and Sobocinski, 1977). Administration of LEM results in several important metabolic alterations: (1) release of amino acids from muscle to the bloodstream; (2) stimulation of amino acid uptake by the liver; (3) alteration of the pattern of protein synthesis in liver away from albumin synthesis and toward manufacture of "acute-phase reactant" proteins; (4) enhancement of uptake and sequestration of iron, zinc, and other trace metals into liver macrophages; and (5) increased serum levels of insulin and glucagon (Powanda, 1977).

Release of LEM may thus mediate both the catabolic and anabolic response to infection. The effect of LEM on muscle in experimental studies in animals is a decrease of local protein synthesis and initiation of breakdown of contractile proteins with release of amino acids (Powanda, 1977). Experiments using rats indicate that muscle protein constitutes a labile pool of nitrogen for protein synthesis elsewhere in the body as amino acids released into the bloodstream are utilized in large part by the liver (Beisel, 1977b; Wannemacher *et al.*, 1972). Studies in man utilizing 3-methyl histidine as a marker for turnover of muscle protein appear to validate the detailed studies performed thus far in the animal models (Long *et al.*, 1975). These metabolic shifts do not result in an increase of blood amino acid concentration; in fact, serum levels of amino acids fall during infection. There are also particular changes in the patterns of amino acids, for example an increased phenylalanine/tyrosine ratio (Powanda, 1977; Wannemacher, 1977). Liver uptake of amino acids and hepatic protein synthesis are simultaneously enhanced by LEM (Wannemacher *et al.*, 1975). In the past, anabolic activity in the liver was considered to be a diversion of nutrients and a functional wastage. On the basis of more recent information, Powanda (1977) has suggested a possible beneficial role for most of the acute-phase reactant proteins (Table I). In this context hepatic new protein synthesis might be considered in the same category as increased turnover of phagocytic cells, lymphocyte proliferation, consumption and turnover of the complement system proteins, and production of immunoglobulins. Although these biosynthetic activities are obviously important to the survival of the host, because they draw upon scarce precursor nutrients and substrates, there must be a concomitant decrease in synthesis, or even degradation of other proteins. In growing children this is certainly accomplished at the expense of visceral protein, manifested both by the decrease in albumin synthesis and the cessation of growth. Moreover, to the extent that these newly made cells and proteins are discharged into exudates or pass to the exterior through epithelial cell mucosal surfaces, the process contributes to the absolute loss of nutrients during infection. This is not mitigated by the fact that there may be purpose to the loss: immediate survival of the host.

The same argument pertains to energy metabolism during infection, with particular reference to deamination of amino acids (principally alanine) for gluconeogenesis, the manufacture of metabolic fuels in the liver. This meta-

Table I. Acute-Phase Reactant Proteins in Infection[a]

Protein	Effect of infection on serum concentration	Possible role in host defense
α-1-Acid glycoprotein	↑ Typhoid fever, Rocky Mountain spotted fever (RMSF)	↑ Wound healing
α-1-Antitrypsin	↑ Typhoid, RMSF, malaria, hepatitis	↓ Tissue damage
Haptoglobin	↑ or ↓, depending on stage of infection	Improved RES function by removal of free hemoglobin; carrier for cathepsin B inhibitor
Ceruloplasmin	↑ (except ↓ in syphilis)	↑ Oxidation of catecholamines
α-2-Macroglobin	? ↑ Turnover	↑ Granulopoiesis; ↑ acute-phase globulin synthesis; macrophage activator

[a] Adapted from Powanda (1977).

bolic alteration during infection is true wastage of protein stores; the deaminated nitrogen is excreted as urea and the carbon skeleton is oxidized to CO_2. When energy balance is studied in septic patients, the resting consumption of energy is found to be elevated above normal. This is generally expressed as resting metabolic expenditure (RME) rather than basal metabolic rate (BMR) (Long, 1977). The reason for this is that BMR is energy consumption during zero work by a fasting subject in a neutral thermal environment. Because these conditions are not met in the febrile septic patient, it is possible to determine only the RME, the expenditure of energy measured by indirect calorimetry under actual conditions in the hospital environment. Studies by Long and colleagues indicate a 25–40% increase in RME during sepsis in surgical patients (Long, 1977; Long *et al.*, 1971). In contrast, uncomplicated elective surgery has rather minimal effects on RME (Long *et al.*, 1971). During sepsis the extra RME is required to fuel fever as well as the various anabolic responses. However, septic patients often eat little, because of anorexia and culturally determined proscription of food, and therefore they must generate fuel endogenously. Body stores of glucose and glycogen in extracellular fluid are estimated to be around 300 g (Long, 1977) and liver glycogen amounts to about 70 g, giving a total energy reserve sufficient for only 12–24 hr (Long *et al.*, 1971). During infection both fasting hyperglycemia and increased glucose oxidation occur; these two facts make it clear that glucose is being manufactured (Long *et al.*, 1971). This results from gluconeogenesis after deamination of amino acids, principally alanine (Felig and Wahren, 1971). Increased conversion of alanine to glucose has been demonstrated in septic patients by the use of [^{14}C]alanine, confirming that enhancement of gluconeogenesis during infection does occur (Long *et al.*, 1976). In infected experimental animals the source of alanine is catabolism of contractile proteins in muscle; however, this has not been investigated in man. Gluconeogenesis is augmented during infection also as a result of elevated levels of adrenal steroids and glucagon (Exton *et al.*, 1972; Rayfield *et al.*, 1973; Rocha *et al.*, 1973).

Two other major features of energy balance in sepsis in man are reasonably well established. First, whereas glucose infusion equal to the fasting hepatic synthetic rate normally turns off gluconeogenesis, this is not observed in septic patients (Long, *et al.*, 1976). Infection in some way appears to distort the feedback control loop in the process of causing a marked increase in the pool size of glucose, as well as increased oxidation and turnover of glucose (Long, 1977; Long *et al.*, 1971, 1976). The increase in the rate of glucose synthesis is greater than the increase in the rate of its removal, accounting for the observed hyperglycemia. Even conversion of glucose to fat in glucose-loaded infected patients is apparently increased as evidenced by respiratory quotients greater than 1 (Long *et al.*, 1971). Abnormalities in glucose tolerance in septic patients, including both fasting hyperglycemia and diabetic glucose-tolerance test curves, are not a consequence of decreased glucose oxidation rates.

Second, the septic patient does not appear to turn to ketones for energy as the nonseptic fasting subject does (Beisel, 1977a; Long, 1977; Long *et al.*,

1971, 1976; Neufeld *et al.*, 1976; Powanda, 1977; Ryan *et al.*, 1974; Wannemacher, 1977). Ketosis appears to be blocked, perhaps because of hyperinsulinemia. Whatever the reason, serum ketones are not elevated and the price for this inability to use stored lipid for energy is nitrogen wastage.

5. Effects of Malnutrition on Susceptibility and Response to Infection

Infection in a malnourished host may be different in many respects from that in the well-nourished one. The difference lies in susceptibility to certain pathogens and the response to an established infection. For example, the protozoan *Pneumocystis carinii,* not normally pathogenic, is an opportunistic agent in patients with underlying neoplasms, immunodeficiency diseases, drug-caused immunosuppression (Burke and Good, 1973), and severe malnutrition (Hughes *et al.*, 1974). In contrast, measles virus causes disease in both well-nourished and malnourished populations, but in the latter the infection is substantially more severe and has a greater mortality (Gordon *et al.*, 1965; Morley, 1969; Morley *et al.*, 1963).

The key to the understanding of these two examples, as well as many other infections in malnourished hosts, is the impact of nutritional deprivation on the several host systems that function to defend against infection. These host defenses may be conveniently divided along the lines of the general classification of other human immunodeficiencies: cellular, phagocytic, antibody, and complement. Recent advances in knowledge of the integrity and activity of these systems in poorly nourished patients explain the basis for many previous observations of infection in the malnourished and of morphologic changes of central and peripheral lymphoid organs. It has been known that in people suffering from malnutrition the thymus gland is markedly atrophic, with distortion of the lobular architecture, decrease in Hassall's corpuscles, and marked cytopenia (Boyd, 1932; Mugerwa, 1971; Schonland, 1972; Smythe *et al.*, 1971; Watts, 1969). Additionally, the spleen and lymph nodes show cellular depletion of paracortical and periarteriolar areas with reduction in the number of primary follicles and germinal centers (Mugerwa, 1971; Schonland, 1972; Smythe *et al.*, 1971). Peyer's patches in the gastrointestinal tract may be atrophic with diminished lymphoid follicles (Douglas and Faulk, 1977). These tissues are normally populated with thymus-derived lymphocytes, the T-cells, which are involved in cell-mediated immunity (CMI).

CMI is the critical defense against intracellular pathogens, including viruses, some protozoans, and certain bacteria. These pathogens invade and multiply within host cells, where they are protected from antibody and other circulating defense substances. The tuberculin test is a manifestation of a CMI response in the skin; it mimics the host-tissue reaction in lungs and other organs infected by tubercle bacilli. Skin testing for CMI responses to a number of different antigenic stimuli has revealed a marked suppression of delayed-type hypersensitivity in malnourished subjects (Chandra, 1972, 1974b; Edelman *et al.*, 1973; Jayalakshmi and Gopalan, 1958; Law *et al.*, 1973; Schlesinger and Stekel, 1974; Smythe *et al.*, 1971). Although this abnormality could be due to defects in the skin per se (Edelman *et al.*, 1973), or perhaps related to increased circulating levels of adrenocortical hormones, which are known to occur in these patients (Beitins *et al.*, 1975), it is more likely related to depletion of thymus-derived lymphocytes. T-cells can be identified by their ability to stick to the surface of sheep erythrocytes and thus to form a rosette, that is, a lymphocyte surrounded by firmly adherent sheep red blood cells. All reported studies of the number of T-cells in malnourished hosts show that they are deficient (Bang *et al.*, 1975; Chandra, 1974a; Ferguson *et al.*, 1974; Keusch *et al.*, 1977a; Schopfer and Douglas, 1975). This is confirmed by functional assay of T-cells, for example the ability of these cells to multiply *in vitro* in the presence of certain stimuli, such as the plant protein, phytohemagglutinin (PHA). Lymphocytes obtained from malnourished individuals do not respond normally to PHA (Chandra, 1972, 1974a; Ferguson *et al.*, 1974; Geefhuysen *et al.*, 1971; Law *et al.*, 1973; Schopfer and Douglas, 1975; Smythe *et al.*, 1971). Although this points to a deficit of T-cell function, there is no certainty that this is explainable simply by the decreased number of T-cells; perhaps the function of the remaining individual T-cells is also impaired. Moreover, it is unclear whether these functional deficits *in vivo* (skin tests) and *in vitro* (PHA response) can be corrected by administration of transfer factor (Brown and Katz, 1967; Walker *et al.*, 1975), a protein obtained from white blood cells that can restore CMI in certain patients with congenital cellular immunodeficiency disease.

In contrast, the state of antibody-dependent immunity, mediated by a different type of lymphocyte, the B-cell, is far less affected by malnutrition. It is likely that B-cells differentiate their special functions during a period of influence by the bone marrow during fetal life. The B-cell and the antibody system are important in the response to many infections, both bacterial and viral. For example, antibody can participate in killing of bacteria or in virus neutralization. Antibody may also function to inactivate biologically active microbial products such as toxins.

In most studies of malnourished hosts, serum immunoglobulins are quantitatively normal or actually increased above the normal range (Alvarado and Luthringer, 1971; Chandra, 1972; Keet and Thom, 1969; Neumann *et al.*, 1975; Rosen *et al.*, 1971) as is the *in vivo* response to pokeweed mitogen (Schopfer and Douglas, 1975; Chandra, 1974a), which stimulates B-cells to divide. Elevated immunoglobulin levels may be attributable to the high frequency of infections, which constantly stimulate antibody synthesis, or to the absence of normal T-cell function, which is thought to modulate B-cell response to certain antigens. Although immunoglubulin levels are adequate, it is possible that specific antibodies produced in response to antigenic stimulation may not be (Chandra, 1974a; Faulk and Chandra, 1978; Suskind *et al.*, 1976b). Some experiments have suggested that antibody affinity may be reduced when protein-deprived animals are immunized (Passwell *et al.*, 1974). Recent studies also indicate that the secretory IgA antibody response in man to polio and measles virus may be blunted (Chandra, 1975c). In general, however, the number of B-cells is normal and their function largely intact in malnutrition. Therefore, antibody immunodeficiency does not appear to be a major characteristic of such hosts.

For the most part, all the normal processes of phagocytosis can be shown to occur in phagocytic cells of patients with malnutrition; however, the rate at which they occur may be altered. Abnormalities of phagocyte cell function in these patients are thus subtle. Often they appear to be a consequence of infection per se, rather than of the malnutrition per se (Douglas and Schopfer, 1974; Jose *et al.*, 1975; Keusch *et al.*, 1977a; Rosen *et al.*, 1975; Seth and Chandra, 1972). Even in otherwise normal hosts, infections lead to quantitative defects in phagocytic cell function, perhaps because of mobilization of younger, functionally immature cells (McCall *et al.*, 1969, 1971). Quantitative abnormalities of migration (chemotaxis), metabolism, and microbicidal activity have nevertheless been demonstrated in some of the studies of malnourished hosts, whereas no defect in engulfment or degranulation has yet been found. The magnitude of the defects, however, does not allow the conclusion that cellular dysfunction of phagocytes is of major importance in the response to invading microorganisms. Apparently normal resistance to infection may be present in nonmalnourished subjects with similar moderate impairment of the same cellular mechanisms.

Phagocytic cell function must be considered in the context of the activity of the complement system. Complement activity appears to be deficient in the malnourished individual, and as a result the minor *in vitro* cellular abnormalities may be magnified *in vivo* (Chandra, 1975a; Sirisinha *et al.*, 1973; Suskind *et al.*, 1976a). The consequence of this may be gross failure of the phagocytic defense mechanism in the host himself. Complement fragments generated during the inflammatory response are critical for chemotaxis (cellular migration), opsonization (coating microorganisms to prepare them for ingestion), and altered vascular permeability (perhaps to ease the diapedesis of leukocytes from the bloodstream to the tissue). The system is complex, consisting of 18 proteins which activate and inactivate themselves by two distinct pathways in a sequential but modulated and controlled fashion. Total complement activity (hemolytic complement activity) and immunochemically assayed individual components are reduced in malnourished hosts (Chandra, 1975a; Sirisinha *et al.*, 1973; Suskind *et al.*, 1976a). These components may be qualitatively abnormal as well as quantitatively reduced, for electrophoretically altered complement proteins have been reported in sera from nutritionally deprived hosts (Chandra, 1975a). It is possible, too, that there is consumption of complement *in vivo* in these patients because increased immunoconglutinins and positive direct Coombs' tests with anticomplementary activity have been demonstrated (Chandra, 1975a; Smythe *et al.*, 1971; Suskind *et al.*, 1976a). Few quantitative studies of opsonic activity have been reported; they indicate that there is decreased complement mediated opsonization (Keusch *et al.*, 1977a,b; Tanphaichitr *et al.*, 1973). Malnutrition experimentally produced sharply reduces heat-labile (complement mediated) opsonins within 1 week of a protein-poor diet (Keusch *et al.*, 1977a). It is therefore likely that complement, like T-cell function, is a critically affected host system during malnutrition.

There have been very few studies of immune mechanisms during recovery from protein–calorie malnutrition. Long-term effects may be of particular concern in children with prenatal or neonatal malnutrition. Animal experiments by Newberne and colleagues show persistent abnormalities in the ability to defend against *Salmonella* infections in well-fed offspring of diet-deprived mothers (Gebhardt and Newberne, 1974; Newberne *et al.*, 1970). McGhee *et al.* (1974) have also shown persistently depressed complement levels in rats born to malnourished mothers. The levels remain low even after weaning in spite of ingestion of a high-protein diet. Perhaps the only relevant information about man is the finding of decreased serum IgM in infants fully recovered from neonatal kwashiorkor (Aref *et al.*, 1970).

Short-term dietary therapy of protein–calorie malnutrition allows recovery of CMI as detected by skin testing and the number of T-cells (sheep erythrocyte rosetting cells) and their function as assessed by the PHA response (Chandra, 1974a; Edelman *et al.*, 1973; Geefhuysen *et al.*, 1971; Jayalakshmi and Gopalan, 1958; Law *et al.*, 1973). Similarly, complement component levels return toward normal after dietary repletion (Chandra, 1975a; Sirisinha *et al.*, 1973; Suskind *et al.*, 1976a). Both quality and quantity of diet are important; high protein intake favors correction of the complement defect more than high calorie intake. At the same time, anticomplementary activity of serum, which is detectable during acute malnutrition, disappears with dietary treatment (Chandra, 1975a; Suskind *et al.*, 1976a). Both diminished synthesis and increased consumption of complement may be affecting the levels present during malnutrition.

6. Iron Deficiency

Iron deficiency is one of the most common forms of malnutrition in the world (WHO, 1975); iron deficiency anemia is the most prevalent form of nutritional anemia. In addition to the deleterious effect of anemia in pregnancy, which increases maternal and fetal morbidity and mortality (Roszkowski *et al.*, 1966) and the impairment of work capacity (Viteri and Torun, 1974), iron deficiency with or without significant anemia may also affect resistance to infection (Sussman, 1974). Iron is of importance to the growth and virulence of many microorganisms (Bullen *et al.*, 1974), which often produce iron-binding molecules (siderophores) to compete for available molecular iron (Rodgers and Neilands, 1973). Over 90% of body iron is present in the form of hemoglobin, ferritin, and hemosiderin (Dallman, 1974). Only a small proportion of the essential iron is in nonerythroid tissue, including heme iron compounds other than hemoglobin (myoglobin, cytoglobin, catalase, peroxidase), nonheme iron compounds or metalloflavoproteins (dehydrogenases for NADH, succinate, xanthine, and α-glycerophosphate), and enzymes that do not contain iron but require it as a cofactor (aconitase, tryptophan pyrrolase) (Dallman, 1974). Some of these molecules are involved in host response to infection, for example myeloperoxidase and catalase which are essential consituents of bactericidal mechanisms within phagocytic vacuoles (Klebanoff, 1975). Reduction of leukocyte myeloperoxidase activity has been demonstrated in iron deficiency in man (Higashi *et al.*, 1967) along with impaired ability of these leukocytes to kill bacteria *in vitro* (Arbeter *et al.*, 1971; Chandra, 1973; Chandra and Saraya, 1975; MacDougall *et al.*, 1975) and diminished resistance to *in vivo* bacterial challenge in iron-deficient rats (Baggs and Miller, 1973).

Other potentially adverse effects of iron deficiency on defense mechanisms have been suggested by some investigators. These data indicate that iron deficiency impairs the antibody response to tetanus toxoid, lymphocyte transformation and lymphokine production, and resistance of the skin to colonization by *Candida albicans* (Arbeter *et al.*, 1971; Buckley, 1975; Higgs and Wells, 1972; Joynson *et al.*, 1972; Nalder *et al.*, 1972).

Iron deficiency may be protective when the requirements of microorganisms for free iron are considered (Sussman, 1974). Iron-binding proteins in serum and secretions of man have antibacterial properties *in vitro* when they are sufficiently unsaturated to compete with bacterial siderophores for available iron (Bullen *et al.*, 1974; Sussman, 1974). Bacteria may fail to grow or to produce virulence factors such as toxins when iron is limiting *in vitro*. It is possible, therefore, that the iron-deficiency state per se or marked unsaturation of transferrin may be protective *in vitro* and that this protection may be overcome when iron is administered to produce a sudden excess. Some published data may be taken as being supportive of these concepts; however, a critical evaluation that would consider the various effects of iron on the host (including anorexia and impaired growth) (Buckley, 1975; MacDougall *et al.*, 1975) and on the agent has not been reported.

7. Vitamin A Deficiency

Severe vitamin A deficiency in association with eye lesions, particularly xerophthalmia, is a well-known consequence of infection, especially diarrheal disease and measles (Franken, 1974; McLaren *et al.*, 1965; Scheifele and Forbes, 1972). On the basis of animal experiments there is reason to believe that vitamin A deficiency may contribute to infection by an adverse effect on several host defense mechanisms. Gnotobiotic animals deficient in vitamin A survive for extended periods in contrast to conventionally raised animals of deprived vitamin A (Bieri *et al.*, 1969). Challenge of vitamin-A-deficient and control animals with *E. coli* also reveals an undue susceptibility of the former group (Rogers *et al.*, 1971). Lack of the immunoadjuvant effects of vitamin A or actual decreased lymphocyte numbers or abnormalities of mucous membranes in vitamin-A-deficiency states may contribute to this (Bang *et al.*, 1972; Dresser, 1968).

8. Diarrheal Disease

The importance of acute diarrheal disease in malnourished populations was clearly stated by Gordon in 1964: "As a killing disease, the diarrheas far overshadow upper respiratory illnesses and through their after-effects they aggravate numerous serious diseases, especially nutritional disorders. In large parts of the world, deaths from diarrheal disease in the general population outnumber those from any single cause. In all areas, they are a regular and prominent feature of deaths among infants and young children. The effect of the disease on growth and development is long lasting, on the people and on the general economy of the region."

When acute diarrheal disease is considered as an epidemiological entity, endemic disease affecting young children can be separated from other endemic diarrheas of older children, adults, and travelers in terms of its impact on nutritonal status and associated mortality (Gordon, 1964). Generally, in developing countries, morbidity and mortality are both rather low during the first 6 months of life, corresponding to the period of exclusive breast feeding (Gordon *et al.*, 1964a; Mata *et al.*, 1969). Following this apparent grace period however, few escape (Gordon, 1964; Gordon *et al.*, 1964b). There occurs in all populations studied a rising incidence of diarrheal disease and a corresponding increase in deaths as well (Gordon *et al.*, 1964a,b). The peak occurrence varies with specific populations from the latter part of the first year of life to the second year of life and sometimes even later still (Gordon *et al.*, 1963, 1964b). The most important determinant is weaning, which varies from culture to culture (Gordon *et al.*, 1963). Weaning begins when exogeneous foods are introduced to breast-fed infants, and it ends when breast feeding ceases. Weanling infants are increasingly affected by acute diarrhea, the peak incidence of which usually occurs near the end of weaning and continues for several months thereafter (Gordon *et al.*, 1963; Mata *et al.*, 1976). Because this relationship is really dependent on environmental and social factors and not on age, Gordon *et al.* (1963) defined it as an epidemiological entity, weanling diarrhea. It is related to a "nutritionally inadequate and a bacteriologically unsafe environment during a period of high susceptibility" (Gordon, 1964). Clinical observations document that, in contrast to the short course of diarrhea and prompt recovery typical of well-nourished children, an acute illness of longer duration is followed by a chronic and irregular low-grade diarrhea, with periodic recurrent attacks of acute diarrhea (Dingle *et al.*, 1956; Gordon, 1964; Hodges *et al.*, 1956). Because the affected children are already malnourished, the consequence of this is a progressive deterioration in the nutritional state and health.

In rural areas of the developing world there are also periodic increases in both the incidence of and mortality from diarrheal disease (Gordon *et al.*, 1964a). These represent epidemic increases, at times related to emergence of certain pathogens, as for example recently *Shigella dysenteriae* in Central America (Mata *et al.*, 1970). At other times, however, the usual mixture of enteric pathogens is responsible. Epidemics in malnourished populations evolve slowly, spread by contact, and have a protracted course measured in months to years (Gordon, 1964; Gordon *et al.*, 1964a,b). This is in contrast to the "classical" epidemic diarrheal disease characterized by common source outbreaks due to point contamination of water, milk, or food, with a rapid rise in the number of cases, short duration of the disease, and abrupt termination (Hardy, 1959). Even when the entire age spectrum of the population is affected, as happened in the Central American epidemic of 1969–1971, mortality is concentrated in young infants and children, following the pattern of weanling diarrhea already described (Gangarosa *et al.*, 1970).

Although acute diarrheal disease is an example of infection that is thought to occur more frequently and with greater severity in malnourished than well-nourished individuals, there have been few quantitative analyses to evaluate these beliefs. Gordon *et al.* (1964b) evaluated data collected in Guatemala according to the Gomez classification of nutritonal status based on determination of weight for age. These data show a definite increase in frequency of diarrheal disease in malnourished children compared to those who were less than 10% deficient in weight for age. Severity of clinical illness, short of death, is more difficult to define. Gordon *et al.* (1964b) classified simple diarrhea as mild when it was shorter than 4 days in duration and as moderate when it lasted 4 days or more. Disease was considered to be more severe when there was blood or mucus in the stool, irrespective of duration. Severe diarrhea occurred more commonly in malnourished children than in those approximating normality with respect to weight for age. However, it is not certain that these criteria for severity are entirely valid, and in any event, the small number of normally or near-normally nourished children in the study precluded assessment of statistical significance.

Much has been learned in recent years about etiology and pathogenesis of acute diarrhea (Grady and Keusch, 1971). Of major importance have been the separation of enterotoxigenic (choleralike) (Gorbach and Khorana, 1972; Sack, 1975) and invasive (shigellalike) (Guerrant *et al.*, 1975) *Escherichia coli* infections from the previously designated group of "enteropathogenic *E. coli*" (EPEC) strains (DuPont *et al.*, 1971; Sack *et al.*, 1971) and the recognition of the human rotavirus, previously known as orbivirus or duovirus agents of infantile diarrhea (Flewett *et al.*, 1974; Kapikian *et al.*, 1976; Middleton *et al.*, 1974; Wyatt and Kapikian, 1977). EPEC organisms are a group of nine *E. coli* O-antigen serotypes, for which typing antisera are readily available, orginally determined to be important causes of epidemic diarrheal disease of the newborn. EPEC are found regularly unrelated to diarrheal disease in older infants, children, and adults (Goldschmidt and DuPont, 1976; Neter, 1975).

Except for the EPEC strains that infect neonates, virulence of *E. coli* depends on one of two specific attributes: (a) production of enterotoxin, and (b) invasion of intestinal epithelial cells. Toxin synthesis is determined by information in the DNA contained in transmissible extrachromosomal plasmids, in contradistinction to chromosomal DNA, which governs the serotype. Toxigenic *E. coli* can therefore be of any serotype as long as it contains the appropriate plasmid. This explains the lack of correlation of EPEC serotypes with acute diarrhea. The toxigenic organisms, prevalent in developing countries where malnutrition abounds, have recently been shown to be among the principal causes of travelers' diarrhea (Gorbach *et al.*, 1975; Merson *et al.*, 1976; Shore *et al.*, 1974). They also account for a variable proportion of the previously etiologically undefined acute undifferentiated diarrheas in the tropics. Unless toxigenicity is specifically determined by a laboratory test, there is nothing that would distinguish the virulent strains from commensals in the intestinal tract. Invasive *E. coli* produce a diarrhea/dysentery similar to shigella infection. A limited number of serotypes, very closely related to shigella, causes this disease. However, these invasive serotypes are not found among the classical EPEC strains and they are consequently undetectable by use of the EPEC serotyping sets.

Like the toxigenic and invasive *E. coli*, HRVL agent diarrhea is difficult to detect. Until recently, the virus was identified only by immunoelectron microscopy and hence was missed in routine investigations of etiology of diarrheal disease. The recent development of a complement fixation serological test will greatly aid proper identification of disease due to the HRVL viruses. When these two techniques were applied to a population admitted to Children's Hospital National Medical Center in Washington, D.C., half of the children with acute diarrhea admitted over an 18-month period were infected with HRVL (Kapikian *et al.*, 1976). These infections were during the colder months of the year, the peak incidence being in December; no cases were found from May through October. HRVL agents have also been found in malnourished children in Guatemala and Costa Rica (Mata *et al.*, 1977b). Their importance as agents of diarrheal disease and their propensity to cause nutritional deterioration have not yet been determined, although appropriate studies are in progress.

9. Vaccination in Malnutrition

In view of the impairment of host defenses in the malnourished, it is necessary to consider consequences of immunizations in such children. This consideration must include not only the question of vaccine efficacy in the malnourished population, but also the potential dangers of vaccinations. When one deals with inactivated vaccines, which result in an immune response to the antigenic mass injected, but which depend on nonreplicating antigens, the major consideration is vaccine efficacy in hosts who may not respond adequately. On the other hand, live viral vaccines must be considered active infections, and the consequences of these replicating agents in hosts unable to contain replication of viruses may perhaps be dangerous. It is possible to examine what has occurred in a vast malnourished population, because numerous children have received vaccines as part of the general immunization programs in various parts of the world. There are, however, no reported studies of any undesirable consequences of these vaccines in malnutrition, even as analyses have shown the effectiveness of the vaccines in disease prophylaxis.

The specific studies that have been done have usually viewed the administered vaccines as probes of the host response to the antigen and not as potentially harmful agents. A review of these indicates that some have been carried out as long ago as 1919 (Zilva, 1919), but the more critical ones began at the conclusion of World War II. Thus, in 1945, Cannon showed a poor response of malnourished rabbits to a variety of antigens. In 1947, Wissler showed that malnourished rats had a lesser response to the pneumococcal polysacharides than well-nourished controls. Unfortunately, both of these studies are difficult to interpret, because of certain imprecisions that were unavoidable at the time they were done but which rendered them invalid in the light of more contemporary knowledge (Balch, 1950). In 1948, Metcoff *et al.* infected a number of malnourished and well-nourished rats with *Salmonella typhimurium* and noted that the malnourished population had a greater mortality. Both groups, however, had comparable levels of antibodies. A subsequent report by Wohl *et al.* (1949) described antibody response to typhoid vaccine among profoundly malnourished postsurgical patients. This study, also, is not interpretable in the light of current clinical review.

Reddy and Srikantia (1964) reported a differential response to the typhoid vaccine among malnourished Indian children fed higher and lower protein diets, respectively. Those given more protein developed significantly higher levels of circulating antibodies. Suskind *et al.* (1976b) similarly detected failure of response to typhoid vaccine by Thai children with PCM. A group of 10 such children had no H-antibody response, in contrast to nutritionally rehabilitated children, who did. Among kwashiorkor children who were undergoing intensive nutritional rehabilitation, Pretorius and de Villiers (1962) found no failure of response to typhoid vaccine. Brown and Katz (1966) found that children with kwashiorkor, whom they inoculated with the 17D strain of yellow fever vaccine, failed to develop antibodies, in contrast to the well-nourished children comparably treated. The same investigators demonstrated failure of antibody production in 10 of 17 children who became infected with the attenuated oral polio virus vaccine (Katz *et al.*, 1966). Subsequently, Chandra (1975b) showed that in a group of fetally growth retarded children (who were therefore products of long-term malnutrition) polio vaccine did take, but the levels of antibodies resulting were lower than those in normal children. Moreover, the same investigator (Chandra, 1975c) determined in a group of 20 malnourished children in India that live attenuated poliovirus vaccine failed to induce secretory IgA antibody to the same extent that it did in control subjects.

A review of these reported humoral antibody responses indicates that there is no uniformity among the studies. In some instances malnourished children do develop antibodies, and in others they either fail to produce antibodies or produce them in reduced amounts. An explanation for these discrepancies is suggested by the studies of Hoffenberg *et al.* (1966) and Cohen and Hansen (1962), who examined turnover rates of albumin and globulin in malnourished children. These authors cautioned that their studies were not carried out in a steady state of their subjects, but they did indicate that the albumin pool was decreased in the malnourished children and it returned to normal levels after nutritional rehabilitation. The globulin pool, on the other hand, did not show any differences between the state of malnutrition and the state of rehabilitation from the deficiencies. These investigators postulated that the available protein was spared for the purpose of globulin synthesis, at the expense of the circulating albumin and tissue proteins. They speculated further that individuals in marginal, pre-kwashiorkor states of malnutrition may develop kwashiorkor at the time of infection because the demand on the available proteins stores for production of gamma globulin may deplete them to the point of precipitation of kwashiorkor.

These findings may also explain the reason for the discrepancies among the studies of humoral antibodies in malnutrition. It follows that children moderately malnourished may have sufficient available protein to produce immunoglobulins. Indeed, many children with kwashiorkor have very high levels of immunoglobulins. In severe malnutrition, however, there may not be sufficient protein pool from which amino acids can be transferred for the synthesis of gamma globulins, and such children will fail to develop new antibodies and indeed are likely to die.

Very little information is available about the development of cell-mediated immune response in vaccinated malnourished children. Harland (1965) reported that moderately malnourished children who received BCG vaccine had generally less intense reactions to PPD than did the well-nourished controls. He reported no complications of the vaccine. BCG can cause disseminated disease in patients with neoplasia and altered CMI (Aungst *et al.*, 1975).

It is impossible in the present state of our knowledge to predict whether the inadequate response to live viral vaccines in malnourished children may result in latent infections. Several diseases, notably those of the central nervous system, develop years or decades after exposure to a particular virus, which apparently remains dormant in the host. For example, children with subacute sclerosing panencephalitis tend to have their measles infection early in life, before 2 years of age, but develop their neurological disease in late childhood or in the teens. It can be assumed that during the initial infection with the measles virus these young children are in a state of some immunological unresponsiveness, or impaired responsiveness, that may be analogous to what happens in malnourished children. If that were true, one could anticipate complications years later that may not *appear* to be related to the administration of the live virus vaccine.

Our knowledge of action of the vaccines in malnourished children is still quite limited. The few available data do not permit us to calculate the risk of administration of the vaccine. When adequate data are accumulated, a rational judgment about the relative risks and benefits of vaccines will be possible.

10. References

Alford, C. A., Stagno, S., and Reynolds, D. W., 1975, Diagnosis of chronic perinatal infections, *Am. J. Dis. Child.* **129**:455.

Alvarado, J., and Luthringer, D. G., 1971, Serum immunoglobulins in edematous protein-calorie malnourished children, *Clin. Pediatr.* **10:**174.
Arbeter, A., Echevarri, L., Franco, D., Munson, S., Valez, M., and Vitale, J. J., 1971, Nutrition and infection, *Fed. Proc.* **30:**1421.
Aref, G. H., Badr El Din, M. K., Hassan, A. I., and Araby, I. I., 1970, Immunoglobulins in kwashiorkor, *J. Trop. Med. Hyg.* **73:**186.
Arroyave, G., 1975, Nutrition in pregnancy in Central America and Panama, *Am. J. Dis. Child.* **129:**427.
Atkins, E. A., and Bodel, P., 1972, Fever, *New Engl. J. Med.* **286:**27.
Aungst, C. W., Sokal, J. E., and Jager, B. V., 1975, Complications of BCG vaccination in neoplastic disease, *Ann. Int. Med.* **82:**666.
Baggs, R. B., and Miller, S. A., 1973, Nutritional iron deficiency as a determinant of host resistance in the rat, *J. Nutr.* **103:**1554.
Balch, H. R., 1950, Relation of nutritional deficiency in man to antibody production, *J. Immunol.* **64:**397.
Bang, B. G., Bang, F. B., and Foard, M. A., 1972, Lymphocyte depression induced in chickens on diets deficient in vitamin A and other components, *Am. J. Pathol.* **68:**147.
Bang, B. G., Mahalanabis, D., Mukherjee, K. L., and Bang, F. B., 1975, T and B lymphocyte rosetting in undernourished children, *Proc. Soc. Exp. Biol. Med.* **149:**199.
Beisel, W. R., 1975, Metabolic response to infection, *Ann. Rev. Med.* **26:**9.
Beisel, W. R., 1977a, Magnitude of the host nutritional response to infection, *Am. J. Clin. Nutr.* **30:**1236.
Beisel, W. R., 1977b, Résumé of the discussion concerning the nutritional consequences of infection, *Am. J. Clin. Nutr.* **30:**1294.
Beitins, I. Z., Kowarski, A., Migeon, C. J., and Graham, G. G., 1975, Adrenal function in normal infants and in marasmics and kwashiorkor, *J. Pediatr.* **86:**302.
Bieri, J. G., McDaniel, E. G., and Rogers, W. E., Jr., 1969, Survival of germfree rats without vitamin A, *Science* **163:**574.
Bishop, R. F., Davidson, G. P., Holmes, I. H., and Ruck, B. J., 1973, Virus particles in epthelial cells of duodenal mucosa from children with acute non-bacterial gastroenteritis, *Lancet* **2:**1281.
Bistrian, B. R., 1977, Interaction of nutrition and infection in the hospital setting. *Am. J. Clin. Nutr.* **30:**1228.
Bistrian, B. R., Blackburn, G. L., Hallowell, E., and Heddle, R., 1974, Protein status of general surgical patients, *J. Am. Med. Assoc.* **230:**858.
Bistrian, B. R., Blackburn, G. L., Scrimshaw, N. S., and Flatt, J. P., 1975, Cellular immunity in semistarved states in hospitalized adults, *Am. J. Clin. Nutr.* **28:**1148.
Bistrian, B. R., Blackburn, G. L., Vitale, J. J., Cochran, D., and Naylor, J., 1976, Prevalence of malnutrition in general medical patients, *J. Am. Med. Assoc.* **235:**1567.
Boyd, E., 1932, The weight of the thymus gland in health and in disease, *Am. J. Dis. Child.* **43:**1162.
Brown, R. E., and Katz, M., 1966, Failure of antibody production to yellow fever vaccine in children with kwashiorkor, *Trop. Geogr. Med.* **18:**125.
Brown, R. E., and Katz, M., 1967, Passive transfer of delayed hypersensitivity reaction to tuberculin in children with protein calorie malnutrition, *J. Pediatr.* **70:**126.
Buckley, R. H., 1975, Iron deficiency anemia: its relationship to infection susceptibility and host defense (editorial), *J. Pediatr.* **86:**993.
Bullen, J. J., Rogers, H. J., and Griffiths, E., 1974, Bacterial iron metabolism in infection and immunity, in: *Microbial Iron Metabolism: A Comprehensive Treatise* (J. B. Neilands, ed.), p. 517, Academic Press, Inc., New York.
Burke, B. A., and Good, R. A., 1973, *Pneumocystis carinii* infection, *Medicine* **52:**23.
Cannon, P. R., 1945, The importance of proteins in resistance to infection, *J. Am. Med. Assoc.* **128:**360.
Capparelli, E., and Mata, L. J., 1975, Microflora of maize prepared as tortillas, *Appl. Microbial.* **29:**802.

Chandra, R. K., 1972, Immunocompetence in undernutrition, *J. Pediatr.* **81**:1194.
Chandra, R. K., 1973, Reduced bactericidal capacity of polymorphs in iron deficiency, *Arch. Dis. Child.* **48**:864.
Chandra, R. K., 1974a, Rosette-forming T lymphocytes and cell mediated immunity in malnutrition, *Br. Med. J.* **3**:608.
Chandra, R. K., 1974b, Interactions of infection and malnutrition, in: *Progress in Immunology*, Vol. 4 (L. Brent and J. Holborow, eds.), p. 355, North-Holland Publishing Co., Amsterdam.
Chandra, R. K., 1975a, Serum complement and immunoconglutinin in malnutrition, *Arch. Dis. Child.* **50**:225.
Chandra, R. K., 1975b, Fetal malnutrition and postnatal immunocompetence, *Am. J. Dis. Child.* **129**:450.
Chandra, R. K., 1975c, Reduced secretory antibody response to live attenuated measles and poliovirus vaccines in malnourished children, *Br. Med. J.* **2**:583.
Chandra, R. K., and Saraya, A. K., 1975, Impaired immunocompetence associated with iron deficiency, *J. Pediatr.* **86**:899.
Cohen, S., and Hansen, J. D. L., 1962, Metabolism and albumin and γ-globulin in kwashiorkor, *Clin. Sci.* **23**:351.
Copeland, E. M., MacFayden, B. V., Rapp, M. A., and Dudrick, S. J., 1975, Hyperalimentation and immune competence in cancer, *Surg. Forum* **26**:138.
Curry, C. R., and Quie, P. G., 1971, Fungal septicemia in patients receiving parenteral alimentation, *New Engl. J. Med.* **185**:1221.
Dallman, P. R., 1974, Tissue effects of iron deficiency, in: *Iron in Biochemistry and Medicine* (A. Jacobs and M. Worwood, eds.), p. 437, Academic Press, Inc., New York.
Dingle, J. H., McCorkle, L. P., Badger, G. F., Curtiss, C., Hodges, R. G., and Jordan, W. S., Jr., 1956, A study of illness in a group of Cleveland families. XIII. Clinical description of acute non bacterial diarrhea, *Am. J. Hyg.* **64**:368.
Douglas, S. D., and Faulk, W. P., 1977, Immunological aspects of protein-calorie malnutrition, in: *Recent Advances in Clinical Immunology* (R. A. Thompson, ed.), pp. 15-39, Churchill-Livingstone, London.
Douglas, S. D., and Schopfer, K., 1974, Phagocyte function in protein-calorie malnutrition, *Clin. Exp. Immunol.* **17**:121.
Dresser, D. W., 1968, Adjuvanticity of vitamin A, *Nature* **217**:527.
DuBois, E. F., 1937, The mechanism of heat loss and temperature regulation, Lang medical lectures, Stanford University Press, Stanford, Calif.
DuPont, H. L., Formal, S. B., Hornick, R. B., Snyder, M. J., Libonati, J. P., Sheahan, D. G., LaBrec, E. H., and Kalas, J. P., 1971, Pathogenesis of *Escherichia coli* diarrhea, *New Engl. J. Med.* **285**:1.
Edelman, R., Suskind, R., Olson, R. E., and Sirisinha, S., 1973, Mechanisms of defective delayed cutaneous hypersensitivity in children with protein-calorie malnutrition, *Lancet* **1**:506.
Exton, J. H., Friedman, N., Wong, E. H., Brineaux, J. P., Corbin, J. D., and Park, C. R., 1972, Interaction of glucocorticoids with glucagon and epinephrine in the control of gluconeogenesis and glycogenolysis in the liver and of lipolysis in adipose tissue, *J. Biol. Chem.* **247**:3579.
Faulk, W. P., and Chandra, R. K., 1979, Nutrition and Resistance to infection, in: *CRC Handbook of Food and Nutrition*, CRC Press, Cleveland, in press.
Felig, P., and Wahren, J., 1971, Amino acid metabolism in exercising man, *J. Clin. Invest.* **50**:2703.
Ferguson, A. C., Lawler, G. J., Neumann, C. G., Oh, W., and Stiehm, E. R., 1974, Decreased rosette-forming lymphocytes in malnutrition and intrauterine growth retardation, *J. Pediatr.* **85**:717.
Flewett, T. H., Bryden, A. S., and Davies, H., 1974, Diagnostic electron-microscopy of faeces. I. The viral flora of the faeces as seen by electron microscopy, *J. Clin. Pathol.* **27**:603.
Franken, S., 1974, Measles and xerophthalmia in East Africa, *Trop. Geogr. Med.* **26**:39.
Gangarosa, E. J., Perera, D. R., Mata, L. J., Mendizabal-Morris, C., Guzman, G., and Reller, L. B., 1970, Epidemic *Shiga* bacillus dysentery in Central America. II. Epidemiologic studies in 1969, *J. Infect. Dis.* **122**:181.

Gebhardt, B. M., and Newberne, P. M., 1974, Nutrition and immunological responsiveness: T-cell function in the offspring of lipotrope and protein-deficient rats, *Immunology* **26**:489.

Geefhuysen, J., Rosen, E. U., Katz, J., Ipp, P., and Metz, J., 1971, Impaired cellular immunity in kwashiorkor with improvement after therapy, *Br. Med. J.* **4**:527.

Goldmann, D. A., Martin, W. T., and Worthington, J. W., 1973, Growth of bacteria and fungi in total parenteral nutrition solutions, *Am. J. Surg.* **126**:314.

Goldschmidt, M. C., and DuPont, H. L., 1976, Enteropathogenic *Escherichia coli*. Lack of correlation of serotype with pathogenicity, *J. Infect. Dis.* **137**:153.

Gorbach, S. L., and Khorana, C. M., 1972, Toxigenic *Escherichia coli*. A cause of infantile diarrhea in Chicago, *New Engl. J. Med.* **287**:791.

Gorbach, S. L., Kean, B. H., Evans, D. G., Evans, D., and Bessudo, D., 1975, Travelers' diarrhea and toxigenic *Escherichia coli, New Engl. J. Med.* **292**:933.

Gordon, J. E., 1964, Acute diarrheal disease, *Am. J. Med. Sci.* **248**:345.

Gordon, J. E., Chitkara, I. D., and Wyon, J. B., 1963, Weanling diarrhea, *Am. J. Med. Sci.* **245**:345.

Gordon, J. E., Behar, M., and Scrimshaw, N. S., 1964a, Acute diarrheal disease in less developed countries. 1. An epidemiological basis for control, *Bull. WHO* **31**:1.

Gordon, J. E., Guzman, M. A., Ascoli, W., and Scrimshaw, N. S., 1964b, Acute diarrheal disease in less developed countries. 2. Patterns of epidemiological behavior in rural Guatemalan villages, *Bull. WHO* **31**:9.

Gordon, J. E., Jansen, A. A., and Ascoli, W., 1965, Measles in rural Guatemala, *J. Pediatr.* **67**:779.

Gordon, J. E., Wyon, J. B., and Ascoli, W., 1967, The second year death rates in less developed countries, *Am. J. Med. Sci.* **254**:357.

Grady, G. F., and Keusch, G. T., 1971, Pathogenesis of bacterial diarrhea, *New Engl. J. Med.* **285**:834.

Guerrant, R. L., Moore, R. A., Kirschenfeld, P. M., and Sande, M. A., 1975, Role of toxigenic and invasive bacteria in acute diarrhea of childhood, *New Engl. J. Med.* **293**:567.

Hardy, A. V., 1959, Diarrheal diseases of infants and children. Mortality and epidemiology, *Bull. WHO* **21**:309.

Harland, P. S. E. G., 1965, Tuberculin reactions in malnourished children, *Lancet* **2**:719.

Hellman, L. M., and Pritchard, J. A., 1971, *Williams Obstetrics*, 14th ed., p. 336, Appleton Century Crofts, New York.

Higashi, O., Sato, Y., Takamura, H., and Oyama, M., 1967, Mean cellular peroxidase (MCP) of leukocytes in iron deficiency anemia, *Tohoku J. Exp. Med.* **93**:105.

Higgs, J. M., and Wells, R. S., 1972, Chronic mucocutaneous candidiasis. Associated abnormalities of iron metabolism, *Br. J. Dermatol.* **86** (suppl. 8):88.

Hodges, R. G., McCorkle, L. P., Badger, G. F., Curtiss, C., Dingle, J. H., and Jordan, W. S., Jr., 1956, A study of illness in a group of Cleveland families. XI. The occurrence of gastrointestinal symptoms, *Am. J. Hyg.* **64**:349.

Hoffenberg, R., Black, E., and Brock, J. F., 1966, Albumin and γ-globulin tracer studies in protein depletion states, *J. Clin. Invest.* **45**:143.

Hughes, W. T., Price, R. A., Sisko, F., Havion, W. S., Kafatos, A. G., Schonland, M., and Smythe, P. M., 1974, Protein-calorie malnutrition. A host determinant for *Pneumocystis carinii* infection, *Am. J. Dis. Child.* **128**:44.

Jayalakshmi, V. T., and Gopalan, C., 1958, Nutrition and tuberculosis. I. An epidemiological study, *Indian J. Med. Res.* **46**:87.

Jose, D. G., Shelton, M., Tauro, G. P., Belbin, R., and Hosking, C. S., 1975, Deficiency of immunological and phagocytic function in aboriginal children with protein-calorie malnutrition, *Med. J. Austral.* **62**:699.

Joynson, D. H. M., Jacobs, A., Walker, D. M., and Dolby, A. E., 1972, Defect of cell-mediated immunity in patients with iron deficiency anemia, *Lancet* **2**:1058.

Kampschmidt, R. F., Upchurch, H. F., Eddington, C. L., and Pulliam, L. A., 1973, Multiple biological activities of a partially purified leukocytic endogenous mediator, *Am. J. Physiol.* **224**:530.

Kapikian, A. Z., Kim, H. W., Wyatt, R. G., Rodriquez, W. J., Ross, S., Cline, W. L., Parrott, R. H., and Chanock, R. M., 1974, Reovirus-like agent in stools: association with infantile diarrhea and development of serologic tests, *Science* **185:**1049.

Kapikian, A. Z., Kim, H. W., Wyatt, R. G., Cline, W. L., Arrobio, J. O., Brandt, C. D., Rodriguez, W. J., Sack, D. A., Chanock, R. M., and Parrott, R. H., 1976, Human reovirus-like agent as the major pathogen associated with "winter" gastroenteritis in hospitalized infants and young children, *New Engl. J. Med.* **294:**965.

Katz, M., Brown, R. E., and Plotkin, S. A., 1966, Antibody Production in Kwashiorkor, Response to Viral Infection, Proc. VII Int. Congr. Nutr. (abstr.), Hamburg.

Keet, M. P., and Thom, H., 1969, Serum immunoglobulins in kwashiorkor, *Arch. Dis. Child.* **44:**600.

Keusch, G. T., 1977, The consequences of fever, *Am. J. Clin. Nutr.* **30:**1211.

Keusch, G. T., Urrutia, J. J., Fernandez, R., Guerrero, O., and Castaneda, G., 1977a, Humoral and cellular aspects of intracellular bacterial killing in protein-calorie malnutrition, in: *Malnutrition and the Immune Response* (R. Suskind, ed.), Raven Press, New York.

Keusch, G. T., Urrutia, J. J., Guerrero, O., and Castaneda, G., 1977b, Deficient serum opsonic activity in acute protein-calorie malnutrition, unpublished data.

Keusch, G. T., Urrutia, J. J., Guerrero, O., Castaneda, G., and Douglas, S. D., 1977c, Rosette forming lymphocytes in children with protein-calorie malnutrition in Guatemala, in: *Malnutrition and the Immune Response* (R. Suskind, ed.), Raven Press, New York.

Keusch, G. T., Douglas, S. D., Hammer, G., and Braden, K., 1978, Macrophage antibacterial functions in experimental protein-calorie malnutrition. II. Cellular and humoral factors for chemotaxis, phagocytosis and intracellular bactericidal activity, *J. Inf. Dis.* **138:**134.

Klebanoff, S. J., 1975, Antimicrobial mechanisms in neutrophilic polymorphonuclear leukocytes, *Semin. Hematol.* **12:**117.

Kluger, M. J., Ringler, D. H., and Anver, M., 1975, Fever and survival, *Science* **188:**166.

Lasky, R. E., Lechtig, A., Delgado, H., Klein, R. E., Engle, P., Yarbrough, C., and Martorell, R., 1975, Birth weight and psychomotor performance in rural Guatemala, *Am. J. Dis. Child.* **129:**566.

Latham, M. C., 1975, Nutrition and infection in national development, *Science* **188:**561.

Law, D. K., Dudrick, J. J., and Abdou, N. I., 1973, Immunocompetence of patients with protein calorie malnutrition. The effects of nutritional repletion, *Ann. Int. Med.* **79:**545.

Law, D. K., Dudrick, S. J., and Abdou, N. I., 1974, The effects of protein calorie malnutrition on immune competence, *Surg. Gynecol. Obstet.* **139:**257.

Lechtig, A., Delgado, H., Lasky, R., Yarbrough, C., Klein, R. E., Habicht, J. P., and Behar, M., 1975, Maternal nutrition and fetal growth in developing countries, *Am. J. Dis. Child.* **129:**553.

Logan, W. S., 1972, Vitamin A and keratinization, *Arch. Dermatol.* **105:**748.

Long, C. L., 1977, Energy balance and carbohydrate metabolism in infection and sepsis, *Am. J. Clin. Nutr.* **30:**1301.

Long, C. L., Spencer, J. L., Kinney, J. M., and Geiger, J. W., 1971, Carbohydrate metabolism in man: effect of elective operations and major injury, *J. Appl. Physiol.* **31:**110.

Long, C. L., Haverberg, L. N., Young, V. R., Kinney, J. M., Munro, H. N., and Geiger, J. W., 1975, Metabolism of 3-methylhistidine in man, *Metabolism* **24:**929.

Long, C. L., Kinney, J. M., and Geiger, J. W., 1976, Nonsuppressability of gluconeogenesis by glucose in septic patients, *Metabolism* **25:**193.

McCall, C. E., Katayama, I., Cotran, R. S., and Finland, M., 1969, Lysosomal and ultrastructural changes in human "toxic" neutrophils during bacterial infection, *J. Exp. Med.* **129:**267.

McCall, C. E., Caves, J., Cooper, R., and DeChatelet, L., 1971, Functional characteristics of human toxic neutrophils, *J. Infect. Dis.* **124:**68.

MacDougall, L. G., Anderson, R., McNab, G. M., and Katz, J., 1975, The immune response in iron-deficient children: impaired cellular defense mechanisms with altered humoral components, *J. Pediatr.* **86:**833.

McGhee, J. R., Michalek, S. M., Ghanta, V. K., and Stewart, G., 1974, Complement levels in malnourished animals: Quantification of serum complement in rat dams and their offspring, *J. Reticuloendoth. Soc.* **16:**204.

McLaren, D. S., Shirajian, E., Tchalian, M., and Khowry, G., 1965, Xerophthalmia in Jordan, *Am. J. Clin. Nutr.* **17:**117.

Mapes, C. A., and Sobocinski, P. Z., 1977, Differentiation between endogenous pyrogen and leukocytic endogenous mediator, *Am. J. Physiol.* **232:**C15.

Mata, L. J., 1978, *The Children of Santa Maria Cauque. A Prospective Field Study of Health and Growth,* MIT Press, Cambridge, Mass.

Mata, L. J., Urrutia, J. J., Garcia, B., Fernandez, R., and Behar, M., 1969, Shigella infection in breast-fed Guatemalan Indian neonates, *Am. J. Dis. Child.* **117:**142.

Mata, L. J., Gangarosa, E. J., Caceras, A., Perera, D. R., and Mejicanos, M. L., 1970, Epidemic *Shiga* bacillus dysentery in Central America. I. Etiologic investigations in Guatemala, 1969, *J. Infect. Dis.* **122:**170.

Mata, L. J., Urrutia, J. J., Albertazzi, C., Pellecer, O., and Arellano, E., 1972a. Influence of recurrent infections on nutrition and growth of children in Guatemala, *Am. J. Clin. Nutr.* **25:**1267.

Mata, L. J., Urrutia, J. J., Caceres, A., and Guzman, M. A., 1972b, The biological environment in a Guatemalan rural community, in: *Proc. Western Hemisph. Nutr. Congr. III* (P. L. White and N. Selvey, eds.), p. 257, Futura Publishing Co., Inc., New York.

Mata, L. J., Kronmal, R. A., Urrutia, J. J., and Garcia, B., 1975a, Antenatal events and postnatal growth and survival of children. Prospective observation in a rural Guatemalan village, in: *Proc. Western Hemisph. Nutr. Congr. IV* (P. L. White and N. Selvey, eds.), p. 107, Publ. Sciences Group, Acton, Mass.

Mata, L. J., Urrutia, J. J., Kronmal, R. A., and Joplin, C., 1975b, Survival and physical growth in infancy and early childhood. Study of birth weight and gestational age in a Guatemalan Indian village, *Am. J. Dis. Child.* **129:**561.

Mata, L. J., Kronmal, R. A., Garcia, B., Butler, W., Urrutia, J. J., and Murillo, S., 1976, Breast-feeding, weaning and diarrhoeal syndrome in a Guatemalan Indian village, in: *Acute Diarrhoea in Childhood* (K. Elliot and J. Knight, eds.), Ciba Foundation Symposium 42 (new series), p. 311, Elsevier/Excerpta Medica/North-Holland, Amsterdam.

Mata, L. J., Kronmal, R. A., Urrutia, J. J., and Garcia, B., 1977a, Effect of infection on food intake and nutritional status: Perspectives as viewed from the village, *Am. J. Clin. Nutr.* **30:**1215.

Mata, L. J., Urrutia, J. J., Serrato, G., Mohs, E., and Chin, T. D. Y., 1978b, Viral infections during pregnancy and in early life, *Am. J. Clin. Nutr.* **30:**1824.

Merson, M. H., Morris, G. K., Sack, D. A., Wells, J. G., Feeley, J. C., Sack, R. B., Creech, W. B., Kapikian, A. Z., and Gangarosa, E. J., 1976, Travelers' diarrhea in Mexico. A prospective study of physicians and family members attending a congress, *New Engl. J. Med.* **294:** 1299.

Metcoff, J., Darling, D. B., Scanlon, M. H., and Stare, F. J., 1948, Nutritional status and infection response. I. Electrophoretic Circulating plasma protein, hematologic, hematopoietic, and immunologic responses to *Salmonella typhimurium (Bacillus aetrycke)* infection in the protein-deficient rat, *J. Lab. Clin. Med.* **33:**47.

Middleton, P. J., Szymanski, M. T., Abbott, G. D., Bartolussi, R., and Hamilton, J. R., 1974, Orbivirus acute gastroenteritis of infancy, *Lancet* **1:**1241.

Morley, D., 1969, Severe measles in the tropics, *Br. Med. J.* **1:**297.

Morley, D., Woodland, M., and Martin, M. J., 1963, Measles in Nigerian children. A study of the disease in West Africa, and its manifestations in England and other countries during different epochs, *J. Hyg.* **61:**115.

Mugerwa, J. W., 1971, The lymphoreticular system in kwashiorkor, *J. Pathol.* **105:**105.

Nalder, B. N., Mahoney, A. W., Ramakrishnan, R., and Hendricks, D. G., 1972, Sensitivity of the immunological response to the nutritional status of rats, *J. Nutr.* **102:**535.

Neter, E., 1975, Enteropathogenicity of *Escherichia coli.* Occurrence in acute diarrhea of infants and children, *Am. J. Dis. Children* **129:**668.

Neufeld, H. A., Pace, J. A., and White, F. E., 1976, Effect of bacterial infections on ketone concentrations in rat liver and blood and/or free fatty acid concentrations in rat blood, *Metabolism* **25:**877.

Neumann, C. G., Lawler, G. J., Jr., Stiehm, E. R., Swendseid, M. E., Newton, C., Herbert, J., Ammann, A. J., and Jacob, M., 1975, Immunologic responses in malnourished children, *Am. J. Clin. Nutr.* **28:**89.

Newberne, P. M., Wilson, R. B., and Williams, G., 1970, Effects of severe and marginal maternal lipotrope deficiency on response of post-natal rats to infection, *Br. J. Exp. Pathol.* **51:**22.

Niswander, K. R., and Gordon, M., 1972, *Collaborative Perinatal Study. The Women and Their Pregnancies,* W. B. Saunders Company, Philadelphia.

Olson, J. A., 1972, The biological role of Vitamin A in maintaining epithelial tissue, *Isr. J. Med. Sci.* **8:**1170.

Passwell, J. H., Steward, M. W., Soothill, J. F., 1974, The effects of protein malnutrition on macrophage function and the amount and affinity of antibody response, *Clin. Exp. Immunol.* **17:**491.

Powanda, M. C., 1977, Changes in body balances of nitrogen and other key nutrients: Description and underlying mechanism, *Am. J. Clin. Nutr.* **30:**1254.

Pretorius, P. J., and deVilliers, L. S., 1962, Antibody response in children with protein malnutrition, *Am. J. Clin. Nutr.* **10:**379.

Rayfield, E. J., Curnow, R. T., George, D. T., and Beisel, W. R., 1973, Impaired carbohydrate metabolism during a mild viral illness, *New Engl. J. Med.* **289:**618.

Reddy, V., and Srikantia, S. G., 1964, Antibody response in kwashiorkor, *Ind. J. Med. Res.* **52:**1154.

Rocha, D. M., Santeusanio, F., Faloona, G. R., and Unger, R. H., 1973, Abnormal pancreatic alpha-cell function in bacterial infections, *New Engl. J. Med.* **288:**700.

Rodgers, G. C., and Neilands, J. B., 1973, Microbial iron transport compounds, in: *Handbook of Microbiology* (A. Laskin and H. Lechevalier, eds.), p. 823, CRC Press, Cleveland.

Rogers, W. E., Jr., Bieri, J. G., and McDaniel, E. G., 1971, Vitamin A deficiency in the germfree state, *Fed. Proc.* **30:**1773.

Rosen, E. U., Geefhuysen, J., and Ipp, I., 1971, Immunoglobulin levels in protein calorie malnutrition, *S. Afr. Med. J.* **45:**980.

Rosen, E. U., Geefhuysen, J., Anderson, R., Joffe, M., and Rabson, A. R., 1975, Leukocytes function in children with kwashiorkor, *Arch. Dis. Child.* **50:**220.

Roszkowski, I., Wojcicka, J., and Zaleska, K., 1966, Serum iron deficiency during the third trimester of pregnancy: maternal complications and fate of the neonate, *Obstet. Gynecol.* **28:**820.

Rowell, L. B., 1974, Human cardiovascular adjustments to exercise and thermal stress, *Physiol. Rev.* **54:**75.

Rudoy, R. C., and Nelson, J. D., 1975, Enteroinvasive and enterotoxigenic *Escherichia coli.* Occurrence in acute diarrhea of infants and children, *Am. J. Dis. Child.* **129:**668.

Ryan, N. T., Blackburn, G. L., and Clowes, G. H. A., Jr., 1974, Differential tissue sensitivity to elevated endogenous insulin levels during experimental peritonitis in rats, *Metabolism* **23:**1081.

Sack, R. B., 1975, Human diarrheal disease caused by entertoxigenic *Escherichia coli, Ann. Rev. Microbiol.* **29:**333.

Sack, R. B., Gorbach, S. L., Banwell, J. G., Jacobs, B., Chatterjee, B. D., and Nutra, R. C., 1971, Enterotoxigenic *Escherichia coli* isolated from patients with severe cholera-like disease, *J. Infect. Dis.* **123:**378.

Scheifele, D. W., and Forbes, C. E., 1972, Prolonged giant cell excretion in severe African measles, *Pediatrics* **50:**867.

Schlesinger, L., and Stekel, A., 1974, Impaired cellular immunity in marasmic infants, *Am. J. Clin. Nutr.* **27:**615.

Schonland, M., 1972, Depression of immunity in protein–calorie malnutrition: A post mortem study, *J. Trop. Pediatr.* **18:**217.

Schopfer, K., and Douglas, S. D., 1975, *In vitro* studies of lymphocytes from children with kwashiorkor, *Clin. Immunol. Immunopathol.* **5:**21.

Seth, V., and Chandra, R. K., 1972, Opsonic activity, phagocytosis, and bactericidal capacity of polymorphs in undernutrition, *Arch. Dis. Child.* **47:**282.

Shore, E. G., Dean, A. G., Holik, K. J., and Davis, B. R., 1974, Enterotoxin-producing *Escherichia coli* and diarrheal disease in adult travelers: A prospective study, *J. Infect. Dis.* **129:**577.
Sirisinha, S., Edelman, R., Suskind, R., Charupatana, C., and Olsen, R. E., 1973, Complement and C3-proactivator levels in children with protein-calorie malnutrition and effect of dietary therapy, *Lancet* **1:**1016.
Smythe, P. M., Schonland, M., Brereton-Stiles, G. G., Coovadia, H. J., Loening, W. E. K., Mafoyane, A., Parent, M. A., and Vos, G. H., 1971, Thymolymphatic deficiency and depression of cell-mediated immunity in protein-calorie malnutrition, *Lancet* **2:**939.
Stiehm, E. R., 1975, Fetal defense mechanisms, *Am. J. Dis. Child.* **129:**438.
Suskind, R., Edelman, R., Kulapongs, P., Pariyanonda, A., and Sirisinha, S., 1976a, Complement activity in children with protein-calorie malnutrition. *Am. J. Clin. Nutr.* **29:**1089.
Suskind, R., Sirisinha, S., Vithayasi, V., Edelman, R., Damrongsak, D., Charupatana, C., and Olsen, R. E., 1976b, Immunoglobulins and antibody response in children with protein-calorie malnutrition, *Am. J. Clin. Nutr.* **29:**836.
Sussman, M., 1974, Iron and infection, in: *Iron in Biochemistry and Medicine* (A. Jacobs and M. Worwood, eds.), p. 649, Academic Press, Inc., New York.
Tanphaichitr, P., Mekanandha, V., and Valyasevi, A., 1973, Impaired plasma opsonic activity in malnourished children, *J. Med. Assoc. Thai.* **56:**116.
Urrutia, J. J., Mata, L. J., Trent, F., Cruz, J. R., Villatoro, E., Alexander, R. E., 1975, Infection and low birth weight in a developing country. A study in an Indian village of Guatemala, *Am. J. Dis. Child.* **125:**558.
Viteri, F., and Torun, B., 1974, Anemia and work capacity, *Clin. Haematol.* **3:**609.
Walker, A. M., Garcia, R., Pate, P., Mata, L. J., and David, J. R., 1975, Transfer factor in the immune deficiency of protein-calorie malnutrition: a controlled study with 32 cases, *Cell Immunol.* **15:**372.
Wannemacher, R. W., Jr., 1977, Key role of various amino acids in host response to infection, *Am. J. Clin. Nutr.* **30:**1269.
Wannemacher, R. W., Jr., Pekarek, R. S., and Beisel, W. R., 1972, Mediator of hepatic amino acid flux in infected rats, *Proc. Soc. Exp. Biol. Med.* **139:**128.
Wannemacher, R. W., Jr., Pekarek, R. S., Thompson, W. L., Curnow, R. T., Beall, F. A., Zenser, T. V., Derubertis, F. R., and Beisel, W. R., 1975, A protein from polymorphonuclear leukocytes (LEM) which affects the rate of hepatic amino acid transport and synthesis of acute phase globulins, *Endocrinology* **96:**651.
Watts, T., 1969, Thymus weights in malnourished children, *J. Trop. Pediatr.* **15:**155.
WHO, 1975, Control of Nutritional Anaemia With Special Reference to Iron Deficiency, *WHO Tech. Rep. Ser. 580,* Geneva.
Wissler, R. W., 1947, The effects of protein-depletion and subsequent immunization upon the response of animals to pneumococcal infection, *J. Infect. Dis.* **80:**250.
Wohl, M. G., Reinhold, J. G., and Rose, S. B., 1949, Antibody response in patients with hypoproteinemia, *Arch. Int. Med.* **83:**402.
Wyatt, R. G., and Kapikian, A. Z., 1977, Viral agents associated with acute gastroenteritis in humans, *Am. J. Clin. Nutr.* **30:**1857.
Zilva, S. S., 1919, The influence of deficient nutrition on the production of agglutinins, complement and amboceptors, *Biochem. J.* **13:**172.

11

Nutrition in Dental Development and Disease

Juan M. Navia

1. Introduction

Nutrition has a major role in the development of dental tissues and is a primary contributor to the establishment of an oral environment compatible with dental health. Dietary factors, together with microbial ones, influence oral disease processes to a large extent, depending on quality, quantity, and frequency of foods consumed. While the effect of nutrition on dental development and oral environment is generally accepted, the specific effects of malnutrition on development of oral tissues and the true magnitude of this factor in human oral disease has not been sufficiently measured or investigated.

To oral tissues, nutrition is of special relevance, not only because of their own nutrient requirements, but because nutrients come in contact with these tissues twice: once when foods are masticated and ingested, thus contributing to the oral millieu, and second, after foods are digested and nutrients absorbed, these return to nourish the tissue via the circulatory system. Of these two effects of nutrients, the latter one is most critical for teeth before they erupt into the oral cavity, as this is the time when the dentition undergoes development and active mineralization. Local effects of nutrients are important later, after eruption of the tooth, when adequate nutrition can modify the oral disease challenge.

The nutrient composition of foods in the diet, therefore, can influence dental tissues at two distinct periods.

1. The *preeruptive period* spans the developmental time from the initial bell-shaped stages where the organic matrix of the tooth predominates, until the mineralized tooth is ready to break through the oral epithelium. During

Juan M. Navia • Institute of Dental Research, School of Dentistry, University of Alabama in Birmingham, Birmingham, Alabama.

this period the developing tooth must obtain all nutrients necessary for its formation and maturation from its trophic environment. Failure to obtain these nutrients at this time may affect the structure and chemical composition, as well as the eruption time, of the tooth.

2. The *posteruptive period* begins as the newly formed tooth emerges into the oral cavity. The maturation process of the enamel, which terminates when the enamel is fully calcified, continues for some time after the tooth is exposed to the oral environment. Nutrients in the diet, as well as mineral components of saliva, contribute to this mineralization process. Saliva also contributes to the formation of an organic film, the acquired pellicle, which may have protective as well as other as yet unknown functions for the enamel surface. Posteruptively, nutrients in the diet have a major influence on bacterial colonization on the different ecological niches in the oral cavity. The composition of foods and the frequency with which they are consumed will contribute to the selection and implantation of these bacteria, and stimulate the metabolic activity of the bacterial masses or plaque accumulating normally on the tooth enamel surface. While some of the components of plaque may be innocuous, some have definite pathogenic potential and give rise to the two most common bacterial plaque-dependent diseases: dental caries and periodontal disease.

Specifically, therefore, nutrients in the diet can influence oral disease in the following ways:

1. Changing the chemical environment of cells responsible for the formation of tissues, such as enamel, and changing preeruptively the tooth structure, composition, and its physiocochemical properties.
2. Influencing, either independently or together with hormones, the cellular enzyme systems involved in mineralization processes.
3. Altering protein–synthetic reactions and thus modifying the nature of the calcifying organic matrix of mineralized tissues.
4. Modifying the rate of flow, the quantity, or the physical, chemical, or immunological properties of saliva.
5. Enhancing or inhibiting the remineralization process taking place normally on the tooth surface of erupted teeth.
6. Influencing the multiplication, implantation, and metabolism of the plaque flora (Navia, 1976).

The scope of this chapter will be limited to a discussion of the understanding existing today about the effects of malnutrition on the developing tooth, particularly protein–calorie and vitamin A deficiencies, and how malnutrition affects tooth formation, eruption, and susceptibility to specific oral diseases such as dental caries. The discussion will therefore center primarily on dental preeruptive effects, although effects of malnutrition on other tissues, such as salivary glands, which condition and determine to a large extent the oral environment of the erupted tooth, will also be discussed.

2. Amelogenesis and Chronology of Human Teeth Eruption

Before the interrelationship between malnutrition and dental development is discussed, it is essential to review briefly some of the major events taking place during the formation of enamel, the dental tissue exposed to the initial caries attack. In addition, the normal chronology of tooth eruption will be presented.

Amelogenesis (Sicher, 1966) is characterized by two distinct processes: (a) the organic matrix formation and (b) its subsequent mineralization. Human enamel formation by ameloblasts has been studied by several investigators (Scott, 1972; Angmar-Mänsson, 1971a,b; Gwinnett, 1967; Swancar *et al.*, 1971; Nylen, 1967; Meckel *et al.*, 1965), who concluded that the mineralization process of this tissue differs from that of other hard tissues, such as bone. In amelogenesis, the ameloblasts begin to secrete the organic matrix in a thin layer along the dentin, shortly after dentin deposition has started. Mineralization begins to take place in the matrix segments and the interprismatic substance soon after this secretory activity has started, and it is characterized by the appearance of numerous long, thin crystals of apatitelike material (Frazier, 1968). This initial mineralization (Nylen, 1964) is quickly followed by a second stage or maturation, during which crystals increase in thickness (Fig. 1). This process spreads from the dentinoenamel junction toward the enamel surface, from the cuspal areas it extends to the fundus of the fissures, and also progresses cervically toward the cementoenamel junction. The advancing mineralization front is at first parallel to the dentinoenamel junction and later to the outer enamel surface, and therefore the incisal and occlusal regions reach maturity before the cervical areas of teeth. The maturation process continues for sometime after eruption. Understanding of this developmental process is clinically important because, during the maturation process, the hydroxyapatite crystals attain their maximum size. Depending on atomic size and chemical properties, various inorganic ions are either incorporated into the crystal lattice or into the hydration layer surrounding the apatite crystals, thus determining to a large extent the structure and chemical organization of enamel (Miles, 1967). The oral trophic conditions influence the end result of this process either to form a mature enamel surface, high in density and low in reactivity, or enamel that is immature, highly reactive, and capable of allowing diffusion of various ionic and molecular species.

If nutritional insults are imposed early in the formation of the organic mixture, the clinical expression may be hypoplasia of enamel, characterized by pitting, furrowing, or absence of enamel. If the stress is imposed later, when the maturation process is the primary activity, the result may be a hypocalcification, manifested by opaque or chalky areas, surrounded by normal-looking enamel (Pindborg, 1970). Both hypoplasia and hypocalcification can be induced by other environmental stresses (Molnar and Ward, 1974) such as febrile episodes, chemical intoxications, and genetic defects. Mineralization defects of the primary teeth from children who have experienced intrauterine

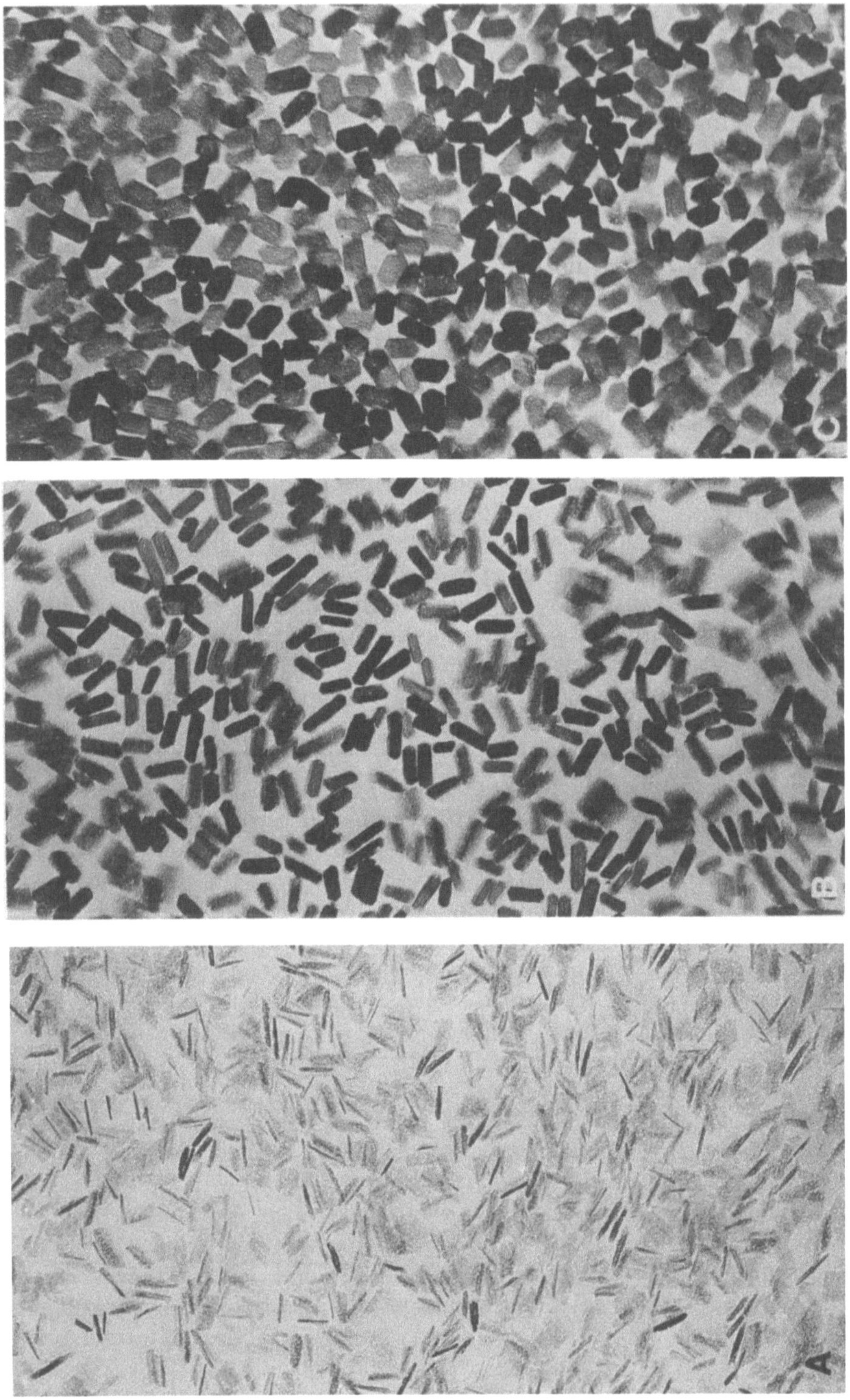

Fig. 1. Sequential stages in enamel crystal growth. Note the increase in both thickness of the cross-cut crystals and the concomitant reduction in the intercrystalline spaces. A, initial crystals deposited; B and C, more advanced stages of enamel mineralization. × 160,000. From Nylen (1964).

undernutrition were investigated by Grahnén *et al.* (1972). The material studied consisted of 52 "small-for-date" children, 26 dysmature children, and 56 normal controls. The children's average age at the time of the dental examination was 28 months. Defects such as external and internal enamel hypoplasia (opacities) and internal pigmentation were recorded and found to be equal for these three groups. These results suggested that intrauterine undernutrition does not always seem to play a definite role in the etiology of enamel hypoplasia of primary teeth, either because matrix formation and mineralization has not yet started or because other stresses may also cause mineralization defects.

Enamel represents for the tooth what the skin is to the body. Both tissues are ectodermal in origin, and each serves as a protective barrier between the systemically controlled internal environment and the variable and potentially pathologic environment. Like the skin, enamel has a bacterial load which may harbor pathogens capable of disrupting the integrity of this tissue, thus penetrating the underlying tissues. For these reasons, a sound enamel surface is critical, as it will determine to a large extent whether the tooth will be healthy or carious. Adequate nutrition can allow normal enamel formation to take place, while malnutrition will interfere with normal amelogenesis. While malnutrition is undesirable for all tissues, it is even worse for enamel. This tissue, once the tooth erupts into the oral cavity, has no cellular mechanism to repair whatever developmental damage has taken place, and therefore the lesion is to a large extent irreversible. It is true that the enamel surface is constantly exposed to the reparative effects of saliva and its components, but this can be interfered with and completely neutralized by the destructive activity of bacterial plaque.

Inspection of Table I, where the chronology of human dental development is tabulated (Sognnaes *et al.,* 1966; Lunt and Law, 1974a,b) indicates that the various surfaces for different teeth in the primary or the permanent dentition are at risk at different times. Enamel matrix formation begins for primary, maxillary central incisors a few months after conception. For permanent third molars, the last teeth formed, enamel matrix formation starts 8–10 years later, and therefore the first 10–14 years of age encompass critical periods for different teeth whose enamel is being formed at various times.

It is important to notice that there is an extended period of time between the completion of enamel formation and the time of eruption when the enamel can undergo extensive changes in the mineralization of the surface layer, and acquire elements such as fluoride, which will contribute to the formation of large, stable crystals of hydroxyapatite-containing fluoride on selected sites on the crystal surface (Brown and König, 1977). Some of these trace elements will tend to enhance maturation of the enamel and thus reduce the caries susceptibility of the tooth after eruption. Further research is necessary to understand the mechanisms of preeruptive maturation of enamel in humans, as well as in experimental animals, such as the rat, which have, in contrast to the long human amelogenesis, a 20-day period between beginning of amelogenesis and eruption of the tooth into the oral cavity.

Table I. Chronology of Human Dentition[a]

Tooth	Formation of enamel matrix and dentine begins[b]	Amount of enamel matrix formed at birth	Enamel completed	Eruption (range for age containing ±1 S.D.)	Root completed (yr)
Primary dentition					
Maxillary					
Central incisor	13–16 wk	Five-sixths	1½ mo.	8–12 mo	1½
Lateral incisor	14⅔–16½ wk	Two-thirds	2½ mo	9–13 mo	2
Canine	15–18 wk	One-third	9 mo	16–22 mo	3¼
First molar	14½–17 wk	Cusps united[c]	6 mo	13–19 mo	2½
Second molar	16–23½ wk	Cusp tips still isolated[d]	11 mo	25–33 mo	3
Mandibular					
Central incisor	13–16 wk	Three-fifths	2½ mo	6–10 mo	1½
Lateral incisor	14–18 wk	Three-fifths	3 mo	10–16 mo	1½
Canine	16–20 wk	One-third	9 mo	17–23 mo	3¼
First molar	14½–17 wk	Cusps united[e]	5½ mo	14–18 mo	2¼
Second molar	17–19½ wk	Cusp tips still isolated	10 mo	23–31 mo	3
Permanent dentition					
Maxillary					
Central incisor	3–4 mo	—	4–5 yr	7–8 yr	10

Lateral incisor	10–12 mo	—	4–5 yr	8–9 yr	11
Canine	4–5 mo	—	6–7 yr	11–12 yr	13–15
First premolar	$1\frac{1}{2}$–$1\frac{3}{4}$	—	5–6 yr	10–11 yr	12–13
Second premolar	2–$2\frac{1}{4}$ yr	—	6–7 yr	10–12 yr	12–14
First molar	At birth	Sometimes a trace	$2\frac{1}{2}$–3 yr	6–7 yr	9–10
Second molar	$2\frac{1}{2}$–3 yr	—	7–8 yr	12–13 yr	14–16
Third molar	7–9 yr	—	12–16 yr	17–21 yr	18–25
Mandibular					
Central incisor	3–4 mo	—	4–5 yr	6–7 yr	9
Lateral incisor	3–4 mo	—	4–5 yr	7–8 yr	10
Canine	4–5 mo	—	6–7 yr	9–10 yr	12–14
First premolar	$1\frac{3}{4}$–2 yr	—	5–6 yr	10–12 yr	12–13
Second premolar	$2\frac{1}{4}$–$2\frac{1}{2}$ yr	—	6–7 yr	11–12 yr	13–14
First molar	At birth	Sometimes a trace	$2\frac{1}{2}$–3 yr	6–7 yr	9–10
Second molar	$2\frac{1}{2}$–3 yr	—	7–8 yr	11–13 yr	14–15
Third molar	8–10 yr	—	12–16 yr	17–21 yr	18–25

[a] Lunt and Law (1974).
[b] Primary dentition = age expressed from fertilization time *in utero*. Permanent dentition = age expressed from birth.
[c] Occlusal completely calcified plus one-half to three-fourths crown height.
[d] Occlusal incompletely calcified; calcified tissue covers one-fifth to one-fourth crown height.
[e] Occlusal completely calcified.

3. Dental Dysplasias and Malnutrition

3.1. Epidemiology of Dental Dysplasias

Several vitamin deficiencies, especially vitamins A and D, have been implicated together with protein–calorie deficiencies in dental abnormalities in malnourished populations. For a long time it has been observed that children in underpriviledged groups present dental dysplasias which have been ascribed to inadequate nutrition during tooth development. Baume and Meyer (1966) studied the dental condition and nutritional practices of several thousand school children in French Polynesia and described two types of dental dysplasias possibly related to malnutrition: an odontoclasia in the deciduous dentition and a "yellow teeth" condition seen in permanent teeth. Both defects were highly susceptible to rampant decay. These authors described a third dysplastic condition, referred to as "infantile melanodontia," which was occasionally described in the deciduous teeth of Chinese children. Other alterations, such as mottled enamel, were observed in children living in distant islands, who seemed to be caries-resistant. Baume (1968, 1969) made an extensive study of these epidemiological findings and suggested that to some extent these enamel alterations were related to malnutrition experienced early in life when the teeth were undergoing active mineralization. The enamel of these carious "yellow teeth" contained half the fluoride concentrations of the slightly mottled teeth from caries-resistant counterparts (Baume and Vulliémoz, 1970). Microradiograph studies (Baume and Vulliémoz, 1972) using teeth from malnourished individuals showed extensive areas of hypomineralized enamel, which they referred to as "developmental hypomineralization."

In many parts of the world other defects have been observed. Sweeney *et al.* (1971) described in Guatemalan children a linear hypoplasia of the deciduous incisor teeth. The timing of occurrence for these dysplasias seems to correspond to the birth or postnatal period (Fig. 2). This defect is similar to the one described by Nicholls *et al.* (1961) in Asiatic underprivileged children, the condition referred to as "bar decay." Jelliffe and Jelliffe (1971) have called attention to the fact that in rural Jamaica (Jelliffe *et al.*, 1954) and also in Haiti (Jelliffe and Jelliffe, 1961) it is frequent to find in 3- to 6-year-olds a characteristic "carved-out" erosion of deciduous incisors affecting adjacent sides, particularly in the maxillary teeth. The same defect, with even a higher prevalence (31%), was reported in 1- to 4-year-old Cuna Indians in the San Blas Islands off the Caribbean coast of Panama (Jelliffe *et al.*, 1961).

From these epidemiological studies it is apparent that a linear hypoplasia resulting in increased susceptibility to dental decay (provided the cariogenic challenge is present) is a common occurrence in young children in areas where malnutrition is rampant. The exact etiological factor for these lesions is difficult to ascertain; however, epidemiological data suggest that malnutrition, or hypovitaminosis A synergistically combined with infections (Sweeney and Guzmám, 1966; Sweeney *et al.*, 1969), has a profound influence on the development of teeth.

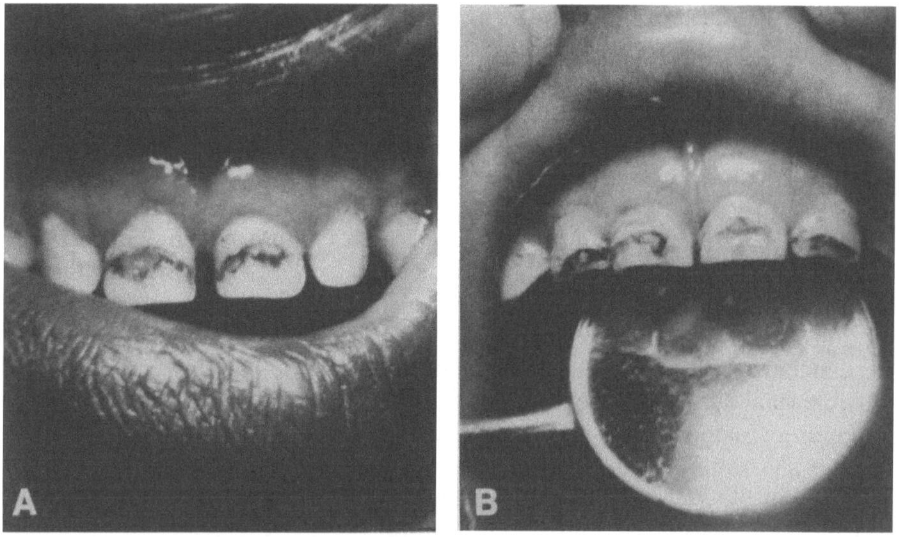

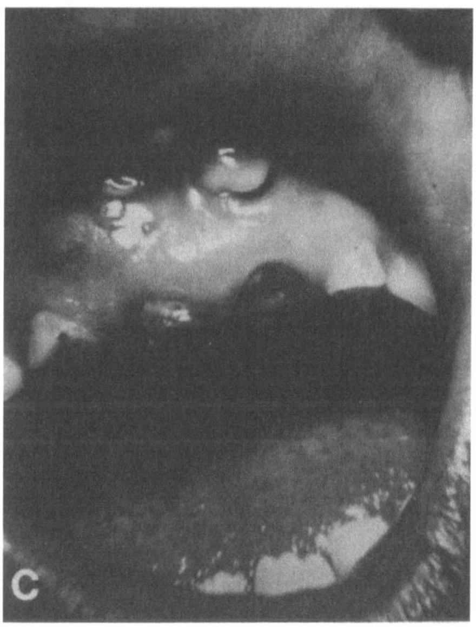

Fig. 2. Linear hypoplasia of the deciduous incisor teeth of Guatemalan children. From Sweeney *et al.* (1969).

3.2. Experimental Studies Related to Vitamin A Deficiency and Dental Development

Vitamin A is considered to be one of the most frequently diagnosed vitamin deficiencies among preschool children in both technically underdeveloped countries and in the United States (Ten-State Nutrition Survey, 1968-1970; HANES, 1974). The effects of such deficiencies imposed early in life on the subsequent development and function of the organs affected is largely unknown in the case of vitamin A. The biochemical role of vitamin A in organs such as the eye has been elucidated, but its role in bone and tooth formation is yet to be defined. However, because vitamin A deficiency is widespread among children in many parts of the world and because of its importance for ectodermal tissues, it is of interest to review some of the experimental studies performed to study its effect on tooth development.

More than 50 years ago Wolbach and Howe (1925) reported the histopathologic changes observed in young rats fed a vitamin-A-deficient diet consisting of casein, starch, McCollum's salt mixture, lard, and Brewer's yeast. Gross changes in teeth were not noted during the experimental period, but eruption rate seemed to be slower and microscopic changes were reported. These included irregular down growths of odontoblasts which were associated with abnormal deposits of dentin. Another interesting finding was the presence of islands of osteoid tissue, incorporated and surrounded by osteoblasts in the pulp. As the deficiency became more severe, enamel formation ceased. The ameloblasts were either shrunken and atrophic or replaced by a narrow layer of stratified, nonkeratinized epithelium.

One of the first investigators to point out the possible relation between several nutritional deficiencies and the development of structural defects in teeth, which may alter their susceptibility to diseases such as dental caries, was Mellanby (1928). She used young dogs which were fed a variety of diets deficient in one or more nutrients to produce all grades of defects in teeth from pigmentation of enamel to induction of interglobular spaces in highly irregular and disorganized dentin. Most of the work was related to vitamin D and calcium/phosphorus intake and has been reviewed by Shaw and Sweeney (1973). Their results and those of other earlier investigators, such as McCollum and Toverud, suggest that changes in the dentition of animals deficient in vitamin A may lead to increased susceptibility to caries.

The concept of the influence of vitamin A deficiency on tooth development (Mellanby, 1941) and caries susceptibility was not readily accepted at that time. Bloch (1931) evaluated the dentition of 45 blind children who had developed xerophthalmia during the first year of life. He could not detect a clear relation between vitamin A deficiency and caries. The sample was limited and the understanding of the carious process was quite incomplete, so no valid conclusions could be drawn from such early research attempts.

Wolbach and Howe (1933) reported on the effects of vitamin A deficiency on the incisor teeth of albino rats and guinea pigs. These teeth are continuously erupting and therefore subject at any stage of development to the influence of dietary levels of vitamin A. In the incisor, the initial effect of vitamin A deficiency was found to be on the enamel organ. The ameloblasts responded earliest by atrophy, followed by atrophy and depolarization of odontoblasts. These dentin-forming cells were reported to survive longest on the side where they are in apposition to the enamel organ (labial). The final stages of this nutritionally induced syndrome in teeth involved metaplasia, calcification, and ossification of the pulp, especially in the guinea pig.

Some of these changes were observed in some human cases reported at the time. Boyle (1933) reported the histological finding in the tooth germ of an undernourished 3½-month-old infant who presented atrophy of the enamel organ. The ameloblasts and the stellate reticulum had been replaced by a nonkeratinized layer of squamous epithelium. The dentin was abnormally calcified and the predentin was excessively wide and showed capillary and cell inclusions (Boyle and Bessey, 1941). Dinnerman (1933), working with postmortem material from five infants aged 3–7 months, also observed hypoplasia of enamel and poorly calcified dentin, but a clear relationship between these defects and vitamin A deficiency could not be conclusively established.

The classic work of Wolbach and Howe (1925, 1933) and some of the other reports relating case histories in infants stimulated other noted investigators, such as Schour *et al.* (1941), to evaluate further the changes in the incisor teeth of albino rats brought about by a vitamin A deficiency. Rats were placed at weaning (21 days) on a diet deficient in vitamin A for a period of 9–81 days. The rats were considered to be depleted when their body weight became stationary and they showed early signs of xerophthalmia (21–26 days). Some of the rats were assigned to a group receiving replacement therapy consisting of additions of definite amounts of alfalfa or cod liver oil.

In general, their results agreed with those of Wolbach and Howe (1933). The primary effect of vitamin A deficiency detected by Schour and co-workers seemed to be on the histodifferentiation of the odontogenic epithelium, which gives rise to severe morphologic and functional alterations. Concomitant with this lack of differentiation, there was a continuation of the proliferative activity of the odontogenic epithelium, characterized by an invasion of the pulp by epithelial cords. The end result of these disturbances of the cellular organization was a distinct distortion of the morphologic outline of the teeth. Replacement therapy resulted in the reestablishment of the rate of dentine apposition and the quick differentiation of the peripheral pulpal cell into odontoblasts.

Mellanby (1941a,b), aware of those studies where vitamin A deficiency had clearly affected the growing incisor and conscious of the fact that nutritional deficiencies could be imposed through the mother to the offspring, decided to study such early effects of vitamin A deficiency. Mellanby used a natural-type diet which was marginal in vitamin A. After feeding this diet for some time (12–34 weeks), deficient female rats were bred with stock males. The young born of these matings were killed at various ages, depending on their nutritional status. Dams which had pups after being fed the experimental diet for 12–13 weeks produced offspring with normal dentition, notwithstanding that their liver vitamin A store was practically depleted. If the feeding period was longer (15–19 weeks), the litter presented some minor tooth alterations. If the deficient diet was fed even longer (24–25 weeks), the pups presented gross abnormalities in the shape of incisors and retardation in their development. It was then clear that irreversible alterations in tooth histology and structure in the young could be induced by manipulating the nutritional status of the rat mother.

Similar results were obtained by maintaining the rats under a mild but chronic vitamin A deficiency for periods up to 1 year (Burn *et al.*, 1941). Incisor teeth presented loss of pigmentation, ridging, constrictions, and misalignment. Focal areas of degeneration were observed among ameloblasts, and lingual odontoblasts became atrophic and finally disappeared. An interesting finding was the development of large odontomas in older animals, half of which also developed tumors in both jaws. The odontomas were described as being of embryonic-type connective tissue, similar in appearance to that of the pulp tissue.

In spite of the work done up to that time in experimental animals, as well as using postmortem tissues from humans, enamel hypoplasia in humans could not be principally associated with vitamin A deficiency and therefore the effects of vitamin A malnutrition remained essentially limited to the oral epithelium, where the deficiency may result in hyperkeratotic changes of the oral mucosa. The relationship between caries and vitamin A deficiency was still obscure and the only suggestion mentioned (Schour and Massler, 1945) was the possibility that salivary lysozyme might be lowered under conditions of vitamin A deficiency and thus might result in a loss of bacteriostatic action.

Mellanby (1947) suggested that vitamin A was essential to osteoblasts and osteoclasts to be able to remodel bone in an orderly manner, and Irving (1949) investigated the effects of avitaminosis A upon the alveolar bone and upper incisor teeth of rats. The odontoblasts were observed to produce an excessive amount of abnormal dentin on the labial side of the incisor and occasionally osteoid masses in the pulp. The changes seen in this study were principally in the dentin and odonblasts, which seem to be retarded rather than accelerated.

Not only vitamin A deficiency, but other manipulations, including irradiation, high fever, or alloxan administration, arrested the activity of ameloblasts and produced macroscopic and microscopic abnormalities in the enamel. Paynter and Grainger (1956) decided to examine macroscopic changes which might be produced in teeth through nutritional manipulations imposed during tooth development. Analysis of measurements of the upper first molar of weanling rats, born and nursed by rat dams whose diet was deficient in vitamin A, resulted in small teeth which could be related to a concomitant smaller body size in the animals.

Realizing that vitamin A deficiency not only affected teeth but also salivary glands, where squamous metaplasia of the epithelial components is produced by the deficiency, Salley *et al.* (1959) conducted a study with the premise that salivary gland degeneration due to vitamin A deficiency changed the oral conditions enough to increase the caries susceptibility of the molars. Weanling albino hamsters were assigned to an experimental group receiving a synthetic, vitamin-A-free diet, and a control group receiving the same diet, but supplemented with an oral dose of 250 IU of vitamin A ester weekly. Because the experiment lasted for an extremely long time (233 days for the depleted group and 366 days for the control), maintenance doses of 5 IU of vitamin A were given orally to hamsters in the experimental group. Results for this experiment showed that hamsters in the deficient group had a total of three times more tooth structure destroyed by caries than those in the control group. These investigators also detected major abnormalities in salivary glands, particularly complete disappearance of secretory acini, replacement of gland parenchyma with infiltrates of acute and chronic inflammatory cells and squamous metaplasia of ductal epithelium, and the presence of large keratotic plugs in the large collecting ducts. There is a strong possibility, therefore, that these salivary gland changes were responsible for the increased caries susceptibility of deficient animals that were not due to enamel changes. The question of the effect on caries susceptibility of developmental enamel defects due to vitamin A deficiency would have to be answered in other types of experimental designs where the nutritional insult is imposed during development and the caries challenge is instituted posteruptively under adequate vitamin A status.

Recent studies related to vitamin A and dental tissues have tried to incorporate new methodology and approaches into experiments designed to answer some of the questions asked by earlier investigators, but little progress has been made in understanding the biochemical changes taking place in the composition of enamel and dentin that are associated with the well-described morphological alterations. Very few human studies have been undertaken to try to describe the relationship of possible alterations induced in teeth during development on caries susceptibility. Baume *et al.* (1972) have investigated some aspects of this problem, as they realize that vitamin A deficiency, a widespread problem around the world, may give rise to enamel defects that may be unable to withstand the cariogenic stress of the dietary habits in affluent societies. They were particularly interested in the effects on the offspring mediated by depriving the mother. Their results suggest that dental development in the newborn is to a large extent dependent on a sufficient maternal supply of vitamin A. Further studies are necessary to establish and describe possible relationships between the irreversible enamel alterations brought about by vitamin A deficiency and caries. Research is needed to evaluate whether malnutrition imposed during tooth development constitutes a weakening factor contributing to the increased susceptibility to caries seen in malnourished populations changing to new life-styles and food habits (Baume and Meyer, 1966). Answers to some of these questions should help to elucidate the role of nutritional deficiencies imposed during development on the integrity of these dental structures and their susceptibility to diseases such as dental caries.

3.3. Fluoride, Dental Development, and Caries

Although the evidence supporting the nutritional essentiality of fluoride is not yet sufficiently clear, the role of this element in dental development and health has to be stressed. Fluoride administration is unquestionably the single most significant deterrent of dental caries. Its provision in water supplies at a level around 1 ppm has been found to be effective in reducing caries of people who drink this fluoridated water (Arnold *et al.*, 1962; Ast *et al.*, 1956). Other forms of fluoride delivery and application have been developed, which have also been shown to be effective cariostatic procedures. Provision of fluoride at levels about 1–2 ppm in the drinking water may give rise to an enamel defect commonly referred as fluorosis or mottling, characterized by hypoplasia and discoloration of tooth enamel. This enamel defect is commonly seen in children from many parts of the world, such as certain regions in India and Africa where the drinking water may contain 8–12 ppm of fluoride. The mechanism of action for this cariostatic element (Brown and König, 1977), provided at optimal levels, is still under active investigation. Its exact mode of action, optimal time–dose relationship, and interactions with other mineral elements and nutrients have not been sufficiently clarified. Some of the mechanisms suggested to explain its cariostatic properties include: (1) alteration of tooth crown morphology; (2) formation of large, perfect crystals of apatite in enamel; (3) stimulation of remineralization processes at the enamel surface; (4) decreased solubility of enamel; and (5) decreased bacterial enzymatic activity in dental plaque. It is possible that more than one of these mechanisms is responsible for the cariostatic effect of fluoride. Identification of which ones are most important would help in determining the optimal level and timing for fluoride administration.

Fluoride metabolism has been reviewed in monographs published by WHO (1970), the National Academy of Sciences (1971), and in publications by Hodge and Smith (1970), Jenkins *et al.* (1970), and Underwood (1971). Fluoride from inorganic sources is rapidly absorbed and distributed throughout the body. It accumulates in bone and tooth apatite (Eanes and Posner, 1970), where it substitutes for hydroxyl groups at selected sites in the apatite lattice. Body fluids such as saliva, bile, and blood contain fluoride concentrations ranging from 0.1 to 0.2 ppm. These concentrations are influenced by fluoride ingestion from diet and drinking water. Mammary glands and placenta seem to act as barriers for the fluoride ion, so the fetus and the suckling child are protected from an excessive maternal fluoride intake.

Carlos *et al.* (1962) and Horowitz and Heifetz (1967) studied several cohorts of children who had different patterns of prenatal exposure followed by continuous exposure to fluoridated water. Their results indicated that there was no significant additional benefits from maternal ingestion of fluoridated water if the offspring also was provided with fluoridated water from birth.

Because development of the primary and permanent dentitions of humans takes place over a long period of time (Table I), the question of determining the critical dose and period of time of fluoride administration that would avoid mottling of teeth and yet facilitate the most beneficial cariostatic effect of fluoride is fundamental. The previously mentioned studies of Carlos *et al.* (1962) and Horowitz and Heifetz (1967) suggest that administration of fluoride prenatally does not confer special advantage to be deciduous dentition. Epidemiological studies (Arnold *et al.*, 1962) have shown that fluoride administration for long periods of time to erupted teeth is beneficial in the control of caries. It remains to be investigated and clearly determined whether provision of fluoride at some specific dose from birth to the appearance of the first molar (around 6 years of age) is also desirable.

Margolis *et al.* (1975) conducted a 10-year longitudinal prospective study to compare the effect of fluoride on dentitions of children from infancy through 10 years of age. In Kalamazoo, Michigan (fluoridated), and Oneida, New York (nonfluoridated), parallel groups of children were given a fluoride–vitamin supplement either from infancy or from age 4 years and compared with children consuming fluoridated or nonfluoridated water. Incidence of new caries activity in both deciduous and permanent teeth was measured by determining mean number of new decayed or filled teeth. Prevalence of caries was also studied in 6-year molars. Their results seemed to indicate a reduction of caries activity for deciduous teeth when fluoride was provided from infancy. The prevalence of decayed and filled permanent 6-year molars indicated that provision of fluoride–vitamin supplements from infancy to children in both communities, compared to controls not receiving fluoride, was significantly beneficial. Provision of fluoride–vitamin supplement from 4 years of age was found to be also significantly cariostatic for Oneida children compared to controls receiving no fluoride. The number of children was small in this particular study of the permanent dentition, and therefore other comparisons did not show statistical significance. The question of how much fluoride and when to start its administration needs further clinical research.

Navia *et al.* (1976) studied experimentally in rats the caries effect of providing fluoride preeruptively and posteruptively. When the fluoride was offered posteruptively there was an increase in fluoride level in the molars, no signs of mottling, and a significant decrease (20–54%) in caries in all molar surfaces. However, preeruptive administration of fluoride solutions by gavage to suckling rat pups, at doses ranging from 0 to 105 μg of F/10 g of body weight did not seem to confer caries protection. It was interesting to find that in rats assigned to groups given the higher fluoride treatments, there were hypoplastic changes in the appearance of the enamel of molars and an increase in susceptibility to caries.

It is recognized that the mineralization of the rat pup molars is an extremely fast process which takes place in approximately 3 weeks, while in the human it may take several years. Such differences preclude direct extrapolation of these experimental results to the human situation. In view of the scarcity of data related to the actual fluoride intake of infants and their caries experience in subsequent years, it is important to conduct clinical studies to determine the fluoride dose that will provide caries protection without causing mottling and thus compromising dental health.

4. Protein–Calorie Malnutrition and Dental Caries

4.1. Epidemiology of Malnutrition and Dental Caries

Epidemiological studies to determine the pattern of distribution of dental caries do not seem to correspond always with that of malnutrition. In view of what was discussed in the previous sections, this may seem to be a contradiction or a negation of the role of nutrition on dental health. To understand this seemingly complicated paradox it is necessary to discuss some of the etiological characteristics of dental caries which may help in understanding these differences in distribution.

Russell (1963), and recent data compiled by WHO (Barmes, 1977), show that caries is widespread, although its prevalence varies in different parts of the world. The variation is quite large and it is possible to find clusters of people with extremely low levels of caries and just a few miles away another group of people presenting extensive dental decay (Schamschula *et al.*, 1972; Barmes *et al.*, 1970). Notwithstanding this type of variation, attempts have been made (Russell, 1963) to group populations living in different nations into areas of high, intermediate, and low prevalence of caries (Table II). Interestingly enough, populations that are reported to be poorly nourished seemed to have a low level of caries and conversely people living in areas where overt malnutrition is rare seem to have high caries prevalence and severity.

Efforts to correlate the factors responsible for this disease with the patterns and variations in severity of the disease are difficult and frustrating because of the multiple etiological factors (related to diet, bacteria, and host) that affect caries and the long time it takes for the disease to develop. One factor that stimulates caries prevalence in a population is the abundant and frequent consumption of fermentable carbohydrates (Nizel, 1973). An important protective factor would be the availability of fluoride in the drinking water. These two factors, among others, make interpretation of the effects of malnutrition during tooth formation difficult to define and quantitate. Comparison between the nutritional status of a group of people at a specific time when a nutritional survey is conducted with the accumulated lifetime number of decayed, missing, and filled (DMF) teeth has not clearly yielded evidence that optimum nutrition decreases the susceptibility to caries. Even though in

Table II. Populations Living in Regions with Relatively High, Intermediate, and Relatively Low Prevalences of Dental Caries[a]

Relatively high	Intermediate	Relatively low
Most populations in:	Fluoride areas in:	Ethiopia
North America	United States	
South America	Ecuador	
Europe	Boliva	Burma, Thailand,
New Zealand	Greece	Viet Nam, India
Australia	Israel	China, Taiwan
Tahiti	Lebanon	
Hawaii	Egypt	
		Remote areas of
		Alaska
Urban Alaska	Malaya	
	Indonesia	Jordan
Trinidad		
		New Guinea

[a] From Russell (1963).

Alaska, Ethiopia, Viet Nam, and Thailand nutritional deficiencies in vitamin A, thiamine, and riboflavin have been documented, the people in these countries exhibit low caries levels (Russell, 1963).

Such epidemiological data, and the one generated by the studies (Sognnaes, 1948) on wartime reduction in dental caries in European children, should be carefully and cautiously interpreted in the light of new understanding of the disease, its etiological factors, and interrelationships with dietary habits. The consumption of fermentable carbohydrate at frequent intervals is a powerful stimulator of the implantation, colonization, and metabolic activity of cariogenic bacteria in tooth plaque. It can be shown experimentally in rats that some cariostatic treatments and manipulations can be effectively neutralized by feeding *ad libitum* a diet containing 67% sucrose from the time of tooth eruption, which in the rat starts to take place approximately at 16–17 days of age (Menaker and Navia, 1973c). The intensity of the caries-promoting potential of the high-sugar diet fed under these *ad libitum* conditions can overwhelm the beneficial effects of even adequate nutrition during tooth development. Conversely, there might be protective factors in certain foods which may retard caries and contribute to the maintenance of oral health (Nizel, 1973).

In this context, two aspects should also be considered. One is the specific and characteristic distribution of food within a family unit, depending on sex, economic status, and rural or urban life-style. Depending on these factors, children in the family may be exposed to varying influences, resulting in different nutritional status at a time when tooth development is most active. Therefore, the true contribution of nutritional status early in development to oral diseases will be difficult to assess. Another important aspect relates to the previously mentioned frequency of consumption of snack foods containing fermentable carbohydrates. Malnourished populations not only consume foods with low nutritional quality, but they also eat less frequently than those living in technically developed societies. Therefore, a person working in a rural environment in a technically developing country may have only one or two meals a day, supplemented by consumption of beverages such as coffee, while in an affluent urban economy, a person may have two or three meals plus a series of snacks, thus nibbling six or more extra times during the day. It is therefore obvious that while overall nutritional requirements may be better fulfilled in the latter situation, the challenge to oral health is greater, and the end result is that these individuals have a higher prevalence of dental caries than their undernourished counterparts in the rural community.

Overall, in human populations, nutritional factors probably have a greater importance on dental caries prevention than can be surmised from experimental studies. These studies are usually conducted using diets fed *ad libitum* and containing high levels of fermentable sugars, and with animals inoculated with highly virulent strains of cariogenic organisms. Under these experimental situations, studies on the influence of malnutrition usually yield limited information, for the severity of the cariogenic challenge overwhelms and confounds whatever beneficial effects might be contributed by an adequate nutrition provided during oral tissue development.

Fluoride availability is another factor which has been clearly shown to interfere with the course of the disease, and its availability in certain regions also confounds the interpretation of epidemiological studies. Water drinking patterns and proportions of fluids such as tea, milk, water, or fruit juices consumed by people around the world vary enormously. Whereas in some areas young children are weaned early and start to consume water, fruit juices, or even beverages such as tea with a high fluoride content, in other parts of the world the consumption of milk is continued for long periods of time to the practical exclusion of water. It is well known that while the nutritional value of milk is excellent, a child raised on water containing fluoride, or tea, may obtain the benefit of a higher fluoride intake, while a child who is relatively better nourished on milk will have a lower fluoride intake. A child raised on milk may be better nourished overall, but the protective value of fluoride on caries incidence will be absent. Therefore, total fluoride intakes of diet and drinking water in different areas should be determined to evaluate the contribution of this parameter to the disease picture.

There is a strong probability that adequate nutrition will contribute to the prevention of the disease in a large proportion of the world population. Nutrition, however, will not influence the caries experience of the two extreme portions of the population who either have a very low cariogenic challenge (and therefore will have low levels of caries) or the extreme situation of people who are exposed to a highly virulent challenge. Between these two extremes there will be a large number of people who will benefit from having adequate nutrition to allow for normal development of teeth, particularly enamel.

We hypothesize that there is an inverse relationship between good nutrition during oral tissue development and caries susceptibility. Furthermore, we feel that in the case of the tooth, just as with many other tissues, there is a critical time when the necessary nutrients have to be available in their trophic environment. This will enable formation of the kind of enamel that would be able to withstand better the pathological stresses of the life-style chosen by people living in technically developed societies.

Undernutrition imposed early in life can affect the following:

1 The development of the tooth, particularly enamel.
2. The secretion of saliva, particularly its flow and composition.
3. The secretory antibody formation and secretion by salivary glands into the oral cavity.

Changes in any of these factors can determine an increase in the susceptibility to oral diseases such as dental caries, and need to be further understood to evaluate their contribution to the overall disease process.

4.2. Experimental Studies Related to Protein Malnutrition and Caries

Epidemiological studies are useful in identifying those factors in a population that may have a significant correlation with certain diseases, and therefore help in identification and understanding of variables to be studied under controlled experimental conditions. In some cases such laboratory studies yield information that enables epidemiologists to return to field studies and reevaluate the data under a new hypothesis. The study of the preeruptive effects of protein malnutrition on caries susceptibility is an example of the latter.

Holloway *et al.* (1961) conducted an investigation on the various effects of different dietary sucrose/casein ratios on the teeth and supporting structures of rats. Purified diets containing varying sucrose/casein ratios (i.e., 67:16, 19:64; 83:0) were fed to female rats for a stabilizing period of 28 days and then through pregnancy and lactation. Offspring of females fed the high sucrose/low protein diet grew more slowly and their molar teeth erupted later than rats born to dams fed a control diet with adequate protein content. Molars of the offsprings from protein-deficient dams were significantly smaller than those of the controls, mainly due to a reduction in the distance between the outer borders of the dentin. Many of the third molars from the malnourished rats had missing cusps. Similar findings were reported by Shaw and Griffiths (1963) when they fed rats a low-protein diet during reproduction. These findings were studied in more detail by Shaw (1963, 1969, 1972), who investigated the influence of diets with various sucrose/protein ratios fed at different times during the perinatal period on third molar structure and eruption. He reported that marginal protein deficiency in female rats during pregnancy and lactation resulted in reduced weaning weights, delayed eruption, and abnormal cuspal patterns of third molars. When a complete protein deficiency was imposed for 5 days at different times during lactation, third-molar eruption was delayed proportionally to how late the protein deficiency was imposed during the lactation period.

These and other studies reviewed by Larson (1964) and Shaw (1970) clearly suggested that changes in the structure, enamel morphology, and eruption of teeth could take place when nutritional insults were imposed prior to the emergence of the tooth into the oral cavity. However, many experimental studies on the effects of malnutrition and undernutrition on tooth development and caries susceptibility were inconclusive because: natural diets of unknown nutritional composition had been used; some studies were carried out during postdevelopmental periods, after weaning when the first and second molars had already erupted into the oral cavity; no distinction had been made between the different stages of tooth development; and protein ingredients in the diet had been substituted at the expense of sucrose, so caries effects could be due to the increase in the new ingredient or to the reduction in sucrose.

Using special manipulation of diets, foster rat dams and rat pup interchanges among litters, Navia *et al.* (1970) were able to start a series of experimental studies to investigate the effect of undernutrition imposed to rat dams during pregnancy and/or lactation on body growth, incisor and molar weight, and caries susceptibility of their offspring. In one of their experiments, 5-day-pregnant rats were assigned to four groups (A to D). During gestation rat dams in groups A and B were fed an agar gel diet 425, containing 25% protein, while rats in groups C and D were given 8% protein diet 408 (Table III). Compen-

Table III. Composition of Gel Diets Fed to Dams During Gestation and Lactation[a]

Ingredients	Diet 408 (8% protein diet)[b]	Diet 425 (25% protein diet)[b]
Casein	480	1500
l-Methionine	18	18
Dextrose	1000	1000
Sucrose	1002	1002
Dextrin	1000	1000
Cornstarch	1078	58
Corn oil	900	900
Salt mix (Harper)	240	240
Vitamin mix[c]	60	60
Choline	24	24
Agar	210	210
Distilled water (liters)	6	6

[a] Menaker and Navia (1973a).
[b] Grams per 6000-g dry diet.
[c] Composed of the following (in grams): vitamin A acetate and D_2, 1.00 (325,000 USP IU A/32,500 D_2); vitamin E acetate (25%), 40.00; vitamin K, 0.50; thiamine hydrochloride, 1.00; riboflavin, 2.00; niacin, 5.00; vitamin C, 20.00; pyridoxine, 1.00; *p*-aminobenzoic acid, 10.00; biotin, 0.050; calcium pantothenate, 5.00; folic acid, 0.20; inositol, 20.00; vitamin B_{12} (0.1%), 5.00; and sucrose, 890.27.

sation was made for the reduction of protein in the diet by an increase in the cornstarch component. The sucrose and dextrose were kept at the same level in both diets to ensure a similar sugar supply in each diet.

At birth, dams in group A continued to receive the 25% protein diet, and dams in group D continued on the 8% protein diet; therefore, they were either well fed (A) or underfed (D) both during gestation and lactation. Dams were exchanged between the litters from group B, which received diet 425, and the litters in group C. Rat dams in groups B and C were continued on their respective diet regimes. All litters were adjusted to eight pups and kept at this number by adding pups from spare litters. These spare pups were not used in the final evaluation. Feeding a low-protein diet to the dams limits the amount, but not the quality, of the milk, and therefore rat pups in group B were undernourished only during lactation. The offspring of rats in group C were undernourished only during gestation. At 16 days, when the first and second molars start to erupt, rats in all groups were fed caries-promoting diet 305 (Navia, 1977) and provided with distilled water until the termination of the experiment at 36 days of age.

Mean tibia weight and caries scores for buccal and sulcal surfaces are presented in Table IV. These data show that rats undernourished during lactation (which corresponds with the time of active mineralization of enamel) were retarded in their growth and also had highest caries scores. Body weights and incisor weights of progeny were not affected by feeding diet 408 to rat dams during pregnancy. Severe caries was not observed in molars of rats nutritionally stressed during gestation. It was not possible to discern from this study whether the latter lack of influence was due to protection of the fetus by the mother or whether this state of tooth development is not sensitive to the marginal effects of undernutrition used in this study.

It is obvious that while a late period in tooth development had been identified as being affected by undernutrition, further experiments (DiOrio *et al.*, 1973) were needed to separate the effects of protein and calorie malnutrition on the development of teeth. The approach used was to feed either a diet adequate in protein (groups A and B) or a low-protein diet (groups C, D, and E) to a total of five groups of rat litters. Pups in group A were not intubated, while those in group B were intubated with distilled water from days 1 to 19 of age. Pups in groups C and D were intubated with isocaloric dietary supplements simulating rat milk in composition, with group C pups receiving a supplement containing 10% protein, while those in group D, a supplement containing no protein, and therefore were subject to a true protein malnutrition. Pups in group E were supplemented with distilled water and experience both protein and calorie deficits. The basis of this experimental approach was to produce a protein–calorie malnutrition in rat pups by reducing the milk output of rat dams through feeding a low-protein diet. The pups were then selectively supplemented with either protein and calories or calories alone in a simulated rat milk formulation which also provided minerals and vitamins.

Results of these controlled animal studies indicated that when a specific protein deficiency was imposed on suckling pups during the period of active mineralization for first and second molars, tooth development and eruption times were severely retarded. Administration of a protein supplement was able to correct these developmental alterations in the underfed pups, but supplementation of calories did not correct the effect of the undernutrition produced in the pups by the limited amount of milk available from the rat dams fed a low-protein diet.

These experimental results prompted the next question, which was to determine whether a specific protein malnutrition imposed during late stages of tooth development would increase the susceptibility to caries. This type of study is difficult to perform with humans, as malnourished populations seldom

Table IV. Mean Tibia Weight and Buccal- and Sulcal-Caries Scores of Rat Pups Nutritionally Stressed During Gestation or Lactation or Both[a]

	Diet treatment		Mean tibia	Caries scores (mean ± S.E.)	
Group[b]	Gestation	Lactation	Wt. ± S.E.[c] (mg)	Buccal	Sulcal
A	25	25[d]	122.5 ± 4.4	6.1 ± 1.5	6.4 ± 0.6
B	25	8	60.9 ± 2.8	14.2 ± 1.5	12.1 ± 0.6
C	8	25	106.4 ± 2.8	5.3 ± 1.1	7.6 ± 0.4
D	8	8	62.4 ± 2.6	8.9 ± 1.4	10.8 ± 0.6

[a] From Navia *et al.* (1970).
[b] Sixteen rat pups per group.
[c] Percent of casein in the purified agar diet.
[d] S.E., standard error.

display deficiencies limited to a single nutrient during a specific period, and also because the cariogenic stress that different groups of people experience varies in intensity and timing, and thus caries experiments conducted under these conditions would not be meaningful. Menaker and Navia (1973a) designed an experiment similar to that done previously by DiOrio *et al.* (1973). Two groups (1A and 1B) of rat dams with their randomized, adjusted litters of eight pups were fed gel diet 425, containing 25% protein, while four groups (2A, 2B, 2C, 2D) were fed a diet containing 8% protein (408). Pups in groups 1A and 2D received no supplements by intubation; those in groups 1B and 2C received distilled water; those in group 2A were given by intubation-simulated rat milk supplement containing 10% protein supplement; and those in group 2B an isocaloric-simulated rat milk supplement without protein. All supplementation was stopped at 19 days of age, when the rat pups were weaned. All rats were then fed a mildly caries-promoting diet 305 (Navia, 1977). Results of these experiments are presented in Table V. Specific protein malnutrition imposed during tooth mineralization resulted in increased caries susceptibility under these experimental conditions. The addition of protein alone was sufficient to improve body weight gains and to restore the low level of caries observed in the control rats receiving adequate nutrition. In this study, protein and not calories was the primary factor in overcoming the adverse effects of undernutrition on caries susceptibility of the rats.

Because a difference in tooth eruption pattern between adequate and undernourished rat pups could to some extent account for these results of protein malnutrition, Menaker and Navia (1973c) studied the influence of the 1-day lag in eruption observed between these two groups. Results of these experiments indicated that the difference in susceptibility to dental caries between rats suckled by dams receiving either the 8% or the 25% protein diet could not be due to this difference in tooth eruption, which would therefore affect the time of exposure to the cariogenic stress.

Table V. Mean Caries Scores for First and Mandibular Molars from 35-Day-Old Rats[a,b]

Group	Buccal enamel	Sulcal: Enamel	Sulcal: Dentinal (slight)	Proximal enamel
1, A[c]	8.7 ± 0.9[1d]	9.6 ± 0.3[1]	6.5 ± 0.3[1]	0.5 ± 0.2[1]
1, B	8.2 ± 0.7[1]	10.8 ± 0.3[1]	8.0 ± 0.3[1]	0.9 ± 0.3[1]
2, A	7.2 ± 0.9[1]	10.9 ± 0.5[1]	7.9 ± 0.4[1]	0.5 ± 0.2[1]
2, B	13.4 ± 0.9[2]	14.2 ± 0.4[2]	10.7 ± 0.4[2]	1.2 ± 0.3[1]
2, C	13.5 ± 0.6[2]	13.5 ± 0.4[2]	10.2 ± 0.4[2]	1.3 ± 0.3[1]
2, D	14.3 ± 0.7[2]	13.4 ± 0.4	10.3 ± 0.4[2]	1.1 ± 0.3[1]

[a] Scores are expressed as mean ± standard error.
[b] From Menaker and Navia (1973a).
[c] Minimum of 21 pups per group.
[d] Groups identified with same superscript (1 or 2) are not significantly different ($p > 0.05$).

One additional aspect needed to be evaluated in this experimental model, and that was the effect of protein malnutrition on salivary glands. Menaker and Navia (1973b) followed the alterations in some basic biochemical parameters that are associated with protein biosynthesis in the submandibular salivary glands of rats stressed nutritionally during the perinatal period. It is common clinical knowledge that patients with salivary glands that were removed surgically or treated with radiotherapy undergo a rapid and severe breakdown of tooth structure due to caries. The exact mechanism for this protective behavior of saliva is not clearly understood, but it is assumed that it is related to posteruptive enamel remineralization, the formation of protective enamel pellicle, the production of salivary secretory immunoglobulins, and enzymes such as lysozyme, which in some way help maintain the integrity of the tooth. Results of these salivary gland studies indicated that undernutrition imposed on the rats under identical nutritional manipulations to those used to produce increased caries susceptibility also affects many of the essential components of protein biosynthetic pathways in rat submandibular salivary gland. DNA, RNA, protein concentration, and total net weight of glands from undernourished rats did not return to the levels recorded in control rats after intubation with the simulated rat milk supplement containing no protein. However, administration of a supplement containing 10% protein during the suckling period completely restored these biochemical parameters to normal levels. It was also of interest to study some of the changes taking place in saliva, so whole saliva volume and its protein content were evaluated by Menaker and Navia (1974). This study compared pups suckling from dams that were fed diet 408 or 425 up to 16 days of age. Results show that in rats undernourished during development, total salivary volume and protein content are severely reduced. This may also contribute to the increased susceptibility to caries observed in rats undernourished during the postnatal period.

In another study Menaker and Navia (1976) investigated which period during active mineralization would make teeth more susceptible to the effects of malnutrition. The 18-day suckling period was divided into three equal parts and by using the fostering technique, litters were interchanged between rat dams receiving either a 25% (425) or an 8% (408) protein diet. A factorial experiment was designed to test all combinations of 6-day periods of malnutrition during the suckling period. At weaning (day 19), all pups received a moderately caries promoting diet (305). On day 38 the rats were sacrificed and the molars scored for caries. Statistical analysis suggested that early malnutrition (1–6 days of age) significantly increased sulcal caries susceptibility. Rats malnourished later (9–12 days) presented significantly high increases in both sulcal and buccal molar surfaces, and those challenged just prior to tooth eruption (13–18 days) showed increased caries on sulcal, buccal, and proximal molar surfaces. From these results it can be concluded that malnutrition imposed at different stages of molar development affects the caries susceptibility of different tooth surfaces, with the time just before eruption being most sensitive to the effects of malnutrition on caries susceptibility.

4.3. *Protein–Calorie Malnutrition and Oral Immunology*

Oral health is not only promoted through the adequate development of tissues in the oral cavity, but also through appropriate development of the reticuloendothelial system. McGhee *et al.* (1974) studied the complement levels in malnourished rats by quantifying the serum complement levels in rat dams and their offspring. Rat dams were either fed a purified diet containing 25% or 8% casein and the serum complement from the dams and their pups were assayed with radial hemolysis in gel. When rat mothers in the two groups were compared, no significant differences were found in serum complement concentration. However, differences in the serum complement level in malnourished and normal offspring were apparent at each time of sampling (25, 30, 35 days of age) after weaning. A threefold difference was observed at termination of the experiment (35 days) between the two groups, thus suggesting that undernutrition during early stages of rat development results in damage to the tissues responsible for the synthesis of complement.

Michalek *et al.* (1974) evaluated milk complement level in rat dams fed the two diets differing in protein content. As expected, the amount of milk collected from rat dams fed a low-protein diet (8% casein) was significantly less (0.5–5.0 ml/hr) than that collected from rats fed a diet with normal protein (25% casein) (2.0–10 ml/hr). In addition to this low milk production, rat dams fed the low-protein diet produced milk that contained less complement (2.3–3.5 CH_{50}/ml) than milk from well-fed rats (4.3–7.9 CH_{50}/ml) throughout lactation.

Michalek *et al.* (1975) have extended these studies to include a quantitation of immunoglobulin levels in colostrum and milk from adequate and malnourished rat dams during gestation and lactation. Serum and saliva of their offspring were also quantitated using radial immunodiffusion and purified anti-rat α, γ, and μ. Colostral IgA in malnourished dams was not significantly different from that of adequately fed dams. There were also no differences in the level of serum IgA and IgM and salivary IgA and IgG_{2a} of their offspring. However, IgG_{2a} in colostrum and milk from malnourished rat dams was approximately twofold lower throughout the entire period of lactation. This difference was also reflected in the level of IgG_{2a} in the serum of rat pups suckling from rat dams fed a low-protein diet which was always 1.5- to 2-fold less than adequately fed controls. The low levels of IgG_{2a} in the offspring's serum may be an important aspect of the lowered resistance to infection associated with malnutrition, and further research is needed to clarify the role of IgG in the maintenance of oral health.

The decrease in both humoral and cell-mediated immunity documented in humans and experimental animals subjected to protein deficiency (Kenny *et al.*, 1968; Jose and Good, 1971; Chandra, 1972; Edelman, 1975) prompted McGhee *et al.* (1976) to study the influence of protein malnutrition on the implantation and cariogenicity of *S. mutans*, a bacterium found in dental plaque and considered to be one of the important microbial etiological agents for dental caries. Previous rat studies (Tanzer *et al.*, 1973; Taubman, 1973;

Taubman and Smith, 1974) conducted under gnotobiotic and conventional conditions had suggested that rats locally injected with *S. mutans* in the salivary gland region and challenged with the homologous bacteria elicit a protective immune response against dental caries. Michalek *et al.* (1975) injected either saline or whole killed *S. mutans* 6715 in the region of the submandibular gland of rat pups which suckled from dams fed either low-protein diet 408 or adequate protein diet 425. They were subsequently infected with a 24-hr culture of *S. mutans* 6715. A third set of two groups of rats were neither injected nor infected with *S. mutans* and fed either diet 408 or 425. Infected rat pups developed serum and salivary agglutinins against the *S. mutans*. When rats which were locally injected with *S. mutans* were compared for titers of agglutinins, levels in malnourished rats were found to be similar to those observed in rats which had suckled from dams fed a nutritionally adequate diet. At the end of 45 days caries scores were determined for these groups of rats, with the following results (Table VI); (1) both groups of immunized rats and those subsequently infected with the cariogenic *S. mutans* had lower mean caries scores than infected, nonimmunized rats in all molar surfaces examined; and (2) malnourished rats, not immunized, but infected with *S. mutans,* had significantly more severe caries than well-fed nonimmunized rats. As expected, both dietary groups of noninfected rats had few carious lesions. These results indicate that (1) caries developed in these rats as a result of the oral infection with *S. mutans*; (2) protein-malnourished rats developed higher levels of caries, as shown previously (Navia *et al.*, 1970); and (3) even a nutritionally compromised host can elicit an immune response that may reduce the severity of *S. mutans*-induced caries. The exact nature and mechanism of action of such an immune response acting in the oral cavity remains to be clearly elucidated.

Table VI. Mean Caries Scores: Effect of Immunization with S. mutans in Protein-Malnourished and Normal Rats[a]

		Mean caries scores[c]				
		Buccal	Sulcal		Proximal	Mean body
Diet	Group[b]	Enamel	D_s	D_m	Enamel	wt (g)[d]
408	A (immunized, infected)	3.9 ± 0.6	10.4 ± 0.5	7.6[e] ± 0.6	0.2[e] ± 0.1	104.0 ± 2.6
	B (infected only)	8.6 ± 0.6	12.3 ± 0.6	10.5 ± 0.5	3.1 ± 0.5	100.5 ± 3.2
	C (control—no infection)	0.8 ± 0.5	1.2 ± 0.7	0.8 ± 0.1	0.0 ± 0.0	101.7 ± 5.0
425	D (immunized, infected)	1.2 ± 0.4	8.7 ± 0.6	4.7[e] ± 0.8	0.3[e] ± 0.2	144.6 ± 3.9
	E (infected only)	4.0 ± 0.9	10.5 ± 0.6	7.9 ± 0.8	1.6 ± 0.4	154.5 ± 6.5
	F (control—no infection)	0.7 ± 0.2	1.8 ± 0.1	0.3 ± 0.1	0.0 ± 0.0	142.6 ± 5.0

[a] From Michalek *et al.* (1976).
[b] Rat pups in groups A, B, D, and E challenged with 5.6×10^8 viable *S. mutans* 6715 in 50 μl.
[c] Evaluated by the Keyes procedure (1958); E, penetration into enamel; D_s, slight penetration into dentin; D_m, moderate penetration into dentin. Score ± standard error [represents 31 (A), 22 (B), 15 (C), 23 (D), 16 (E), and 23 (F) rats].
[d] Values as mean ± standard error.
[e] Values significantly lower than infected controls ($p \leq 0.01$).

5. Conclusions

Oral tissues in their development have critical periods when nutrients from the trophic environment have to be provided to ensure normal expression of their genetic information. Failure to obtain such required nutrients results in developmental dysplasia and defects which, to a large extent, are irreversible. This is especially true for tissues such as enamel which is formed prior to tooth eruption and loses its ameloblasts once development is completed. In this tissue, therefore, there is no cellular repair mechanism once its eruption into the oral cavity is completed. The tooth, then, must rely on the reparative remineralizing action of saliva and other protective factors present in food and water, such as fluoride, to maintain its integrity (Navia, 1972). Saliva also provides proteins such as immunoglobulins, which may play an important role in the maintenance of oral health. Adequate nutrition during salivary gland development is therefore also critical to ensure proper function of these important glands, which secrete into the oral cavity.

Nutritional adequacy is of primary importance to oral tissues pre- and postnatally, particularly in the availability of nutrients such as protein, vitamin A, and the mineral elements. This is not to infer that other nutrients are not necessary, but to stress that these nutrients are the ones which have been studied experimentally and have been shown to play a significant role in the development of oral structures and in the maintenance of oral health. At this time the greatest need is for clinical studies to explore the mechanisms and consequences of these deficiencies on oral health. Such information would facilitate understanding of the contribution that malnutrition, imposed early in development, has on diseases such as dental caries, compared with the effects of an improper diet after the eruption of the teeth. The answer to these questions will then enable health professionals to determine which nutritional and dietary measures are essential to prevent and control oral disease.

ACKNOWLEDGMENT. The research for this chapter was supported by NIH–NIDR Grant DE02670-09 and Training Grant DE07020-01.

6. References

Angmar-Mänsson, B., 1971a, A polarization microscopic and micro x-ray diffraction study on the organic matrix of developing human enamel, *Arch. Oral Biol.* **16:**147.

Angmar-Mänsson, B., 1971b, A quantitative microradiographic study on the organic matrix of developing human enamel in relation to the mineral content, *Arch. Oral Biol.* **16:**135.

Arnold, F. A., Likins, R. C., Russel, A. L., and Scott, D. B., 1962, Fifteenth year of Grand Rapids fluoridation study, *J. Am. Dent. Assoc.* **65:**780.

Ast, D. B., Smith, D. J., Wachs, B., and Cantwell, K. T., 1956, Newburgh-Kingston caries-fluorine study. XIV. Combined clinical and roentgenographic dental findings after 10 years of fluoride experience, *J. Am. Dent. Assoc.* **52:**314.

Barmes, D. E., 1977, Global problems of oral disease, *J. Dent. Res.* **56:**C9.

Barmes, D. E., Adkins, B. L., and Schamschula, R. G., 1970, Etiology of caries in Papua-New Guinea-Association in soil, food and water, *Bull WHO* **43:**769.

Baume, L. J., 1968, Report on a dental survey among the school population of French Polynesia, *Arch. Oral Biol.* **13:**787.

Baume, L. J., 1969, Caries prevalence and caries intensity among 12,344 schoolchildren of French Polynesia, *Arch. Oral Biol.* **14:**181.

Baume, L. J., and Meyer, J., 1966, Dental dysplasia related to malnutrition with special reference to melanodontia and odontoclasia, *J. Dent. Res.* **45:**726.

Baume, L. J., and Vulliémoz, J. P., 1970, Dietary fluoride uptake into the enamel of caries-susceptible "yellow" permanent teeth and of caries-resistant permanent and primary teeth of Polynesians, *Arch. Oral Biol.* **15:**431.

Baume, L. J., and Vulliémoz, J. P., 1972, Variations in the mineral content of the enamel of Polynesian teeth, *Int. Dent. J.* **22:**193.

Baume, L. J., Franquin, J. C., and Kömer, W. W., 1972, The prenatal effects of maternal vitamin A deficiency on the cranial and dental development of the progeny, *Am. J. Orthod.* **62:**447.

Bloch, C. E., 1931, Effects of deficiency in vitamins in infancy, *Am. J. Dis. Child.* **42:**263.

Boyle, P. E., 1933, Manifestation of vitamin A deficiency in a human tooth-germ, *J. Dent. Res.* **13:**39.

Boyle, P. E., and Bessey, O. A., 1941, The effect of acute vitamin A deficiency on the molar teeth and paradontal tissues, with a comment on deformed incisor teeth in this deficiency, *J. Dent. Res.* **20:**236.

Brown, W. E., and König, K. G., 1977, Cariostatic Mechanisms of Fluoride, Proceedings of ADA-NIDR Conference, *Caries Res.* **11**(suppl. 1):1–327.

Burn, C. G., Orten, A. U., and Smith, A. H., 1941, Changes in the structure of the developing tooth in rats maintained on a diet deficient in vitamin A, *Yale J. Biol. Med.* **13:**817.

Carlos, J. P., Gittelsohn, A. M., and Haddon, W., 1962, Caries in deciduous teeth in relation to maternal ingestion of fluoride, *Public Health Rep.* **77:**658.

Chandra, R. K., 1972, Immunocompetence in undernurtition, *Trop. Pediatr.* **81:**1194.

Dinnerman, M., 1933, Vitamin A deficiency in unerupted teeth of infants, *Oral Surg., Oral Med., Oral Pathol.* **13:**1024.

DiOrio, L. P., Miller, S. A., and Navia, J. M., 1973, The separate effects of protein and calorie malnutrition on the development and growth of rat bones and teeth, *J. Nutr.* **103:**856.

Eanes, E. D., and Posner, A. S., 1970, Structure and chemistry of bone mineral, in: *Biological Calcification: Cellular and Molecular Aspects* (H. Schraer, ed.), p. 1, Appleton-Century-Crofts, New York.

Edelman, R., 1975, Cell-mediated immunity in protein-calorie malnutrition, in: *Protein-Calorie Malnutrition* (R. E. Olson, ed.), pp. 377–381, Academic Press, Inc., New York.

Frazier, P. D., 1968, Adult human enamel: An electron microscopic study of crystallite size and morphology, *J. Ultrastruct. Res.* **22:**1.

Grahnén, H., Holm, A., Magnusson, B., and Sjölin, S., 1972, Mineralisation defects of the primary teeth in intra-uterine undernutrition, *Caries Res.* **6:**224.

Gwinnett, A. J., 1967, The ultrastructure of the "prismless" enamel of permanent human teeth, *Arch. Oral Biol.* **12:**381.

HANES, 1974, Health and Nutrition Examination Survey, United States (1971–1972), *Public Health Serv. DHEW (HRA) 74-1219-1.*

Hodge, H. C., and Smith, F. A., 1970, Minerals: Fluorine and dental caries, *Adv. Chem. Sem.* **94:**93.

Holloway, P. J., Shaw, J. H., and Sweeney, E. A., 1961, Effects of various sucrose: Casein ratios in purified diets on the teeth and supporting structure of rats, *Arch. Oral Biol.* **3:**185.

Horowitz, H. S., and Heifetz, S. B., 1967, Effects of prenatal exposure to fluoridation on dental caries, *Publ. Health Rept.* **82:**297.

Irving, J. T., 1949, The effects of avitaminosis and hypervitaminosis A upon the incisor teeth and incisal alveolar bone of rats, *J. Physiol.* **108:**92.

Jelliffe, D. B., and Jelliffe, E. F. P., 1961, The nutritional status of Haitian children, *Acta Trop.* **18:**1.

Jelliffe, D. B., and Jelliffe, E. F. P., 1971, Linear hypoplasia of deciduous incisor teeth in malnourished children, *Am. J. Clin. Nutr.* **24**:893.
Jelliffe, D. B., Williams, L. L., and Jelliffe, E. F. P., 1954, A nutrition survey in rural Jamaica, *J. Trop. Med. Hyg.* **57**:27.
Jelliffe, D. B., Jelliffe, E. F. P., Garcia, I., and DeBarrios, G., 1961, The children of the San Blas Indians of Panama, *J. Pediatr.* **59**:271.
Jenkins, G. N., Venkateswarku, P., and Zippin, I., 1970, Physiological effects of small doses of fluoride, World Health Organ. Monogr. 59, Chap. 6, p. 163.
Jose, D. G., and Good, R. A., 1971, Absence of enhancing antibody in cell-mediated immunity to tumor heterografts in protein-deficient rats, *Nature (London)* **231**:323.
Kenney, M. A., Roderuck, C. E., Armich, L., and Piedad, F., 1968, Effect of protein deficiency on the spleen and antibody formation in rats, *J. Nutr.* **95**:173.
Keyes, P. H., 1958, Dental caries in the molar teeth of rats. II. A method for diagnosis and scoring several types of lesions simultaneously, *J. Dent. Res.* **37**:1088.
Larson, R. H., 1964, Effect of prenatal nutrition on oral structures, *Am. Dietet. Assoc.* **44**:368.
Lunt, R. C., and Law, D. B., 1974a, A review of the chronology of calcification of deciduous teeth, *J. Am. Dent. Assoc.* **89**:599.
Lunt, R. C., and Law, D. B., 1974b, A review of the chronology of eruption of deciduous teeth, *J. Am. Dent. Assoc.* **89**:872.
McGhee, J. R., Michalek, S. M., Ghanta, V. K., and Stewart, G., 1974, Complement levels in malnourished animals: quantification of serum complement in rat dams and their offspring, *J. Retinculoendothel. Soc.* **16**:204.
McGhee, J. R., Michalek, S. M., Navia, J. M., and Narkates, A. J., 1976, Effective immunity to dental caries: studies of active and passive immunity to *Streptococcus mutans* in malnourished rats, *J. Dent. Res.* **55**:C206.
Margolis, F. J., Reames, H. R., Freshman, E., Macauley, J. C., and Mehaffey, H., 1975, Fluoride-term-year prospective study of deciduous and permanent dentition, *Am. J. Dis. Child.* **129**:794.
Meckel, A. H., Griebstein, W. J., and Neal, R. J., 1965, Structure of mature human dental enamel as observed by electron microscopy, *Arch. Oral Biol.* **10**:775.
Mellanby, E., 1941a, Nutrition in relation to bone growth and the nervous system, *Proc. Roy. Soc. (Lond.)* **B132**:28.
Mellanby, E., 1941b, Skeletal changes affecting the nervous system produced in young dogs by diets deficient in vitamin A, *J. Physiol.* **99**:467.
Mellanby, E., 1947, Vitamin A and bone growth: the reversibility of vitamin A deficiency changes, *J. Physiol.* **105**:382.
Mellanby, H., 1941, The effect of maternal dietary deficiency of vitamin A on dental tissues in rats, *J. Dent. Res.*, **20**:489.
Mellanby, M., 1928, The influence of diet on the structure of teeth, *Physiol. Rev.* **8**:545.
Menaker, L., and Navia, J. M., 1973a, Effect of undernutrition during the perinatal period on caries development in the rat: IIa. Caries susceptibility in underfed rats supplemented with protein or caloric additions during the suckling period, *J. Dent. Res.* **52**:680.
Menaker, L., and Navia, J. M., 1973b, Effect of undernutrition during the perinatal period on caries development in the rat: IIIb. Effect of undernutrition on biochemical parameters in the developing submandibular salivary gland, *J. Dent. Res.* **52**:688.
Menaker, L, and Navia, J. M., 1973c, Effect of undernutrition during the perinatal period on caries development in the rat: IVc. Effects of differential tooth eruption and exposure to a cariogenic diet on subsequent dental caries incidence, *J. Dent. Res.* **52**:692.
Menaker, L., and Navia, J. M., 1974, Effect of undernutrition during the perinatal period on caries development in the rat: V. Changes in whole saliva volume and protein content, *J. Dent. Res.* **53**:592.
Menaker, L., and Navia, J. M., 1976, A critical period for caries development in the undernourished rat, *Proc. Int. Assoc. Dent. Res. Abstr. 255,* p. 125B.
Michalek, S. M., McGhee, J. R., and Ghanta, V. K., 1974, Complement levels in malnourished

animals: Measurement of complement in rat milk during the period of lactation, *J. Reticuloendothel. Soc.* **16:**213.

Michalek, S. M., Rahman, A. F. R., and McGhee, J. R., 1975, Rat immunoglobulins in serum and secretions: comparison of IgM, IgA and IgG in serum, colostrum, milk and saliva of protein malnourished and normal rats, *Proc. Soc. Exp. Biol. Med.* **148:**1114.

Michalek, S. M., McGhee, J. R., Navia, J. M., and Narkates, A. J., 1976, Effective immunity to dental caries: protection of malnourished rats by local injection of *Streptococcus mutans, Infect. Immun.* **13:**782.

Miles, A. E. W., 1967, *Structural and Chemical Organization of Teeth,* Vol. 1, Academic Press, Inc., New York.

Molnar, S., and Ward, S. C., 1974, Mineral metabolism and microstructural defects in primate teeth, *Am. J. Phys. Anthrop.* **43:**3.

National Academy of Sciences, 1971, *Biological Effects of Atmospheric Pollutants: Fluoride,* Div. Med. Sci., National Research Council, Washington, D.C.

Navia, J. M., 1972, Prevention of dental caries: Agents which increase tooth resistance to dental caries, *Int. Dent. J.* **22:**427.

Navia, J. M., 1976, Nutrition and oral disease, in: *Preventive Denistry* (R. Stallard and R. Caldwell, eds.), W. B. Saunders Company, Philadelphia.

Navia, J. M., 1977, *Animal Models in Dental Research,* University of Alabama Press, University, Ala.

Navia, J. M., DiOrio, L. P., Menaker, L., and Miller, S., 1970, Effect of undernutrition during the perinatal period on caries development in the rat, *J. Dent. Res.* **49:**1091.

Navia, J. M., Hunt, C. E., First, F. B., and Narkates, A. J., 1976, Fluoride metabolism-effect of pre-eruptive or posteruptive fluoride administration on rat caries susceptibility, in: *Trace Elements in Human Health and Disease* (A. S. Prasad, ed.), Vol. II, pp. 249–268, Academic Press Inc., New York.

Nicholls, L., Sinclair, H. M., and Jelliffe, D. B., 1961, *Tropical Nutrition and Dietetics,* 4th ed., Ballière, Tindall & Cox, London.

Nizel, A. E., 1973, Nutrition and oral problems, *World Rev. Nutr. Dietet.* **16:**226.

Nylen, M. U., 1964, Electron microscope and allied biophysical approaches to the study of enamel mineralization, *J. R. Micro. Soc.* **85:**135.

Nylen, M. U., 1967, Recent electron microscopic and allied investigations into the normal structure of human enamel, *Int. Dent. J.* **17:**719.

Paynter, K. J., and Grainger, R. M., 1956, The relation of nutrition to the morphology and size of rat molar teeth, *J. Can. Dent. Assoc.* **22:**519.

Pindborg, J. J., 1970, *Pathology of the Dental Hard Tissues,* W. B. Saunders Company, Philadelphia.

Russell, A. L., 1963, International nutrition surveys: a summary of preliminary dental findings, *J. Dent. Res.* **42:**233.

Salley, J. J., Bryson, W. F., and Eshleman, J. R., 1959, The effect of chronic vitamin A deficiency on dental caries in the syrian hamster, *J. Dent. Res.* **38:**1038.

Schamschula, R. G., Barnes, D. E., and Adkins, B. L., 1972, Caries etiology in Papua-New Guinea-Associations of tooth size and dental arch width, *Aust. Dent. J.* **17:**188.

Schour, I., and Massler, M., 1945, The effects of dietary deficiencies upon the oral structures, *Physiol. Rev.* **25:**442.

Schour, I., Hoffman, M. M., and Smith, M. C., 1941, Changes in the incisor teeth of Albino rats with vitamin A deficiency and the effects of replacement therapy, *Am. J. Pathol.* **17:**529.

Scott, D. B., Simmelink, J. W., Swancar, J. R., and Smith, T. J., 1972, Apatite crystal growth in biological systems, in: *Structural Properties of Hydroxyapatite and Related Compounds* (W. F. Brown and R. A. Young, eds.), Gordon and Breach, Science Publishers, Inc., New York.

Shaw, J. H., 1969, Influence of marginal and complete protein deficiency for varying periods during reproduction on growth, third-molar eruption, and dental caries in rats, *J. Dent. Res.* **48:**310.

Shaw, J. H., 1970, Preeruptive effects of nutrition on teeth, *J. Dent. Res.* **49:**1238.

Shaw, J. H., and Griffiths, D., 1963, Dental abnormalities in rats attributable to protein deficiency during reproduction, *J. Nutr.* **80:**123.
Shaw, J. H., and Sweeney, E. A., 1973, Nutrition in relation to dental medicine, in: *Modern Nutrition in Health and Disease,* 5th ed. (R. S. Goodhard and M. E. Shils, eds.), Lea and Febiger, Philadelphia.
Sicher, H. 1966, *Orban's Oral Histology and Embryology,* The C. V. Mosby Co., Saint Louis.
Sognnaes, R. F., 1948, Analysis of wartime reduction of dental caries in European children, *Am. J. Dis. Child.* **75:**792.
Sognnaes, R. F., Scott, D. B., Mylen, M. U., and Yaeger, J. A., 1966, Enamel in: *Orban's Oral Histology and Embryology* (H. Sicher, ed.), The C. V. Mosby Company, St. Louis, Mo.
Swancar, J. R., Scott, D. B., and Njemirovskij, Z., 1971, Studies on the structure of human enamel by the replica method, *J. Dent. Res.* **49:**1025.
Sweeney, E. A., and Guzmán, M., 1966, Oral conditions in children from three highland villages in Guatemala, *Arch. Oral Biol.* **11:**687.
Sweeney, E. A., Cabrera, J., Urrutia, J., and Mata, L., 1969, Factors associated with linear hypoplasia of human deciduous incisors, *J. Dent. Res.* **48:**1275.
Sweeney, E. A., Saffir, A. J., and DeLeon, R., 1971, Linear hypoplasia of deciduous incisor teeth in malnourished children, *Am. J. Clin. Nutr.* **24:**29.
Tanzer, J. M., Hageage, G. J., and Larson, R. H., 1973, Variable experience in immunization of rats against *Streptococcus mutans*-associated dental caries, *Arch. Oral Biol.* **18:**1425.
Taubman, M. A., 1973, Role of immunization in mental disease, in: *Comparative Immunology of the Oral Cavity* (S. Mergenhagen, ed.), pp. 138–158, HEW-NIH Publ. No. 73–438.
Taubman, M. A., and Smith, D. J., 1974, Effect of local immunization with *Streptococcus mutans* on induction of salivary immunoglobulin A antibody and experimental dental caries in rats, *Infect. Immun.* **9:**1079.
Ten-State Nutrition Survey in the United States, 1968–1970, U.S. Department of Health, Education and Welfare, Health Services and Mental Health Administration, DHEW Publ. No. (HSM) 72-8134.
Underwood, E. J., 1971, *Trace Elements in Human and Animal Nutrition,* 3rd ed., Academic Press, Inc., New York.
WHO, 1970, Fluorides and human health, *WHO Monogr. Sem. 59.,* Geneva.
Wolbach, S. B., and Howe, P. R., 1925, Tissue changes following deprivation of fat-soluble A vitamin, *J. Exp. Med.* **42:**753.
Wolbach, S. B., and Howe, P. R., 1933, The incisor teeth of Albino rats and Guinea pigs in vitamin A deficiency and repair, *Am. J. Pathol.* **9:**275.

12

Pediatric Nutrition: Potential Relationship to the Development of Atherosclerosis

Charles J. Glueck and Reginald C. Tsang

1. Introduction

The concept that nutritional patterns in childhood might relate to atherosclerosis is relatively new, as is the suggestion that the pathologic and biochemical genesis of atherosclerosis is in childhood (Holman, 1961; Atherosclerosis Study Group, 1970; American Academy of Pediatrics, 1972; Mitchell *et al.*, 1972; deHaas, 1973; Mitchell and Jesse, 1973; Kannel and Dawler, 1972; Drash and Hengsteinberg, 1972; Fallat *et al.*, 1974; Glueck *et al.*, 1974a; Enos *et al.*, 1955; Holman *et al.*, 1958; McGill, 1968; McMillan, 1973; McNamara *et al.*, 1971; Vlodaver *et al.*, 1969). The purpose of this chapter is to review the evidence, hypotheses, and speculation that pediatric nutrition might play a role in the longitudinal progression of atherosclerosis. The relationship between the type of dietary fat intake and plasma lipid and lipoprotein levels in children and young adults will be examined. Nutritional data referable to obesity and hypertension in children will be reviewed, since both may directly or additively potentiate atherosclerosis risk. The distinction between "genetic" and acquired hyperlipidemias in children will be emphasized, and the degree of urgency relative to nutritional modification will be viewed within this frame of reference. Finally, as emphasized by Fomon (1974),

> There are two mechanisms leading to development of premature atherosclerotic disease in individuals and populations: (1) the *pathologic mechanism* that begins to find expression in late childhood and accelerates during young adult life and (2) the

Charles J. Glueck • General Clinical Research Center and Lipid Research Center, University of Cincinnati College of Medicine, Cincinnati, Ohio. ***Reginald C. Tsang*** • Fels Division of Pediatric Research, Children's Hospital Research Foundation, Cincinnati, Ohio.

behavioral mechanism that has its origins much earlier when lifelong patterns begin to develop. In the United States . . . behaviors leading to increased risk of premature atherosclerotic disease become firmly established for large numbers of individuals during childhood and early adolescence. This is particularly true of food preferences leading to high intakes of animal fats and cholesterol.

There is no uniform agreement and there is considerable controversy as to the effectiveness, safety, and advisability of nutritional modification for all children as a pediatric approach to primary prevention of atherosclerosis (American Academy of Pediatrics, 1972; Mitchell *et al.*, 1972; Mitchell and Jesse, 1973; North, 1975; Bergman, 1975; Davidson, 1975; Schubert, 1973). Although "genetic" and acquired hyperlipidemias are associated with accelerated coronary heart disease (CHD) (Kannel *et al.*, 1971; Tamir *et al.*, 1972; Glueck *et al.*, 1974b; Chase *et al.*, 1974; Blumenthal *et al.*, 1975), there still is a serious lack of consistent, unequivocal evidence that (primary) amelioration of hyperlipidemia antecedent to a clinical or pathologic CHD event will prevent or retard development of atherosclerosis. Hence, critics of *population-wide dietary modification* aimed at reducition of cholesterol and saturate intake in childhood can suggest that little should be done until "smoking pistol" evidence is available to validate the lipid hypothesis (Mitchell *et al.*, 1972; North, 1975; Bergman, 1975; Davidson, 1975; Schubert, 1973). Such evidence will become indirectly available only through careful long-term blinded studies in young adults or middle-aged adults (currently under way in the Lipid Research Centers). The unequivocal answers in regard to pediatric populations will, in addition, require at least several generations for proof.

In this chapter we will review the rapidly expanding body of information about the relationship of diet to lipid disorders, obesity, and hypertension to indicate potentially useful avenues available prior to receipt of firmer evidence of the efficacy of primary atherosclerosis prevention.

2. Anatomic Studies: Pediatric Precursors of Mature Atherosclerotic Lesions

As early as 1961, Holman (1961) speculated that the genesis of atherosclerosis might well be a pediatric one, perhaps related to excess intake of cholesterol and/or saturated fats in childhood. Evidence for some atherosclerosis was present in at least 30% of autopsy studies of young American soldiers (Enos *et al.*, 1955; Holman *et al.*, 1958; McGill, 1968; McMillan, 1973; McNamara *et al.*, 1971). Most pathologic evidence also suggests that the adult atherosclerotic lesion evolves from the ubiquitous fatty streak in infancy (Holman, 1961; Holman *et al.*, 1958; McGill, 1968; McMillan, 1973). Although this evidence is short of rigid proof (being cross-sectional rather than longitudinal), it is uniform, persuasive, and broadly based (Glueck *et al.*, 1974a; Enos *et al.*, 1955; Holman *et al.*, 1958; McGill, 1968; McMillan, 1973; McNamara *et al.*, 1971; Vlodaver *et al.*, 1969; Keys *et al.*, 1958). Fatty streaks are probably not the only pediatric precursors of mature atherosclerotic plaques. Vlodaver *et al.* (1969) demonstrated that aortic muscular elastic development and eccentric intimal tissue with increased collagen were more marked in infants from Israeli ethnic groups with higher prevalences of premature coronary atherosclerosis and myocardial infarction as compared to groups with lower prevalences.

3. *Distribution of Serum Lipids in Children*

Although there is uncertainty about the probable mechanism of progression of the atherosclerotic lesion, a series of epidemiologic, geographic, and national population studies have prospectively shown that a high serum cholesterol concentration is one of the most readily identifiable risk factors in the development of coronary heart disease (Kannel and Dawler, 1972; Kannel *et al.*, 1971; Keys *et al.*, 1958; Johnson *et al.*, 1965; Sherwin, 1974; Keys, 1970; Goldrick *et al.*, 1969; Kimura, 1956; Connor, 1961). Eight to 25% (depending on the definition of abnormal) of free-living children in industrial societies have elevated plasma cholesterol levels, closely related to national and community dietary habits. By age 3, plasma cholesterol levels in most American children are as high as those of Japanese men of middle age (Sherwin, 1974; Berenson *et al.*, 1974). In a group of 8- to 12-year-old American children, 46% had one risk factor (hypercholesterolemia, hypertension) for coronary heart disease; 14% had two or more risk factors (Wilmore and McNamara, 1974). By age 20, most American men are hypercholesterolemic by worldwide norms (Keys, 1970; Schilling *et al.*, 1964).

In Iowa teenagers, mean plasma cholesterol was 157 ± 36 mg/dl and cholesterol was above an arbitrary "normal limit" of 200 mg/dl in 20% of children (Hodges and Krehl, 1965). Thirteen percent of adolescent children in Vermont had cholesterol levels above 200 mg/dl (Clarke *et al.*, 1970). Six percent of California school children, ages 6–14, had cholesterol levels >220 mg/dl (Starr, 1971). Coronary heart disease risk factors in 4829 school children in Muscatine, Iowa, have been carefully evaluated (Lauer *et al.*, 1975). Twenty-four percent of children had levels greater than or equal to 200 mg/dl, 9% were greater than or equal to 220 mg/dl, 3% were greater than or equal to 240 mg/dl, and 1% were greater than or equal to 260 mg/dl. Fifteen percent of the children had serum triglyceride levels of 140 mg/dl or more. In the age group 14–18 years, 9% had systolic blood pressures greater than or equal to 140 mm of mercury, and 12.2% had diastolic blood pressures greater than or equal to 90 mm of mercury (Lauer *et al.*, 1975). In 4.4% both pressures were at or above these levels. At ages 6–9, 20% of the children were moderately obese and 5% were severely obese. The authors concluded "these data indicated that a considerable number of school-age children have risk factors which in adults, are predictive of coronary heart disease" (Lauer *et al.*, 1975).

Pediatric hypercholesterolemia is also relatively common in middle-class children of Australia and South Africa (Godfrey *et al.*, 1972; Duplessis *et al.*, 1967). Godfrey *et al.* (1972) reported that 2½% of Australian school children had plasma cholesterol levels over 238 mg/dl. Duplessis *et al.* (1967) noted that

9% of white South African children ages 7–11, and 14% of those ages 12–15, had cholesterol levels greater than 260 mg/dl, whereas only 1% of the various groups of nonwhite children were similarly affected.

Thus, by the middle of the second decade of life, children from industrial urban societies have plasma cholesterol levels in the 160- to 180-mg/dl range, and 5–20% have levels in excess of 200 mg/dl (Lauer *et al.*, 1975; Wilmore and McNamara, 1974; Berenson *et al.*, 1974; Whyte and Yee, 1958; Margolis, 1975; Hodges and Krehl, 1965; Clarke *et al.*, 1970; Starr, 1971; Godfrey *et al.*, 1972; Duplessis *et al.*, 1967). Children from underdeveloped, nonindustrial, and rural populations usually have cholesterol levels 50 mg/dl lower than their Westernized counterparts (Lauer *et al.*, 1975; Whyte and Yee, 1958; Duplessis *et al.*, 1967. Cholesterol levels are further often aggregated in families and children in both the high and low ends of the cholesterol distributions, and tend to remain in these sections of the distribution when followed longitudinally (Johnson *et al.*, 1965; Sherwin, 1974; Deutscher *et al.*, 1966).

What is the (longitudinal) significance of the relatively high prevalence of hypercholesterolemia in children from urban, industrial countries? The Framingham study indicated a continuous gradation of risk for coronary heart disease for adults with serum cholesterol levels above 180 mg/dl (Kannel *et al.*, 1971). In countries in which the incidence of coronary heart disease is very low, mean serum cholesterol concentrations of adults range from 140 to 160 mg/dl (Keys, 1970; Goldrick *et al.*, 1969; Kimura, 1956; Connor, 1961), and the 95th percentile for plasma cholesterol is about 200 to 220 mg/dl. If one accepts that a level of 200 mg/dl may represent a "threshold value" above which there is significant risk for atherosclerosis (Kannel *et al.*, 1971; Keys, 1970; Goldrick *et al.*, 1969; Kimura, 1956; Connor, 1961), the 24% of Muscatine, Iowa, school children with plasma cholesterol ≥ 200 mg/dl, and the 9.2% ≥ 200 mg/dl are putatively at considerably higher risk for the future development of coronary heart disease (Lauer *et al.*, 1975). Since after the age of 24 years in industrialized countries there may be a progressive rise in the serum cholesterol level for the next 30 years, plasma cholesterol would be ≥ 220 mg/dl in most Iowa school children as they become young adults (Lauer *et al.*, 1975). They would then exceed the 220-mg/dl level, which is the "trip-wire" point recommended for treatment by the Food and Nutrition Board of the National Academy of Sciences (Council of Foods and Nutrition, 1972) and by the Committee on Diet and Heart Disease of the National Heart Foundation of Australia (1974).

4. Childhood Diet and Serum Lipids

Nutritional factors are powerful governors of plasma lipids and lipoproteins in infancy, childhood, and young adulthood, and probably account for the fact that children from urban industrial societies have considerably higher cholesterol levels than children from rural, "Third World" cultures (Atherosclerosis Study Group, 1970; Keys *et al.*, 1958; Whyte and Yee, 1958; Duplessis *et al.*, 1967; Scrimshaw *et al.*, 1957; Golubjatnikov *et al.*, 1972; Schilling *et al.*, 1964). As shown by Duplessis *et al.* (1967), dietary cholesterol and saturated fat intake correlate significantly with plasma cholesterol levels in both South African whites and blacks. Although cord blood and year-1 cholesterol levels are similar in urban Australian and rural, "nonwestern" New Guinean infants, the children's cholesterol levels then diverge (Whyte and Yee, 1958). Concomitant with ingestion of a saturate and cholesterol-rich diet, Australian pediatric cholesterol levels by late adolescence are much higher than those in New Guinean children whose serum cholesterol *declines* after age 1 (Whyte and Yee, 1958). Similarly, mean plasma cholesterol in boys from a Guatemalan urban upper-income group (with higher cholesterol intake) was 187 mg/dl, from an urban lower-income group was 143 mg/dl, and from a rural lower-income group was 121 mg/dl (Scrimshaw *et al.*, 1957). The mean serum cholesterol level of rural Mexican children was approximately 100 mg/dl, nearly one-half that of mean levels, 187 mg/dl for Wisconsin school children (Golubjatnikov *et al.*, 1972). Striking differences in the cholesterol levels in these two populations led the authors to conclude that population-wide diet modification should be started in the preschool and school-age period to achieve maximum success (Scrimshaw *et al.*, 1957; Golubjatnikov *et al.*, 1972). In a comparable study in the United States, McGandy reported that plasma cholesterol levels in Eastern U.S. school children were in the 180-290 mg/dl range (McGandy, 1971). He concluded that if atherosclerosis prevention was ever to achieve its maximal effect (McGandy, 1971), attention must be turned to much younger groups. As recently demonstrated in young adults (Hill and Wynder, 1976), reduction of cholesterol intake to <250 mg/day and increasing the P/S ratio from 0.3 to 1.2 significantly lowered serum cholesterol levels. Similar results have been achieved in 500 teenagers at a boarding school, where substantial cholesterol lowering was effected by reduction of fat from 39 to 33% of calories and increasing the P/S ratio from 0.2 to 1.0 (Stare and McWilliams, 1973). There is reason to believe that prudent and simple dietary changes similar to those suggested by Hill and Wynder (1976) and Stare and McWilliams (1973) would successfully and safely lower plasma cholesterol in a majority of American children.

5. *Neonatal and Infantile Cholesterol Levels and Their Relationship to Diet*

In all populations, plasma cholesterol increases sharply within days after birth and is markedly affected in the first years of life by fat and cholesterol content of diet. Generally higher mean cholesterol concentrations have been reported in breast-fed infants as compared to infants receiving vegetable oil-rich, cholesterol-poor infant formulae (Tsang and Glueck, 1975; Fomon and Bartels, 1960; Fomon *et al.*, 1970; Goalwin and Pomeranze, 1962; Sweeny *et al.*, 1961; Fomon, 1974). During the first 12–24 hours, neonatal plasma cholesterol values increase significantly with maximum levels reached by 4–5 days

(Persson and Grentz, 1966; Rafestedt, 1955; Senn and McNamara, 1937). Plasma cholesterol at birth is about 65 mg/dl, increases to 80–90 mg/dl in 24 hours, and to about 120–140 mg/dl in several days (Tsang and Glueck, 1975; Persson and Grentz, 1966; Rafestedt, 1955; Senn and McNamara, 1937). The postpartum increase in cholesterol is predominantly in low-density lipoprotein cholesterol (C-LDL) with the ratio of high-density lipoprotein cholesterol (C-HDL)/C-LDL falling sharply from birth to 10 days (Rafestedt, 1955). In premature infants, plasma cholesterol levels rise in a fashion similar to full-term infants during the first few weeks of life, but the quantitative increments are much less marked (Rafestedt, 1955; Jezerniczky and Csorba, 1971).

Breast and cow's milk have more cholesterol and saturated fat and much less polyunsaturated fat than most commercially prepared proprietary formulas (Stare and McWilliams, 1973; Tsang and Glueck, 1975; Fredrickson and Breslow, 1973). Plasma cholesterol levels at 6–8 weeks of life in infants receiving milk with 35% of the calories as butterfat were 136 mg/dl, notably higher than 88 mg/dl in infants who received vegetable oil formulations which contained 20- to 30-fold more lineoleic acid than the butterfat milk preparation (Sweeny *et al.*, 1961). Similar findings have been reported in comparison of a evaporated milk formula and polyunsaturate-rich propietary formula (Pomeranze *et al.*, 1960). At 2 weeks of age, mean plasma cholesterols were 98 and 84 mg/dl, respectively, at 4 weeks 128 and 98 mg/dl, at 8 weeks 162 and 123 mg/dl, and at 12 weeks 187 and 127 mg/dl (Pomeranze *et al.*, 1960). Comparable results have been reported for premature infants (Gyorgy *et al.*, 1963). Premature infants on cow's milk (at 2 weeks of age) had mean plasma cholesterol levels of 144 mg/dl as compared with 95–108 mg/dl on vegetable oil formula, soy formula, or soy with coconut formula. At 3 months, mean cholesterol in infants on cow's milk was 145 mg/dl compared with 95–122 mg/dl in infants on formula. At 6 months of age, the respective values were 163 mg/dl, compared with 119 to 139 mg/dl (Gyorgy *et al.*, 1963).

Breast and cow's milk cholesterol content may be relatively similar, but breast milk is considerably richer in linoleic acid (Insull *et al.*, 1959). Breast milk polyunsaturate content may be appreciably increased by augmented intake of polyunsaturates by the lactating mother (Insull *et al.*, 1959). Feeding cows formalin-treated grains which allow the polyunsaturates to escape the saturating effects of the first rumen has a similar effect on cow's milk (Nestel *et al.*, 1974).

6. Longitudinal Effects of Infant Feeding on Plasma Cholesterol Levels

The studies of Hahn and Kirby (1973), Kubat *et al.* (1967), Sidelman *et al.* (1972), and Rieser (1971) have examined the hypothesis (Reiser, 1971; Fomon, 1971; McBean and Speckman, 1974) that the intake of dietary cholesterol early in life may initiate enzyme systems responsible for serum cholesterol degradation later in adulthood. Male, but not female rats, and swine, when fed high-cholesterol diets during infancy, had lower serum cholesterol levels when grown than did control groups initially fed low-cholesterol diets. Kubat *et al.* (1967) and Hahn and Kirby (1973) evaluated effects of premature weaning versus normal weaning on lipid concentrations in rats. Rats weaned prematurely to a low-cholesterol diet had significantly higher serum cholesterol levels as adults than did normally weaned littermates consuming dam's milk higher in cholesterol content for a longer period. These studies inferred that dietary cholesterol in early infancy somehow induced enzyme systems responsible for the normal function of a feedback mechanism relating the biosynthesis and catabolism of cholesterol. The hypothesis has been reviewed in some depth by Fomon (1974). The studies have been done in relatively few animals and merit further exploration (Fomon, 1974), especially in primates, to determine their generalizability.

This "early-cholesterol-intake" hypothesis (Hahn and Kirby, 1973; Kubat *et al.*, 1967; Sidelman, 1972; Reiser, 1971; Fomon, 1971; McBean and Speckman, 1974) has been examined in humans by Glueck *et al.* (1972, 1976) and by Friedman and Goldberg (1975). In both normal and hypercholesterolemic infants Glueck *et al.* (1972, 1976) failed to demonstrate a relationship between "low" and moderate intakes of cholesterol during early infancy and plasma concentration of cholesterol at age 1. Friedman and Goldberg (1975) studied 1-year-old infants who had been fed either a human milk or a prepared formula (cholesterol free) from birth to 4 months of age. The formula-fed children had significantly *lower* serum cholesterol levels at 1 year of age. By 18–24 months of age the two groups had similar serum cholesterol levels. Friedman and Goldberg (1975), like Glueck *et al.* (1972, 1976), could not therefore demonstrate any deleterious effect of initial low cholesterol feeding in humans. Three- to five-year follow-up studies of some of the infants initially evaluated by Glueck *et al.* (1976) have subsequently failed to show a delayed increase in plasma cholesterol following early cholesterol restriction. Hence, studies in children fail to provide any evidence which indicates that early low-cholesterol intake may provide a "rebound" phenomenon later in life which would lead to diminished ability to catabolize cholesterol (Hahn and Kirby, 1973; Kubat *et al.*, 1967; Sidelman, 1972; Reiser, 1971; Fomon, 1971; McBean and Speckman, 1974). The data in humans, in fact, suggests the opposite, in that low-cholesterol intakes, particularly in children with familial hypercholesterolemia, may facilitate lower plasma cholesterol levels later in life (Glueck *et al.*, 1972, 1976; Friedman and Goldberg, 1975).

7. Nutritional Approaches to Treatment of Pediatric Familial Hyperlipoproteinemias

A considerable proportion of premature atherosclerosis can be accounted for by the presence of the "monogenic" familial hyperlipoproteinemias which can be diagnosed in childhood: familial hypercholesterolemia, familial hypertriglyceridemia, and familial combined hyperlipoproteinemia (Tamir *et al.*, 1972; Glueck *et al.*, 1974b; Chase *et al.*, 1974; Blumenthal *et al.*, 1975; Stone *et al.*, 1974; Kwiterovich *et al.*, 1974; Glueck *et al.*, 1973a; Kwiterovich and Margolis, 1973; Kwiterovich and Farah, 1975; Goldstein *et al.*, 1973; Glueck *et al.*, 1973b,c; James *et al.*, 1974, 1975). Treatment of these familial hyperlipoproteinemias in childhood requires a more stringent dietary approach than that for acquired hyperlipoproteinemias.

Four recent studies (Tamir *et al.*, 1972; Glueck *et al.*, 1974b; Chase *et al.*, 1974; Blumenthal *et al.*, 1975) have testified to the close relationship between familial hyperlipidemias, atherosclerosis, and pediatric diagnosis of hyperlipidemia. Tamir *et al.* (1972) studied hypercholesterolemic Israeli families where the proband had sustained myocardial infarction prior to age 42. Thirty of 110 children born to these kindreds were found to have elevated lipoprotein levels. Glueck *et al.* (1974b) evaluated 70 kindreds from the United States where the proband had a myocardial infarction prior to age 50. Thirty-one percent of 223 children born to these 70 probands had hyperlipoproteinemia, with type II-A patterns in 15%, type II-B in 4%, and type IV in 11%. Chase *et al.* (1974) evaluated 179 offspring of 71 probands who had sustained myocardial infarction before age 50, and found blood lipid elevations in 29% of these children. Relatively similar findings were made by Blumenthal *et al.* (1975). These studies indicate that the concept of screening for hyperlipidemia can be very profitably applied to children of parents suffering myocardial infarction before age 50.

7.1. Familial Hypercholesterolemia: Diagnosis and Dietary Therapy

Familial hypercholesterolemia, inherited as an autosomal dominant trait, is characterized by major elevations in total and LDL cholesterol and occasionally by modest increments in triglyceride and VLDL cholesterol (Fredrickson and Breslow, 1973; Stone *et al.*, 1974; Kwiterovich *et al.*, 1974). Homozygotes have extraordinary elevations of cholesterol (800–1000 mg/dl) and often have arteriosclerotic cardiovascular disease before age 10 (Glueck *et al.*, 1974a). At the cellular level, the homozygous state is distinctively characterized by a lack of feedback responsiveness of cultured human fibroblast to LDL due to deficient or absent LDL receptors on the cell wall and a basal 40- to 60-fold increment in synthesis of HMG CoA reductase molecules (Brown and Goldstein, 1975).

Familial hypercholesterolemia is probably the most commonly diagnosed form of familial hyperlipoproteinemia in children (Glueck *et al.*, 1974a; Kwiterovich *et al.*, 1974). Approximately 50% of children born to matings of familial hypercholesterolemia times normal will have hypercholesterolemia (Kwiterovich *et al.*, 1973, 1974). Familial hypercholesterolemia can be ascertained by measurement of cord blood cholesterol or LDL cholesterol (Tsang and Glueck, 1975; Fredrickson and Breslow, 1973; Kwiterovich *et al.*, 1973; Goldstein *et al.*, 1974; Darmady *et al.*, 1972; Greten *et al.*, 1974; Tsang *et al.*, 1974b, 1975) coupled with exhaustive family study. The prevalence of neonatal familial hypercholesterolemia in free-living unselected populations appears to be approximately one in 250 live births in Cincinnati and Seattle (Goldstein *et al.*,

1974; Tsang *et al.*, 1974b, 1975). Greten *et al.* (1974) found a somewhat higher prevalence of nearly 1% in Germany. Kwiterovich *et al.* (1973) reported that half of the children born to parents with well-documented familial hypercholesterolemia had elevations of cord blood LDL cholesterol at birth. At follow-up at age 1, most children with neonatal familial hypercholesterolemia on moderate to high cholesterol intake have distinctive elevations of cholesterol. A single dissenting study, however, has been published by Darmady *et al.* (1972). They concluded that cord blood cholesterol determinations had little utility in the diagnosis of familial hypercholesterolemia, and found poor correlation between cord blood cholesterol and plasma cholesterol at age 1.

At 1-year follow-up, in Tsang and Glueck's study, using rigourous criteria for the diagnosis of familial hypercholesterolemia (three-generation vertical transmission of hypercholesterolemia and/or the presence of tendon xanthomas) eight neonates with hyperlipidemia had evidence of dominant inheritance (0.44%) (Tsang *et al.*, 1974b). Only three of these eight still had hypercholesterolemia at 1 year. In four others, normal plasma cholesterol levels were thought to be due to cholesterol and saturate-poor, polyunsaturate-rich diets which modified continued expression of the phenotype (Tsang *et al.*, 1974b). In a study by Kwiterovich *et al.* (1973), 11 of 12 infants who had been abnormal as neonates remained hypercholesterolemic at 1–2 years of age. The one infant with normal cholesterol at follow-up had been on a cholesterol- and saturated-fat-restricted diet. None of the seven infants who were normal at birth and born of hypercholesterolemic parents developed hyperlipidemia later.

In infants with familial hypercholesterolemia, plasma cholesterol levels at 6–24 months of age are lower on low-cholesterol diets than on moderate- to high-cholesterol intake (Tsang *et al.*, 1974b, 1975). Plasma cholesterols of hypercholesterolesterolemic infants on a low-cholesterol diet are often indistinguishable from those of normal infants receiving *ad libitum* diets (Kwiterovich *et al.*, 1973; Greten *et al.*, 1974; Tsang *et al.*, 1974b, 1975). In older hypercholesterolemic children, low-cholesterol diets have reduced plasma cholesterol levels by 24% under strict hospital conditions (Segal *et al.*, 1970). In hypercholesterolemic children, ages 7–21, on a low-cholesterol diet as outpatients, plasma cholesterol levels were normalized in 11 of 36 children (Glueck *et al.*, 1973b). Six of 10 children who were not normalized on diet alone attained normal cholesterol levels when cholestyramine resin was added to the diet (Glueck *et al.*, 1973b).

Between ages 2 and 7 in children heterozygous for familial hypercholesterolemia, diet had variable effect (Glueck *et al.*, 1976; Larsen *et al.*, 1974; Glueck and Tsang, 1972). Effects of a cholesterol and saturate-poor, polyunsaturate-rich diet on plasma total and low-density lipoprotein cholesterol (C-LDL) have been assessed in 23 children, ages 2–7, heterozygous for well-documented familial hypercholesterolemia (Glueck *et al.*, 1976). Sixteen of the 23 children, whose mean age at inception of diet was 4.8 years, sustained 10.5 and 11.3% reductions in total and LDL cholesterol after 6 months on the diet, $p < 0.001$. Six of these 23 children with familial hypercholesterolemia, all 2 years old, had been previously maintained on low-cholesterol diets since age

1 or earlier. At age 2, 3 of the 6 had normal plasma cholesterol levels while on diet. After 1 year of follow-up, plasma cholesterol was normal in 5 of these 6 children. Dietary therapy in 2- to 7-year-old children heterozygous for familial hypercholesterolemia provided a mean overall reduction of total and LDL cholesterol of 6–15%, and appeared to be most effective when instituted in children at ages 1 or 2.

7.2. *Familial Hypertriglyceridemia*

Familial hypertriglyceridemia (Glueck *et al.*, 1973c) is less commonly diagnosed in childhood than is familial hypercholesterolemia (Glueck *et al.*, 1974a), probably because the genetic disorder is less penetrant. Diagnosis of familial hypertriglyceridemia cannot be made by measurement of cord blood triglyceride levels, since cord blood triglycerides primarily reflect external stressful environmental conditions relative to the delivery of the fetus (Tsang *et al.*, 1974a; Fosbrooke and Wharton, 1973). It is, however, important to diagnose pediatric familial type IV because of exceptional sensitivity to diet. Approximately 20–25% of children at genetic risk for hypertriglyceridemia have elevated triglyceride. Weight reduction and the NIH type IV diet were uniformly effective in reducing triglyceride levels to normal (Glueck *et al.*, 1973c).

7.3. *Familial Combined Hyperlipidemia*

Goldstein *et al.* (1973) have delineated a disorder, familial combined hyperlipidemia, genetically distinct from familial hypercholesterolemia. Relatives of probands had hypercholesterolemia, hypertriglyceridemia, or mixed elevations of both cholesterol and triglyceride. Familial combined hyperlipidemia appeared to be transmitted as a simple Mendelian autosomal dominant trait.

Cord blood studies by Goldstein *et al.* (1974) suggested that this disorder also could be diagnosed at the time of birth, but adult studies indicated that the trait was not fully expressed until adulthood (Goldstein *et al.*, 1973). In the studies of Glueck *et al.* (1973a), however, the disorder was highly penetrant in childhood. Not enough experience with children with the type II-A, type II-B, or type IV lipoprotein phenotype from these kindreds has, however, been accumulated to determine if their response to diet or diet plus cholestyramine differs from that of children from kindreds with familial hypercholesterolemia.

8. *Obesity*

Another area of pediatric nutrition which relates to the development of atherosclerosis may well be overnutrition and obesity. Obesity is linked with arteriosclerotic cardiovascular disease and probably is an additive cardiac risk factor along with hypertension, hyperlipidemia, and diabetes (Fiser and Fisher, 1975; Dawber *et al.*, 1962). Considerable evidence has been accrued which indicates a relationship between accretion of excess body weight in childhood and obesity in adult life (Dawber *et al.*, 1962; Abraham *et al.*, 1971; Abraham and Nordsieck, 1960; Eid, 1970; Cheek *et al.*, 1974; Hirsch, 1974; Bjorntorp and Sjorstrom, 1971; Salans *et al.*, 1973). Pediatric overnutrition also may lead to eventual life-long obesity by setting a persistent overeating format (Mellander *et al.*, 1959; Fomon *et al.*, 1971). Inasmuch as obesity increases the size and number of adipocytes and makes them relatively insulin-resistant, obesity may also become self-perpetuating (Cheek *et al.*, 1974; Hirsch, 1974; Bjorntorp and Sjorstrom, 1971; Salans *et al.*, 1973). Obesity also has a strong genctic component (Weil, 1974).

Recent changes in infant feeding habits may also predispose to obesity. Bottle-fed infants often ingest excessive calories as compared to breast-fed infants (Fomon, 1974). Since the dietary intake of infants is predominantly volume-limited and not calorie-limited, infants are particularly susceptible to overfeeding by bottle (Fomon *et al.*, 1964). With bottle feeding, there is maternal pressure on the infant to "finish the bottle" (Fomon, 1974; Shukla *et al.*, 1972; Oates, 1973), which often provides excessive caloric intake. In addition, the introduction of solid foods very early, while the infant still is taking a full milk diet, leads to excess calorie and excess protein intake (Fomon *et al.*, 1970). Overall, there are many vectors active in infant nutritional patterns which might predispose to obesity. Prudent approaches to prevent the development of obesity in infants and children would be very valuable lest the obesity, once developed, become self-sustaining.

9. Hypertension

Although the lineal connection of hypertension with infant nutrition is probably more tenuous than the connections for hypercholesterolemia or obesity, it is important to examine this relationship since hypertension is responsible for at least 25% of cardiovascular deaths which occur before age 60 in the U.S. population. As shown by Dahl (1961), early feeding of salt to genetically sensitive rats predisposes them to later hypertension. Familial aggregation of blood pressure may well include children (Johnson *et al.*, 1965). Certain epidemiological studies suggest a linear relationship between salt intake (on an average) and prevalence of hypertension (Dahl, 1972). After reviewing the problem in detail, the Committee on Nutrition of the American Academy of Pediatrics concluded "there is a reasonable possibility that a low salt intake begun early in life may protect, to some extent, persons at risk (from developing hypertension)" (American Academy of Pediatrics, 1974).

The salt requirements of infants have been calculated to be about 6–8 meq/day (sodium), yet 6-month-old infants often receive from 25–75 meq/day (Dahl, 1961). The relatively high salt intake in infancy probably reflects the common use of salt in the infants' cereals, meats, and vegetables (Guthrie, 1968; Dahl, 1968).

Evaluation of the salt intake of infancy, as well as familial aggregations of hypertension in families (Johnson *et al.*, 1965), may be very important, since hypertension may be a self-perpetuating phenomenon. Persistent hypertension may produce changes in arterioles, leading to increased peripheral resistance and to a higher threshold for "barostats." Factors which initiate hypertension may then be rather different from those which maintain it, further stressing the importance of recognizing hypertension, or its precursors, in childhood.

10. Safety and Nutritional Adequacy of Fat- and Cholesterol-Modified Diets

An effective *general* program of dietary modification which begins in childhood must be carefully scrutinized for potential hazards of modifying traditional dietary practices. Concerns about the safety of diets low in cholesterol for infants and children are based on the following considerations.

10.1. Hypothesis 1

A relatively high cholesterol intake (breast milk) has been speculated to be necessary in the first year of life to provide satisfactory induction of cholesterol catabolic systems which would otherwise be impaired later in life. Although evidence has been proffered in rats and swine that early cholesterol feeding potentiates catabolic systems later in life (Hahn and Kirby, 1973; Kubat *et al.*, 1967; Sidelman, 1972; Reiser, 1971; Fomon, 1971; McBean and Speckman, 1974), no such data have been observed in studies of normal and hypercholesterolemic infants and children (Glueck *et al.*, 1972, 1976; Friedman and Goldberg, 1975).

10.2. Hypothesis 2

Cholesterol might be an essential nutrient in life and central nervous system myelination may be affected by low cholesterol intake in early infancy. Normal brain maturation progresses at a critical rate during the perinatal period and first 2 years of infant life (Davison, 1973) and cholesterol is required for optimal central nervous system myelination. However, brain cholesterol is practically entirely derived from endogenous biosynthesis (Tsang and Glueck, 1975; Chevallier, 1967; Morris and Cahikoff, 1961), and thus it would appear unlikely that dietary cholesterol would play any important role. The data of Plotz *et al.* (1968) indicate that in the brain of the human fetus, cholesterol is synthesized *in situ* from glucose. After birth, exogenous cholesterol appears to be of little importance in myelination.

10.3. Hypothesis 3

Cholesterol lowering modification of diets might result in deficiencies in essential nutrients. Cholesterol-modified diets would not be restricted in lean meat, fish, poultry, vegetable sources of protein, or skim milk (Fomon, 1974). The usual pediatric protein intake is much greater than recommended, and moderate reduction in protein intake (restriction in eggs and dairy products rich in cholesterol) would be of trivial significance (Fomon, 1974). On strict low-cholesterol diets, iron requirements might not be met and might necessitate iron supplementation (Larsen *et al.*, 1974).

10.4. Hypothesis 4

Cholesterol-modified diets might elevate plasma phytosterol levels in infancy and childhood. A fourth potential effect of long-term low-cholesterol, vegetable-oil-rich diets would be to elevate plasma phytosterol levels (Mellies *et al.*, 1976a,b). Low-cholesterol diets are usually high in polyunsaturates and plant sterols, the phytosterols (Mellies *et al.*, 1976a). Three- to fivefold increases in plasma phytosterols in children on cholesterol-lowering diets and in infants taking vegetable-oil-rich commercial formulas have been demonstrated (Mellies *et al.*, 1976a). The implications of such elevations over time are unknown, although infants on phytosterol-rich diets rapidly accrue phytosterols in normal aortic tissue (Mellies *et al.*, 1976b).

10.5. Hypothesis 5

There might be an enhanced incidence of gallstones by virtue of a polyunsaturate-rich, low-cholesterol diet. This potential problem related to diet was outlined by Sturdevant *et al.* (1973), who reported an enhanced incidence of gallstones in a group of adult males taking a polyunsaturate-rich, low-cholesterol diet, when compared to a control group. Sturdevant *et al.* speculated that a diet with a high dietary plant sterol/cholesterol ratio or a high dietary polyunsaturate/saturate ratio or both might be associated with increased risk of gallstone formation. Confirmatory evidence of these findings is awaited.

Finally, there is a vague and unspecified uneasiness that proposed dietary changes might be followed by currently unanticipated adverse consequences. The utility of making dietary changes must be balanced against this anxiety, and against the supposition that the diet would have significant beneficial effects. At the current stage of the art, it appears that the benefits will more than likely overshadow the known and unanticipated risks.

11. Conclusion

The development of atherosclerosis relates to a cobweb of tangled risk factors which include prominently the following: elevated cholesterol and triglyceride, hypertension, diabetes, cigarette smoking, obesity, and many other lesser factors. Nutrition in childhood may well relate to the eventual lipid levels in adulthood and to development of atherosclerosis. Obesity is probably partially determined by patterns of childhood diet and weight gain. Precursors to hypertension may lie in high salt intake in infancy and childhood. The ability to recognize aggregations of hypertension in childhood and longitudinal follow-up will be helpful. Cigarette smoking habits usually are fixed in the teens.

Initiation of prudent preventive measures directed at the primary amelioration of these risk factors should be in children since the pathologic mechanisms of atherosclerosis are present in children and are accelerated in the second and third decade, and because certain lifelong patterns may be formed in behavior early in childhood, particularly food preferences. Prudent attention could be focused on cardiovascular risk factors in the total pediatric population. In the small number of children with well-defined monogenic hypercholesterolemia or hypertriglyceridemia, known to be at exceptionally high risk as adults, intervention at an early age is especially important.

It is not yet clear whether effective ablation or amelioration of risk factors will retard and/or prevent the development of atherosclerosis. Provided that such interventions are without substantial risk, recognition of factors in pediatric nutrition which relate to the development of atherosclerosis is important as a pediatric approach to the prevention of atherosclerosis.

ACKNOWLEDGMENTS. A portion of this work was carried out during Dr. Glueck's tenure as an Established Investigator of the American Heart Association. A portion of this work was supported by the General Clinical Research Center, Grant RR 00068-15.

12. References

Abraham, S., and Nordsieck, M., 1960, Relationship in excess weight in children and adults, *Public Health Rep.* **75**:263–273.

Abraham, S., Collins, G., and Nordsieck, M., 1971, Relationships of childhood weight status to obesity in adults, *HSMHA Health Rep.*, **86**:273–283.

American Academy of Pediatrics, Committee on Nutrition, 1972, Childhood diet and coronary heart disease, *Pediatrics* **49**:305–307.

American Academy of Pediatrics, Committee on Nutrition, 1974, Salt intake and eating patterns of infants and children in relation to blood pressure, *Pediatrics* **53**:115–212.

Atherosclerosis Study Group and Epidemiology Study Group of the Inter-Society Commission for Heart Disease Resources, 1970, Primary prevention of the atherosclerotic disease, *Circulation* **42**:A55–A95.

Berenson, G. S., Pargaonkar, P. B., Srinivasan, S. R., Dalfereo, E. R., Jr., and Radhakrishnamurthy, B., 1974, Studies of serum lipoprotein concentrations in children, *Clin. Chim. Acta* **56**:65–74.

Bergman, S. G., 1975, The atherosclerosis problem—pediatric and dietary aspects, *Clin. Pediatr.* **14**:61–65.

Bjorntorp, P., and Sjorstrom, L., 1971, Number and size of adipose tissue fat cells in relation to metabolism in human obesity, *Metabolism* **20**:703–713.

Blumenthal, S., Jesse, M. J., Hennekens, C. H., Klein, T. E., Ferrer, P. L., and Gourley, J. E., 1975, Risk factors for coronary artery disease in children of affected families, *J. Pediatr.* **87**:1187–1192.

Brown, M. S., and Goldstein, J. L., 1975, Regulation of activity of low density lipoprotein receptor in human fibroblasts, *Cell* **6**:307–316.

Chase, H. P., Oquin, R. J., and O'Brien, D., 1974, Screening for hyperlipidemia in childhood, *J. Am. Med. Assoc.* **230**:1535–1537.

Cheek, D. B., Pania, A., and White J., 1974, Overnutrition, in: *Overnutrition in Fetal and Postnatal Cellular Growth* (D. B. Cheek, ed.), pp. 453–474, John Wiley & Sons, Inc., New York.

Chevallier, F., 1967, Dynamics of cholesterol in rats, studied by the isotopic equilibrium method, *Adv. Lipid Res.* **5**:209–239.

Clarke, R. P., Merrow, S. B., Morse, E. H., and Keyser, D. E., 1970, Interrelationships between plasma lipids, physical measurements and body fatness of adolescents in Burlington, Vermont, *Am. J. Clin. Nutr.* **23**:754–763.

Committee on Diet and Heart Disease of the National Heart Foundation of Australia, 1974, Dietary fat and coronary heart disease, a review, *Med. J. Aust.* **1**:575–579, 616–620, 663–668.

Connor, W. E., 1961, Dietary cholesterol and the pathogenesis of atherosclerosis, *Geriatrics* **16**:407.

Council of Foods and Nutrition, 1972, Diet and coronary heart disease—a council statement, *J. Am. Med. Assoc.* **222**:1647.

Dahl, L. K., 1961, Effects of chronic excess salt feeding; induction of self-sustaining hypertension in rats, *J. Exp. Med.* **114**:231–236.

Dahl, L. K., 1968, Salt in processed baby foods, *Am. J. Clin. Nutr.* **21**:787–792.

Dahl, L. K., 1972, Salt and hypertension, *Am. J. Clin. Nutr.* **25**:231–244.

Darmady, J. M., Fosbrooke, A. S., Lloyd, J. K., 1972, Prospective study of serum cholesterol levels during first year of life, *Br. Med. J.* **2**:685–688.

Davidson, M., 1975, Obesity and the question of cholesterol reduction in infancy and childhood, *Pediatr. Ann.* **4**:101–111.

Davison, A. N., 1973, Biology of normal brain development, *Pediatr. Res.* **7**:48.

Dawber, T. R., Kannel, W. B., Revotskie, N., and Kagan, A., 1962, The epidemiology of coronary heart disease—The Framingham Enquiry, *Proc. R. Soc. Med.* **55**:265–271.

deHaas, J. H., 1973, Primary prevention of coronary heart disease, *Overdruk Hart Bull.* **4**:3–11.

Deutscher, S., Epstein, F. H., and Kjelsberg, M. O., 1966, Familial aggregation of factors associated with coronary heart disease, *Circulation* **33**:911–924.

Drash, A., and Hengsteinberg, F., 1972, The identification of risk factors in normal children in the development of atherosclerosis, *Ann. Clin. Lab. Sci.*

Duplessis, J. P., Viver, F. R., and Delange, D. J., 1967, The biochemical evaluation of nutrition status of urban school children aged 7–15 years; serum cholesterol and phospholipid levels and serum and urinary amylase activities, *S. Afr. Med. J.* **41**:1212–1216.

Eid, E. E., 1970, Follow-up study of physical growth of children who had excessive weight gain in first six months of life, *Br. Med. J.* **2**:74–76.

Enos, W. F., Beyer, J. C., and Holmes, R. H., 1955, Pathogenesis of coronary disease in American soldiers killed in Korea, *J. Am. Med. Assoc.* **158**:912–914.

Fallat, R. W., Tsang, R. C., and Glueck, C. J., 1974, Hypercholesterolemia and hypertriglyceridemia in children, *Prev. Med.* **3**:390–405.

Fiser, R. H., Jr., and Fisher, D. A., 1975, Current understanding of the pathogenesis of obesity, *South. Med. J.* **68**:931–933.

Fomon, S. J., 1971, A pediatrician looks at early nutrition, *Bull. N.Y. Acad. Med.* **47**:569–578.

Fomon, S. J., 1974, *Infant Nutrition,* W. B. Saunders Company, Philadelphia.

Fomon, S. J., and Bartels, D. J., 1960, Concentrations of cholesterol in serum of infants in relation to diet, *Am. J. Dis. Child.* **99**:27–30.

Fomon, S. J., Owen, G. M., and Thomas, L. N., 1964, Milk or formula volume ingested by infants fed ad libitum, *Amer. J. Dis. Child.* **108**:601.

Fomon, S. J., Filer, J., Jr., Thomas, L. N., 1970, Growth and serum values of normal breastfed infants, *Acta Paediatr. Scand.* **59**:(suppl. 202):1–14.

Fomon, S. J., Thomas, L. N., Filer, L. J., Jr., Ziegler, E. E., and Leonard, M. T., 1971, Food consumption and growth of normal infants fed milk-based formulas, *Acta Paediatr. Scand.* **60**:(suppl. 223):1–36.

Fosbrooke, A. S., and Wharton, B. A., 1973, Plasma lipids in umbilical cord blood from infants or normal and low birth weight, *Biol. Neonate* **23**:330–338.
Fredrickson, D. S., and Breslow, J. L., 1973, Primary hyperlipoproteinemia in infants, *Ann. Rev. Med.* **24**:315–324.
Friedman, G., and Goldberg, S. J., 1975, Concurrent and subsequent serum cholesterol of breast-fed and formula-fed infants, *Am. J. Clin. Nutr.* **28**:42–45.
Glueck, C. J., and Tsang, R. C., 1972, Pediatric familial type II hyperlipoproteinemia: Effects of diet on plasma cholesterol in the first year of life, *Am. J. Clin. Nutr.* **25**:224–230.
Glueck, C. J., Tsang, R., Balistreri, W., and Fallat, R. W., 1972, Plasma and dietary cholesterol intake on subsequent response to increased dietary cholesterol, *Metabolism* **21**:1181–1192.
Glueck, C. J., Fallat, R. W., Buncher, C. R., Tsang, R. C., and Steiner, P. M., 1973a, Familial combined hyperlipoproteinemia: studies in 91 adults and 95 children from 33 kindreds, *Metabolism* **22**:1403–1428.
Glueck, C. J., Fallat, R. W., and Tsang, R. C., 1973b, Pediatric familial type II hyperlipoproteinemia: therapy with diet and cholestyramine resin, *Pediatrics* **52**:669–679.
Glueck, C. J., Tsang, R. C., Fallat, R. W., Buncher, C. R., Evans, G., and Steiner, P. M., 1973c, Familial hypertriglyceridemia: Studies in 130 children and 45 siblings of 36 index cases, *Metabolism* **22**:1287–1309.
Glueck, C. J., Fallat, R. W., and Tsang, R. C., 1974a, Hypercholesterolemia and hypertriglyceridemia in children. A pediatric approach to primary atherosclerosis prevention, *Am. J. Dis. Child.* **128**:569–577.
Glueck, C. J., Fallat, R. W., Tsang, R. C., and Buncher, C. R., 1974b, Hyperlipidemia in progeny of parents with myocardial infarction before age 50, *Am. J. Dis. Child.* **127**:70–75.
Glueck, C. J., Tsang, R. C., Fallat, R. W., and Mellies, M. J., 1976, Effects of diet in children, heterozygous for familial hypercholesterolemia, *Am. J. Dis. Child.* **131**:162–166.
Goalwin, A., and Pomeranze, J., 1962, Serum cholesterol studies in infants, *Arch. Pediatr.* **79**:58–62.
Godfrey, R. C., Stenhouse, N. S., Cullen, K. J., and Blackman, U., 1972, Cholesterol and the child studies of cholesterol levels of Busselton school children and their parents, *Aust. Pediatr. J.* **8**:72–78.
Goldrick, R. B., Sinne, H. P. F., and Whyte, H. M., 1969, An assessment of coronary heart disease and coronary risk factors in a New Guinea highland population, Atherosclerosis: Proceedings of the Second International Symposium, 366.
Goldstein, J. L., Schrott, H. G., Hazzard, W. R., Bierman, E. L., and Motulsky, A. G., 1973, Hyperlipidemia in coronary heart disease II. Genetic analysis of lipid levels in 176 families and delineation of a new inherited disorder, combined hyperlipidemia, *J. Clin. Invest.* **52**:1544–1568.
Goldstein, J. L., Albers, J. J., Schrott, H. G., Hazzard, W. R., Bierman, E., and Motulsky, A. G., 1974, Plasma lipid levels and coronary heart disease in adult relatives of newborns with normal and elevated cord blood lipids, *Am. J. Hum Genet.* **26**:727–735.
Golubjatnikov, R. T., Paskey, T., and Inhorn, S. L., 1972, Serum cholesterol levels of Mexican and Wisconsin school children, *Am. J. Epidemiol.* **96**:36–39.
Greten, H., Wagner, M., and Schettle, G., 1974, Early diagnosis and incidence of familial type II hyperlipoproteinemia—analysis of umbilical cord blood from 1323 newborns, *Deut. Med. Wochschr.* **99**:2553–2557.
Guthrie, H. A., 1968, Infant feeding practices—a predisposing factor in hypertension, *Am. J. Clin. Nutr.* **21**:863–867.
Gyorgy, P., Rose, C. S., and Chu, E., 1963, Serum cholesterol and lipoproteins in premature infants, *Am. J. Dis. Child.* **106**:165–169.
Hahn, P., and Kirby, L., 1973, Immediate and late effects of premature weaning and of feeding a high fat or high carbohydrate diet to weanling rats, *J. Nutr.* **102**:690–696.
Hill, P., and Wynder, E. L., 1976, Dietary regulation of serum lipids in healthy, young adults, *J. Am. Dietet. Assoc.* **68**:25–30.
Hirsch, J., 1974, Cell number and size as a determinant of subsequent obesity, in: *Childhood Obesity* (M. Winick, ed.), p. 15, Wiley-Interscience, New York.

Hodges, R. E., and Krehl, W. A., 1965, Nutrition status of teenagers in Iowa, *Am. J. Clin. Nutr.* **17**:200–210.

Holman, R. L., 1961, A pediatric nutrition problem? *Am. J. Clin. Nutr.* **9**:565–659.

Holman, R. L., McGill, H. C., Jr., Strong, J. P., and Geer, J. C., 1958, The natural history of atherosclerosis. The early aortic lesions as seen in New Orleans in the middle of the 20th century, *Am. J. Path.* **34**:209–235.

ICHD Statement, 1979, in press.

Insull, W., Jr., Hirsch, J., James, T., and Ahrens, E. H., Jr., 1959, The fatty acids of human milk, II. Alterations produced by manipulation of caloric balance and exchange of dietary fats, *J. Clin. Invest.* **38**:443–450.

James, F., Glueck, C. J., Fallat, R. W., Moulton, R., and Millett, F., 1974, Maximal exercise electrocardiography in children with familial hypercholesterolemia, *Circulation* **49, 50**(suppl. III): 266.

James, F., Glueck, C. J., Fallat, R. W., Millett, F., and Kaplan, S., 1975, Exercise electrocardiograms in normal and hypercholesterolemic children (Meeting Abstract), *Pediatr. Res.* **9**:267.

Jezerniczky, J., and Csorba, S., 1971, Serum lipides in prematures during the first three months of life, *Acta Pediatr. Acad. Sci. Hung.* **12**:43–47.

Johnson, B. C., Epstein, F. H., and Kjelsberg, M. D., 1965, Distributions and familial studies of blood pressure and serum cholesterol levels in a total community, Tecumsah, Michigan, *J. Chronic Dis.* **18**:147–160.

Kannel, W. B., and Dawler, T. R., 1972, Atherosclerosis as a pediatric problem, *J. Pediatr.* **80**:544–554.

Kannel, W. B., Castelli, W. P., Gordon, T., and McNamara, P. M., 1971, Serum cholesterol, lipoproteins, and the risk of coronary heart disease, *Ann. Int. Med.* **74**:1–12.

Keys, A., 1970, Coronary heart disease in seven countries, *Circulation* **41**(suppl. 1):I-1–I-211.

Keys, A., Kimura, N., Kusukawa, A., Bronte-stewart, B., Larsen, N., and Keys, M. H., 1958, Lessons from serum cholesterol studies in Japan, Hawaii and Los Angeles, *Ann. Int. Med.* **48**:83–94.

Kimura, N., 1956, *Analysis of 10,000 Postmortem Examinations in Japan, World Trends in Cardiology, I. Cardiovascular Epidemiology,* p. 22, Harper & Row, Inc., New York.

Kubat, K., Hahn, P., and Koldovsky, O., 1967, Cited by P. Hahn and O. Koldovsky, Utilization of nutrients during postnatal development, *Int. Ser. Monogr. Pure Appl. Biol. Div. Zool.* **33**:154.

Kwiterovich, P. O., Jr., and Farah, J. R., 1975, Type V hyperlipoproteinemia biochemical defect of childhood (Meeting Abstr.) *Pediatr. Res.* **9**:352.

Kwiterovich, P. O., Jr., and Margolis, S., 1973, Type IV hyperlipoproteinemia, *Clin. Endocrinol. Metabol.* **2**:41–71.

Kwiterovich, P. O., Jr., Levy, R. I., and Fredrickson, D. S., 1973, Neonatal diagnosis of familial type II hyperlipoproteinemia, *Lancet* **1**:118–121.

Kwiterovich, P. O., Jr., Fredrickson, D. S., and Levy, R. I., 1974, Familial hypercholesterolemia (one form of familial type II Hyperlipoproteinemia), *J. Clin. Invest.* **53**:1237–1249.

Larsen, R., Glueck, C. J., and Tsang, R. C., 1974, Special diet for familial type II hyperlipoproteinemia, *Am. J. Dis. Child.* **128**:67–72.

Lauer, R. M., Connor, W. E., Leaverto, P. E., Reiter, M. A., and Clarke, W. R., 1975, Coronary heart-disease—risk factors in school children, Muscatine Study, *J. Pediatr.* **86**:697–706.

McBean, L. D., and Speckman, E. W., 1974, An interpretive review: Diet in early life and the prevention of atherosclerosis, *Pediatr. Res.* **8**:837–842.

McGandy, R. B., 1971, Adolescence and the onset of atherosclerosis, *Bull. N.Y. Acad. Med.* **47**:590–600.

McGill, H. C., Jr. (ed.), 1968, The geographic pathology of atherosclerosis, *Lab. Invest.* **18**:465–467.

McMillan, G. C., 1973, Development of arteriosclerosis, *Am. J. Cardiol.* **31**:542–546.

McNamara, J. J., Molot, M. A., Stremple, J. F., and Cutting, R. T., 1971, Coronary artery disease in combat casualties in Vietnam, *J. Am. Med. Assoc.* **216**:1185–1187.

Margolis, S., 1975, Hyperlipoproteinemias in adolescence, *Med. Clin. North Am.* **59:**1359–1369.
Mellander, O., Vahlquist, B., and Mellbin, T., 1959, Breast feeding and artificial feeding, *Acta Paediatr. Scand.* **48**(suppl. 116):11–108.
Mellies, M., Glueck, C. J., Sweeney, C., Fallat, R. W., Tsang, R. C., and Ishikawa, T. T., 1976a, Plasma and dietary phytosterols in children, *Pediatrics* **57:**60–67.
Mellies, M. J., Ishikawa, T. T., Glueck, C. J., Bove, K., and Morrison, J., 1976b, Phytosterols in aortic tissue in adults and infants, *J. Lab. Clin. Med.* **88:**914–921.
Mitchell, S. C., and Jesse, M. J., 1973, Risk factors of coronary heart disease—their genesis and pediatric implication. *Amer. J. Cardiol.* **31:**588–590.
Mitchell, S. C., Blount, S. G., Jr., Blumenthal, S., Jesse, M. J. K., and Weidman, W. H., 1972, The pediatrician and atherosclerosis, *Pediatrics* **49:**165–168.
Morris, M. D., and Chaikoff, I. L., 1961, Concerning incorporation of labelled cholesterol, fed to mothers, into brain cholesterol of 20 day old suckling rats, *J. Neurochem.* **8:**226–229.
Nestel, P. J., Havenstein, N., Scott, T. W., and Cook, L. J., 1974, Polyunsaturated ruminant fats and cholesterol metabolism in man, *Aust. N. Z. J. Med.* **4:**497–501.
North, A. F., 1975, Should pediatricians be concerned about children's cholesterol levels? *Clin. Pediatr.* **14:**439–444.
Oates, R. K., 1973, Infant feeding practices, *Br. Med. J.* **2:**762–764.
Persson, B., and Grentz, J., 1966, The pattern of blood lipids glycerol and keton bodies during the neonatal period, infancy and childhood, *Acta Paediatr. Scand.* **55:**353–362.
Plotz, E. J., Kabara, J. J., Davis, M. E., LeRoy, G. V., and Gould, R. G., 1968, Studies on the synthesis of cholesterol in the brain of the human fetus, *Am. J. Obstet. Gynecol.* **101:**534–538.
Pomeranze, J., Gaulwin, A., and Slobody, L., 1960, Infant feeding practices and blood cholesterol levels, *Am. J. Clin. Nutr.* **8:**340–343.
Rafestedt, S., 1955, Studies on serum lipids and lipoproteins in infancy and childhood, *Acta Paediatr. Scand.* **102**(suppl.):1–109.
Reiser, R., 1971, Control of adult serum cholesterol by the nutrition of the suckling, *Circulation* **43**(suppl. II):3.
Salans, L. B., Cushman, W. W., and Weismann, R. E., 1973, Adipose cell size and number in non-obese and obese patients, *J. Clin. Invest.* **52:**929–941.
Schilling, F. J., Christakis, G. J., Bennett, N. J., and Coyle, J. F., 1964, Studies of serum cholesterol in 4244 men and women: an epidemological and pathogenetic interpretation, *Am. J. Public Health* **54:**461–476.
Schubert, W., 1973, Fat nutrition and diet in childhood, *Am. J. Cardiol.* **31:**581–587.
Scrimshaw, N. S., Balsam, A., and Arroyave, G., 1957, Serum cholesterol levels in school children from three socio-economic groups, *Am. J. Clin. Nutr.* **5:**629–633.
Segal, M. M., Fosbrooke, A. S., Lloyd, J. K., and Wolff, O. H., 1970, Treatment of familial hypercholesterolemia in children, *Lancet* **1:**641–644.
Senn, J. J. E., and McNamara, H., 1937, The lipids of the blood plasma in the neonatal period. *Am. J. Dis. Child.* **53:**445–454.
Sherwin, R. M. B., 1974, The epidemiology of atherosclerosis and coronary heart disease, *Postgrad. Med.* **56:**81–87.
Shukla, A., Forsyth, H. A., Anderson, C. M., and Marwah, S. M., 1972, Infantile overnutrition in the first year of life. A field study in Worchestershire, *Br. Med. J.* **4:**507–515.
Sidelman, Z., 1972, The influence of milk cholesterol ingested by suckling rats on serum cholesterol concentrations of the adult rat, *Dairy Sci. Abstr.* **34:**820.
Stamler, J., 1967, Blood pressure as risk for atherosclerosis, in: *Lectures on preventive cardiology*, p. 277, Grune & Stratton, Inc., New York.
Stare, F. J., and McWilliams, M., 1973, in: *Living Nutrition,* John Wiley & Sons, Inc., Inc., New York.
Starr, P., 1971, Hypercholesterolemia in school children, *Am. J. Clin. Pathol.* **56:**515–522.
Stone, M. J., Levy, R. I., Fredrickson, D. S., and Verter, J., 1974, Coronary artery disease in 116 kindred with familial type II hyperlipoproteinemia, *Circulation* **49:**476–488.

Sturdevant, R. A. L., Pearce, M. L., and Dayton, S., 1973, Increased prevalence of cholelithiasis in men ingesting a serum-cholesterol-lowering diet, *New Engl. J. Med.* **288**:24–27.

Sweeny, M. J., Etteldrof, J. N., Dobbins, W. T., Somervill, B., Fischer, R., and Ferrell, C., 1961, Dietary fat and concentrations of lipid in the serum during the first six to eight weeks of life, *Pediatrics* **27**:765–771.

Tamir, I., Bojamower, Y., Levtow, O., Heldenberg, D., Dickerman, Z., and Werbin, B., 1972, Serum lipids and lipoproteins in children from families with early coronary disease, *Arch. Dis. Child.* **47**:808–810.

Tsang, R. C., and Glueck, C. J., 1975, Perinatal cholesterol metabolism, *Clin. Perinatol.* **2**:275–294.

Tsang, R. C., Glueck, C. J., Evans, G., and Steiner, P. M., 1974a, Cord blood hypertriglyceridemia, *Am. J. Dis. Child.* **127**:78–82.

Tsang, R. C., Fallat, R. W., and Glueck, C. J., 1974b, Cholesterol at birth and age 1: comparison of normal and hypercholesterolemic neonates, *Pediatrics* **53**:459–470.

Tsang, R. C., Glueck, C. J., Fallat, R. W., and Mellies, M. J., 1975, Neonatal familial hypercholesterolemia, *Am. J. Dis. Child.* **129**:83–91.

Vlodaver, A. H., Kahn, H. A., and Neufeld, H. N., 1969, The coronary arteries in early life in three different ethnic groups, *Circulation* **39**:541–550.

Weil, W. B., Jr., 1974, Infantile obesity, in: *Childhood Obesity* (M. Winick, ed.), p. 61, Wiley-Interscience, New York.

Whyte, H. M., and Yee, I. L., 1958, Serum cholesterol levels of Australians and natives of New Guinea from birth to adulthood, *Aust. Ann. Med.* **7**:336–339.

Wilmore, J. H., and McNamara, J. J., 1974, Prevalence of coronary heart disease risk factors in boys 8–12 years of age, *J. Pediatr.* **84**:527–533.

13

Iron Deficiency: Behavior and Brain Biochemistry

Rudolph L. Leibel, Daryl B. Greenfield, and Ernesto Pollitt

"Gold is for the mistress—silver for the maid—
Copper for the Craftsman cunning at his trade."
"Good!" said the Baron, sitting in his hall,
"But Iron—Cold Iron—is master of them all."

—Rudyard Kipling, "Cold Iron"

1. Introduction

That severe degrees of undernutrition are associated with behavioral changes in animals and humans has been known since antiquity. The twentieth century has seen the accelerating development of distinctions between behavioral aberrations associated with different types of gross nutritional disturbances (e.g., marasmus and kwashiorkor) as well as critical factor (vitamins, minerals) deficiencies. Beginning with Goldberger's description in 1915 of the dementia caused by a specific nutritional deficiency (later shown to be niacin) and building upon the insights gained through the neurologic impact of various "inborn errors of metabolism," nutrition science has reached a point where even specific dietary amino acids are being examined for their impact on behavior and its apparent neurochemical substrates (Fernstrom, 1976; Fernstrom and Lytle, 1976). Along with this burst of interest in potential neurotransmitters and their dietary precursors, there has been a renewal of interest in various "trace elements" and their role in normal somatic and neural

Rudolph L. Leibel • Rockefeller University, New York, New York. ***Daryl B. Greenfield*** • Boston College, Boston, Massachusetts. ***Ernesto Pollitt*** • School of Public Health, University of Texas, Houston, Texas.

physiology (Burch and Sullivan, 1976). Iron, named *sideros* (star) by the Greeks and believed by them to be a special gift sent to earth by the god Ares, has long been attributed, in both lay and medical thinking, a definite, although often nondescript role in the individual's physical and mental well-being. From the Berus Papyrus (1500 B.C.) to the present-day hyping of Geritol[R] and raisins, mankind has been convinced of the salubrity of iron (Fairbanks *et al.*, 1971).

Iron is the fourth most abundant (4.8% by weight) element in the earth's crust and is probably the major component of the core, which alone accounts for some 32% of the mass of the earth (Harris, 1972). However, its metabolic importance throughout the plant, microbial, and animal kingdom (Harris, 1972) is not simply a biochemical reflection of iron's environmental ubiquity. Rather, iron's crucial role in the energy metabolism of all living cells, through its inclusion in redox enzymes and oxygen-carrying proteins, is based upon physicochemical properties (several stable valence states, ability to bind molecular oxygen, a chemistry strongly influenced by ligands) characterizing the transition metals and permitting their functional role in these proteins (Malmstrom, 1970). As Weinberg (1974) has emphasized, animal, plant, and microbial cells require similar quantities of iron (0.3–1.8 μM) for normal growth. In plants, iron is required for chlorophyll synthesis and is a component of chloroplast and mitochondrial cytochromes and ferredoxin (Devlin and Barker, 1971). Bacterial growth rates and exotoxin synthetic capacities (i.e., virulence) are markedly influenced by ambiant iron availability (Weinberg, 1974). In humans, iron plays important roles in oxygen transport, mitochondrial redox reactions, immunocompetence, and possibly in neurotransmitter synthesis and catabolism and collagen synthesis.

It is widely accepted that iron deficiency is presently mankind's most prevalent, although certainly not most important nutritional disturbance (WHO, 1975). *Severe* iron deficiency and the resultant *severe anemia* may be associated with a number of adverse, in some cases life-threatening, effects (e.g., cerebral hypoxemia, congestive heart failure). Here the impact reflects a generic effect of reduced hemoglobin mass and attendant reduced tissue oxygen delivery, and apparently is not, to any significant extent, the result of tissue iron deficiency per se. Less certain is the impact, if any, of iron deficiency with or without attendant mild to moderate anemia on the general health status of the individual of any age. Aside from a reduction in maximal muscular work capacity which attends relatively minor decrements in hemoglobin concentration (e.g., to 10 g/dl) (WHO, 1975), there are no proven, significant somatic or neurologic disturbances accompanying moderate reductions in hemoglobin mass (due to iron deficiency or blood loss) in otherwise healthy individuals. On the other hand, there are nutritional anemias such as that occurring in B_{12} deficiency in which major neuropsychiatric abnormalities may precede the onset of even mild degrees of anemia (Smith, 1960), and persist long after any associated anemia is remediated. In such instances, it is clear that the factor in question plays an important role in several major organ systems.

The iron-deficient state has been subjected to intense scrutiny regarding just such a multiplicity of physiologic functions for the metal separate from its crucial role in systemic oxygen transport. In addition to the alleged influences on various aspects of behavior and affective states to be discussed in this chapter, iron has been implicated in various aspects of the host–bacterial pathogen interaction (Weinberg, 1974; Ganzoni and Puschmann, 1975; Chandra, 1976; Macdougall *et al.*, 1975; Srikantia *et al.*, 1976; Beisel, 1976); gut mucosal function (Naiman *et al.*, 1964); disorders of lipid metabolism, including both elevations in the plasma triglycerides of pups of iron-deficient rats (Guthrie *et al.*, 1974) and reduced cholesterol levels in patients with low hemoglobins (Elwood *et al.*, 1970); prematurity (MacGregor, 1963); and somatic growth failure (Judisch *et al.*, 1966; Anonymous, 1966). It is important to reiterate that none of these apparent effects of iron have been demonstrated conclusively to be of clinical import. The aim of this chapter is to present a general background for the contention that relatively subtle alterations in nutritional status may alter behavior and then to examine the nature of the behavioral–affective changes attributed to iron deficiency and to indicate potential biochemical mechanisms for these putative clinical effects.

2. Basic Aspects of Iron Metabolism and the Physiology of Anemia

Mammalian iron metabolism is described in great detail in a number of excellent reviews (Harris and Kellermeyer, 1970; Jacobs and Worwood, 1974; Hallberg *et al.*, 1970). For our purposes here, it is necessary only to emphasize certain fundamental aspects of the normal distribution of iron and the ontogeny of its depletion in man. Table I indicates the major compartmental locations of iron in the iron-replete adult male. The enzyme iron, which represents such a small percentage of the body's total iron content, includes iron-containing enzymes and cofactors which are crucial in oxidative phosphorylation, liver-mediated catabolic processes, and catecholamine biosynthesis and degradation. It is among the metabolic pathways presided over by these enzymes that the search for a nonhemoglobin impact of iron deficiency on behavior has centered. It had been taught, more on teleological grounds than anything else, that these enzymes and cofactors represented an inviolate sanctuary even in

Table I. Functional Compartments of Body Iron (Adult Male)

Site	Iron (mg)	Total iron content (%)
Hemoglobin	2500	67
Storage (ferritin hemosiderin)	1000	27
Myoglobin	130	3.5
Labile pool	80	2.2
Enzymes/cytochromes	8	0.2
Transport	3	0.08

the face of the most profound degrees of iron deficiency. Beutler (1957) was one of the first to demonstrate the fallacy of this argument, showing that in rats, liver and kidney cytochrome c were more susceptible to iron depletion than was the circulating hemoglobin mass.

Figure 1 (Fairbanks *et al.*, 1971) represents the current view of the ontogeny of iron deficiency, emphasizing that iron stores are depleted, transferrin undersaturated, and a significant portion of the enzyme compartment adversely affected before any hematologic derangements (fall in hemoglobin mass, mean red cell volume, or mean red cell hemoglobin) are discerned. Recently, Cook *et al.* (1976) reaffirmed that iron-related biochemistries may be significantly deranged in the face of a "normal" hemoglobin level. Even individuals with concomitant signficant reductions in transferrin saturation and ferritin levels and elevations in free erythocyte protoporphyrin levels—all biochemical indicators of iron deficiency—had only a 63% prevalence of anemia (hemoglobin < 13 g/dl in adult males; < 12 g/dl in others 5 years and older). The hemoglobin mass is, then, a crude and end-stage reflection of the actual status of body iron.

In any population, the hemoglobin and iron biochemistry values of functionally anemic and nonanemic individuals are likely to overlap considerably,

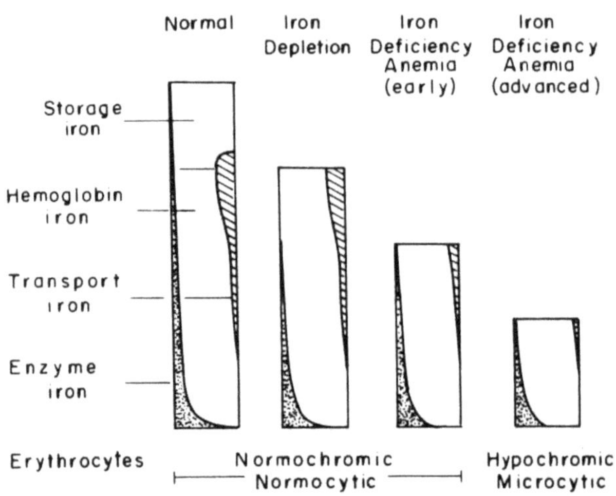

Fig. 1. Stages in the progression of iron deficiency. The earliest stage, iron depletion, is characterized by loss of storage iron. Some diminution in tissue (enzyme) iron and in transport (serum) iron may occur at this stage, but hemoglobin concentration and transport (serum) iron concentration are in the normal range. The next stage (not illustrated here) has been defined as iron deficiency without anemia. Its characteristics are similar to those of iron depletion, except that the serum iron concentration and transferrin saturation are low. When iron-deficiency anemia is mild (early), the correct diagnosis often cannot be made except by evaluation of iron stores; the anemia is commonly normochromic and normocytic or microcytic, and the serum iron concentration may also be normal. Finally, when about half the normal body content of iron has been lost (advanced), the classic features of a moderate-to-severe, hyposideremic, hypochromic, microcytic anemia appear, and the correct diagnosis is usually quite easily made. From Fairbanks *et al.* (1971), p. 164, with permission of the publisher and author.

thus confounding epidemiologic efforts at separating these two subpopulations (Cook *et al.*, 1976; Jacobs and Worwood, 1974). Oski (1973) has emphasized that the definition of anemia as a reduction in red cell or hemoglobin mass fails to consider the most important (and discontinuous) covariant, tissue oxygenation. The ability of various organic phosphates (DPG, ATP, ADP) and hydrogen ions to influence the molecular respiration of the hemoglobin molecule by reducing its affinity for oxygen (Adamson and Finch, 1975) provides a powerful compensatory mechanism assuring sufficient tissue oxygen delivery in the face of marked reductions in circulating hemoglobin mass (Lovric *et al.*, 1975). For this reason, the cardiovascular compensations (increased cardiac output, tachycardia with exercise, reduced pulmonary vascular resistance, shifts in regional blood flow) characteristic of *chronic* anemia are generally not invoked at hemoglobin concentrations above 7–8 g/dl. Likewise, elevations in the ratio of blood lactate to pyruvate (a reflection of tissue hypoxia) are rarely seen at hemoglobins greater than 6 g/dl (Siebert and Ebough, 1967).

The purpose here is to emphasize the related points that hemoglobin is a poor index of an individual's body iron status and of the potential impact of that status on tissue oxygenation, and that any behavioral/affective alterations noted in iron-deficient patients with hemoglobin concentrations above 8 g/dl are unlikely to be the result of tissue hypoxemia secondary to reduced oxygen delivery via hemoglobin. Finch (1970) has described the relative efficacy of various other methods of assessing body iron status. Although bone marrow iron, serum ferritin, transferrin level and saturation, and free erythrocyte protoporphyrin all reflect, with decreasing degrees of accuracy, the status of body iron stores, none provides a reliable index of the status of the iron compartment of most interest in the context of putative behavioral changes in iron deficiency—the enzyme/cofactor compartment. In most clinical studies of iron and behavior, it is only some combination of the above body-iron-dependent parameters which are measured and correlated with the dependent variable in question.

Bainton and Finch (1964) showed that a transferrin saturation less than 15% was associated with exhausted iron stores (by marrow aspiration) in adults. However, Scott and Pritchard (1967) reported that although a transferrin saturation of less than 15% was reliably associated with absent marrow iron, the converse was not the case; negligible iron stores may exist with transferrin saturation greater than 15% in up to two-thirds of cases. More recently, measurements of serum ferritin (Jacobs and Worwood, 1975; Siimes *et al.*, 1974; Cook *et al.*, 1974) have proved useful in providing a more direct and reliable indicator of the presence of reduced or absent iron stores. The diurnal variation of serum iron (Speck, 1968) and the significant number of disorders aside from iron deficiency which can perturb the transferrin level (Harris and Kellermeyer, 1970) render the transferrin saturation a less reliable method than serum ferritin measurement. If, however, such confounding medical situations can be ruled out, the transferrin level itself appears to be inversely proportional (in a linear fashion) to serum ferritin over a wide range of normal and subnormal ferritin values (Lipschitz *et al.*, 1974). The usefulness of transferrin or ferritin in assessing iron overload states is much less clear (Wands *et al.*, 1976).

These iron-related chemistries have been described in some detail because they are used clinically to categorize patients being scrutinized for somatic or neurological sequelae of varying degrees of iron deficiency. Table II gives the criteria for iron deficiency recommended by the AMA's Committee on Iron Deficiency (1968). When compared to the hematologic values in known iron-replete children (Moe, 1965; Marner, 1969) (Table III), it is clear that the AMA figures would tend to underestimate the extent of iron deficiency in a community because the lower-limit Hb is too low by up to 2.5 g/dl (from mean values of Marner and Moe).

The reader must realize, then, when examining the data regarding the possible impact of iron deficiency on behavior, that:

1. Despite their frequent use, the hemoglobin and hematocrit do not provide a direct measurement of the sufficiency of tissue oxygenation.
2. Hematologic measurements fail to provide insight into subtle derangements of iron metabolism.
3. Iron biochemistries (serum iron, transferrin, ferritin, erythrocyte protoporphyrin) are variously sensitive to body iron stores but do not directly assess the status of enzyme/cofactor iron in the patient. As indicated in Fig. 1, there is evidence that tissue iron changes occur relatively early in the development of iron deficiency. However, the extent of the decrement and its impact, if any, on subcellular oxidation and other functions may be the product of complex interactions in which the actual enzyme content of a tissue is only one factor.

Table II. AMA Criteria for Iron Deficiency[a]

1. Iron depletion: decreased or absent bone marrow hemosiderin.
2. Iron deficiency without anemia: absent bone marrow hemosiderin, transferrin saturation less than or equal to 15%.
3. Iron deficiency with anemia: changes per No. 2, plus anemia, defined as a value less than those listed as follows.

	Hb (g/dl)	Hct (%)
6 mo–4 yr	11	33
5–9 yr	11.5	34.5
10–14 yr	12	36
Men	14	42
Women	12	36
Pregnant women	11	33

[a] From the Committee on Iron Deficiency (1968).

Table III. Hematocrit and Hemoglobin Values of Presumably Iron-Replete Children[a]

Age (yr)	Hct	S.D.	Hb	S.D.
1½–3	38.9	2.20	11.8	0.53
3	38.0	2.00	13.0	0.58
4	40.0	2.37	13.5	0.79
5	40.0	2.33	13.8	0.87
6	41.0	2.78	14.0	0.81

[a] Data from Marner (1969) and Moe (1965).

4. So-called "normal" hematologic and iron–biochemical values are arbitrary in nature, and, in general, although useful in nutritional demographic analyses, do not provide useful information about individual patients (Hegsted, 1970).

3. *The Prevalence of Iron Deficiency*

As was mentioned previously, iron deficiency is probably the most common nutritional disturbance in the world. It is estimated that from 20 to 60% of individuals in various populations have latent or overt iron deficiency. Highly efficient means for the capture and reuse of iron from senescent red cells render basal iron requirements rather low in all age groups. Table IV enumerates the sites and average rates of loss for iron in the adult male.

Children (whose iron losses are proportional to body surface area) require the absorption of an additional 0.5–1.0 mg of iron/24 hr to permit optimal somatic iron content (45 mg of iron/kg weight increment) and menstruating females need an extra 0.8 mg of iron absorbed per day to compensate for the iron loss in a 50-ml monthly menstrual blood flow. In the last 2 months of pregnancy, about 2.7 mg of absorbed iron daily is necessary just to meet the fetal demand for iron and to compensate for parturition blood loss. The extensive literature relating to iron requirements in man has recently been reviewed in depth (Bowering and Sanchez, 1976).

Obviously, iron deficiency may develop from a derangement in dietary intake, absorption, intermediary handling, or from excessive loss. Looked at from a worldwide perspective, the majority of instances of iron deficiency are

Table IV. Iron Loss per 24 Hours in Healthy Adult Male[a]

Urine	0.1 mg
Sweat and desquamating skin	0.2 mg
Shedding of intestinal mucosa and bile secretion	0.2 mg
Normal GI blood loss (1 ml/24 hr)	0.5 mg

[a] Data from Pollycove (1972).

secondary to suboptimal dietary intake. The amount of iron available from the diet is not simply a direct correlate of its concentration, since the efficiency of absorption is affected by a number of factors, including age, status of iron stores, the type of food in which the iron is contained, and the types of food comprising the diet as a whole. The elemental iron in cereals and vegetables is, for example, less well absorbed than the heme-bound iron in animal flesh. Furthermore, the addition of amino acids as meat to vegetable iron enhances the absorption of the latter. Thus, in assessing an individual's iron requirements, one needs to consider not only age and sex but also the nature of the diet which is to be the source of the iron (Table V) (Beaton, 1974, p. 507).

Given the preceding information regarding the age groups and medical situations in which iron requirements are greatest plus the generally direct relationships between social class and quality/quantity of diet with regard to protein and iron content, it should not be difficult to predict the results of population studies of iron status. Those at highest risk of suboptimal iron store status—infants, children, and women of childbearing age—are generally found to have the worst iron status relative to the remainder of the community, regardless of how this is defined (hemoglobin, transferrin saturation, ferritin, marrow iron). Within age groups, as expected, individuals of lower socioeconomic status have a higher frequency of iron deficiency than those from upper socioeconomic groups (Department of Health, Education and Welfare, 1968–1970).

It should be mentioned, however, that not all prevalence studies report data in conformity to this anticipated pattern. McFarlane *et al.* (1967), in a study which included information on 98 well women reflecting the national socioeconomic distribution of Scotland, found a 2% prevalence of iron-deficiency anemia (hemoglobin <12 g/dl; transferrin saturation <16%) and a 16.3% prevalence of iron deficiency without anemia (hemoglobin >12 g/dl; transferrin

Table V. Recommended Daily Intakes (mg) of Iron[a]

		Recommended intake according to diet		
Age/sex group	Absorbed iron required[b]	Animal foods below 10% of calories	Animal foods 10–25% of calories	Animal foods above 25% of calories
Infants 5–12 mo[c]	1.0	10	7	5
Children 1–12 yr	1.0	10	7	5
Boys 13–16 yr	1.8	18	12	9
Girls 13–16 yr	2.4	24	18	12
Menstruating women	2.8	28	19	14
Men and postmenopausal women	0.9	9	6	5

[a] From Beaton (1974, p. 507), with permission of the publisher and author. Data from the Report of the FAO/WHO Expert Group on Requirements for Ascorbic Acid, Vitamin D, Folic Acid, Vitamin B_{12} and Iron (1970).
[b] The requirements are estimates of "upper limits" of requirements in normal individuals; they do not take into account any additional requirement imposed by unusual environmental stress, by disease, or by parasitic infection.
[c] For infants 0–4 mo, breast feeding is assumed to be adequate.

saturation <16%). There was, however, no apparent influence of menstrual or socioeconomic status on these findings. Scott and Pritchard (1967) studied 113 Caucasian female college students who had no history of pregnancy or menstrual disorders and found a surprisingly high incidence of depleted marrow iron (one-fourth had no stainable iron; only one-third had more than "1 plus" iron). Using quantitative phlebotomy, they estimated that two-thirds of these women had 350 mg (the amount present in a pint of whole blood) or less of available storage iron. It is interesting to note that the mean hemoglobin of those with no stainable marrow iron was 13 g/dl (compared to 13.7 g/dl in those with "2–4 plus" iron), while their mean transferrin saturation was 17% (30% in those with "2–4 plus" iron). While reinforcing the concept of severe iron deficiency without anemia, this paper also indicates the very high prevalence rates of iron deficiency which can occur even in individuals of presumably high socioeconomic standing.

Many surveys have been conducted examining the prevalence of anemia and/or iron deficiency in different segments of various populations. Virtually all these studies suffer from the types of inherent difficulties enumerated in Section 6.2: (1) it is impossible to choose a physiologically meaningful lower limit of normal for hemoglobin; (2) serum iron has a significant diurnal variation which it may be difficult to control in large surveys, and (3) a portion of the difference in hemoglobin between blacks and whites is apparently not due to socioeconomic/dietary differences between these two groups (Garn *et al.*, 1975). Consequently, prevalence rates for "anemia" and its biochemical correlates are subject to a host of semantic and methodologic problems tending to give each a unique bias and rendering interstudy comparisons hazardous. For this reason it makes little sense to recapitulate the findings of these studies. Instead, and in order to give the reader an idea of recent findings vis-à-vis the U.S. population, we have tabulated the results of several large, current population studies. Tables VI and VII indicate the percentage of persons having "deficient" or "low" hemoglobin values from the Ten State Nutrition Survey (Department of Health, Education and Welfare, 1968–1970). Table VIII presents hemoglobin determinations for three age ranges and for white and black samples from preliminary findings of the first Health and Nutrition Examination Survey (Department of Health, Education and Welfare, 1973).

Studies focusing on prevalence rates of anemia and/or sideropenia in infants and children less than 2 years old are not as numerous as studies of older populations. This very young group is most susceptible to both iron deficiency itself and to the possible adverse effects of this state on the maturing nervous system. Table IX enumerates the findings of a number of surveys performed in this age group.

4. Developmental and Ecological Factors in the Impact of Nutritional and Other Insults on Cognition

Given the high prevalence rates of reduced body iron in infants and young children indicated in Section 3 and given the fact that significant brain devel-

Table VI. Percent of Persons Having Deficient[a] or Low[b] Hemoglobin Values by Age, Sex, and Ethnic Group for Low-Income-Ratio States[c]—Ten-State Nutrition Survey

	Ethnic Group								
	White			Black			Spanish-American		
		Percent with:			Percent with:			Percent with:	
Age (yr)	*N*	Deficient[a] values	Low[b] values	*N*	Deficient[a] values	Low[b] values	*N*	Deficient[a] values	Low[b] values
2, all	52	3.8	15.4	262	19.8	15.6	16	6.3	6.3
2–5, all	313	3.2	9.9	1060	11.2	22.8	158	5.7	3.7
6–12, all	876	1.0	12.9	2435	3.9	31.3	572	2.6	7.2
13–16, males	197	5.1	16.8	543	19.0	30.6	126	11.1	14.3
13–16, females	193	0.5	5.7	628	2.9	23.7	156	1.3	9.6

[a] Deficient hemoglobin values (g/dl) are defined as follows for both males and females: 2 yr, ≤ 9.0; 2–5 yr, ≤ 10; 6–12 yr, ≤ 10. At 13–16 yr for males is ≤ 12; for females is ≤ 10.0.

[b] Low hemoglobin values (g/dl) are defined as follows for both males and females: 2 yr, 9.0–9.9; 2–5 yr, 10.0–10.9; 6–12 yr, 10–11. At 13–16 yr for males it is defined as 12.0–12.9, and for females 10.0–11.0.

[c] There were five low-income-ratio states: Kentucky, Louisiana, South Carolina, Texas, West Virginia. In these states more than half of the families surveyed were living at below poverty level.

Table VII. Percent of Persons Having Deficient[a] or Low[b] Hemoglobin Values by Age, Sex, and Ethnic Group for High-Income-Ratio States[c]—Ten-State Nutrition Survey

	Ethnic Group								
	White			Black			Spanish-American		
		Percent with:			Percent with:			Percent with:	
Age (yr)	*N*	Deficient values	Low values	*N*	Deficient values	Low values	*N*	Deficient values	Low values
2, all	181	3.9	10.5	87	10.3	14.9	103	4.8	8.7
2–5, all	845	1.5	6.4	305	3.3	15.7	330	4.2	11.8
6–12, all	2650	0.1	4.3	996	1.3	20.3	832	1.0	12.0
13–16, males	539	3.3	9.5	206	16.0	25.2	151	6.6	21.9
13–16, females	507	0.0	2.6	221	5.0	16.7	167	1.2	14.4

[a] Deficient hemoglobin values (g/dl) are defined as follows for both males and females: 2 yr, ≤ 9.0; 2–5 yr, ≤ 10; 6–12 yr, ≤ 10. At 13–16 yr for males it is defined as ≤ 12.0, and for females ≤ 10.0.

[b] Low hemoglobin values (g/dl) are defined as follows for both males and females: 2 yr, 9.0–9.9; 2–5 yr, 10.0–10.9; 6–12 yr, 10.0–11.0. At 13–16 yr for males it is defined as 12.0–12.9, and for females 10.0–11.0.

[c] There were five high-income-ratio states: California, Michigan, Massachusetts, New York, and Washington.

Table VIII. Hemoglobin Determinations (g/dl) for Three Age Ranges and for White and Black Samples from Preliminary Findings of the First Health and Nutrition Examination Survey

	Mean values			Percent of low values[a]		
Age ranges	Total	White	Black	Total	White	Black
1–5 yr	12.67	12.79	12.01	2.65	1.66	8.23
6–11 yr	13.33	13.44	12.66	2.59	1.41	9.39
12–17 yr	14.29	14.44	13.40	4.68	2.48	17.78

[a] Low values are less than 10.0, 11.5, and 13.0 g/dl for the 1–5, 6–11, and 12–17 age ranges, respectively.

opment occurs in the human organism during this time period, it is important to understand what behavioral derangements might be expected to result from iron deficiency given our current understanding of the factors influencing the effects of early central nervous system (CNS) trauma (e.g., nutritional, asphyxial, emotional) on later cognitive development.

The hypothesis that early traumas result in later impairment of development and behavior can be traced back to the nineteenth century (Little, 1862). Thereafter, and especially in the last 30 years, much research has been conducted regarding long-term behavioral effects of environmental stress in early life (Caputo and Mandell, 1970; Gottfried, 1973; Hunt, 1976a). Recent work by a series of investigators, however, has raised doubts about many forms of early brain insult as a source of detectable sequelae in later childhood (Sameroff, 1975; McCall, 1976; Scarr-Salapatek, 1976).

Two recently published, thorough reviews (Gottfried, 1973; Sameroff and Chandler, 1975) of the literature report that the intelligence test performance of preschool children with a history of perinatal anoxia is inferior to that of controls. However, this is not the case with children in later childhood and adolescence. As development progresses, early signs of intellectual deviancy tend to disappear. By approximately 10 years of age children with a history of mild perinatal anoxia perform as well as normal controls (Gottfried, 1973). Gottfried emphasizes that although anoxic children as a group are not mentally retarded, their probability of being mentally retarded may be increased by having experienced this result.

Another situation relevant to the issue of early trauma and later deviancy is early protein–calorie malnutrition. CNS biochemical, physiological, and histological derangements associated with severe undernutrition in early life have been well documented in laboratory animals (Wurtman and Wurtman, 1977a,b). Moreover, malnourished children present dramatic attentional deficits (Lester, 1975). However, the evidence regarding intellectual impairment in later life is inconclusive. Only in cases of severe and prolonged undernutrition associated with multiple other biosocial deprivations is one likely to find permanent signs of mental impairment (Pollitt and Thomson, 1977). Severe but acute episodes may, but generally do not, leave measurable deficits in intellectual function. The weight of the evidence supports the conclusion that mental impairments will not result if early nutritional trauma due to protein–calorie deficiency occurs independent of social and economic deprivation (Stein *et al.*, 1975; Lloyd-Still *et al.*, 1974).

Low birth weight has been associated with intellectual deficits during later school years. Here again, however, the data on cognitive development are also suspect. There are well-controlled studies yielding no differences in intelligence testing between children with and without a history of low birth weight (Robinson and Robinson, 1965). A review of the relevant studies published in 1970 concluded that the deficits reported are generally associated with very low birth weights—that is, about 1500 g (Caputo and Mandell, 1970). Studies of children with milder birth-weight deficits show minimal—if any—IQ deficits at school age. The Collaborative Perinatal Project of the National Institutes of Neurological Diseases and Stroke (Broman *et al.*, 1975), with a sample size of over 30,000 individuals, showed that in a regression analysis including the predictor variables studied, birth weight made no significant contribution to IQ at 4 years of age. Low birth weight is often associated with many pre- and postnatal stress factors (Braine *et al.*, 1966; Drorbaugh *et al.*, 1975) and any sequelae could be, to some extent at least, the result of exposure to multiple adverse conditions rather than of the low birth weight alone. Thus, as a whole, it is unjustified to claim that the literature on the consequences of low birth weight supports the early trauma/later deviancy hypothesis. Intellectual derangements seem to be associated only with very low birth weights, usually below 1500 g.

Table IX. Prevalence Values of Anemia and Sideropenia from Four Published Studies

			Hemoglobin level (g/dl)		Transferrin
Study	SES	Age of subjects	<9.9 (%)	10–10.9 (%)	saturation <17% (%)
Haughton (1963)	Low income	<12 mo, N = 46	41.3		
		12–36 mo, N = 240	22.6		
Fuerth (1971)	Middle income	12 mo, N = 526, 1959	6.3	15.3	
		9 mo, N = 315, 1969	3	11.0	
Haddy *et al.* (1974)	Low income	4–24 mo, N = 70, 1970	11	33	60
Owen *et al.* (1971)	70–80%, Low income	12–24 mo, N = 349, 1971	6.5	14	N = 259, 49

In connection with multiple perinatal stress, it should be specified that its effects have also been shown to be modified by social–environmental conditions (Werner *et al.,* 1971). Ten-year-old children with a history of perinatal stress (e.g., low birth weight and mild convulsive disorder) but growing up in an emotionally supportive and organized environment were found to show no behavioral inadequacies or intellectual deficits. Conversely, children with the same history but exposed to a nonsupportive environment had high probabilities of showing behavioral derangements or intellectual deficits by age 10.

In conclusion, the early trauma/later deviancy hypothesis is not well supported by much of the data currently available. It is apparent that as development progresses, the probability of detecting intellectual deviancy decreases. In some cases this shift depends on the existence of a favorable supportive environment, but in other cases it seems to depend on the developmental process itself.

Jerome Kagan (1975) recently made the following observation in his review of a report on the Collaborative Perinatal Project from the National Institutes of Neurological Diseases and Stroke.

> There are many beliefs about the effects of pre- and perinatal trauma on childhood intelligence, but these are based, in the main, more on hunch and prejudice than firm information. This monograph is important because the size of its sample and the breadths of its predictive variables dwarf all earlier investigations on this theme. Although the results do not provide new insights regarding the functional relation between early biological distress and cognitive development, they clearly disconfirm the notion that maternal infection, asymmetry of reflexes or mild apnea in a newborn makes either a simple or dramatic contribution to the level of intellectual functioning of the preschool child.
>
> The elegant treatment of such a large corpus of data presented in a coherent manuscript has moved the truth status of hypotheses about the relation of trauma to intelligence from "perhaps true" to "unlikely." That is scientific progress! (In Broman *et al.,* 1975, pp. vi–vii)

In contrast to the uncertainty regarding long-term effects of early biological trauma on intelligence, there is now conclusive evidence that the social environment has profound effects on cognitive growth. Data from the Collaborative Perinatal Project has shown that maternal education and socioeconomic status explain more intelligence test variance at 4 years of age than any other social or biological predictor variable. This statistical association was confirmed for males and females in the samples of both black and white subjects. The long-term effects of the environment are also documented in the epidemiological study on the Island of Kauai already cited (Werner *et al.,* 1971). In this study, home education stimulation was the most potent predictor variable of achievement, IQ, language, and perceptual functioning. Early ratings of high or very high educational stimulation at home were seldom, if ever, associated with cognitive malfunction by age 10. Conversely, 62% of the children who received very low ratings of home stimulation had difficulties in school or IQs below 85. It should be noted that parental IQ was excluded from analysis.

The excellent epidemiological study of the effects of the Dutch famine of

1943 (Stein *et al.*, 1975) has contrasted the effects of one early biological traumatic episode and the social environment on intelligence in early adulthood. Pre- or early postnatal exposure to a 7-month restriction of calorie intake bore no relationship to the prevalence of mental retardation by age 18. Conversely, social class was a good predictor of the level of intellectual function. The children of men involved in manual labor were more likely to manifest mild or moderate mental retardation in later life than were the children of men involved in more intellectually demanding work.

Under certain circumstances environmental influences may play a remedial role in intellectual development. Specific signs of intellectual derangement associated with early biological trauma may disappear due to particular favorable qualities of the child's environment.

The proposition that the psychosocial environment makes a large contribution to the ways in which a child uses his intellect and relates to his social surroundings is largely unquestioned. Moreover, there is evidence that programs of psychoeducational intervention partly compensate for the adverse effects of a financially and psychosocially impoverished environment (Lewis, 1976). At issue here is the question of the environment's capacity to remediate apparently organic deficits in cognitive development.

A recently published study on the fate of poor and severely malnourished Korean children illustrates further the corrective potential of the environment (Winick *et al.*, 1975). The intellectual competence of these children showed no signs of derangement after years of adoption by middle-class American families. All adverse effects of their early traumatic experiences seemed to be eliminated following the remarkable environmental change. The new families apparently fostered intellectual development, remediating a neuropsychological effect which would probably have been unchanged had the children remained in their original milieu. Again, it may be reemphasized that only in cases of severe and prolonged undernutrition associated with multiple other biosocial deprivations is one likely to find signs of permanent mental impairment.

The findings in the Korean children may be questioned because the study was retrospective. This methodological approach makes it impossible to determine the validity of the description of the children's antecedent psychosocial and nutritional status. However, its agreement with the findings from another retrospective study conducted in Jamaica enhances the validity of the data (Richardson, 1976). This latter investigation found that the probabilities of rehabilitation following severe malnutrition in early life depended on environmental characteristics. In those cases where the environment reinforced intellectual competence—even under extreme conditions of poverty—it was unlikely that the affected children would show signs of impairment. Conversely, an unfavorable environmental background and a history of severe undernutrition resulted in serious intellectual deficits.

Recent studies on premature infants suggest further that physical-environmental manipulation may also have salutary effects on sensorimotor development (Scarr-Salapatek and Williams, 1973). Early stimulation of different sensory modalities tends to improve the premature's performance on tests of sensorimotor development. Also, as compared with nonstimulated premature infants, index cases tend to excell in motor development, muscle tone, and responsiveness to auditory stimuli (Cornell and Gottfried, 1976). A few studies have even shown that the stimulated infant gains weight faster than do control cases (White and Labarba, 1976).

Salubrious environmental influences on biological trauma are therefore not limited to the social domain. The environment may also have remedial effects through its physical dimensions. Thus, effects may be produced not only in language and symbol-formation abilities but also on more primitive sensorimotor skills.

It follows from the previous discussion that the social ambience appears to be a critical variable in the calculus of cognitive development. On the one hand, an impoverished psychosocial setting may severely impair the intellectual development of an organically normal child, while on the other hand an optimal environment may protect against and, in some instances, actually remediate the cognitive impairments which appear to accompany the various nutritional/asphyxial central nervous system insults.

These observations argue strongly for an extreme plasticity in the growing human brain, permitting correction and/or compensation for a variety of adverse events provided that the environment is supportive in the sense of supplying adequate emotional and intellectual stimulation. If this thesis is correct, behavioral derangements observed in children with discrete nutritional alterations (e.g., iron deficiency) may be codetermined by the biochemical state and its psychosocial setting. The long-term impact of such a nutritional state would perhaps depend as much on the child's subsequent psychosocial ambience as on the nature of the biochemical abnormality itself.

5. *Iron Deficiency and Behavior*

In this section we review the literature on the effects of iron deficiency ± anemia on certain aspects of behavior and mental development. Only a small number of studies have dealt with behaviors classified as "cognitive," and these will be examined in detail. Reports dealing with affective states and pica will not be reviewed in detail since much of the relevant literature is in the form of anecdotal reports. Readers interested in the topic of iron-deficiency anemia and physical performance are referred to a number of earlier papers (Read, 1973; Pollitt and Leibel, 1976) which cover this topic adequately.

Most of the research to date on the behavioral correlates of iron deficiency is flawed by methodological errors in research design, compounded by failures both to accurately pinpoint the mental process under question and to choose an experimental task which could isolate that process. For these reasons the conclusions that have been advanced concerning the role of iron deficiency anemia on mental development should be viewed only as tentative hypotheses which still require formal testing.

5.1. Affective States and Pica

Although a large number of clinical symptoms have been associated with iron deficiency anemia, it is not unusual to find patients, even with severe anemia, who are asymptomatic. Fatigue, headache, weakness, lightheadedness, and irritability (to name but a few) have often been reported, but a strong correlation between the extent of symptoms and the severity of anemia does not appear to exist (Fairbanks *et al.,* 1971). Wood and Elwood (1966), for example, studied 295 patients, 56 of whom were anemic, and found no significant correlation between either the presence or severity of clinical symptoms and the concentration of hemoglobin. Other studies (Berry and Nash, 1954; Elwood *et al.,* 1969; McFarlane *et al.,* 1967) have also failed to find a significant relationship between symptom severity and the degree of anemia.

A number of studies have failed to demonstrate symptomatic improvement in sideropenic subjects following iron therapy, despite significant increases in hemoglobin. Morrow *et al.* (1968), for example, studied 20 women (Hb concentration 11.7 g/dl or higher; transferrin saturation less than 16%) with various symptoms, including tiredness, headaches, irritability, and depression. The women received 3 months of iron therapy as well as 3 months of placebo treatment, with neither the women nor the experimenter having any knowledge as to the order of treatment (the order was random). No significant difference in symptomatic improvement between the iron and placebo treatment was observed despite a 1.0–1.3 g/dl increase in Hb while on iron.

Results contradicting the findings above were reported by Beutler *et al.* (1960a), who studied 44 iron-deficient, nonanemic subjects (hemoglobin levels above 12.0 g/dl) complaining of chronic fatigue. The women received 3 months of iron therapy and 3 months of placebo treatment, with the evaluator and subject again unaware of the order of treatment. When the subjects were subsequently divided into groups as a function of initial level of iron depletion, 13 of the 19 most depleted subjects showed more symptomatic improvement with iron than with placebo therapy. The remaining 5 subjects showed the opposite trend. Of the 11 least depleted subjects, 4 improved more with iron therapy, 5 improved more with placebo therapy, and 2 remained the same. The differences were statistically significant.

The symptoms which have been studied are virtually impossible to define objectively and measure. This alone may have obscured any true relationship apparent in the data. Both vagueness of the symptoms and the presence of a concerned and attentive physician enhance the likelihood of a placebo effect. Small sample sizes may also have contributed to the lack of any correlational effect. In the study by Wood and Elwood (1966) only 56 patients were anemic, and very few more than moderately so. In the study of Elwood *et al.* (1969), all subjects had normal hemoglobin levels (13.5 g/dl or higher). Choosing a restricted range or only nonanemic subjects may have masked any effects, especially with sample sizes in the 40- to 60-person range. At present the role of iron deficiency anemia or iron deficiency alone in the etiology of symptoms such as headache, irritability, weakness, or fatigue is unclear.

Pica may be defined as a perversion of appetite characterized by the craving for and compulsive ingestion of food or nonfood substances. This type of behavior has been associated with iron deficiency (with and without anemia) (Roselle, 1970; Crosby, 1971). Although some have argued that the iron-deficient state is the result of pica, there are reports providing impressive demonstrations of the rapid alleviation of these cravings following iron therapy (McDonald and Marshall, 1964; Crosby, 1976). Perhaps more impressive is the recent evidence that a distinctive form of pica, pagophagia (compulsive ice-eating), is specifically associated with iron deficiency (Reynolds *et al.*, 1968; Coltman, 1969). This symptom is also rapidly alleviated by the introduction of iron-repletion therapy. The pathophysiology of these behaviors is obscure. Suggested mechanisms have ranged from iron depletion of brain "appetite centers" (Van Bondorff, 1977) to deficiencies of iron-dependent enzymes in the buccal mucosa (Coltman, 1969).

5.2. Learning: Animal Studies

Very little work has been done with the laboratory animal in exploring the role of iron deficiency in learning. Two studies (Bernhardt, 1936; Scarpelli, 1959) experimentally manipulated iron level at two different points in development and found different results when the deficient rats were compared to normal rats on a maze-learning task. Bernhardt (1936) found that rats who were made iron deficient after weaning performed less well on a maze-learning task than did rats with normal iron levels. Scarpelli (1959), on the other hand, found that the offspring of rats fed iron-deficient diets during pregnancy showed no maze-learning deficit when tested at 42 days of age.

These studies do not necessarily contradict one another, since the period of onset may have been a critical factor. However, the validity of both results can be questioned on the grounds that both investigators used food as a positive reinforcer in the maze-learning task. Since after an experience of food deprivation early in life, hungry animals may show an exaggerated response to food (Barnes *et al.*, 1968; Dobbing and Smart, 1973), it is conceivable that the iron-deficient and normal animals differed in motivational level as well as iron status. This confounding of iron level with motivation renders unclear the respective roles of each variable in the animals' subsequent performance.

5.3. Mental Development: Human Studies

5.3.1. Infants

Only two studies to date have investigated possible adverse effects of iron-deficiency anemia on mental development in an infant population. In a sample of 61 children (Cantwell, 1974), 32 developed iron-deficiency anemia between 6 and 18 months of age (Hb 6.1–9.5 g/dl). Twenty-nine infants received neonatal iron dextran injections and were not anemic (Hb 11.5–12.9 g/dl) during this age period. Neurological evaluation was done at 6–7 years of age with the examiner having no knowledge of the presence or absence of prior anemia. The group that was formerly anemic was described as having a higher incidence of "soft neurological signs" (clumsiness in balancing on one foot, in tandem walking, and repetitive hand or foot movements) and as being more inattentive and hyperactive. Stanford-Binet IQ of the normal and anemic group were 98 and 92, respectively. Cantwell, unfortunately, did not report the results of any statistical tests nor did he indicate how attention and hyperactivity were defined and measured in these latter two cases and whether or not the IQ difference and higher incidence of "soft signs" were actually statistically significant.

Oski and Honig (1978) reported a study involving 24 iron-deficient anemic infants ranging in age from 9 to 26 months. The infants were randomly divided into two groups, half receiving an intramuscular injection of an iron dextran complex, half receiving a placebo also injected intramuscularly. The infants were tested on the Bayley Scales of Infant Development both before and 5–8 days after treatment.

Analysis of the results indicated that both groups displayed statistically insignificant increments in performance on the Physical Development Index. Both groups also showed an increase on the Mental Development Index. Only the increase in the experimental group (13.6 points), however, proved significant. The authors concluded that their findings support the hypothesis that developmental alterations in infants produced by iron deficiency can be rapidly reversed with iron therapy.

These interesting findings were derived from a well-controlled study design. One should note, however, that the Bayley Scales, although widely used, do not correlate very well with later measures of intellectual development. In addition, test–retest correlations for the scales are not high. Finally, the sample size was small and the age range tested fairly wide. The results should therefore be viewed as tentative. Hopefully, these findings will encourage additional research with a larger sample, a more restricted age range, and additional measures of infant development.

5.3.2. Preschool Children

Two studies, each of fairly large scope, have concerned low-income black children 3–5 years of age.

Howell (1971) studied 8744 low-income black children 3–5 years of age enrolled in a Headstart program in Philadelphia. When a hemoglobin level of less than 10 g/dl and a hematocrit of less than 31% were used to define anemia, a low prevalence (1.9%) of anemia was found. When the cutoff value was raised to 10.5 g/dl, the prevalence of anemia was 5%.

For those who fell in the anemic (Hb less than 10.5 g/dl) group, a complete medical history was taken and a physical exam performed in order to eliminate those who were anemic for reasons other than iron deficiency. In addition, any child with a history of a bleeding diathesis or evidence of infection at the time of the study, as well as children with any hematologic disorder other than hypochromic microcytic anemia, were excluded from further study. Eighty-three children with hemoglobin ranging from 9 to 10.5 g/dl ($\bar{x}$ = 9.75) met these selction criteria.

These children were then randomly divided into two groups. The first group (42 children) received intramuscular iron at a dose calculated to bring their Hb level to 12 g/dl. The other group received the same volume of physiological saline intramuscularly. A battery of psychological tests was administered before as well as 2 and 4 months after this injection.

The investigators hypothesized that attentional processes would be more affected by iron level than intellectual potential. Intellectual potential was measured by the Stanford-Binet scales and the Goodenough Draw-A-Man Test. Two aspects of attentiveness—field articulation and scanning—were tested. (Field articulation is a technique of finding an embedded figure within a picture; scanning shows the ability of the child to inspect a total picture and evaluate differences.) The authors indicate that the attention tasks were designed to measure initiative, resourcefulness, goal directedness, and impulse control. The authors also indicate that results were subjected to factor analysis, analysis of variance, and t-tests. No statistical information is reported, however.

No difference was evident between anemic and nonanemic children in the intelligence quotients obtained. Performance on the attentional tasks was subjected to a factor analysis. A different structuring of factors emerged for anemic and nonanemic children with respect to the controlling of impulses during intentional behavior.

A final task utilized an unstructured situation in which the child's attention was not specifically directed to any one aspect of the stimulus array. In a subsequent analysis of performance on this task, several differences emerged between anemic and nonanemic children. Anemic girls displayed more aimless manipulation, a narrower attention span, and less complex purposeful activity than the nonanemic girls. Anemic boys "perceived significantly fewer stimulus objects in the visual field than nonanemic boys when a dominant stimulus was present." The authors interpreted this to mean that the anemic boy is more passive and less able to respond to nondominant features of the environment when a dominant stimulus overshadows the visual field.

The authors indicate that the study should be regarded as incomplete, with much additional work being required, but emphasize that children with mild iron deficiency had normal IQ but showed markedly decreased attentiveness, more aimless manipulation, less complex and purposeful activity, narrower attention span, and perceived fewer stimuli in the presence of dominant stimuli.

The results, even though considered tentative by the investigators, are suspect to an even greater extent for a number of reasons. First, no statistical information is provided. The authors, for example, state that the psychological tests were given before, 2 and 4 months after iron treatment. No mention of this time variable, however, appears anywhere in the study. Thus, it is not clear whether or not the two groups were actually equivalent in performance at the initial time of testing and whether or not performance at the end of 2 months was the same as performance at the end of 4 months. Second, the psychological battery is not described in great detail, and it is therefore difficult to assess whether or not the tests actually measured what the authors allege. Measures of initiative, resourcefulness, goal directedness, and impulse control are categorized as attentional variables. Certain of these behaviors appear to have processes other than attention as their main determinant. Motivation, for example, may play a greater role in initiative or impulse control in a situation where many processes are operating simultaneously.

It is also not clear that optimal behaviors were chosen for study. Anemic boys, for example, were said to have "perceived significantly fewer stimulus objects in the visual field than nonanemic boys when a dominant stimulus was present" (p. 68). It may be the case, however, that looking at nondominant stimuli when a dominant one is present may not be a more developmentally advanced behavior in all situations. One can certainly conceive of instances where such behavior might be less desirable and the child labeled "easily distracted." Although the authors indicate between group differences, it is not obvious what these differences signify. Unless the behaviors in question relate to some known developmental sequence or to some signficant area of cognition in which nonanemic children show superior performance, the differences obtained are not particularly meaningful.

Impulse control was reported to differentiate anemic from nonanemic children. This same factor, however, differentiates boys from girls, regardless of iron status. Since the ratio of boys to girls in each group is not reported, it is unknown whether or not this effect is present for the iron variable when sex is partialled out.

Finally, the response of the "anemic" children in this study to parenteral iron is of interest. Of the 42 tested, 10 showed no rise in hemoglobin level after 4 months and an additional 20 showed a rise of only 1 g/dl. Given the fact that "normal" hemoglobin levels in blacks 3–5 years of age are 0.6–1.1 g/dl below those of socioeconomically matched whites (Garn *et al.*, 1975), and given the nonexistent or minimal response of nearly 75% of the "anemic" subjects, it seems likely that a number of these children were neither iron-deficient nor anemic. Without further iron-related biochemistries it is not possible to ascertain the exact size of that group. Obviously, the inclusion of a sizable number of normals in the anemic groups would seriously confound efforts to demonstrate behavioral differences between anemic and nonanemic children.

In another study involving preschool children, Sulzer *et al*, (1973) examined over 300 four- to five-year-old black children enrolled in a Headstart program in New Orleans. Testing took place at five Headstart centers. Since not all children were tested on every task (see below for reasons), the number of subjects in each group varied for different tasks. Using a hemoglobin of less than 10 g/dl to define anemia, approximately 30 of the 300 children per task were classified as anemic. Additional analysis using 10.5 g/dl as the separation point increased the group classified as anemic in each task to approximately 75 children.

The testing situation was far from ideal. There was insufficient time to establish rapport with the child before the testing began, and the physical setting was noisy. Testing was terminated if the child became uncooperative, highly distracted, or excessively bored or shy. Although this information was noted in the child's record, no further mention of this factor is made. It would have been of interest to note whether or not these behaviors distributed themselves equally among anemic and nonanemic children, since if the anemic group had a greater percentage of incomplete tests, children with the greater deficits (assuming lack of cooperation may have been caused by difficulty with the task) may not have been tested, possibly reducing to some extent the performance differences obtained between the anemic and nonanemic groups.

The behavioral tests were subdivided into two batteries, one measuring cognitive ability and the other consisting of measures of endurance, memory, attention, reaction time, learning, and transfer of training. The first battery consisted of two IQ tests, the Kahn Intelligence Test (KIT) and the Van Alstyne Picture Vocabulary Test (VAPT), a test of moral development, and a measure of classification or grouping behavior. The second battery was comprised of three reaction-time tasks, a short-term memory task, and a test of physical endurance.

Performance on both batteries was initially analyzed as a function of age. This was done to demonstrate that the tasks without established age norms were developmentally sensitive and could function as meaningful diagnostic tools. If proficiency on a task increases as a function of age (the assumption here is that the average older child is at a higher developmental level than the average younger child), differences in ability at the same age may also be viewed as developmental differences. Thus, poorer performance may be interpreted as evidence for a lag in development of the cognitive process being tested.

All tasks of specific processes and behaviors, except two, produced significant age effects in the expected direction. Both the sorting task and the moral judgement tasks proved too difficult and failed to produce age effects. Also, the overall low level of performance on the memory task seems to indicate that even though age trends were present, the task was probably too difficult for the majority of subjects and thus inappropriate for detecting difference in memory as a function of iron level. Reference to these three tasks will no longer be made since in practice they do not function as a meaningful part of the test battery.

Using a hemoglobin of less than 10 g/dl to define anemia, performance on the two test batteries was analyzed to determine if differences were present between the anemic and nonanemic groups. In the first test battery, anemic children scored lower on both IQ tests, but only one difference proved statistically reliable. In the second battery some tests showed trends in the expected direction, but performance between anemic and nonanemic children did not differ statistically on any test.

A second analysis was conducted with the separation point between deficient and normal children raised to 10.5 g/dl. This resulted in a reliable statistical difference between normal and deficient children on both IQ tests as well as on a reaction-time task in which associative learning was involved.

In the initial analysis of age trends described earlier, both IQ tests showed decreases in the rate of mental development with age. In order to determine whether or not a greater relative decline in IQ with age occurred in the anemic group, both groups were divided into three age categories (53.7–57.3 months, 57.4–61.7 months, 61.8–66.7 months) for additional analysis. All three anemic groups showed poorer performance on the VAPT, but on the KIT only the anemic children in the middle age group displayed significant performance deficits relative to nonanemic controls. The authors also indicate that both tests seem to show greater "decline" in IQ for the iron-deficient subjects (p. 89) but no significant interaction is reported, and inspection of the differences in IQ points between nonanemic and anemic children in the three age groups on both the KIT (+1.4, −5.1, −1.0, respectively) and the VAPT (−5.1, −5.8, −5.3, respectively) indicates that no such trend is evident. The authors' conclusion that IQ decreases with age to a greater extent in the anemic group appears unwarranted.

The associative reaction task, the only task in battery 2 in which differences between the anemic and nonanemic groups occurred, was subjected to the same age analysis outlined above. Of particular interest is the relative improvement over the four sequentially learned associations, since to the extent that previous learning is retained, the task becomes easier on subsequent trials. The nonanemic children in all age groups displayed the predicted progression. Reaction time steadily decreased from 1.69 sec on the first association to 1.34 sec on the fourth association, averaged over all groups. Mean differences in reaction time between the age groups did occur with older children reacting faster, but the decrease over associations was equivalent in the three groups. The anemic group, however, showed somewhat more erratic behavior. The oldest anemic group performed as well as their nonanemic counterpart. The two younger groups, however, performed more poorly than the remaining four groups, and although some improvement occurred over associations, the improvement was irregular, with increased reaction times occurring on some subsequent associations.

The purpose of the associative reaction-time task was to compare learning ability in the anemic and nonanemic groups. In addition to learning ability differences, however, poorer performance may have resulted from poorer attention, less motivation, or greater fatigue in the anemic children. This issue is not easily resolved, for cogent arguments exist both for rejecting and accepting these factors (attention, motivation, fatigue) as having a contributing role in performance on the task. One can argue that since anemic and nonanemic subjects did not differ on the first two reaction-time tasks which included all the processes above, save learning, these processes do not differ in the two groups. In addition, the endurance task which had a strong motivational component was performed for a longer time by the anemic children. If anything, then, motivation was higher in the anemic group, and thus could not account for the difference in the associative reaction task. If, however, learning ability deficits controlled performance, one would predict a regular decrease in reaction time with additional associations for the anemic groups, similar to the normal group but with a smaller slope. The erratic performance of the two younger-age anemic groups do not conform to this prediction. One can argue that even though motivation, attention, and fatigue were the same in the two easier and earlier tasks, the anemic children may have become more fatigued, less motivated, or less attentive as the task progressed (the associative task was given last) and became more difficult.

In an attempt to separate these confounding variables, a number of follow-up studies were conducted (Sulzer, 1971). Although the results tended to support the contention that attentional and motivational factors played a greater role in determining performance, fatigue and learning ability could not be ruled out.

The results of the studies by Sulzer seem to indicate, then, that intelligence and either learning ability, or more likely, motivation or attention, may be adversely affected by iron-deficiency anemia. Since the study was retrospective in design and did not experimentally manipulate the independent variable (iron level), it is necessary to demonstrate, before forming any firm conclusions, that the groups were, indeed, equivalent with regard to any moderating variable that might in itself account for the obtained differences.

In order to test the role of past nutritional status (irrespective of the presence of anemia) on performance, children were divided into groups according to their height (the assumption being that shorter stature is an indication of poorer past nutrition), and performance differences were analyzed, controlling for age. The analysis indicated that differences in height, presumably reflecting past nutritional status, could account for the noted variation in performance.

To separate the possible contributions of past nutritional and present nutritional status on the performance difference, four groups were formed from all possible combinations of past (normal nutrition—tall, poor nutrition—short) and present (nonanemic, anemic) nutritional status. The four groups were comprised of children with both past and present nutritional deficiencies (short, anemic), children with only a present nutritional deficiency (tall, anemic), children with only a past nutritional deficiency (short, nonanemic), and children normal at both times (tall, nonanemic).

Statistical comparison of the performance of these four groups yielded group differences on only the two tests of intelligence. The results of the analysis indicated that the adverse effects of past and present nutritional deficiency combined in an interactive fashion. When compared to the group normal at both times, neither past (1.9-point deficit on VAPT; 3.6-point deficit on KIT) nor present (2.1-point deficit on VAPT; no deficit on KIT) deficiency alone led to a large deficit in IQ. In combination (both past and present deficiency), however, large IQ deficits occurred (11.5-point deficit on VAPT; 9.3-point deficit on KIT).

Although slightly larger deficits occurred in the group with past deficiencies only than occurred in those with normal growth but currently anemic, the actual IQ deficits are small (an average of 1.9 points, which is probably within the range of the standard error of these tests), as is the difference between the two groups (an average of 1.7 points). In addition, the sample is not large, with less than 20 subjects in half of the groups. The tentative conclusion that one can draw is that these results are consistent with the hypothesis that either a mild past nutritional deficiency (given the population in question, we must assume that we are dealing with mild forms of undernutrition) or a mild present nutritional deficiency (anemia) alone are weakly associated with mental development as measured by the IQ test. It is only when these two conditions are combined that significant deficits are encountered. Of course, no *causal* relationship between either of the nutritional measures and mental development can be inferred. Covariants, such as socioeconomic status, maternal education, etc., may have been much more important determinants of the variance noted.

5.3.3. *Adolescents*

Webb and Oski (1973a) studied 12- to 14-year-old junior high school students living in an economically deprived, virtually all black community in Philadelphia. After a hematologic survey of 1807 students, 193 were selected to participate in the study. Ninety-two of these students were classified as anemic (Hb ranged from 10.1 to 11.4 g/dl), with the remaining 101 students serving as a normal (Hb ranged from 14.0 to 14.9 g/dl) control group. The specific criteria for selecting the control subjects and the reason for the larger size of this group are not given.

The dependent variable under investigation was scholastic performance as measured by the composite score on the Iowa Tests of Basic Skills, Levels A-F/Form 3, a test routinely given each year in the Philadelphia school system. The composite score is derived from 11 subtests, which test vocabulary, reading comprehension, spelling, capitalization, punctuation and usage, map-graph table reading, knowledge and use of reference material, arithmetic concepts, and problem solving. These subscales were not analyzed separately.

The students were divided into 12 groups according to their sex, age to the nearest year (12, 13, or 14 years old), and hematologic status. Their composite test scores were then analyzed for differences. Results indicated that anemic students performed less well than the nonanemic groups. In addition, a sex by age by hematologic status interaction occurred. This triple interaction may be stated as follows. Anemic girls performed less well than nonanemic girls at all ages by approximately the same degree (the difference in test score means between nonanemic and anemic 12-, 13-, and 14-year-old girls was −0.50, −0.59, and −0.54 grade levels, respectively). Anemic boys, however, showed no deficit at 12 years of age but greater deficits in performance with increasing age. The difference in test score means between nonanemic and anemic 12-, 13-, and 14-year-old boys was +0.34, −0.25, and −1.65 grade levels, respectively.

As is the case with a number of the studies already reviewed, the present study did not experimentally manipulate the independent variable but rather was retrospective in design. In light of the confounding effect of past nutritional status in the Sulzer study, the results of the present study may be questioned on the same grounds. The investigators themselves point out that the obtained difference in performance may not have resulted directly from the anemic condition but may have been mediated by a more generalized nutritional derangement or concomitant social factors.

In a subsequent brief report, Webb and Oski (1973b) indicate that these same children were tested on a visual after-image task. The anemic group took significantly longer (4.08 sec) than nonanemic children (1.81 sec) in reporting a visual after-image. The authors indicate that this finding, along with the one discussed above, and teacher evaluations which indicated that anemic males displayed significantly more conduct problems than did nonanemic males (Webb and Oski, 1974) suggest that iron deficiency leads to disturbances in attention and perception which may be manifest as deficits in scholastic ability. Although their results are consistent with this hypothesis, the evidence is very far from being conclusive. The problems of interpreting the results of the scholastic ability test have already been discussed. The details of the after-image task are not given, and it is not clear what other processes may have been operating in the task. The authors also suggest that cerebral metabolism was likely altered, accounting for the results of the after-image task. After-images, however, depend not only on cerebral function, but peripheral aspects of the visual system as well. Their findings certainly do not contradict their hypothesis, but neither do they provide strong support for it.

5.3.4. Adults

Elwood and Hughes (1970) conducted a study of 47 women 20 years and older living in a Welsh mining valley, who had Hb levels below 10.5 g/dl. These women were randomly divided into a treatment (150 mg of Fe as ferrous carbonate daily) and a no-treatment (tablets of similar appearance containing no iron) condition in a 4:3 ratio (no reason for this ratio is given). A battery of psychological tests was administered at home, and following 8 weeks of treatment the psychological tests and hemoglobin determinations were repeated. The experimenters were blind to the treatments. All subjects did not receive all tests, owing to either lack of cooperation or technical difficulty.

The battery consisted of five tests which the investigators said covered a range "from an almost pure test of intellectual function to a simple test of manual dexterity."

After analysis of the follow-up blood test, the women were further divided into three groups: no iron (21 subjects); received iron but Hb increment was less than 2 g/dl (12 subjects); and received iron and increment in Hb was greater than 2 g/dl (14 subjects). Changes in performance following 8 weeks of treatment were then analyzed. In none of the post-test comparisons were the differences between the means significant.

The authors offer a series of explanations to account for these results. One explanation provided is that iron supplementation has no beneficial effect on psychomotor function in women with iron-deficiency anemia. An alternative explanation is that subjects might not have been sufficiently iron deficient for an effect to occur. The authors deemed both of these interpretations to be plausible. Two alternative explanations, however, were rejected : (1) the tests covered a limited area of psychomotor function and thus may have been inappropriate for detecting changes in the iron-supplemented group. The authors rejected this because of what they deemed the wide range of functions tested; and (2) the tests were not sensitive enough to detect an improvement. The authors reject the second explanation, arguing that since the group not treated with iron showed improvement, the tests detected learning effects and therefore would most likely detect any beneficial effect of iron therapy.

Upon closer scrutiny these last two explanations are not as easily eliminated as the authors indicate. The test that the authors deem a "pure test of intellectual function involving concentration, arithmetic reasoning and short-term memory" is one in which the subject repeatedly subtracts seven from each subsequent number beginning with 100. The dependent measure appearing in the analysis table is the time taken for 14 subtractions, regardless of errors. The appropriateness of such a task as a "pure measure of intellectual function" may be questioned. Previous education, aptitude in math, and strategy may have differed in the women and may have played a role in the results. Some subjects may have tried to go as fast as possible, disregarding error, some may have tried to eliminate errors, some may have adopted an intermediate strategy. There is clearly a trade-off between speed and accuracy, and it is difficult to assess if all subjects adopted the same strategy. The remaining four tasks consisted of crossing out e's in a text; tracing a maze; sorting playing cards by suit; and taking pegs out of a board, turning them upside down, and replacing them. It seems that these four tasks are highly similar and probably overdetermined to a large extent by manual dexterity. The authors did not analyze the data in such a manner (e.g., by a multivariate technique) as to determine whether or not these tests, indeed, tapped a variety of psychomotor functions. It seems reasonable, then, that only a limited area of psychomotor function may have been tested and that the tests may have been inappropriate for detecting differences among the groups.

The authors also reject the hypothesis that the tests were not sensitive enough to detect an improvement. This conclusion may also be contested. The improvement in performance could be easily explained as a generalized learning-set phenomenon, familiarity with the tests, or any number of effects associated with practice, since improvement did occur in four of five tests (which further indicated that the test might have had some strong common component). The criterion for determining whether or not the tests were sensitive is not this improvement across testing times but the level of performance. If performance for all groups was very good, one might conclude that the test was too easy. If performance was very poor, one might conclude that again the tasks were inappropriate, this time because the tasks were too hard. If performance were at some intermediate level and no difference occurred, one could contend with greater confidence that performance on these tasks was unaffected by iron treatment.

The lack of a control group with normal iron levels from the outset poses more serious problems than those outlined above. Even if the psychomotor tests were appropriate for detecting differences, no difference between a treated and nontreated anemic group has extremely different consequences, depending upon the performance of these groups in relation to a normal control group. If normal women performed significantly better than the anemic groups at both the initial time of testing and also after iron repletion of the latter individuals, one might conclude that anemia has a marked and irreversible effect on these functions. If the normal group performed the same as the anemic groups at both testing times, one might conclude that these functions are unaffected by iron status initially, so that any improvement would not be expected. Without this additional group the present data are difficult to interpret.

6. Biochemical Substrates for Behavioral Derangements

6.1. Tissue Morphologic and Biochemical Changes

The search for nonhematologic impacts of iron deficiency was prompted by the repeated observation of clinicians that patients with iron-deficiency anemia displayed more somatic and psychic symptomatology (weakness, irritability, pica, etc.) than individuals with equivalent reductions in Hb from other causes (Fairbanks *et al.,* 1971; Harris and Kellermeyer, 1970; Brown *et al.,* 1950). In many instances, these symptoms were alleviated or cured within days of the initiation of iron therapy, before any substantial augmentation of hemoglobin mass could occur. Additional impetus was provided by the fact that iron-deficient patients appeared to be susceptible to a number of mucosal and cutaneous disturbances which did not correlate well with the extent of reduction of the hemoglobin mass per se. Table X indicates the gross and microscopic changes which occur in various body tissues during iron deficiency. Some of these (glossal atrophy, ozena) are rarely seen today and probably reflect the mucocutaneous impact of a concatentation of nutritional disturbances. Koilonychia is probably the only cutaneous finding definitely related to iron deficiency and has been noted in both anemic and nonanemic iron-deficient patients (Hogan and Jones, 1970). Recently, iron has been shown to play a role in systemic cellular (both phagocytic and lymphocytic) immune mechanisms (Chandra, 1976) as well as in mucosal defense against *Candida albicans* (Fletcher *et al.,* 1975). Impairments in cutaneous delayed hypersensitivity, lymphocyte DNA synthesis with PHA stimulation, rosette-forming thymus-dependent lymphocytes, and polymorphonuclear bactericidal capacity

Table X. Tissue Changes[a]

A.	Skin and mucous membranes
	Angular stomatitis, atrophic oral mucosa, atrophic tongue (smooth glossal papillae), ozena (nasal mucosa atrophy), post-cricoid webs (Plummer-Vinson), koilonychia (concave, thin, brittle nails), poor hair growth, dry and fissured skin.
B.	Gastrointestinal
	Gastric and esophageal atrophy, reduced acid and intrinsic factor secretion, blunting of duodenal villae, occult blood and protein loss, fat and protein malabsorption.
C.	Red blood cells
	Reduced number and survival time.
D.	Muscle
	1. Skeletal: no gross or microscopic changes. 2. Cardiac: hypertrophy secondary to increased mitochondrial size.
E.	Bone
	Hyperplasia of erythroid elements.
F.	Brain
	Rarely, intracranial hypertension ± papilledema; usually, no gross or microscopic changes.

[a] Based on data in Hallberg *et al.* (1970, pp. 370–382) and Fairbanks *et al.* (1971, p. 128).

have all been noted in the "latent" iron-deficient state and are rapidly repaired by iron administration (Macdougall *et al.*, 1975; Chandra and Saraya, 1975; Chandra, 1973). These gross clinical and cellular-function findings all tend to imply that iron deficiency affects more than just the individual's Hb mass and that many of the symptoms/signs of this disorder are related to a direct tissue effect of the iron-deficient state. Buckley (1975) has emphasized the fact that these biochemical derangements may be reflections of concomitant nutritional and/or infectious disturbances in these children and of limited, if any, functional significance in host resistance. Some authors (Masawe *et al.*, 1974; Editorial, *Lancet*, 1974) have, in fact, reported reduced rates of bacterial infection in iron-deficient children. Nonetheless, the studies demonstrating rapidly reversible (with iron administration) functional derangements in apparently iron-dependent biochemical systems related to immune function, regardless of their physiologic significance, reinforce the concept of enzyme susceptibility to modest degrees of iron deficiency.

For reasons mentioned earlier, it is likely that tissue iron content is the key to the putative behavioral derangements in iron deficiency. In its role here, iron is a component of both myoglobin and various important enzyme systems. In these systems, iron participates in several forms (Table XI).

1. *Heme proteins* (Williams, 1974). These proteins contain an iron-prophyrin prosthetic group within which the ferrous iron may (cytochromes) or may not (Hb) be readily and reversibly oxidized. Most heme proteins participate in oxidative metabolic pathways.

2. *Metalloflavoproteins* (Hall *et al.*, 1974). In these proteins the iron is associated with flavin prosthetic groups. Like the cytochromes, these compounds function largely in the electron-transfer process. The isoalloxazine ring of the riboflavin moiety participates in oxidoreduction reactions by being reversibly reduced. Also, like the cytochromes, most of the metalloflavoproteins are located in mitochondria. Nonheme compounds account for far more mitochondrial iron than do cytochromes.

3. *Cofactor* (Hall *et al.*, 1974). In these instances iron is necessary for the activity of an enzyme, perhaps by virtue of a catalytic action at the enzyme's active site.

4. *Nucleic acid metabolism* (Robbins and Pederson, 1970; Hoffbrand *et al.*, 1974, 1976; Siimes and Dallman, 1974; Hershko *et al.*, 1970). Iron apparently plays a role in the mitotic process. The interphase nucleolus of HeLa cells is a repository for iron which is subsequently transferred to the metaphase chromosomes. Removal of iron from the culture medium selectively inhibits DNA synthesis (regardless of the stage in the mitotic cycle) without immediately influencing RNA or protein synthesis. In various studies reduced [^{3}H]thymidine and ^{32}P incorporation into DNA in tissues from iron-deficient animals have been noted. Ribonucleotide reductase, which is concerned in the reduction of ribonucleotides to deoxyribonucleotides, is an iron-containing enzyme whose activity is decreased *in vitro* by iron deficiency. There is also evidence of impaired folate metabolism in iron deficiency (Vitale *et al.*, 1966).

Every mammalian cell contains iron in at least one of these forms. One would anticipate that those tissues with the most rapid cellular turnover rates would, on the one hand, be most susceptible to enzyme depletion in iron-deprivation states and, on the other, would show the most rapid repletion following the administration of iron. This does, in fact, turn out to be the case, and thus gut epithelium is more susceptible to, for example, cytochrome c

Table XI. Nonstorage Iron

1. Heme-containing proteins: myoglobin, hemoglobin, cytochromes, catalase, peroxidase, tryptophan pyrrolase.
2. Metalloflavoproteins: succinate dehydrogenase, NADH dehydrogenase, monoamine oxidase, aldehyde oxidase, α-glycerophosphate dehydrogenase, xanthine oxidase.
3. Cofactor: aconitase, protocollagen-proline hydroxylase, phenylalanine hydroxylase, tryptophan hydroxylase, ribonucleotide reductase, (?) THFA formiminotransferase.

depletion than is skeletal muscle (Dallman and Schwartz, 1965a), but returns to normal levels within 2 days of iron therapy, while 40 days are required for the achievement of normal levels in muscle (Dallman and Schwartz, 1965a). Dallman (1974) has extensively reviewed the various tissue enzyme changes in iron deficiency.

Table XII is a summary of some of the enzyme changes found in various tissues. Several points should be emphasized.

1. Certain of these reductions have been noted to occur in iron-deficient, but not anemic, organisms. Beutler (1957) demonstrated marked reductions in liver cytochrome c in iron-deficient nonanemic rats. Dagg *et al.* (1966) reported reduced cytochrome oxidase in buccal mucosa biopsies of patients with latent iron deficiency; enzyme levels returned to normal within 48 hr after the institution of parenteral iron therapy. In this study no correlation between enzyme levels and buccal mucosa atrophic changes could be demonstrated. Many of the *in vitro* derangements in immunologic function ascribed to iron deficiency are noted in patients with latent iron deficiency (Macdougall *et al.*, 1975). Bactericidal function returns to normal in iron-treated patients within 10–12 days (Chandra, 1973), implying a direct cellular impact of iron (possibly on leukocyte myeloperoxidase levels).

Table XII. Biochemical Changes in Iron Deficiency

A. *Heme proteins*
   1. Hb—decreased.
   2. Mb—decreased.
   3. Cytochromes—decreased cytochrome c in skeletal muscle, intestine, liver, and kidney of rat. Little or no change in brain or heart. Depletion proportional to work load (44% of control in skeletal, 79% in diaphragm, 89% in heart). Reduced cytochrome oxidase in buccal and intestinal mucosa of human.
   4. Reduced catalase (RBC).

B. *Metalloflavoproteins*
   1. Monoamine oxidase—decreased in liver of rat and in platelets of human. Normal in rat brain.
   2. Aldehyde oxidase—decreased in rat brain mitochondria. Normal in liver and heart (homogenates).
   3. α-Glycerophosphate dehydrogenase—decreased in skeletal muscle of iron-deficient, nonanemic rat. Normal in brain.
   4. Succinate dehydrogenase—decreased in heart and kidneys of rats. Normal in liver and brain.
   5. Xanthine oxidase—(?)decreased in liver and duodenal mucosa of iron-deficient rat.

C. *Cofactor*
   1. Aconitase—decreased in human blood (most marked in WBC). Decreased in kidneys (but not liver or brain) of rat.
   2. Protocollagen-proline hydroxylase—hydroxylation of proline and lysine in conversion of protocollagen to collagen. Decreased collagen synthesis in iron-deficient rat.

D. *Nucleic acid metabolism*—impaired DNA synthesis and mitosis. Reduced ribonucleotide reductase activity.

E. *Folate and/or B_{12} metabolism*—(?)reduced THFA formimino transferase activity; (?)decreased B_{12} absorption.

2. The same enzyme may be affected differently in different organs of the same animal and the same organ of different animals. For example, succinate dehydrogenase activity is reduced in the heart and kidneys but not in the liver of iron-deficient rats. Catalase is reduced in the iron-deficient pig liver but not in the iron-deficient mouse liver. This same enzyme may be reduced in human red cells and increased in pig erythrocytes in the iron-deficient state (Fairbanks *et al.*, 1971).

3. The work load on an organ appears to influence its susceptibility to iron depletion. Cytochrome c, for example, may be reduced to 45% of control values in the skeletal muscle of the iron-deficient rat, while the myocardium of the same animal remains near (89%) control levels (Dallman and Schwartz, 1965b). Likewise, there is a hierarchy of myoglobin depletion in these animals, with myocardium being most resistant (80% control), diaphragmatic muscle intermediate, and skeletal muscle (45% control), most adversely affected (Dallman and Schwartz, 1965b).

4. Because of the influence of cellular multiplication rate and the rate of replacement of subcellular organelles on tissue susceptibility, the organ impact of iron deficiency may depend upon the timing of the insult. The postweanling rat, with a nearly completely developed brain, shows great resistance to the CNS biochemical impact of iron deficiency. Thus, there is no significant reduction in brain cytochrome c in the postweanling iron-deficient rat (Dallman and Schwartz, 1965b). A recent report (Chepelinsky and de Lares Arnaiz, 1970) has confirmed the rapid rise in rat brain mitochondrial cytochromes in the immediate postnatal period, with adult levels achieved by 30 days of age. The hepatic cytochromes (b_5, P_{450}), important in various detoxification reactions, rise even more rapidly (Dallner *et al.*, 1966). The extended postpartum development of the human brain, a significant portion of which occurs during periods characterized by a high likelihood of somatic iron deficiency, provides the biological basis for a possibly long term neurobehavioral effect of iron deficiency in infancy and early childhood (Dobbing, 1974; Burman, 1973).

6.2. Brain Iron: Ontogeny and Distribution

Hallgren and Sourander (1958) studied a total of 98 brains of patients of all ages (none with cerebrovascular or neuropsychiatric disturbance) and reported the development and distribution of nonheme iron by both chemical and histopathologic methods. In patients 30–100 years of age, the iron content was highest in the phylogenetically oldest structures. The globus pallidus had the highest concentration (21.3 mg/100 g of fresh weight of brain). This was considerably more than the liver (13.4 mg/100 g), a major storage site for iron. In the cortex, the highest concentrations (5.03 mg/100 g) were found in the motor cortex with smaller amounts in the occipital, parietal, and prefrontal cortices. The whole adult brain was estimated to contain about 60 mg of nonheme iron, approximately 20% of the content of an iron-replete liver. Although in general the iron content of the brain appeared independent of that present in the liver, in patients who had experienced severe gastrointestinal bleeding, a somewhat lowered iron content was noted in the extrapyramidal structures. Since the amount of iron present in the brain is greater than that which can be accounted for by known enzymes and cofactors, and since one-third of the nonheme iron is extractable by buffer solutions and bound to a protein similar to ferritin, it can be inferred that the brain harbors reserve iron depots which are, to a large extent, independent of somatic iron status.

All areas of the brain except the medulla oblongota (constant at 1-2 mg/100 g from birth) showed an increase in nonheme iron with advancing age. Although the exact rates of increase vary with the brain part, all show maximal increments in the first 20 years of life. As a generalization, levels are at about 10% of adult values at birth, 50% by age 10 years, and maximal between ages 20 and 50 years. Dallman (1974) has pointed out that this gentle rise parallels neither the developmental pattern of brain cytochromes, which increase to adult levels concurrent with myelinization, nor the fluctuations of total body iron stores. The subcellular location and biochemical function of this nonheme iron in the human brain is largely unknown. Fractionation of homogenized dog brain (2, 5, and 7 years of age) revealed the mitochondria and microsomes to contain a total of about 40% of the nonheme iron (Hallgren and Sourander, 1958).

As was indicated earlier, the developing rat brain appears resistant to the effects of iron deficiency as far as cytochrome c levels are concerned (Dallman and Schwartz, 1965b). However, recently Dallman *et al.* (1975) have demonstrated that a transient period of severe iron deficiency in the young rat results in a permanent deficit of nonheme brain iron despite the subsequent provision of an adequate iron intake and restoration of hematologic and liver ferritin iron status to normal. Rats were deprived of dietary iron from days 10–28 and days 10–48 of life, resulting in nonheme iron concentrations in the brain 27% and 22% below control values, respectively. Body weight, liver weight, and hematocrit, but not brain weight, were significantly different between experimental and control animals. Intramuscular iron (5 mg) corrected the difference in the first three measures within 14 days, but 14–45 days after this treatment the nonheme iron in the brain remained 19% (early corrected) to 29% (late corrected) below the control means. Ferritin iron in the brain also remained depressed 33–42% below the control means (early and late repair, respectively). This failure to ameliorate iron deficiency in the brain is apparently the result of a very slow turnover rate for brain iron compounds rather than of a developmental decrease in the permeability of the blood–brain barrier to iron (Dallman and Spirito, 1977).

It is not known at present whether transient iron deficiency can in any way affect the short- or long-term biochemical or morphological characteristics of the human brain. The studies of Cantwell and Oski (see Section 5.3.1) provide very oblique evidence that such a situation may occur. What is clearer is that the human infant has a relative somatic iron deficiency at the time of maximal brain growth (Burman, 1973). Minor alterations in this tenuous balance could, theoretically, impair the deposition of brain iron and hence the synthesis of enzymes and cofactors for several critical biochemical pathways.

Dallman (1974) has emphasized that in such a situation subsequent restoration of brain chemistry to normal might not necessarily imply restoration of function, since subtle secondary morphologic abnormalities could persist. Thus, even studies which demonstrate restoration of various biochemical parameters should not be viewed as demonstrating an *absence* of permanent adverse impact.

6.3. *Biochemical Changes Relevant to CNS Function*

Despite an awareness of the role of iron in a series of pathways important to central nervous system function, very little research has focused directly on the possible impact of iron deficiency on brain-related biochemical processes.

Enumerated below are the available biochemical studies of systems relevant to CNS function. Section 6.4 then provides a summary of the attempts to assess the physiologic pertinence of the various biochemical changes which have been noted or predicted.

6.3.1. *Respiratory Enzymes and Mitochondrial Cytochromes*

Aconitase concentrations (Beutler, 1959) and succinate dehydrogenase and α-glycerophate oxidase activity (Mackler *et al.*, 1978) have been reported normal in the brain of iron-deficient rats. Cytochrome concentrations and mitochondrial oxidative phosphorylation remain normal in the rat brain despite severe iron deficiency (Dallman and Schwartz, 1965a; Mackler *et al.*, 1978).

6.3.2. *Phenylalanine Hydroxylase*

Phenylalanine hydroxylase (Fig. 2), a microsomal enzyme (present only in liver, pancreas, and kidney), catalyzes the conversion of phenylalanine (an essential amino acid) to tyrosine, which is subsequently used (in part) for neurotransmitter synthesis. Iron, apparently in the ferric form, is essential for this enzyme's activity (Hall *et al.*, 1974). Complete, and in some instances partial, deficiencies of phenylalanine hydroxylase result in hyperphenylalaninemia (PKU), a systemic metabolic disturbance associated with deranged brain biochemistry and mental retardation (Scriver and Rosenberg, 1973). Although the etiologies of the mental disturbances associated with that disorder are not known, the protean secondary biochemical abnormalities (e.g., reduced synthesis of serotonin, GABA, and catecholamines) represent a fertile field of possibilities. We are unaware of any studies regarding *in vivo* phenylalanine hydroxylase function or blood amino acid patterns in iron-deficient animals or patients. Subtle perturbations in plasma amino acid patterns and hence brain amino acid transport, resulting from iron-deficiency-related impairments in this enzyme function, could contribute to the behavioral changes noted.

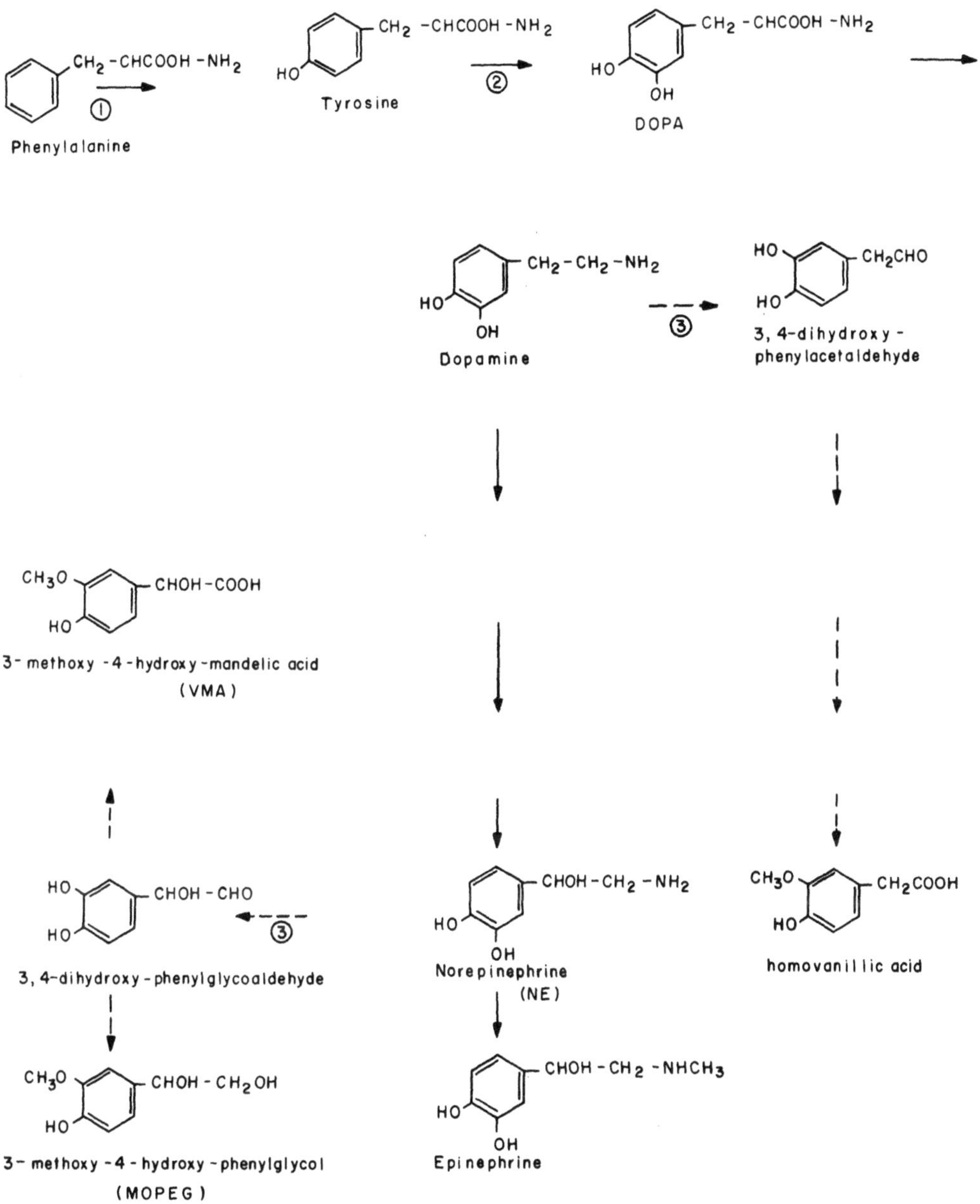

Fig. 2. CNS catecholamine metabolism. In the CNS reduction of 3,4-dihydroxyphenylglycoaldehyde predominates and therefore MOPEG is the predominant NE catabolite. Little, if any, VMA is formed in the CNS. (——→) Synthesis; (- - -→) catabolism; (1) phenylalanine hydroxylase; (2) tyrosine hydroxylase; (3) monoamine oxidase.

6.3.3. Tyrosine Hydroxylase

Tyrosine hydroxylase (unique to catecholamine-containing neurons and chromaffin cells) (Fig. 2) is the first and rate-limiting enzyme in catecholamine biosynthesis and requires ferrous iron as a cofactor. Iron's role in this enzyme's function is unknown, but selective iron chelators can inhibit its activity *in vitro* (Cooper *et al.*, 1974). No determinations of its activity in the brains of iron-deficient animals have been reported, but the activity of this enzyme has been found to be increased in the sympathectomized adrenal glands of iron-deficient rats (Quik and Sourkes, 1977).

6.3.4. Tryptophan Hydroxylase

Serotonin (5-hydroxytryptamine) is an indolealkylamine which is an apparent CNS neurotransmitter whose derangements have been implicated in a number of psychoaffective disturbances (Cooper *et al.*, 1978). Tryptophan hydroxylase (see Fig. 3), which may require iron as a cofactor (Youdim *et al.*, 1975b), is the rate-limiting enzyme in the synthesis of serotonin. Youdim and Green (1977) have reported that tryptophan hydroxylase activity in homogenates of iron-deficient rat brain stem is the same as that of control rats.

6.3.5. Tryptophan Pyrrolase

Tryptophan pyrrolase (see Fig. 3), a heme-containing enzyme, catalyzes the degradation of tryptophan to *N*-formyl L-kynurenine, thus shunting it away

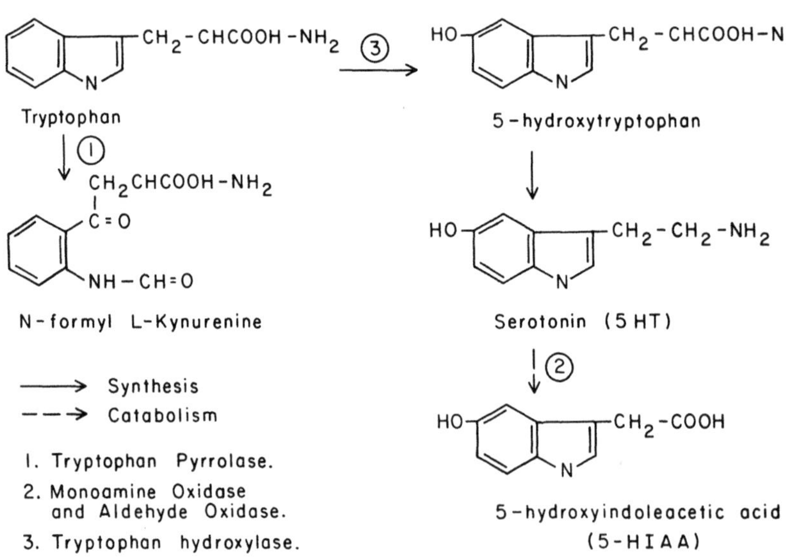

Fig. 3. Tryptophan metabolism.

from serotonin (5-HT) synthesis (Lehninger, 1975). Badawy (1977) reported no influence of mild or severe iron deficiency on the level of tryptophan pyrrolase in the liver of rats.

6.3.6. *Aldehyde Oxidase*

Like monoamine oxidase, aldehyde oxidase is an iron-containing metalloflavoprotein. It functions in the catabolism of serotonin by converting 5-hydroxyindoleacetaldehyde (formed from serotonin by MAO) to 5-hydroxyindoleacetic acid, which is then excreted from the body. Although Youdim and Green (1977) report no decrease in the level of the enzyme in liver or heart homogenates of iron-deficient rats, Mackler *et al.* (1978) have reported a significant decrement in brain mitochondrial aldehyde-oxidase-specific activity; liver mitochondrial aldehyde oxidase was unaffected by iron deficiency. That this change in a serotonin catabolic enzyme may be of some functional significance is indicated by the fact that the concentrations of both tryptophan and 5-hydroxyindoles were increased (albeit by small amounts) in the brain of iron-deficient animals. Both the level of aldehyde oxidase activity and concentrations of 5-hydroxyindole compounds were normalized within one week of parenteral iron therapy. The possible significance of these biochemical findings with regard to CNS functions is discussed in Section 6.4 of this chapter.

6.3.7. *Monoamine Oxidase*

Intracellularly, monoamine oxidase, a metalloflavoprotein, is found (probably exclusively) on the outer mitochondrial membrane and may well exist in different forms (conformational, polymeric) in various tissues (Youdim, 1973). Iron may function as a metal prosthetic group for the enzyme, provide a bond for the enzyme to the mitochondrial membranes, or act as a cofactor for the biosynthesis of the MAO apoenzyme (Sourkes, 1972). In highly purified preparations, the concentration of iron may be as low as 0.5 atom/mol of enzyme (Tipton, 1973). While pig and rat liver MAO appears to contain iron, pig brain MAO has been reported to be iron-free. Whether or not human brain MAO contains iron is not known (Tipton, 1973). In the nervous system MAO is found in glial cells as well as in synaptosomes. The enzyme limits the duration of action of various neurotransmitters (serotonin, dopamine, norepinephrine) by oxidizing them after their release or reuptake into the presynaptic nerve fiber (Snyder, 1976). The enzyme catalyzes the conversion of the amine to its corresponding aldehyde.

Sourkes (1972) has shown that both riboflavin and iron deficiency result in reduced hepatic MAO activity in rats; the abnormality is reversed by repleting the animal with the missing vitamin or mineral. Symes *et al.* (1969) made young male albino rats (40 g) severely iron-deficient (and anemic) by feeding a low-iron diet (for 5 or more weeks). Rats studied after 35–80 days of iron deficiency showed a mean Hb of 4.8 g/dl and a 64% reduction in liver iron (mg of Fe/g of tissue) compared to control animals. The same animals showed an 18–41% reduction in MAO activity as estimated by the rate of O_2 consumption by liver homogenates using isoamylamine, tyramine, and benzylamine as substrates. This report does not describe the body or organ weights of the iron-deficient animals. Since iron-deficient animals are frequently anorectic and lose weight, it is possible that the enzyme "impact" is an artifact of malnutrition in these animals. In a subsequent set of experiments (Symes *et al.*, 1971), the same group showed that nutritional iron deficiency with anemia in the rat resulted in a 30–40% reduction in the *in vivo* rate of metabolism of pentylamine (a substrate for MAO) to CO_2. The iron-deficient rats showed a definite weight decrement versus controls, but dietary iron supplements returned pentylamine metabolism to normal within 6 days. During that period, Hb rose (but not to control levels) but body weight did not approach control levels. Further evidence against a nonspecific effect of the iron deficiency was provided by a demonstration of normal rates of oxidation of *n*-amyl alcohol in the iron-deficient animals. In the same study, riboflavin deficiency was shown to produce a similar reduction (specifically reversible upon 7 days of riboflavin feeding) in *in vivo* MAO activity while copper deficiency produced no change. It is noteworthy that the copper-deficient animals showed no decrease in weight or MAO activity. This study would have been more convincing had animals with weight loss secondary to a nutritionally balanced but hypocaloric diet been used as controls. Nonetheless, the results of both together strongly imply that iron deficiency limits the activity of an enzyme with a major role in neurotransmitter function. The speed of correction with iron repletion is congruent with clinical studies mentioned earlier in which subjective improvement precedes significant increments in Hb. Clearly, however, proof of a deficit in hepatic metabolism does not necessarily imply a similar abnormality in the CNS. And, in fact, both Mackler *et al.* (1978) and Youdim and Green (1977) report no significant change in monoamine oxidase activity in the brain mitochondria or brain homogenates of young rats deficient in iron for 4–5 weeks. Such a short period of iron deficiency may not, however, be an appropriate model for the CNS effects of the long-term nutritional iron deficiencies seen in humans. Furthermore, just as the demonstration of mild whole-brain excesses/deficiencies of neurotransmitter-related compounds implies little or nothing about abnormal brain function, so the demonstration of "normal" levels of these compounds provides no assurance that specific subareas of the brain are not biochemically and functionally deficient.

Platelet monoamine oxidase, which is believed by some to reflect brain levels of the enzyme (Murphy and Wyatt, 1972; Wyatt *et al.*, 1973; Robinson, 1975; Edwards and Chang, 1976), was measured by Youdim *et al.* (1975a) in iron-deficient patients. Because the enzyme is known to have multiple forms, each with different substrate specificities (Youdim, 1973), several substrates were used. Sixteen iron-deficient adults had hemoglobins of 8.79 ± 0.24 g/dl with accompanying marked reductions in transferrin saturation. All showed significant reductions in platelet MAO with tyramine, dopamine, and 5-hydroxytryptamine as substrate. Seven were treated with oral ferrous sulfate for

30 days; those whose serum iron levels had returned to normal had normal platelet MAO activity. Unfortunately, no controls were given a short course of iron to assess its impact on "normal" MAO levels. Subsequently, 37 adults were studied for the relationship between Hb, serum Fe, and platelet MAO activity. When grouped on the basis of serum iron (low < 10.7 μmol/liter for female and 14.3 μmol/liter for male), there was a significant difference in MAO activity between groups, when grouped by Hb, there was no significant difference between MAO activity between individuals with normal or low (< 11.5 g/dl for female, < 13.5 g/dl for male) Hb. This approach is replete with the methodologic pitfalls referred to earlier. Serum iron level alone is a poor indicator of body iron status. Nonetheless, there does appear to be some relationship between body iron status and platelet MAO levels. This, of course, says nothing about CNS status in this regard.

In the developing rat, MAO levels increase at different rates in different organs. Whereas adult levels of renal MAO are achieved at 8 days postpartum, liver MAO increases steadily to adult levels by 120 days. Brain MAO is less than 50% of adult levels at birth and actually shows a transient arrest or actual decline during the first postpartum week followed by a parabolic rise to adult levels of 120 days (Kuzuya and Nagatsu, 1969). Karki *et al.* (1962) correlated brain MAO levels at birth with the functional maturity of CNS, showing that whereas newborn rats and rabbits, who are "born helpless," have brain MAO levels about 30% of adult values, newborn guinea pigs who are more neurologically mature at birth have adult levels of brain MAO. Humans show a steady rise in brain MAO throughout life, roughly paralleling MAO levels in platelets and plasma. The early ontogeny of this enzyme in the human CNS is unknown, but levels approaching those of adults have been reported in 4-year-olds (Robinson, 1975). Relative to adult capabilities, the neonatal human CNS is obviously quite immature, and thus by analogy with the animal studies described, one might anticipate that newborn brain levels would be well below those of adults and increase rapidly during early postnatal life. If this were the case, and assuming that human brain MAO contains iron, it is possible that early nutritional derangements vis-à-vis iron could affect CNS development.

Platelet, plasma, and hindbrain MAO levels appear to rise with advancing age in humans. Hindbrain norepinephrine concentration progressively decreases with age, while 5-hydroxytryptamine and 5-hydroxyindoleacetic acid levels remain unchanged. At all ages, females have a higher mean plasma and platelet MAO than men (Robinson, 1975; Roth *et al.*, 1976) and slightly higher mean brain MAO levels (Robinson, 1975). These findings are not only consistent with the CNS catecholamine depletion theory of depressive illness (higher frequency in older females), but also predict that the male should be more susceptible to functional MAO depletion in the face of iron deficiency. Furthermore, since estrogens appear to elevate peripheral MAO levels (Gilmore *et al.*, 1971; Klaiber *et al.*, 1971), the adult female should be even further protected against the impact of iron deficiency on MAO levels. Such considerations could influence the results obtained in behavioral studies of iron-deficient women at different ages and at different points in their menstrual cycles. They are also congruent with studies showing a differential effect of iron deficiency in male and female adolescents (see Section 6.5).

6.3.8. Nucleic Acid Metabolism

As was indicated earlier, iron deficiency appears to influence various aspects of DNA and protein synthesis and cell division. This may indeed be secondary to a direct effect of iron on these processes (Robbins and Pederson, 1970; Hoffbrand *et al.*, 1974, 1976; Siimes and Dallman, 1974; Hershko *et al.*, 1970; London *et al.*, 1976) but could also be the result, at least in part, of iron's apparent effect on B_{12} and folate metabolism. Yeh *et al.* (1961) reported reduced intestinal B_{12} absorption in iron-deficient rats (possibly due to reduced gastric secretion), and Vitale *et al.* (1966) have reported biochemical and hematologic evidence of folate deficiency in iron-deficient rats, possibly due to an iron-lack-mediated reduction in the activity of formimino transferase. Both B_{12} and folate influence DNA synthesis, and both, in addition, influence CNS function in a fashion which may be separate from the impact of nucleotide metabolism (Das and Herbert, 1976; Reynolds, 1976). For example, B_{12} interacts with a folic acid derivative in the conversion of homocysteine to methionine, while THFA formimino transferase catalyzes an important step in histidine catabolism, the conversion of FIGLU to glutamate (see Fig. 4). It is interesting to note that inborn deficiencies of both these enzyme systems have been described (Rowe, 1978) and are associated with deranged CNS function. Despite these associations it seems unlikely that iron deficiency acts via these pathways to produce its effects on CNS function.

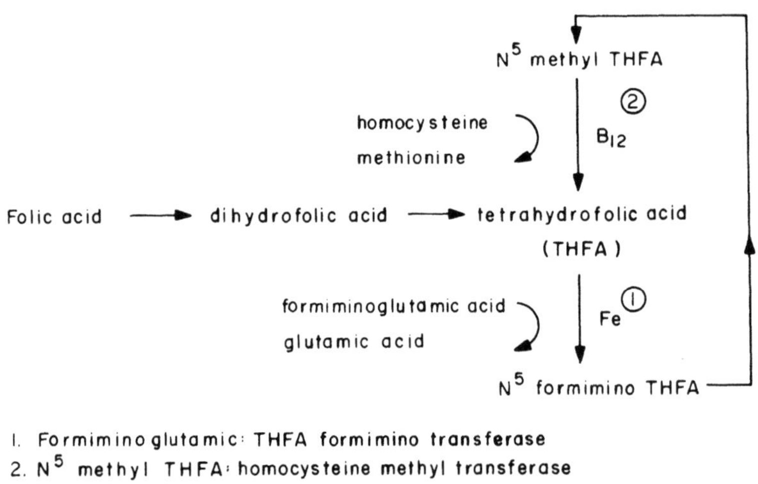

Fig. 4. Interrelationships of folate, B_{12}, and iron metabolism. Folate coenzymes participate in purine and thymidylate biosynthesis as well as the metabolism of several amino acids (glycine, serine, histidine, remethylation of homocysteine to methionine).

6.4. Physiologic Impact of Biochemical Changes

Enumerated in the previous section have been some of the biochemical derangements which characterize the iron-deficient state. In looking for non-hematologic modes of impact of iron deficiency, we now examine the potential effects of these biochemical changes on brain function and behavior. Table XIII summarizes the areas of possible CNS impact of iron deficiency.

6.4.1. Oxidative Metabolism

6.4.1a. Brain. Metabolically, the brain is an extremely active organ. The average adult male consumes about 250 ml of oxygen per minute in the basal state. Of this, about 50 ml/min is accounted for by the brain. Thus, an organ which constitutes some 2% of the total body weight accounts for 20% of the oxygen consumption. Cerebral oxygen consumption is apparently low at birth but rises rapidly during the period of brain growth. In children, maximal brain oxygen consumption is achieved at 6 years of age, at which time the brain uses 5.2 ml of O_2/100 g/min as opposed to 3.5 ml of O_2/100 g/min in an adult. At this rate, the child's brain metabolism accounts for more than 50% of the total body basal oxygen consumption. The cerebral blood flow is also higher at this time (106 ml/100 g/min) than in adulthood (62 ml/100 g/min). Most of this oxygen is used for the oxidation of glucose and the generation of high-energy phosphate bonds. The very high brain oxygen-consumption levels of

Table XIII. Areas of Possible CNS Impact of Iron Deficiency (without Anemia)

General biochemical	Specific sites	Impact
Heme synthesis	1. Porphyrin synthesis (general ↓; ↑ FEP)	Toxic or intracellular deficiency
	2. Mitochondrial cytochromes (c, oxidase)	Respiration oxidative phosphorylation
	3. Microsomal cytochromes (P_{450}, b_5)	Toxic
Krebs cycle	Succinate dehydrogenase (Fe-S flavoprotein), aconitase (Fe cofactor)	Respiration
Fe-S flavoproteins	NADH-ubiquinone reductase, α-GP dehydrogenase	Respiration, oxidative phosphorylation
Nucleic acids	DNA synthesis, mitosis	Brain growth
Catecholamines	Phenylalanine hydroxylase, tyrosine hydroxylase, monoamine oxidase	Neurotransmitter levels
Serotonin	Tryptophan pyrrolase, tryptophan hydroxylase, aldehyde oxidase	Neurotransmitter levels
Folic acid/B_{12}	Formimino transferase, THF methyltransferase	

childhood apparently reflect the large energy requirements of growth and differentiation of the nervous system (Sokoloff, 1976).

Various heme, iron-flavoprotein, and iron-dependent enzymes are components of the Krebs cycle and cytochrome pathway, both of which are, of course, essential in oxidative metabolism. These enzymes and cofactors are located almost exclusively in the inner compartment of the mitochondrion. Dallman and Goodman (1970) have described a marked enlargement of the mitochondrial compartment in heart, muscle, liver, and erythroblast of iron-deficient rats. The mitochondria are increased in size and are more radiolucent than normal; in some organs they lose their elongated shape and become rounded and may have their usual double outer membrane replaced by a single one. Similar mitochondrial morphologic changes have been seen in erythroid precursor cells in iron-deficient children as well as in animals with copper and riboflavin deficiency (Dallman and Goodman, 1970). Despite these changes in morphology and enzyme content (referred to earlier) of the iron-deficient mitochondria, no defect in respiration or oxidative phosphorylation has been detected in isolated mitochondria from iron-deficient rats (Goodman *et al.*, 1970; Wohlrab and Jacobs, 1967). Mackler *et al.* (1978) recently reported that in iron-deficient rats the cytochrome content of brain mitochondria and their phosphorylation capacity were not different from control animals.

Goodman *et al.* (1970) reported morphologic abnormalities in heart mitochondria of iron-deficient rats whose mitochondrial cytochrome concentrations remained normal. In copper-deficient rats, however, the copper-containing cytochrome c oxidase (aa_3) was depressed to less than 50% of control levels in association with mitochondrial morphologic abnormalities similar to those in iron-deficient animals. In both copper- and iron-deficient mitochondria, however, respiration and phosphorylation proceeded at control rates. Thus, mitochondrial oxidative metabolism was apparently preserved despite severe morphologic changes and/or cytochrome depletions. Restoration of mitochondrial morphology occurred after about 16 days of iron therapy, a period of time twice as long as that required to return the hemoglobin level to normal.

Using each patient as her own control, Beutler *et al.* (1960b) were unable to demonstrate any change in systemic oxygen consumption or exercise-induced oxygen debt in four iron-deficient women before and after 2 weeks of iron therapy. The authors concluded that "the symptoms in iron deficiency may arise from a mechanism other than a generalized defect in oxygen transport." Gardner *et al.* (1975) studied 29 adult iron-deficient anemic subjects (secondary to hookworm infestation) with hemoglobins ranging from 4 to 12 g/dl. Various aspects of cardiopulmonary function were compared before and 80 days after intramuscular administration of iron dextran or a saline placebo. No differences in hand grip or shoulder adductor strength, resting pulse rates, or oxygen consumption with exercise were demonstrable as a result of iron therapy. On the other hand, iron administration resulted in a significant reduction in exercise heart rate and minute ventilation as well as postexercise blood lactate levels. Thus, enhanced cardiopulmonary function (increased pulse rate and minute ventilation) had apparently been invoked to compensate for reduced hemoglobin mass. The elevated lactate levels imply that anaerobiosis was occurring despite the compensatory changes noted above, and could reflect impaired mitochondrial function. Finally, Jacobs (1965) reported that while leukocytes from iron-deficient patients metabolize oxygen at the same rate (per cell) as cells from nonanemic individuals, the iron-deficient leukocyte could not increase its respiratory activity upon exposure to dinitrophenol (which dissociates oxidative phosphorylation from the respiratory enzyme system) as greatly as normal cells. Jacobs interpreted this finding as an indication of limited respiratory reserve in the iron-deficient cells. Homogenates of these cells, when incubated with supplemental cytochrome c, showed a greater increase in respiratory activity than did normal cells.

Given the known resistance of the central nervous system to iron-induced deficiencies in cytochromes and Krebs cycle enzymes, and the essentially normal oxygen consumption rates reported in four iron-deficient patients and certain of their presumably iron-deficient cells, it seems unlikely that brain oxidative metabolism is affected via enzymatic/cytochrome changes in iron deficiency. Further, oxygen delivery to this organ is protected by a series of homeostatic adjustments described earlier down to a hemoglobin of 7–8 g/dl. Nonetheless, recent studies (Rosenthal *et al.*, 1976) of the changes in the redox level of cytochrome oxidase in intact cerebral cortex of cats (using dual wavelength reflectance spectrophotometry) have indicated that this cytochrome is at least 20% reduced under resting conditions and has a sharp reductive response to any lowering of inspired oxygen. These findings, which are at variance with the very low reduction levels reported in isolated mitochondria, support the concept that the mammalian cerebral cortex is normally near a state of oxygen insufficiency (Davies and Bronk, 1957). The implication is that relatively small shifts in brain oxygenation and/or cytochrome levels might significantly impair oxidative phosphorylation in that organ. Iron-deficiency anemia, of course, offers an opportunity for derangement in both these areas.

6.4.1b. Muscle. Iron deficiency may adversely affect oxidative metabolism in skeletal muscle by causing a reduction in myoglobin or influencing certain muscle enzymes. The former situation obtains in various animals but has not been studied in humans. The latter possibility has been carefully examined by Finch *et al.* (1976), who studied treadmill work performance in iron-deficient rats whose hemoglobins had been acutely restored by transfusion, thus removing oxygen delivery as a variable. They found that although the concentrations of all cytochromes, as well as myoglobin and α-glycerophosphate dehydrogenase levels, were significantly reduced in the skeletal muscle of iron-deficient animals who showed impaired treadmill performance, only the activity of α-glycerophosphate dehydrogenase rose to nearly (80%) normal levels over the 3- to 4-day period of iron repletion necessary to eliminate the deficit in maximum work performance. There was, in fact, no change in the reduced myoglobin and cytochrome levels during that short period of therapy. α-Glycerophosphate dehydrogenase is a mitochondrial iron metalloprotein which appears to function both in the regeneration of NAD^+ for the glycolytic cycle (via the α-glycerophosphate shuttle) and in the generation ATP by oxidation of α-glycerophosphate. A derangement in this enzyme's function in iron deficiency may explain, at least in part, the muscle weakness and easy fatigability which characterize this disorder and are, to some extent, independent of the hemoglobin level.

6.4.2. Liver Microsomal Drug-Metabolizing Enzyme System

In the liver, along with a flavoprotein electron carrier (NADPH-cytochrome P_{450} reductase), the microsomal cytochrome P_{450} (which acts as a binding site for oxygen and substrate) participates in the detoxification (by hydroxylation and desaturation reactions) of a host of endogenous compounds as well as drugs. In addition, another microsomal cytochrome (b_5) is active in the desaturation of fatty acids (Lehninger, 1975). There is convincing evidence that various metals, including iron, directly influence the cellular content of heme. Increased intracellular levels of iron, for example, result in a decrease in ALA synthetase activity and an increase in heme oxygenase; the net effect is to decrease intracellular heme by reducing its synthesis and accelerating its catabolism. An attendant diminution in cytochrome-P_{450}-dependent chemical metabolism has been demonstrated under these circumstances (Maines and Kappas, 1977). As a corollary to these observations, one might anticipate that iron deficiency would increase intracellular heme, thereby preserving or even augmenting the activity of the microsomal cytochrome system. And, in fact, both the P_{450} and b_5 cytochromes appear to be extremely resistant to any adverse effect of iron deficiency. When, after 3 weeks of iron deficiency, hepatic mitochondrial cytochromes a and c are diminished, levels of P_{450} and b_5 are unchanged. The inducibility of P_{450} with phenobarbital (fourfold rise) was unaltered after 3 weeks of iron deficiency. At 8 weeks there is a slight fall in the levels of P_{450} while b_5 remains unchanged (Dallman and Goodman, 1971). Similar findings were reported by Becking (1972, 1976) who found no change in the hepatic P_{450} or b_5 levels of rats on diets deficient in iron for up to 35 days. In fact, *in vitro* and *in vivo* metabolism of aniline and amino pyrine was actually increased. Becking (1976) has hypothesized that the ferric iron in ferritin may be an inhibitor of certain hepatic enzymes (including cytochrome P_{450}) and that the reduced hepatic ferritin which is a concomitant of iron deficiency may "release" these enzymes, thus augmenting the metabolism of various drugs. An alternative explanation for these findings is, of course, the effect of iron on rates of heme synthesis/catabolism discussed above. Though rats and mice show these iron-deficiency-related increments in drug catabolism, the one study in iron-deficient humans has failed to demonstrate an increased rate of drug metabolism (O'Malley and Stevenson, 1973). If iron deficiency were shown to influence drug metabolism in humans, the high prevalence of the disturbance would render this an important pharmacologic issue.

6.4.3. Heme Dysmetabolism

As was pointed out in Section 6.2, erythrocyte protoporphyrin is increased in iron deficiency. However, its rate of synthesis, along with that of other porphyrins is probably depressed in the iron-deficient patient. The increase in erythrocyte protoporphyrin in iron deficiency is apparently the result of a relatively greater block at the final (ferrochelatase, iron insertion) step than more proximally in the synthetic pathway (Moore and Goldberg, 1974). It should be noted that two other disorders characterized by deranged heme metabolism, lead poisoning and acute intermittent hepatic prophyria (AIP), share certain neurobehavioral characteristics with iron deficiency (Goldberg, 1968; Tschudy *et al.*, 1975). It now seems unlikely that either of these latter two disorders is, in fact, caused by excessive circulating porphyrin levels (Tschudy *et al.*, 1975; Shanley *et al.*, 1975). Acute intermittent hepatic porphyria is the result of an inherited reduction in urophophyrinogen I synthetase in the liver and elsewhere. The resultant reduction in heme synthesis (the apparent negative feedback compound for ALA synthetase) results in the synthesis of excessive amounts of heme precursors in the liver. Since the neurologic derangements characteristic of this disease correlate poorly with circulating porphyrin precursor levels and since hematin infusions cause rapid symptomatic relief in many patients, some have postulated that the neurologic symptoms of AIP may be due to heme lack within the central nervous system (Tschudy *et al.*, 1975) or to dissociation of the heme and apocytochrome component of cytochrome P_{450} in the liver (Watson, 1975). The former situation might influence cerebral oxidative metabolism directly while the latter would influence various detoxification processes. Similar abnormalities might occur in plumbism, and their likelihood in iron deficiency has already been discussed. Becking (1976) has emphasized that normal concentrations of cytochrome P_{450} do not necessarily imply normal function, since specific substrate binding and reduction rates of the complex could be altered in iron deficiency, with attendant alterations in oxidative drug metabolism.

Thus, disturbances in heme metabolism could influence brain function directly by some adverse effect of intracellular heme lack and/or excessive heme precursors, or indirectly by influencing hepatic mixed oxidase function (for which cytochrome P_{450} is a cofactor) with resulting "autointoxication" of the organism. It is interesting to note that although Becking (1976) could demonstrate no significant change in hepatic cytochrome P_{450} or b_5 levels, nor pentobarbital catabolic rates in iron-deficient rats, these same animals exhibited longer sleeping times (86 ± 8 min versus 63 ± 7 min) than controls when pentobarbital was administered. Iron deficiency may, then, influence cerebral sensitivity to this compound or its metabolites by means other than those reflected by the status of heme metabolism in the brain. This serves to reemphasize the point that, although there is little question that iron deficiency may impact certain known iron-containing enzymes, there is no reason to believe that these are the only biochemical pathways affected or that their involvement can account for the putative behavioral abnormalities of iron deficiency.

6.4.4. Cellular Growth

Although iron appears to play a role in various aspects of DNA metabolism and cell mitosis, there is little compelling evidence that these biochemical findings have clinical significance. Judisch *et al.* (1966) studied the records of 156 children under the age of 3 years who had iron deficiency with a hemoglobin less than 9 g/dl. They concluded that iron deficiency was associated with diminished body weight for age. Whether this was due to a direct effect of iron deficiency on somatic growth or was mediated by a secondary anorexia could not be ascertained from the available data, but treatment with iron resulted in accelerated weight gain in the iron-deficient children. It has been pointed out that the significance of these findings is confounded by the large number of low-birth-weight children in this study. This group, whose weights at the time of diagnosis correlated with birth weight, would be expected to have a higher prevalence of low hemoglobins because of their prematurity and to be showing catch-up growth which could be misinterpreted as a response to iron therapy (Anonymous, 1966).

A study by Canale and Lanzkowsky (1970) of the impact of iron deficiency in the postweanling rat (iron-deficient diet introduced at 40 days of age, animals sacrificed at 143 days) revealed a distinct anorectigenic effect of iron deficiency, with secondary, significant decrements in body weight. In addition, by comparing iron-deficient animals (mean hemoglobin 7.8 g/dl) with pair fed/iron supplemented and *ad libitum* fed/iron supplemented animals, iron deficiency per se was shown to be associated with (a) reduced hepatic DNA content (cellular hypoplasia) as well as increased protein/DNA (cellular hypertrophy); (b) increased renal cell protein/DNA; and (c) no significant change in brain total DNA content. Siimes and Dallman (1974) reported a reduced content and synthetic rate of hepatic DNA in iron-deficient rats. They demonstrated that the reduced DNA content was apparently the result of reduced *de novo* synthesis as reflected by depressed incorporation of ^{32}P into DNA. This finding is consistent with the reduced activity of ribonucleotide reductase (an enzyme in the *de novo* DNA synthetic pathway) described by Hoffbrand *et al.* (1976). Dallman *et al.* (1975) also showed that, whereas transient periods of iron deficiency during nursing could produce irremediable deficits in nonheme brain iron in rats, no apparent decrement in brain weight occurred in these animals. Although reversible drops in body and liver weight occurred in those animals exposed to early iron deficiency, the brain appeared to suffer no adverse impact as far as size was concerned. Thus, although the size of the body and certain of its organs may be adversely affected by iron deficiency, the gross growth of the brain seems protected. This is, of course, no guarantee that subtle and permanent morphologic abnormalities are not being induced in the growing, iron-deficient brain. In fact, recent studies have indicated that an iron-related compound, hematin, may play a role in the differentiation of nervous tissue (Ishi and Maniatis, 1978).

6.4.5. *Neurotransmitters*

There is extensive evidence that a number of monoamines (e.g., norepinephrine, dopamine, 5-hydroxytryptamine) are among the most important synaptic transmitters in the central nervous system (Cooper *et al.,* 1974). Shifts in the brain levels of these compounds have been implicated by some in the pathogenesis of various affective disorders. Thus, depressive illness has been viewed as being the result, in some instances, of decrements in the levels of these monoamines secondary to *elevated* monoamine oxidase levels (Sandler *et al.*, 1975). Recently, a population study has indicated a correlation between vulnerability to psychiatric disorders and *reduced* platelet monoamine oxidase activity (Buchsbaum *et al.*, 1976); various hypotheses have linked disordered monoamine metabolism to schizophrenia (Sourkes, 1976). Oral tryptophan, the amino acid precursor of serotonin, has been reported to be as effective as electroshock therapy in the treatment of depression, but to be associated with a deterioration in the mental status of schizophrenics (Copper, 1967; Young and Sourkes, 1974). Monoamine oxidase inhibitors have, of course, been used for many years in the treatment of various types of depression (Byck, 1975). As was pointed out earlier (see Section 6.3), there are a considerable number of iron-containing enzymes involved in various aspects of monoamine metabolism.

Iron deficiency has been shown to be associated with reduced levels of platelet monoamine oxidase (Youdim *et al.*, 1975a; Woods *et al.*, 1977). The only other attempt, to date, to correlate possible alterations in monoamine metabolism with iron deficiency in humans was the study of Voorhess *et al.* (1975), in which the urinary excretion of dopamine, norepinephrine, epinephrine, metanephrine-normetanephrine, and 3-methoxy-4-hydroxymandelic acid (VMA) (see Figure 2) was measured in 24-hr urine samples of 11 iron-deficient children (10 were 10 months–2½ years; 1 was 13 years old) before and after treatment with intramuscular iron. These investigators found norepinephrine excretion to be abnormally high (greater than 2 standard deviations from the mean in 9 of 11) in the iron-deficient children, returning to normal within 1 week of iron therapy. The urinary norepinephrine excretion did not directly correlate with the initial hemoglobin, mean red cell volume, serum iron, or transferrin saturation. VMA excretion was also higher before than after treatment, but in most instances these higher values were still within the range of normal for body size. There was no significant difference between dopamine, epinephrine, and metamephrine-normetanephrine excretion before and after iron therapy. Controls with anemia not due to iron deficiency (e.g., leukemia, thalassemia major) had normal urinary norepinephrine levels before and after transfusion therapy, implying that iron deficiency per se, rather than the reduced hemoglobin, was causally related to the high catecholamine excretion noted in the iron-deficient children. The authors suggest that these findings regarding catecholamine excretion may implicate monoamine oxidase-mediated shifts in central nervous system monoamines in the irritability, inattentiveness, and poor scholastic performance often associated with the iron-deficient state. However, it must be stressed that the source of the excess catecholamines excreted by the iron-deficient children was not demonstrated and that they may well have been derived in large part from the peripheral sympathetic nervous system. Even peripherally derived catechols could, however, influence the central nervous system, as presumably occurs in patients with behavioral derangements associated with pheochromocytomas (Liddle and Melman, 1974). The authors argue that a fall in VMA excretion would be expected if monoamine oxidase levels were reduced from iron deficiency, but no change was noted. Although VMA is the major catabolite of norepinephrine in the peripheral nervous system, very little VMA is found in the central nervous system (see Fig. 2). There, reduction of norepinephrine to its corresponding glycol [3-methoxy-4-hydroxyphenylglycol (MOPEG)] predominates. The sulfate conjugate of MOPEG readily diffuses from the brain into the cerebrospinal fluid and general circulation and might, therefore, reflect the activity of central noradrenergic neurons (Sharman, 1973). Despite the fact that peripheral sources still account for 70–80% of the urine content of this compound, it would seem to be a more reasonable candidate than either VMA or norepinephrine as an indirect reflection of central noradrenergic activity (Cooper *et al.*, 1974).

As noted in Section 6.3, Mackler *et al.* (1978) found reduced brain aldehyde oxidase activity and an attendant rise in the brain concentration of 5-hydroxyindole compounds in iron-deficient rats. This situation was normalized 1 week following 5 mg of intraperitoneal iron dextran. The authors point out the similarity between behaviors (diminished responsivity and learning ability) reported in individuals with elevated brain tryptophan levels and those with iron deficiency. Youdim and Green (1977), on the other hand, found *reduced* brain tryptophan levels in iron-deficient rats but no apparent functional deficit in enzymes concerned with catecholamine and indoleamine synthesis and catabolism. They attributed the diminished serotonin levels to reduced serotonin binding protein in synaptosomes.

Youdim and Green (1977) also demonstrated a reduced physical activity ("movements per minute") response to the administration of catecholamine and serotonin precursors in iron-deficient rats. This reduction disappeared after 8 days of oral iron therapy. Because synthetic/catabolic rates for these compounds were apparently normal [in conflict with Mackler *et al.* (1978)] and because there was also a diminished response to serotonin and dopamine agonists, the authors concluded that the apparent defect in neurotransmission in iron deficiency is located somewhere in the postsynaptic response. However, the study by Finch *et al.* (1976) in which muscle dysfunction (impaired running ability) was correlated with reduced muscle α-glycerophosphate dehydrogenase activity and in which behavioral and biochemical remediation were achieved within 4 days of iron therapy, implies that Youdim and Green may have been observing a peripheral (i.e., muscle tissue) rather than a central nervous system effect of iron deficiency. Likewise, Dallman's studies of iron turnover rates in brain are in apparent conflict with the finding of Mackler *et al.* (1978) regarding short-term remediation of a CNS iron-related biochemical lesion.

7. Conclusion

Iron is a micronutrient with an apparently broad spectrum of biochemical influence. Thus, iron deficiency, a highly prevalent nutritional disturbance, involves more than a potential reduction in hemoglobin mass. In fact, given the various homeostatic adjustments to moderate decrements in circulating hemoglobin, it is unlikely that the alleged psychoaffective morbidity associated with mild-to-moderate iron deficiency can be attributed to diminished oxygen-carrying capacity per se. What remains unsettled are the intertwining issues of (1) whether iron deficiency without anemia results in physiologically significant organ system derangements, and (2) what are the biochemical substrates of any such changes.

The examination of behavioral disturbances associated with iron deficiency illuminates both these issues. Despite strongly held clinical impressions and firm lay-person acceptance, there exists no unequivocal demonstration of an adverse effect of iron deficiency on intelligence, learning, attention, motivation, or general sense of well-being. Likewise, no *functionally significant* derangement in brain growth or brain biochemistry has been demonstrable in iron-deficient animals.

It is important to note, however, that the psychological tests and experimental designs employed to date have been relatively simple and that many of the studies regarding areas of potential CNS biochemical impact have been indirect, looking at effects in nonnervous tissue or measuring only relatively gross CNS biophysical parameters. In fact, the various biochemical derangements discussed in this paper should be regarded more as evidence that critical enzyme systems can be influenced by iron deficiency than as examples of specific pathways etiologically related to the suspected nervous system derangements. It is quite likely that the culpable metabolic process (if there is one) is unknown at present. However, the biochemical ontogeny of the known iron-related neural systems would lead one to anticipate that the impact of iron deficiency should be most striking when imposed on a developing nervous system. Iron-related metabolic derangements occurring during critical growth periods might leave permanent, although subtle, CNS functional sequelae.

The high prevalence of iron deficiency even in otherwise well-nourished populations and the particular predisposition of infants and children to this micronutrient deficiency make the issue of its role in brain development and behavior an important one. Although the effects are likely to be subtle, they are worth searching for both as a paradigm for other metal-deficiency states and as a public health issue.

ACKNOWLEDGMENT. This research was supported in part by Research Grant 1 R22 HDO 9228-01 from the National Institute of Child Health and Human Development.

8. References

Adamson, J. W., and Finch, C. A., 1975, Hemoglobin functions, oxygen affinity and erythropoietin, *Ann. Rev. Physiol.* **37**:351.

Anonymous, 1966, Iron deficiency and growth, *Nutr. Rev.* **24**:330–331.

Badawy, A. A.-B., 1977, The functions and regulation of tryptophan pyrrolase, *Life Sci.* **21**:755.

Bainton, D. F., and Finch, C. A., 1964, The diagnosis of iron deficiency anemia, *Am. J. Med.* **37**:62.

Barnes, R. H., Neely, C. S., Kwong, E., Labadan, B. A., and Frankova, S., 1968, Postnatal nutritional deprivation as determinant of adult rat behavior towards food, its consumption and utilization, *J. Nutr.* **96**:467.

Beaton, G. H., 1974, Epidemiology of iron deficiency, in: *Iron in Biochemistry and Medicine* (A. Jacobs and M. Worwood, eds.), pp. 477–528, Academic Press, Inc., New York.

Becking, G. C., 1972, Influence of dietary iron levels on hepatic drug metabolism *in vivo* and *in vitro* in the rat, *Biochem. Pharmacol.* **21**:1585–1593.

Becking, G. C., 1976, Hepatic drug metabolism in iron-, magnesium- and potassium-deficient rats, *Fed. Proc.* **35**:2480–2485.

Beisel, W. R., 1976, Trace elements in infectious processes, *Med. Clin. North Am.* **60**:831.

Bernhardt, K. S., 1936, Phosphorus and iron deficiencies and learning in the rat, *J. Comp. Psychol.* **22**:273.

Berry, W. T. C., and Nash, F. A., 1954, Symptoms as a guide to anemia, *Br. Med. J.* **1**:918.

Beutler, E., 1957, Iron enzymes in iron deficiency. I. Cytochrome c, *Am. J. Med. Sci.* **234**:517.

Beutler, E., 1959, Iron enzymes in iron deficiency. VI. Aconitase activity and citrase metabolism, *J. Clin. Invest.* **38**:1605.

Beutler, E., Larsh, S. E., and Gurney, C. W., 1960a, Iron therapy in chronically fatigued, nonanemic women: A double blind study, *Ann. Int. Med.*, **52**:378.

Beutler, E., Larsh, S. and Tanzi, F., 1960b, Iron enzymes in iron deficiency. VII. Oxygen consumption measurements in iron-deficient subjects, *Am. J. Med. Sci.* **239**:759–765.

Bowering, J., and Sanchez, A. M., 1976, A conspectus of research on iron requirements of man, *J. Nutr.* **106**:985.

Braine, M. D. S., Heimer, C. B., Wortis, H., and Freedman, A. M., 1966, Factors associated with impairment of the early development of prematures, *Monogr. Soc. Res. Child Dev. 106.*

Broman, S. H., Nichols, P. L., and Kennedy, W. A., 1975, *Preschool IQ: Prenatal and Early Developmental Correlates,* John Wiley & Sons, Inc., New York.

Brown, E. B., Moore, C. V., Reynafarje, C., and Smith, D. E., 1950, Intravenously administered saccharated iron oxide in the treatment of hypochromic anemia. Therapeutic results, potential dangers and indicators, *J. Am. Med. Assoc.* **144**:1084.

Buchsbaum, M. S., Coursly, R. D., and Murphy, D. C., 1976, The biochemical high-risk paradigm: Behavioral and familial correlates of low platelet mono-amine oxidase activity, *Science* **194**:339–341.

Buckley, R. H., 1975, Iron deficiency anemia: Its relationship to infection susceptibility and heart defense, *J. Pediatr.* **86**:993.

Burch, R. E., and Sullivan, J. F., (eds.), 1976, Symposium on trace elements, *Med. Clin. North Am.* **60**:4.

Burman, D., 1973, Iron deficiency in infants and children, *Clin. Haematol.* **2**:257.

Byck, R., 1975, Drugs and the treatment of psychiatric disorders, in: *The Pharmacological Basis of Therapeutics* (L. S. Goodman and A. Gilman, eds.), pp. 180–184. Macmillan Publishing Co., Inc., New York.

Canale, V., and Lanzkowsky, P., 1970, Cellular growth in specific nutritional deficiency states in rats, *Br. J. Haematol.* **19**:579–586.

Cantwell, R. J., 1974, The long-term neurological sequelae of anemia in infancy, *Pediatr. Res.* **8**:342 (abstr.).

Caputo, D. V., and Mandell, W., 1970, Consequences of low birth weight, *Dev. Psychol.* **3**:363.

Chandra, R. K., 1973, Reduced bactericidal capacity of polymorphs in iron deficiency, *Arch. Dis. Child.* **48**:864.

Chandra, R. K., 1976, Iron and immunocompetence, *Nutr. Rev.* **34**:129.

Chandra, R. K., and Saraya, A. K., 1975, Impaired immunocompetence associated with iron deficiency, *J. Pediatr.* **86**:899.

Chepelinsky, A. B., and de Lares Arnaiz, G. R., 1970, Levels of cytochromes in rat-brain mitochondria during post-natal development, *Biochim. Biophys. Acta* **197**:321.

Coltman, C. A., 1969, Pagophagia and iron lack, *J. Am. Med. Assoc.* **207**:513.

Committee on Iron Deficiency, 1968, Iron deficiency in the United States, *J. Am. Med. Assoc.* **203**:407.

Cook, J. D., Lipschitz, D. A., Miles, L. E. M., and Finch, C. A., 1974, Serum ferritin as a measure of iron stores in normal subjects, *Am. J. Clin. Nutr.* **27**:681.

Cook, J. D., Finch, C. A., and Smith, N. J., 1976, Evaluation of the iron status of a population, *Blood* **48**:449.

Cooper, J. R., Bloom, F. E., and Roth, R. H., 1974, *The Biochemical Basis of Neuropharmacology,* 2nd ed., pp. 90–201, Oxford University Press, Inc., New York.

Cooper, J. R., Bloom, F. E., and Roth, R. H., 1978, *The Biochemical Basis of Neuropharmacology,* 3rd ed., pp. 196–222, Oxford University Press, Oxford.

Copper, A., 1967, The biochemistry of affective disorders, *Br. J. Psychiat.* **113**:1237.

Cornell, E. H., and Gottfried, A. W., 1976, Intervention with premature infants, *Child Dev.* **47**:32.

Crosby, W. H., 1971, Food pica and iron deficiency, *Arch. Intern. Med.* **127**:960.

Crosby, W. H., 1976, Pica, *J. Am. Med. Assoc.* **235**:2765.

Dagg, J. H., Jackson, J. M., Curry, B., and Goldberg, A., 1966, Cytochrome oxidase in latent iron deficiency (sideropenia), *Br. J. Haematol.* **12**:331.

Dallman, P. R., 1974, Tissue effects of iron deficiency, in: *Iron in Biochemistry and Medicine* (A. Jacobs and M. Worwood, eds.), pp. 437–475, Academic Press, Inc., New York.

Dallman, P. R., and Goodman, J. R., 1970, Enlargement of mitochondrial compartment in iron and copper deficiency, *Blood* **35**:496.

Dallman, P. R., and Goodman, J. R., 1971, The effects of iron deficiency on the hepatocyte: a biochemical and ultrastructural study, *J. Cell Biol.* **48**:79–90.

Dallman, P. R., and Schwartz, H. C., 1965a, Myoglobin and cytochrome response during repair of iron deficiency in the rat, *J. Clin. Invest.* **44**:1631.

Dallman, P. R., and Schwartz, H. C., 1965b, Distribution of cytochrome c and myoglobin in rats with dietary iron deficiency, *Pediatrics* **35**:677.

Dallman, P. R., and Spirito, R. A., 1977, Brain iron in the rat: Extremely slow turnover in normal rats may explain long lasting effects of early iron deficiency, *J. Nutr.* **107**:1075.

Dallman, P. R., Siimes, M. A., and Manies, E. C., 1975, Brain iron: persistent deficiency following short-term iron deprivation in the young rat, *Br. J. Haematol.* **31**:209.

Dallner, G., Siekevitz, P., and Palade, G. E., 1966, Biogenesis of endoplasmic retriculum membranes, *J. Cell Biol.* **30**:97–117.

Das, K. C., and Herbert, V., 1976, Vitamin B_{12}—folate interrelations, *Clin. Haematol.* **5**:697.

Davies, P. W., and Bronk, D. W., 1957, Oxygen tension in mammalian brain, *Fed. Proc.* **16**:689.

Department of Health, Education and Welfare, 1968–1970, Ten-State Nutrition Survey, Vols. I–V, *DHEW Publ. (HSM) 72-8130.*

Department of Health, Education and Welfare, 1973, First Health and Nutrition Examination Survey.

Devlin, R. M., and Barker, A. V., 1971, *Photosynthesis,* pp. 113–130, Van Nostrand Reinhold Company, New York.

Dobbing, J., 1974, The later growth of the brain and its vulnerability, in: *Scientific Foundations of Paediatrics* (J. A. Davis and J. Dobbing, eds.), W. B. Saunders Company, Philadelphia.

Dobbing, J., and Smart, S. L., 1973, Early undernutrition, brain development and behavior, in: *Clinics in Developmental Medicine,* Publ. 47, Spastic International Medicine Publications, London.

Drorbaugh, J. E., Moore, M. D., and Warram, J. H., 1975, Association between gestational and environmental events and central nervous system function in 7 year old children, *Pediatrics* **56**:529.

Editorial, 1974, Iron and resistance to infection, *Lancet* **2**:325.

Edwards, D. J., and Chang, S. S., 1976, Multiple forms of monoamine oxidase in rabbit platelets, *Life Sci.* **17**:1127–1134.

Elwood, P. C., 1973, Evaluation of the clinical importance of anemia, *Am. J. Clin. Nutr.* **26**:958.
Elwood, P. C., and Hughes, D., 1970, Clinical trial of iron therapy of psychomotor function in anemic women, *Br. Med. J.* **1**:254.
Elwood, P. C., Waters, W. E., Green, W. J. W., Sweetman, P. M., and Wood, M. M., 1969, Symptoms and circulating hemoglobin level, *J. Chronic Diseases* **21**:615.
Elwood, P. C., Mahler, R., Sweetnam, P., Moore, F., and Welsby, E., 1970, Association between circulating haemoglobin level serum cholesterol, and blood pressure, *Lancet* **1**:589.
Fairbanks, V. F., Fahey, J. L., and Beutler, E., 1971, *Clinical Disorders of Iron Metabolism*, Grune & Stratton, Inc., New York.
Fernstrom, J. D., 1976, The effect of nutritional factors on brain amino acid levels and monoamine synthesis, *Fed. Proc.* **35**:1151.
Fernstrom, J. D., and Lytle, L. D., 1976, Corn malnutrition, brain serotonin and behavior, *Nutr. Rev.* **34**:257.
Finch, C. A., 1970, Diagnostic value of different methods to detect iron deficiency, in: *Iron Deficiency, Pathogenesis. Clinical Aspects, Therapy*, pp. 409–416, Academic Press, Inc., New York.
Finch, C. A., Miller, L. R., Inamdar, A. R., Person, R., Seiler, K., and Mackler, B., 1976, Iron deficiency in the rat. Physiological and biochemical studies of muscle dysfunction, *J. Clin. Invest.* **58**:447.
Fletcher, J., Mather, J., Lewis, M. J., and Whiting, G., 1975, Mouth lesions in iron-deficient anemia: Relationship to *Candida albicans* in saliva and to impairment of lymphocyte transformation, *J. Infect. Dis.* **131**:44.
Fuerth, J. H., 1971, Incidence of anemia in full-term infants seen in private practice, *J. Pediatr.* **79**:560.
Ganzoni, A. M., and Puschmann, M., 1975, Another look at iron: Role in host pathogen interaction, *Blut* **31**:313.
Gardner, G. W., Edgerton, R., Barnard, R. J., and Bernauer, E. M., 1975, Cardiorespiratory hematological and physical performance responses of anemic subjects to iron treatment, *Am. J. Clin. Nutr.* **28**:982–988.
Garn, S. M., Smith, N. J., and Clark, D. C., 1975, The magnitude and implications of apparent race differences in hemoglobin values, *Am. J. Clin. Nutr.* **28**:563.
Gilmore, N. J., Robinson, D. S., Nies, A., Sylwester, D., and Ravaris, C. L., 1971, Blood monoamine oxidase levels in pregnancy and during the menstrual cycle, *J. Psychosom. Res.* **15**:215.
Goldberg, A., 1968, Lead poisoning as a disorder of heme synthesis, *Semin. Hematol.* **5**:424–433.
Goodman, J. R., Warshaw, J. B., and Dallman, P. R., 1970, Cardiac hypertrophy in rats with iron and copper deficiency: Quantitative contribution of mitochondrial enlargements, *Pediat. Res.* **4**:244–245.
Gottfried, A. W., 1973, Intellectual consequences of perinatal anoxia, *Psychol. Bull.* **80**:231.
Guthrie, H. A., Froozani, M., Sherman, A. R., and Barron, G. P., 1974, Hyperlipidemia in offspring of iron-deficient rats, *J. Nutr.* **104**:1273.
Haddy, T. B., Jurkowski, C., Brody, H., Kallen, D. J., and Czajka-Narins, D. M., 1974, Iron deficiency with and without anemia in infants and children, *Am. J. Dis. Child.* **128**:787.
Hall, D. O., Cammack, R., and Rao, K. K., 1974, Non-haem iron proteins, in: *Iron in Biochemistry and Medicine* (A. Jacobs and M. Worwood, eds.), pp. 279–334, Academic Press, Inc., New York.
Hallberg, L., Harwerth, H. G., and Vannotti, A., eds., 1970, *Iron Deficiency. Pathogenesis. Clinical Aspects. Therapy*, Academic Press, Inc., New York.
Hallgren, B., and Sourander, P., 1958, The effect of age on the nonhaemin iron in the human brain, *J. Neurochem.* **3**:41.
Harris, J. W., and Kellermeyer, R. W., 1970, *The Red Cell*, pp. 85, 121, Harvard University Press, Cambridge, Mass.
Harris, P., 1972, The composition of the earth, in: *Understanding the Earth* (I. G. Gass, P. J. Smith and R. C. L. Wilson, eds.), 2nd ed., pp. 53–69, MIT Press, Cambridge, Mass.

Haughton, J. G., 1963, Nutritional anemia of infancy and childhood, *Am. J. Public Health* **53:**1121.

Hegsted, D. M., 1970, The recommended dietary allowances for iron, *Am. J. Public Health* **60:**653.

Hershko, C. H., Karsai, A., Eylon, L., and Izak, G., 1970, The effect of chronic iron deficiency on some biochemical functions of the human hemopoietic tissue, *Blood* **36:**321.

Hoffbrand, A. V., Ganeshaguru, K., Tattersall, M. H. N., and Tripp, E., 1974, Effects of iron deficiency on DNA synthesis, *Clin. Sci. Molec. Med.* **46:**12.

Hoffbrand, A. V., Ganeshaguru, K., Hooton, J. W. L., and Tattersall, M. H. N., 1976, Effect of iron deficiency and desferrioxamine on DNA synthesis in human cells, *Br. J. Haematol.* **33:**517.

Hogan, G. R., and Jones, B., 1970, The relationship of koilonychia and iron deficiency in infants, *J. Pediatr.* **77:**1054.

Howell, D., 1971, Significance of iron deficiencies. Consequences of mild deficiency in children, in: *Extent and Meanings of Iron Deficiency in the U.S.,* Summary proceedings of a workshop of the Food and Nutrition Board, National Academy of Sciences, Washington, D.C.

Hunt, J. V., 1976a, Environmental risk in fetal and neonatal life and measured infant intelligence, in: *Origins of Intelligence in Infancy and Early Childhood* (M. Lewis, ed.), p. 223, Plenum Press, New York.

Hunt, J. V., 1976b, Environmental programming to foster competence and prevent mental retardation in infancy, in: *Environments as Therapy for Brain Dysfunction* (R. N. Walsh and W. T. Greenough, eds.), p. 201, Plenum Press, New York.

Ishi, D.-N., and Maniatis, G. M., 1978, Haemin promotes rapid neurite outgrowth in cultured mouse neuroblastoma cells, *Nature* **274:**372.

Jacobs, A., 1965, Leukocyte oxygen consumption in iron deficiency anemia, *Br. J. Exp. Pathol.* **46:**545.

Jacobs, A., and Worwood, M., eds., 1974, *Iron in Biochemistry and Medicine,* Academic Press, Inc., New York.

Jacobs, A., and Worwood, M., 1975, Ferritin in serum. Clinical and biochemical implications, *New Engl. J. Med.* **292:**951.

Judisch, J. M., Naiman, J. L., and Oski, F. A., 1966, The fallacy of the fat iron-deficient child, *Pediatrics* **37:**987–990.

Kagan, J., 1975, Foreword, in: *Preschool IQ Prenatal and Early Developmental Correlates* (S. H. Broman, P. L. Nichols, and W. A. Kennedy, eds.), p. viii, John Wiley & Sons, Inc., New York.

Karki, N., Knutman, R., and Brodie, B. B., 1962, Storage, synthesis and metabolism of monoamines in the developing brain, *J. Neurochem.* **9:**53.

Klaiber, E. L., Kobayaski, Y., Broverman, D. M., and Hall, F., 1971, Plasma monoamine oxidase activity in regularly menstruating women and in amenorrheic women receiving cyclic treatment with estrogens and a progestin, *J. Clin. Endocrinol.* **33:**630.

Kuzuya, H., and Nagatsu, T., 1969, Flavins and monoamine oxidase activity in the brain, liver and kidney of the developing rat, *J. Neurochem.* **16:**123.

Lehninger, A. L., 1975, *Biochemistry,* 2nd ed., pp. 501, 503, Worth Publishers, Inc., New York.

Lester, B. M., 1975, Cardiac habituation of the orienting response to an auditory signal in infants of varying nutritional status, *Dev. Psychol.* **11:**432.

Lewis, M. (ed.), 1976, *Origins of Intelligence,* Plenum Press, New York.

Liddle, G. W., and Melman, K. L., 1974, The adrenals, in: *Textbook of Endocrinology,* 5th ed., pp. 306–307, W. B. Saunders Company, Philadelphia.

Linman, J. W., 1968, Physiologic and pathophysiologic effects of anemia, *New Engl. J. Med.* **279:**812.

Lipschitz, D. A., Cook, J. D., and Finch, C. A., 1974, A clinical evaluation of serum ferritin, *New Engl. J. Med.* **290:**1213.

Little, W. J., 1862, On the influence of abnormal parturition, difficult labor, premature birth, and asphyxia neonatorum on the mental and physical condition of the child, especially in relation to deformities, *Trans. Obstet. Soc. London* **3:**293.

Lloyd-Still, J. D., Hurwitz, I., Wolff, P., and Shwachman, H., 1974, Intellectual development after severe malnutrition in infancy, *Pediatrics* **54**:306.

London, I. M., Clemens, M. J., Ranu, R. S., Levin, D. H., Cherbas, L. F., and Ernst, V., 1976, The role of hemin in the regulation of protein synthesis in erythroid cells, *Fed. Proc.* **35**:2218.

Lovric, V. A., Beal, P. J., and Lammi, A. T., 1975, Iron deficiency anemia: evaluation of compensatory changes, *J. Pediatr.* **86**:194.

MacGregor, M. W., 1963, Maternal anaemia as a factor in prematurity and perinatal mortality, *Scot. Med. J.* **8**:134.

Mackler, B., Person, R., Miller, L. R., Inamdar, A. R., and Finch, C. A., 1978, Iron deficiency in the rat: Biochemical studies of brain metabolism, *Pediatr. Res.* **12**:217.

Maines, M. D., and Kappas, A., 1977, Metals as regulators of heme metabolism, *Science* **198**:1215.

Malmstrom, B. G., 1970, Biochemical functions of iron, in: *Iron Deficiency. Pathogenesis. Clinical Aspects. Therapy* (L. Hallberg, H. G. Harwerth, and A. Vanotti, eds.), pp. 9–18, Academic Press, Inc., New York.

Marner, F., 1969, Haemoglobin, erythocytes and serum iron values in normal children 3–6 years of age, *Acta Paediatr. Scand.* **58**:363.

Masawe, A. E. J., Mundi, J. M., and Swai, G. B. R., 1974, Infections in iron-deficiency and other types of anemia in the tropics, *Lancet* **2**:314.

McCall, R. B., 1976, Toward an epigenetic conception of mental development in the first three years of life, in: *Origins of Intelligence in Infancy and Early Childhood* (M. Lewis, ed.), p. 97, Plenum Press, New York.

McDonald, R., and Marshall, S. R., 1964, Iron therapy in pica, *Pediatrics* **34**:558.

Macdougall, L. G., Anderson, R., McNab, G. M., and Katz, J., 1975, The immune response in iron-deficient children: Impaired cellular defense mechanisms with altered humoral components, *J. Pediatr.* **86**:833.

McFarlane, D., Pinkerton, P., Dagg, J. H., and Goldberg, A., 1967, Iron deficiency with and without anemia in women in general practice. I. Incidence, *Br. J. Haematol.* **13**:790.

Moe, P. J., 1965, Normal red blood picture during the first three years of life, *Acta Paediatr. Scand.* **54**:69.

Moore, M. R., and Goldberg, A., 1974, Normal and abnormal haem biosynthesis, in: *Iron in Biochemistry and Medicine* (A. Jacobs and M. Worwood, eds.), Academic Press, Inc., New York.

Morrow, J. J., Dagg, and Goldberg, A., 1968, A controlled trial of iron therapy in sideropenia, *Scot. Med. J.* **13**:78.

Murphy, D. L., and Wyatt, R. J., 1972, Reduced MAO activity in blood platelets from schizophrenic patients, *Nature* **238**:225–226.

Naiman, J. L., Oski, F. A., Diamond, L. K., Vawter, G. F., and Shwachman, H., 1964, The gastrointestinal effects of iron-deficiency anemia, *Pediatrics* **33**:83.

O'Malley, D., and Stevenson, I. H., 1973, Iron deficiency anemia and drug metabolism, *J. Pharm. Pharmacol.* **25**:339.

Oski, F., 1973, Designation of anemia on a functional basis, *J. Pediatr.* **83**:353.

Oski, F. A., and Honig, A. M., 1977, The effects of therapy on the developmental scores of iron deficient infants, *J. Pediatr.* **92**:21.

Owen, G. M., Lubin, A. H., Garry, P. J., 1971, Preschool children in the United States: Who has iron deficiency? *J. Pediatr.* **79**:563.

Pollitt, E., and Leibel, R., 1976, Iron deficiency and behavior, *J. Pediatr.* **88**(3):372.

Pollitt, E., and Thomson, C., 1977, Protein calorie malnutrition and behavior: A view from psychology, in: *Nutrition and the Brain*, Vol. 2, *Control of Feeding Behavior and Biology of the Brain in Protein Calorie Malnutrition* (R. J. Wurtman and J. J. Wurtman, eds.), Raven Press, New York.

Pollycove, M., 1972, Hemochromatosis, in: *The Metabolic Basis of Inherited Disease* (J. B. Stanbury, J. B. Wyngaarden, and D. S. Fredrickson, eds.), p. 1053, McGraw-Hill Book Company, New York.

Quik, M., and Sourkes, T. L., 1977, The effect of chronic iron deficiency on adrenal tyrosine hydroxylase activity, *Can. J. Biochem.* **55**:60.

Read, M., 1973, Anemia and behavior, *Mod. Prob. Pediatr.* **14**:189.

Reynolds, E. G., 1976, Neurological aspects of folate and vitamin B_{12} metabolism, *Clin. Haematol.* **5**:661.

Reynolds, R. D., Binder, H. J., Miller, H. B., Chang, W., and Horan, S., 1968, Pagophagia and iron deficiency anemia, *Ann. Intern. Med.* **69**:435.

Richardson, S. A., 1976, The relation of severe malnutrition in infancy to the intelligence of school children with differing life histories, *Pediatr. Res.* **10**:57.

Robbins, E., and Pederson, T., 1970, Iron: Its intracellular localization and possible role in cell division, *Proc. Natl. Acad. Sci. USA* **66**:1244.

Robinson, D. S., 1975, Changes in monoamine oxidase and monoamines with human development and aging, *Fed. Proc.* **34**:103–107.

Robinson, N. M., and Robinson, H. B., 1965, A follow up study of children of low birthweight and control children at school age, *Pediatrics* **35**:425.

Roselle, H. A., 1970, Association of laundry starch and clay with anemia in New York City, *Arch. Intern. Med.* **125**:57.

Rosenthal, M., Lamanna, J. C., Jobsis, F. F., Levasseur, J. E., Kontos, H. A., and Patterson, J. L., 1976, Effects of respiratory gases on cytochrome A in intact cerebral cortex: Is there a critical PO_2? *Brain Res.* **108**:143–154.

Roth, J. A., Young, J. G., and Cohen, D. J., 1976, Platelet monoamine oxidase activity in children and adolescents, *Life Sci.* **18**:919–924.

Rowe, P. B., 1978, Inherited disorders of folate metabolism, in: *The Metabolic Basis of Inherited Disease* (J. B. Stanbury, J. B. Wyngaarden, and D. S. Fredrickson, eds.), pp. 430–457, McGraw-Hill Book Company, New York.

Sameroff, A. J., 1975, Early influence on development: Fact or fancy? *Merrill-Palmer Q.* **21**:267.

Sameroff, A. J., and Chandler, M. J., 1975, Reproductive risk and the continuum of caretaking casualty, in: *Review of Child Development Research,* Vol. 4, (F. D. Horowitz, ed.), p. 187, University of Chicago Press, Chicago.

Sandler, M., Carter, S. B., Cuthbert, M. F., and Pare, C. M. B., 1975, Is there an increase in monoamine oxidase activity in depressive illness? *Lancet* **1**:1045–1048.

Scarpelli, E. M., 1959, Maternal nutritional deficiency and intelligence of the offspring (thiamine and iron), *J. Comp. Physiol. Psychol.* **52**:536.

Scarr-Salapatek, S., 1976, An evolutionary perspective on infant intelligence: Species patterns and individual variations, in: *Origins of Intelligence in Infancy and Early Childhood* (M. Lewis, ed.), p. 165, Plenum Press, New York.

Scarr-Salapatek, S., and Williams, M. L., 1973, The effects of early stimulation on low birth weight infants, *Child Dev.* **44**:94.

Scott, D. E., and Pritchard, J. A., 1967, Iron deficiency in healthy young college women, *J. Am. Med. Assoc.* **199**:147.

Scriver, C. R., and Rosenberg, L. E., 1973, *Amino Acid Metabolism and Its Disorders,* pp. 290–337, W. B. Saunders Company, Philadelphia.

Shanley, B. C., Neethling, A. C., Percy, V. A., and Carstens, M., 1975, Neurochemical aspects of porphyria. Studies in the possible neurotoxicity of delta-aminolaevulinic acid, *S. Afr. Med. J.* **49**:576–580.

Sharman, D. F., 1973, The catabolism of catecholamines: Recent studies. *Br. Med. J.* **29**:110–115.

Siebert, D. J., and Ebough, F. G., Jr., 1967, Assessment of tissue anoxemia in chronic anemia by the arterial lactate/pyruvate ratio and excess lactate formation, *J. Lab. Clin. Med.* **69**:177.

Siimes, M. A., and Dallman, P. R., 1974, Iron deficiency: Impaired liver growth and DNA synthesis in the rat, *Br. J. Haematol.* **28**:453.

Siimes, M. A., Addiego, J. E., and Dallman, P. R., 1974, Ferritin in serum: Diagnosis of iron deficiency and iron overload in infants and children, *Blood* **43**:581.

Smith, A. D. M., 1960, Megaloblastic madness, *Br. Med. J.* **2**:1840.

Snyder, S. H., 1976, Catecholamines, serotonin and histamine, in: *Basic Neurochemistry,* 2nd ed. (F. J. Siegel, R. W. Albers, R. Katzman, and B. W. Agranoff, eds.), p. 210, Little, Brown and Company, Boston.

Sokoloff, L., 1976, Circulation and energy metabolism of the brain, in: *Basic Neurochemistry*, 2nd ed. (F. J. Siegel, R. W. Albers, R. Katzman, and B. W. Agranoff, eds.), pp. 388–413, Little, Brown and Company, Boston.

Sourkes, T. L., 1972, Influence of specific nutrients on catecholamine synthesis and metabolism, *Pharm. Rev.* **24:**349.

Sourkes, T. L., 1976, Psychopharmacology and biochemical theories of mental disorders, in: *Basic Neurochemistry*, 2nd ed. (F. J. Siegel, R. W. Albers, R. Katzman, and B. W. Agranoff, eds.), pp. 731–733, Little, Brown and Company, Boston.

Speck, B., 1968, Diurnal variations of serum iron and latent iron binding in normal adults, *Helv. Med. Acta* **34:**231.

Srikantia, S. G., Bhaskaram, C., Prasad, J. S., and Krishnamachari, K. A. V. R., 1976, Anaemia and immune response, *Lancet* **1:**1307.

Stein, Z., Susser, M., Saenger, G., and Marolla, F., 1975, in: *Famine and Human Development: The Dutch Hunger Winter of 1944/45,* Oxford University Press, London.

Sulzer, J. L., 1971, Significance of iron deficiencies: effects of iron deficiency on psychological tests in children, in: *Extent and Meanings of Iron Deficiency in the U.S.,* Summary proceedings of a workshop of the Food and Nutrition Board, National Academy of Sciences, Washington, D.C.

Sulzer, J. L., Wesley, H. H., and Leonig, F., 1973, Nutrition and behavior in head start children: Results from the Tulane study, in: *Nutrition, Development and Social Behavior* (D. J. Kallen, ed.), DHEW Publ. (NIH) 73-242.

Symes, A. L., Sourkes, T. L., Youdim, M. B. H., Gregoriadis, G., and Birnbaum, H., 1969, Decreased monoamine oxidase activity in liver of iron-deficient rats, *Can. J. Biochem.* **47:**999.

Symes, A. L., Missala, K., and Sourkes, T. L., 1971, Iron and riboflavin-dependent metabolism of a monoamine in the rat *in vivo, Science* **174:**153.

Tipton, K. F., 1973, Biochemical aspects of monoamine oxidase, *Br. Med. Bull.* **29:**116.

Tschudy, D. P., Valsamis, M., and Magnussen, C. R., 1975, Acute intermittent porphyria: Clinical and selected research aspects, *Ann. Int. Med.* **83:**851–864.

Underwood, E. J., 1971, *Trace Elements in Human and Animal Nutrition,* 3rd ed., pp. 1–13, Academic Press, Inc., New York.

Varat, M. A., Adloph, R. J., and Fowler, N. O., 1972, Cardiovascular effects of anemia, *Am. Heart J.* **83:**415.

Vitale, J. J., Restrepo, A., Velez, H., Riker, J. B., and Hellerstein, E. E., 1966, Secondary folate deficiency induced in the rat by dietary iron deficiency, *J. Nutr.* **88:**315–322.

Von Bonsdorff, B., 1977, Pica: A hypothesis. *Br. J. Haematol.* **35:**476.

Voorhess, M. L., Stuart, M. J., Stockman, J. A., and Oski, F. A., 1975, Iron deficiency anemia and increased urinary norepinephrine excretion, *J. Pediatr.* **86:**542–547.

Wands, J. R., Rowe, J. A., Mezey, S. E., Waterbury, L. A., Wright, J. R., Halliday, J. W., Isselbacher, K. J., and Powell, L. W., 1976, Normal serum ferritin concentrations in precirrhotic hemochromatasis, *New Engl. J. Med.* **294:**302.

Watson, C. J., 1975, Hematin and porphyria, *New Engl. J. Med.* **293:**605–606.

Webb, T. E., and Oski, F. A., 1973a, Iron deficiency anemia and scholastic achievement in young adolescents, *J. Pediatr.* **82:**827.

Webb, T. E., and Oski, F. A., 1973b, The effect of iron deficiency anemia on scholastic achievement, behavioral stability and perceptual sensitivity of adolescents, *Pediatr. Res.* **8:**294.

Webb, T. E., and Oski, F. A., 1974, Behavioral status of young adolescents with iron deficiency anemia, *J. Spec. Ed.* **8:**153.

Weinberg, E. D., 1974, Iron and susceptibility to infectious disease, *Science* **184:**952.

Werner, E. E., Bierman, J. M., and French, F. E., 1971, *The Children of Kauai: A Longitudinal Study from the Prenatal Period to Age Ten,* University of Hawaii Press, Honolulu.

White, J. L., and Labarba, R. C., 1976, The effects of tactile and kinesthetic stimulation on neonatal development in the premature infant, *Dev. Psychobiol.* **9:**569–577.

Williams, R. J. P., 1974, Haem-proteins and oxygen, in: *Iron in Biochemistry and Medicine* (A. Jacobs and M. Worwood, eds.), pp. 183–219, Academic Press, Inc., New York.

Winick, M., Meyer, K., and Harris, R., 1975, Malnutrition and environmental enrichment by early adoption, *Science* **190:**1173.

Wohlrab, H., and Jacobs, E. E., 1967, Copper-deficient mitochondria, electron transport and oxidative phosphorylation, *Biochem. Biophys. Res. Commun.* **28:**998.

WHO, 1975, Control of Nutritional Anemia with Special Reference to Iron Deficiency, *WHO Tech. Rept. Ser. 580,* pp. 5–9, Geneva.

Wood, M. M., and Elwood, P. C., 1966, Symptoms of iron deficiency anaemia: a community survey, *Br. J. Prevent. Soc. Med.* **20:**117.

Woods, H. F., Youdim, M. B. H., Boullin, D., and Callender, S., 1977, Monoamine metabolism and platelet function in iron-deficiency anemia, in: *Iron Metabolism,* CIBA Foundation Symposium, Vol. 51, p. 227, Elsevier, Amsterdam.

Wurtman, R. J., and Wurtman, J. J. (eds.), 1977a, *Nutrition and the Brain,* Vol. 1, *Determinants of the Availability of Nutrients to the Brain,* Raven Press, New York.

Wurtman, R. J., and Wurtman, J. J. (eds.), 1977b, *Nutrition and the Brain,* Vol. 2, *Control of Feeding Behavior and Biology of the Brain in Protein-Calorie Malnutrition,* Raven Press, New York.

Wyatt, R. J., Murphy, D. L., Balmaker, R., Cohen, S., Donnelly, G. H., and Pollin, W., 1973, Reduced monoamine oxidase activity in platelets, a possible genetic marker for vulnerability to schizophrenia, *Science* **179:**916–917.

Yeh, S. D. J., Gabriel, A. R., and Chow, B. F., 1961, Absorption of vitamin B_{12} by iron deficient rats (abstr.), *Fed. Proc.* **20:**451.

Youdim, M. B. H., 1973, Multiple forms of mitochondrial monoamine oxidase, *Br. Med. Bull.* **29:**120.

Youdim, M. B. H., and Green, A. R., 1977, Biogenic monoamine metabolism and functional activity in iron-deficient rats: Behavioral correlates, in: *Iron Metabolism* (R. Porter and D. Fitzsimmons, eds.), CIBA Foundation Symposium 51, pp. 201–225, Elsevier, Amsterdam.

Youdim, M. B. H., Woods, H. F., Mitchell, B., Grahame-Smith, D. G., and Callender, S., 1975a, Human platelet monoamine oxidase activity in iron-deficient anaemia, *Clin. Sci. Mol. Med.* **48:**289–295.

Youdim, M. B. H., Hamon, M., and Bourgoin, S., 1975b, Preparation of partially purified pig brain stem tryptophan hydroxylase, *J. Neurochem.* **25:**407.

Young, S. N., and Sourkes, T. L., 1974, Antidepressant action of tryptophan, *Lancet* **2:**897–898.

14

Inborn Errors of Metabolism

Donough O'Brien

1. Introduction

Human nutrition science and practise to the outsider, at any rate, appears to emphasize the availability, procurement, preparation, and requirement of foodstuffs in man in both health and disease. Special consideration has, of course, been given to defining needs at various ages, the Recommended Dietary Allowances (RDAs), to the biochemistry of nutrients, and to the health problems of too little and too much food. However, adequate nutrition is only achieved when nutrients complete their roles in providing energy and in intermediate metabolism. Any acquired or genetically determined aberration of these processes can thus be construed as a disorder of nutrition. For the nutritionist the special interest of these conditions undoubtedly lies in the opportunities for dietary manipulation as a means of therapy (American Academy of Pediatrics, 1976a,b).

In order to understand the application of dietary principles to the circumvention of inborn errors of metabolism, it must be remembered that these conditions, like any other genetic abnormality, represent fundamentally a change in one specific protein. The affected protein may be produced in diminished amounts, may not be produced at all, or may exist in a variably altered demesne which involves either its structural role or its ability to act enzymatically in combination with its substrate or cofactor (Fig. 1).

Whether the basic defect is in a protein concerned in transport or in an enzyme acting on some nutrient pathway, therapeutic dietary regimens have a limited number of options. The ideal approach, of course, would be to restore the genetic code, and some laboratories are indeed now conducting research on what is called transgenosis, in which phages containing nucleoprotein de-

Donough O'Brien • Department of Pediatrics, University of Colorado Medical Center, Denver, Colorado.

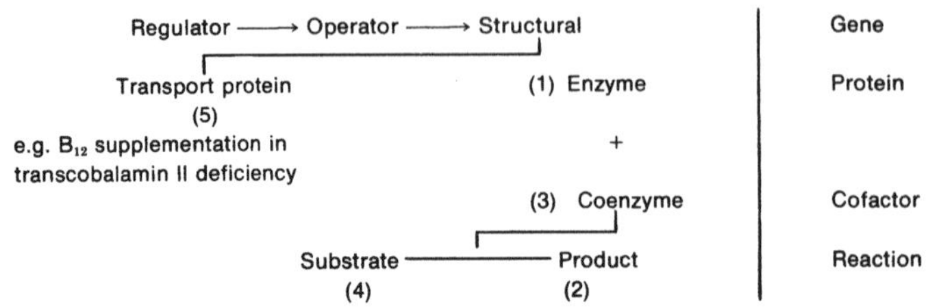

Fig. 1. Schema for nutritional approach to inborn errors of metabolism. (1) Enzyme replacement; (2) product replacement; (3) cofactor supplement; (4) substrate restriction; (5) substrate loading.

rived from *E. coli* are used to try to introduce the ability to synthesize specific missing enzymes into mammalian cells. Exciting as this approach is, it has been tempered by anxieties over the possibly dangerous consequences of uncontrolled experimentation with recombinant DNA. The next best alternative is to try to replace the missing gene product. This, for example, is exactly what is done when antihemophilic globulin (AHG) is given to hemophiliacs or enteric enzymes are given orally to children with cystic fibrosis. There are obvious difficulties in this approach: the protein in question must usually be given parenterally and may not reach the locus where it is required. Repeated pulses are necessary, with the consequent dangers of antibody formation. Placental glucocerebrosidase has been given intravenously to patients with Gaucher's disease, with transient amelioration of biochemical function. Organ transplantation, potentially another way of restoring enzyme activity in a somewhat more physiological way, has been used with occasional success with the liver in Wilson's disease and with the kidney in cystinosis.

Product supplementation is another approach and one which has been rather rarely used. The addition of uridine to the diet of a case of orotic aciduria is a dramatically successful instance of this and so to a lesser degree is the use of substitution therapy in inborn errors of hormone biosynthesis (e.g., thyroxin in inborn errors of tyrosyl iodination). This avenue is clearly only appropriate in situations when the abnormal phenotype is primarily characterized by the absence of the unsynthesizable gene product.

Neither of the preceding two approaches can really be thought of as nutritional. Cofactor supplementation, however, is a third possibility which has been very much involved with nutritional considerations. The supposition is either that the defective enzyme has a structural defect of the cofactor rather than the substrate site which may be overcome in the presence of a substantial excess of cofactor molecules, or that the synthesis of the active form of the cofactor is defective. Some variants of homocystinuria are good examples of a pyridoxine-responsive apoenzyme defect of cystathionine synthase. Likewise, in certain forms of methylmalonic aciduria there is a defect in the synthesis of 5′-deoxyadenosyl-B_{12} as the cofactor for methyl malonyl-CoA mutase. Clinically, this approach has again been rather sparsely useful, but it has provided the scientific basis from which has developed the whole poorly substantiated fabric of "megavitamin therapy" for mental retardation and psychiatric disorders in childhood.

The last and by far the most important approach from the nutrition point of view are those conditions where the unfavorable clinical picture is an expression of the accumulation of a nonmetabolized substrate or its toxic metabolites. Phenylketonuria is the best-known example of such a situation. Simple as such dietary treatment may seem in theory, however, in practice its application has many hazards and at the outset certain criteria must be assured: (1) the untreated disease must be harmful; (2) dietary treatment must be effective in abating these ill effects; (3) steps must be taken to ensure that the treatment is not also harmful to the patient or to others who may inappropriately adopt it; and (4) there must be adequate facilities for an impeccable confirmation of diagnosis and for suitable biochemical monitoring of the course of therapy.

Because of their special practical importance, it is appropriate at this stage to review in greater detail both the principles and practical guidelines of diets where the amino acid levels have to maintain a critical balance between what is toxic and what is insufficient for growth. An example is made of amino acid restriction, but the principles would apply as much to galactose restriction in galactosemia.

2. The Basic Formula

The amount of restricted amino acid provided by diet must be sufficient to meet the metabolic requirements dependent on it, including the requirement for growth; yet its intake must not permit an excess accumulation in body fluids of the amino acid or its derivatives. These requirements for adequate nutrient can be met by providing a semisynthetic diet, derived either from a modified protein hydrolysate or from a mixture of L-amino acids, so that the diet contains either an extremely low amount of the implicated amino acid or is free of it. Other dietary sources of protein can then furnish the implicated amino acid(s) in an amount sufficient to sustain normal metabolism and growth, yet low enough to avoid intoxication. Requirements for other nutrients, calories, fat (including essential fatty acids), carbohydrates, minerals, and vitamins are met either by special dietary formulations incorporated into or added to the amino acid product or by further supplementation with natural foods of known composition.

The need for these special formulations is especially important in early infancy during the most active period of brain myelination. In a number of conditions, the pressure for rigid dietary management diminishes with age and freer use can be made of natural foods. This is not always the case, however. For example, in branched-chain ketoaciduria, the need to restrict leucine, isoleucine, and valine is lifelong: a considerable challenge to sustain with conventional eating patterns.

3. The Progression to More Normal Foods

Infants in the first 6 months of life traditionally receive all or most of their nutrient requirements from breast milk or infant formula. During this period it is relatively easy to meet all the nutritional needs of an infant with an inborn error of amino acid metabolism by providing the kind of semisynthetic dietary product just described. Small amounts of milk may have to be added to meet the requirements of the restricted amino acid(s).

Other foods are introduced into the diet as the infant grows. The composition of these foods are the quantities ingested must be regulated to keep the amino acid composition of the diet under control and to assure provision of other nutritional requirements. Clearly, the lower the basic semisynthetic formulation is the restricted nutrient(s), the greater the permitted variety of natural foods that can be used in balancing the diet. For example, Lofenelac, a complete formula containing 80 mg/100 g of phenylalanine, is suitable for the treatment of phenylketonuria in early infancy. In later months another product, Mead Johnson 3229-A, which has no phenylalaine content, permits a freer introduction to mixed feeding.

4. Monitoring Treatment

Careful monitoring of treatment and its course is essential throughout the period of dietary management. Total nutritional intake, including the micronutrient composition, should be known and monitored in relation to age-appropriate RDAs to be certain the child is receiving a nutritionally adequate diet. The concentration of appropriate amino acids in blood should be determined often enough to assure that the level is adequate to sustain normal protein metabolism but not high enough to be harmful. Even this surveillance is difficult, because levels in plasma of the restricted amino acid may rise not only because of excess intake but also from body protein catabolism when intake is insufficient. The child must be observed frequently to be certain that nutritional deficiencies do not develop.

The degree of sophistication required in laboratory monitoring is such that the American Academy of Pediatrics has stated that these patients should only be managed in specially equipped centers. It is increasingly apparent as the scope for treatment with substrate restricted diets increases that an equal degree of expertise is being demanded of nutritionists. Much can be done to facilitate a rapid, reasonably accurate, and comprehensive index of total nutrition by the use of conventional data-processing equipment. Table I is an example of such a printout.

5. The Team Approach

Dietary constraints of the type required for treatment of the inborn errors of amino acid metabolism can engender difficulties for patient and family.

Table I. Dietary Analysis[a]

Calories/ nutrient	Unit of measure- ment	Mean/30 days	Mean/kg	Recom/kg[b]	g/100 kcal	Percent total calories
Calories	kc	351.8	48.9	108.3		
Protein	g	14.0	1.9	2.3	3.97	15.88
Fat	g	13.5	1.8	0.3	3.84	34.59
Carbon	g	43.6+	5.8	0.3	12.39	49.53
Calcium	mg	926.5+	123.5	360.0		
Phosphate	mg	499.0+	66.5	240.0		
Iron	mg	18.0+	2.4	10.3		
Vitamin A	IU	1096.2−	146.2	1400.3		
Thiamine	mg	0.8+	0.1	0.3		
Riboflavin	mg	1.1+	0.1	0.4		
Niacin	mg	10.0+	1.3	5.0		
Ascorbic acid	mg	28.6−	3.8	35.0		
Tryptophan	mg	271.2+	36.2	22.0		
Threonine	mg	745.8+	99.4	87.0		
Isoleucine	mg	0.0−	0.0	126.0		
Leucine	mg	0.0−	0.0	150.0		
Lysine	mg	1604.6+	213.9	103.0		
Methionine	mg	429.4+	57.3	45.0		
Cystine	mg	474.6+	63.3	0.0		
Phenylalanine	mg	858.8+	114.5	47.0		
Tyrosine	mg	858.8+	114.5	0.0		
Valine	mg	0.0−	0.0	105.0		
Histidine	mg	610.2+	81.4	34.0		
Potassium	mg	902,5+	120.3	0.0		
Sodium	mg	357.2+	47.6	0.0		
Saturated fat	g	0.0	0.0	0.0		
Oleic acid	g	0.0	0.0	0.0		
Linoleic acid	g	0.0	0.0	0.0		
Cholesterol	g	0.0	0.0	0.0		
Zinc	mg	5.6+	0.7	3.0		

Thiamine	2.22 mg/1000 kcal
Riboflavin	3.16 mg/1000 kcal
Niacin equivalent	41.37 mg/1000 kg

Specimen date 4/1/1976. Patient's age is 26.59 weeks
Weight is 7.5 kg, height is 66.0 cm.
MSUD diet formula . . . 6-month-old female.
MSUD = maple syrup urine disease

[a] This is not an actual specimen diet but is a retyped version of a printout of a commonly used branched-chain-deficient formula to which small amounts of natural foods must be added.
[b] Recommended amount per kilogram body weight.

Frequent and open communication between parents and physician in this difficult area can be aided by the recruitment of allied health personnel (e.g., nutritionists, social workers, genetic counselors, and nurses). The competence and availability of these health workers are essential to help parents implement the dietary prescriptions.

Applied to diseases such as phenylketonuria or maple syrup urine disease, these general rules may have a dramatically successful impact. Biochemical control, however, is not always with direct influence on the clinical situation, as shown in the case of disease such as histidinemia and homocystinuria. Substrate reduction in the instances given above involve a specific amino acid and applies to disorders of essential amino acids. However, where the amino acid can be synthesized in the body, such a specific approach is not possible. In these cases there is, nonetheless, an advantage to restricting the total protein in the diet. This is the basis of therapy in the hyperammonemias, and if the needs for special products in these conditions is less immediate, the meticulous demand for careful nutritional monitoring remains.

The genetic defects and their nutritional remedies that have been discussed so far are instructive but decidedly uncommon. Their study has, however, illuminated one basic factor of genetic disease, that of heterogeneity. Put another way, in the production of any one given protein, there is a potential for functional variation that ranges all the way from incompetence to a negligible difference from the normal. The relevance of this to the state of nutrition in a population is obvious: it is that figures such as RDAs can only be approximations, and that there will always be some people who from genetic variation will require more or less of any given nutrient for optimal physiological performance. The understanding of these less severe genetic variants and their nutrition needs, especially in populations with marginal availability of foodstuffs, promises to be an area of interesting future research. It has already been shown to be so in veterinary science.

6. References

American Academy of Pediatrics, Committee on Nutrition, 1976a, Nutritional management in hereditary metabolic disease, *Pediatrics* **40**:289.

American Academy of Pediatrics, Committee on Nutrition, 1976b, Special diets for infant with inborn errors of metabolism, *Pediatrics* **57**:783.

15

Nutritional In-Hospital Management of Chronic Diarrhea in Children

Giorgio Solimano and Sally A. Lederman

1. Introduction

Throughout the world, diarrhea is responsible for a large portion of morbidity and mortality, particularly in children (Amin-Zali, 1971; Kielmann and McCord, 1977; Thomas and Tillett, 1975; Walker and Harry, 1972). Examination of the causes of death in 13 Latin American areas emphasizes the significance of this disease. It was the underlying cause of death in over 40% of the deaths in children under 5 (excluding neonatal deaths). In over 60% of the deaths from diarrhea, moderate or severe malnutrition was an associated cause (Panamerican Health Organization, 1973). If the cure rates in Latin America were comparable to the cure rates in the United States, 98% of the 1969 Latin American deaths from diarrhea would have been prevented.

The economic cost of this disease is enormous. One-fourth to one-third of all medical consultations in Latin America and a similarly high proportion of infant hospitalizations are for diarrhea. If the role of diarrhea in exacerbating malnutrition, and so increasing morbidity and mortality from that condition, is also considered, the importance of a full understanding of its causality, prevention, and treatment is apparent.

Recent increases in knowledge of the mechanisms which produce and perpetuate diarrhea have been paralleled by the development of individualized treatment modalities. Yet this knowledge is not widely used, and therefore many cases are mishandled, at times with fatal consequences.

Based on the previous considerations, the purpose of this chapter is to discuss the nutritional, in-hospital treatment of severe diarrhea in children,

Giorgio Solimano • Institute of Human Nutrition and Center for Population and Family Health, Columbia University College of Physicians and Surgeons, New York, New York. ***Sally A. Lederman*** • Institute of Human Nutrition, Columbia University College of Physicians and Surgeons, New York, New York.

with emphasis on the new diets that have proved useful. To provide a basis from which therapeutic procedures may be understood, the mechanisms which initiate or aggravate diarrhea will be examined, with emphasis on malnutrition and infection as contributory processes.

Diarrhea may accompany many different illnesses. In the normal course of recovery, the diarrhea subsides and no special treatment is necessary. The majority of cases can be effectively treated at home or in a health center. In some instances, however, diarrhea becomes a serious problem. The continued loss of water, electrolytes, and nutrients taxes the body's recuperative ability. At best, diarrhea prolongs recovery from other illnesses; at worst, it can cause death by so depleting the patient, particularly if he is already malnourished, that he can neither throw off the illness nor reverse the changes that diarrhea has caused.

A direct association of nutritional deterioration and subsequent diarrhea has been demonstrated (Ghai and Jaiswal, 1970; Vega-Franco, 1976), but the mechanisms interconnecting these two states are only partially understood (Fig. 1). Malnutrition increases the severity as well as the frequency of diarrheal episodes. In severe and prolonged malnutrition, the gastrointestinal tract shows marked structural and functional changes (Danus *et al.*, 1972), and a decrease in intestinal protein synthesis has been postulated (Ghadimi *et al.*, 1973). The enzymes dipeptide hydrolase and mucosal alkaline phosphatase show reduced activity (Kumar *et al.*, 1971). The activity of lactase (James, 1971; Solimano *et al.*, 1972) and pancreatic lipase (Danus *et al.*, 1970) and amylase (Danus *et al.*, 1970) may also be decreased by malnutrition, and follow-up studies indicate that these impairments are not transient (Oowie *et al.*, 1967; Dahlqvist and Lindquist, 1971; Solimano *et al.*, 1972). Maldigestion

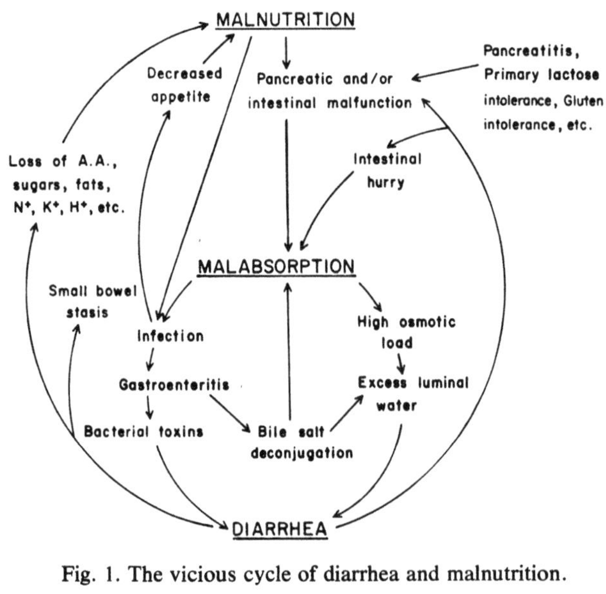

Fig. 1. The vicious cycle of diarrhea and malnutrition.

of fats and disaccharides frequently results from severe malnutrition (James, 1970). Indeed, in malnourished children on a skim milk diet, fat excretion may actually exceed intake (Dutra de Oliveira and Rolando, 1964). Since the efficiency of fat absorption appears to influence the efficiency of nitrogen absorption (Pinter *et al.*, 1964), fat and protein utilization may both be affected.

As structural and functional integrity progressively decline, the intestinal absorptive surface decreases and malabsorption eventually becomes inevitable. Increased amounts of undigested and unabsorbed materials then remain in the lumen, the osmotic pressure rises, and water moves in (Torres-Pinedo *et al.*, 1966). Bacterial fermentation of sugars to organic acids (lactic) can increase water and cation loss (Torres-Pinedo *et al.*, 1971). The resulting dilution of the luminal contents lowers the sodium concentration below that needed for adequate uptake, causing a loss of sodium (Torres-Pinedo *et al.*, 1968). Further along in the bowel, the absorption of water is dependent on Na^+ uptake, and the impairment of Na^+ uptake increases water loss as well (Torres-Pinedo *et al.*, 1966). Absorption of water may be further inhibited by bile acids (chenodeoxycholic and deoxycholic) (Phillips, 1973; Tamer *et al.*, 1974) reaching the colon, where they can decrease absorption and, in higher concentrations, increase secretion (Phillips, 1973). The failure to adequately resorb these substances in the ileum may be due to the increased volume and faster transit, as well as the reduced surface area and function of the intestine. Excessive volume is reached more rapidly in the malnourished child because of the small size of the intestine and its reduced distensibility. Moreover, potassium depletion may become quite serious and this, itself, has been associated with intestinal dilatation, ileus, and alterations in gastric secretions.

Malnutrition per se can encourage infection and its associated diarrhea by impairing the immune response, but it is often difficult to distinguish those changes in immunity which allow infections from those that are caused by infection. It has been reported, however, that the number of sheep erythrocyte rosette-forming cells is reduced in malnutrition (Lifshitz, 1973). These cells are believed to be T-lymphocytes. The T-cell system, associated with cellular immunity, is the most impaired system in severe malnutrition. This system serves as a defense against fungi, viruses, some bacteria, and other pathogens. Moreover, atrophy of thymic lymphoid tissues (Law *et al.*, 1973; Woodruff, 1970), decreased delayed cutaneous hypersensitivity (Law *et al.*, 1973), and other immunological defects (Law *et al.*, 1973; Christie and Ament, 1975) have been associated with protein–calorie malnutrition.

In addition to the immunity changes which may encourage infection generally, malabsorption (which may be associated with malnutrition) is particularly conducive to intestinal infection, since increased availability of undigested sugars in the lumen directly favors bacterial proliferation (Coello-Ramirez and Lifshitz, 1972). Yet malabsorption can be especially detrimental during infection when nutrient needs are increased and appetite is poor.

Whether infection is a primary occurrence or is secondary to malabsorption, it may itself cause diarrhea; diarrhea and infection are frequently found superimposed. Again, all the interrelating mechanisms are not known. Pathological organisms may frequently, but not always, be isolated from patients manifesting gastroenteritis. Altough pathogens may also be isolated from some symptom-free people, it is clear that bacterial or viral infection can cause parenteral or primary gastroenteritis and induce diarrhea. Purified enterotoxin, itself, has been shown to inhibit intestinal tone, contractility, and colonic transit in the dog (Banwell and Sherr, 1973). Staphylococcal enterotoxin reverses net absorption of Na^+, K^+, Cl^-, and water in the upper intestine of the rat and damages mucosal villus crest cells in the monkey (Banwell and Sherr, 1973).

Although the production of toxic materials by the pathogen is probably responsible for many effects, other processes are also involved. For example, some organisms may produce direct cytologic damage by invading the mucosa. Bacterial contamination of the upper intestine may cause increased deconjugation of bile salts (Rodriguez *et al.*, 1974b; Tabaqchali and Booth, 1967). In acute, nonspecific diarrhea, duodenal aspirates have shown increased anaerobic flora and, associated with these, a decrease in primary and conjugated bile acids and reduced glucose absorption (Mastromarino *et al.*, 1974). These changes are not observed in *E. coli* diarrhea, however (Rodriguez *et al.*, 1978). Children with chronic diarrhea also show lower levels of conjugated bile acids and increased duodenal anaerobes (Rodriguez *et al.*, 1974a). *Shigella* diarrhea is associated with different changes; primary bile acids increase and secondary bile acids (deoxycholic and lithocholic) decrease, but the percent of free bile acids is unchanged (Huang *et al.*, 1976).

If the concentration of conjugated bile salts falls below the level required for proper micelle formation, fat absorption can be reduced (Tabaqchali and Booth, 1967). This will reduce the utilization of calories from fats and may significantly affect the types of foods suitable for most rapid rehabilitation.

In addition to malnutrition and infection, other conditions which decrease digestive function, such as hereditary lactose intolerance, pancreatitis, gluten intolerance, and intestinal surgery, can be primary causes of diarrhea because of their associated malabsorption. Early in treatment, then, it is necessary to ascertain that such problems are not present, or to treat them, as well as the diarrhea, if they are.

As shown in Fig. 1, whatever the cause of diarrhea, several processes operate to increase its severity and duration, even in well-nourished patients. Although mouth-to-anus transit time is shorter during diarrhea, small-bowel transit may be delayed. This stasis encourages bacterial growth, with the concomitant damage described above. In addition, after only a few days of diarrhea, lactase activity may be reduced. With a longer or more severe course, maltase and sucrase activity also decline and monosaccharide intolerance, impaired glucose transport (Torres-Pinedo *et al.*, 1966), and reduced vitamin absorption may develop (Araya *et al.*, 1975). In infantile diarrhea, endogenous amino acid loss has also been demonstrated (Ghadimi *et al.*, 1973). These processes further increase the risk of malnutrition and bacterial growth even when these were not contributory causes of the diarrhea. It may be impossible to determine a primary cause.

The consequences of the cycle of diarrhea, malabsorption, and malnutrition can be extremely serious. Gas production may be greatly increased with marked abdominal distension and increased pressure in the intestine. The mucosa can become necrotic under these conditions, and gas can enter the tissues, occasionally producing pneumatosis intestinalis (Aziz, 1973; Coello-Ramirez *et al.*, 1970). Metabolic acidosis, hypokalemia, hyponatremia, and dehydration may occur with a rapid and even fatal course, especially in children who have only small amounts of reserve water. The failure to absorb sugars results in a depletion of glycogen and hypoglycemia may also be severe enough to cause death. With proper treatment, the cycle can be interrupted and the diarrhea stopped.

2. *Diagnosis*

When acute or chronic diarrhea occurs, infection, malnutrition, and malabsorption may all be present as partially overlapping causes and effects. In grave and complicated cases, the most immediate concern is diagnosis of the severity of acidosis, dehydration, and of hypokalemia and hyponatremia, so that lifesaving emergency measures can be instituted immediately. The etiology of the diarrhea must also be determined so that appropriate nutritional treatment can begin as soon as life-threatening problems and infection have been overcome. If diarrhea persists in spite of appropriate emergency treatment, the role of primary or secondary digestive impairment should be examined (see Table I).

Table I. Diagnostic Procedures and Chronic Diarrhea

Condition	Signs (normal diet)	Diagnostic tests
Lactose intolerance	Diarrhea, gas, low stool pH, reducing substances in stool	Lactose intolerance test; disaccharidase activity in biopsies; therapeutic trials
Generalized disaccharide intolerance	Diarrhea, gas, low stool pH, reducing substances in stool	Same as in lactose intolerance, plus other carbohydrate tolerance tests
Monosaccharide intolerance	Diarrhea, gas, low stool pH, reducing substances in stool	Glucose, fructose, galactose tolerance tests
Fat malabsorption	Diarrhea, excessive fecal fat	Stool fat determinations; bile salt in intestinal fluid and in stool; lipase activity in intestinal fluid; tests for coexisting causes (lactose or glucose intolerance or malnutrition)
Protein intolerance	Diarrhea	Peptidase levels in intestinal fluid or biopsy material

Primary malabsorption may be due to congenital absorption defects or to surgery. Secondary malabsorption is associated with several diseases and with the "vicious cycle" described earlier. In either case, recognition of such impairment allows a rational choice of the rehabilitation diet. The duration of the diarrhea cannot be used to determine the severity of the functional impairment, since intermittent bouts of illness may produce progressive malfunction which is not fully corrected between episodes, and an additional short period of diarrhea may add to the prior impairments and leave the patient with villous and pancreatic insufficiency and the inability to tolerate even monosaccharides.

Carbohydrate intolerance can be diagnosed from a low stool pH and from the presence of organic acid, especially lactic, and reducing substances of glucose in the stool (Lifshitz *et al.*, 1971b). The sample analyzed must be freshly excreted, since exogenous bacteria begin fermenting the stool within minutes.

A diagnosis of carbohydrate malabsorption may be supported by carbohydrate-tolerance tests. In these tests, the carbohydrate of interest is given orally and changes in the concentration of blood sugar and in the fecal characteristics are noted (Dahlqvist and Lindquist, 1971). If the blood sugar curve is flat or if large amounts of sugar appear in the stool, especially in association with gas, lowered pH, and the reappearance of diarrhea, intolerance is confirmed. These tests are particularly useful for monitoring the extent of rehabilitation.

Measures of enzymatic activity in intestinal aspirates may aid in evaluating pancreatic function. Determinations of disaccharidase activities in intestinal biopsies (Dahlqvist and Lindquist, 1971), usually combined with histological studies, provide useful information regarding intestinal function (Levin *et al.*, 1970).

Fat malabsorption may be demonstrated by the presence of excessive fat in the stools or in a fecal smear. Measurement of bile salts in the stool and intestinal fluid, and determination of lipase activity in intestinal fluid can then help to define the etiology of fat intolerance. If fat malabsorption is severe or prolonged, as in celiac disease, plasma levels of fat-soluble vitamins may be reduced. Plasma vitamin A determinations are advisable for patients with chronic steatorrhea or chronic gastroenteritis (Araya *et al.*, 1975).

Protein intolerance is less common but can cause extremely persistent diarrhea if not treated properly. Peptidase activity may be determined in intestinal aspirates or in biopsy material. The presence of specific histological changes would enable differential diagnosis in many cases (Solimano *et al.*, 1972).

Therapeutic trials are useful to corroborate a diagnosis of intolerance to fat or protein as well as to carbohydrate; usually, symptoms reappear when the suspected substance is reintroduced into the diet and subside when it is removed. Since secondary malabsorption is transitory, tolerance evaluation should be repeated periodically during the treatment period and a wider variety of nutrients provided as tolerance increases (James, 1970).

3. Nutritional Treatment

The dietetic treatment of diarrheal disease in well-nourished or slightly malnourished children with mild or moderate dehydration generally does not pose serious problems. Basic medical principles should be applied in the management of disorders causing diarrhea, malabsorption, and malnutrition in the pediatric age group (Meneghello, 1972; Nelson *et al.*, 1975). Appropriate treatment of initial episodes of acute diarrhea should be stressed, since these episodes are the usual starting point of repeating and severe forms.

A common mistake in the dietetic treatment of patients with diarrhea has been to restrict excessively the supply of nutrients. Feeding must be resumed as soon as possible according to tolerance, but an excessive load of any form of fat or carbohydrate must be avoided. The importance of breast feeding can hardly be overemphasized. In areas where infant diarrhea is most common and where malnutrition and high infant mortality rates are prevalent, the frequency and intensity of infant diarrhea has been shown to be much greater if the infant is not breast-fed. Breast milk does not merely eliminate the hazards of contaminated bottle feeds but actually provides protection against some diseases, including diarrhea. In some areas, expressed milk administered by intragastric tubes has been used to maintain breast feeding when the infant cannot nurse (Musoke, 1966). If breast feeding is not possible, small volumes of diluted milk formulas should be given as soon as practical. If the tolerance is good, progressive increases in concentration and volume are indicated, even if some abnormalities of the stools are observed for a few days. In acute or chronic diarrhea it is always extremely important to supply adequate amounts of nutrients as quickly as possible, since they are needed not only to improve the nutritional status but to restore the mechanisms and substances (protein, enzymes, etc.) altered or depleted by the diarrhea. Even if stool output increases, more nutrients are absorbed when a larger amount is offered (Dutra de Oliveira and Rolando, 1964).

The degree of tolerance to oral foods determines the type of nutrients to be provided and the route to use to supply them. Where oral foods can be tolerated, they are the choice treatment. When an oral diet is not tolerated, an alternative method must be used to provide nutrients. The main approaches to nutritional management under both of these circumstances and some of the problems of these methods will be considered in the next sections (see Table II).

4. Intravenous Sugar Feeding

Five percent glucose solution is most commonly used for intravenous sugar-feeding solutions. The concentration can be increased to 7.5% and even to 10% in some cases.

Mixtures of glucose and fructose, fructose alone, and ethanol have also

Table II. Nutritional Effects of Various Diets

Diet	Indications	Complications
Normal (breast feeding)	Maintain whenever possible	
IV sugar (glucose or glucose and fructose, or ethanol)	Short-term replacement of oral intake or with carbohydrate-free diets	PCM, if long term; fructose: low renal threshold, hypoglycemia, acidosis; ethanol: hepatotoxicity, inhibition of antidiuretic hormone increases urine volume
Lactose or other disaccharide-free diets	Intolerance to the disaccharide	
Carbohydrate-free diets	Monosaccharide intolerance	Hypoglycemia
Medium-chain triglyceride formulas (MCT)	Fat malabsorption	Hepatic coma; diabetic ketoacidosis
Artificial, oral diets (modular formula, elemental diet)	Weaning from TPN, complex malabsorptive disorders, diagnostic testing	Hypoglycemia; fatty acid deficiency; hypophosphatemic rickets
Total parenteral nutrition	Long-term replacement of oral intake (surgery, severe digestive impairment)	Vitamin deficiencies; phosphate deficiency; phlebitis, venous obstruction, etc.; reduced intestinal function; sepsis

been used. Fructose is claimed to be less irritating to the veins than glucose, less dependent on insulin for its further metabolism, and more rapidly taken up by tissues than glucose is. The renal threshold for fructose is, however, considerable less than that for glucose, and urinary losses of fructose can be substantial when it is infused into children at rapid rates (Harries, 1971). Severe metabolic acidosis and hypoglycemia have also been reported during infusions of fructose. Groups that recommend the use of fructose do indicate, however, that infants should be given at least 50% of the infused carbohydrate as glucose (Harries, 1971).

Ethanol has also been used in conjunction with other caloric sources, but in general most investigators have shied away from it, particularly in view of the relative unpredictability of blood alcohol levels in infants and the likelihood of hepatotoxicity (Heird and Winters, 1975). It can also cause increased losses of urine, owing to its inhibitory action on antidiuretic hormone release (Harries, 1971). Administration of intravenous sugar should be considered a temporary treatment, and provision of other nutrients should begin as soon as they can be tolerated. The replacement of intravenous by oral feeding must be gradual. Metabolic-balance studies of intakes and losses should be performed at short regular intervals, and a close observation for possible complications should be maintained.

5. *Artificial Oral Diets*

The selection of the appropriate formula mixture is most important. Some patients can tolerate increasing amounts of diluted milk formulas without significant deterioration of their general status, even though larger quantities of acidic stools may be observed. Initially, small amounts of powdered semi-skimmed milk diluted to 10% concentration, or whole milk diluted to one-half or one-third of normal concentration, may be used. Increasing the volume and concentration gradually every 2–3 days is recommended if the tolerance is satisfactory. The goal is to reach approximately 150 ml/kg of body weight/day when appetite is good and intestinal absorptive capacity recovers (Lifshitz *et al.*, 1971b).

The modular formula reported by Klish *et al.* (1974) and the elemental diets used by Vega-Franco and Suguihara (1976) in Mexico and Greene *et al.* (1975) in Tennessee constitute new approaches to oral feeding of patients with complex intolerance during severe and prolonged diarrheal episodes. These formulas are useful in the transition from total parenteral nutrition to oral feeding, as well as in the treatment of complex malabsorptive disorders (Christie and Ament, 1975).

The modular formula developed by Klish *et al.* (1974) is structured to provide different nutrients and concentrations sequentially as tolerance improves. The core formula is composed of whole casein and minerals and is calculated to deliver adequate electrolytes when used in a concentration of 3 g of protein/100 ml. To this protein core, fats and sugars can be added as needed and as tolerated by the patient.

A concentration of 1 g/100 ml of the core formula is offered at first to test the patient's tolerance to casein. Modules are added in increments of 1 g/100 ml every 12 hr starting with protein, fat, and then carbohydrates. The kinds of fat most frequently used are corn oil if long-chain triglycerides are desired, or fractionated coconut oil as a source of medium-chain triglycerides. Since fractionated coconut oil is deficient in linoleic acid, 3 ml of safflower oil is added to each liter of formula to supply this essential fatty acid. The final concentration of fat is generally 3.5–4.5 g/100 ml (Klish *et al.*, 1976).

The type and amount of carbohydrate to be added are determined by the gastrointestinal tolerance of the individual patient. Final concentrations range between 5 and 7 g/100 ml. The formula is prepared with sterile tap water, which provides trace elements that are not present in the core, and multivitamins are given as a supplement. When a composition of 3 g/100 ml core mix, 3.5 g/100 ml fat, and 6 g/100 ml carbohydrate, is tolerated for 1 week, the patient is switched to the most closely matching proprietary formula (Klish *et al.*, 1976). If a concentration is reached that causes diarrhea, it is maintained or slightly reduced until intestinal tolerance develops. Some patients use the formula at home since the method of preparation is simple.

This formula has been found useful in the treatment of (1) chronic, protracted diarrhea of infancy, with acquired carbohydrate intolerance, including glucose intolerance; (2) weaning from total parenteral nutrition; and (3) sequential challenge of the intestine for diagnostic purposes.

Long-term use of modular formula is associated with several possible complications, even when it is prepared properly. These include hypoglycemia, essential fatty acid deficiency, and hypophosphatemic rickets. All three are more likely if the formula is being used after cessation of total parenteral nutrition (Paulsrud *et al.,* 1972). Hypoglycemia may be prevented by administering intravenous glucose during weaning from total parenteral nutrition. Fatty acid deficiency may be eliminated by addition of sufficient safflower oil to the formula. Personnel must be alert to the possible development of hypophosphatemia.

The value of the elemental diets (Greene *et al.,* 1975; Vega-Franco and Suguihara, 1976) like that of the modular formula, is that they provide orally nutrients that do not need to be subjected to enzymatic digestion and which can therefore be easily absorbed by patients with different impairments (Greene *et al.,* 1975). They have been used with good results in patients with diarrhea and malnutrition, as well as intestinal fistula, extensive intestinal resections, colon disease, or entities characterized by exaggerated catabolism.

Treatment of 14 infants affected with prolonged diarrhea and malnutrition, 8 of whom were intolerant to disaccharides, showed that this diet resulted in a satisfactory clinical response (Vega-Franco and Suguihara, 1976). There was a significant reduction in the number of stools passed during the week following initiation of the elemental diet, and parenteral fluids were suspended 3 days after initiating the treatment. Weight gain, however, was not observed in most children, except when the elemental diet was administered for a prolonged period at full concentration (Vega-Franco and Suguihara, 1976).

Greene *et al.* (1975) studied changes in intestinal morphology and disaccharidase activities in six infants with protracted diarrhea and malnutrition during treatment with either total intravenous nutrition or oral elemental diets. They concluded that, although the pattern of recovery was very similar in both groups, the period of hospitalization was significantly shorter in those receiving oral elemental diets, suggesting "a more rapid return of intestinal function and of some intestinal enzymes."

6. Medium-Chain Triglyceride Formulas

Recent evidence suggests that the administration of medium-chain triglycerides as the predominant source of fat to patients with different types of malabsorption is of benefit (Graham *et al.,* 1973; Lifshitz, 1973; Pinter *et al.,* 1964). They can also be beneficial during the initial phase of treatment of malabsorption when it is associated with malnutrition, even if other modes of therapy are being used.

The rationale for their use is that they are more easily hydrolyzed than are long-chain triglycerides, requiring less conjugated bile salts and pancreatic enzymes (Lifshitz, 1973). They are also more readily taken up and transported by the mucosal cells, and the medium-chain triglycerides do not require incorporation into chylomicrons or entrance into the lymphatics to complete their absorption (Lifshitz, 1973). Graham *et al.* (1973) reported a higher caloric efficiency using a lactose-free, medium-chain triglyceride formula in malnutrition with severe diarrhea and concluded that "this slight advantage might be of real value to the very sick, severely malnourished infant." Some side effects have been noted, however. High blood levels of medium-chain fatty acids can precipitate hepatic coma in patients with impaired hepatic function, and diabetic kctoacidosis may be induced in diabetes-susceptible individuals (Lifshitz, 1973).

7. Carbohydrate-Free Diets

A significant proportion of severely ill patients do not tolerate milk because of carbohydrate intolerance, or an inability to digest and absorb fats and/or proteins. According to Lifshitz *et al* (1977a), 77% of their children with severe diarrhea had lactose intolerance during the acute stage of the illness. The carbohydrate intolerance was always transient, varying in duration from a few days to several weeks (Lifshitz *et al.*, 1971a). Eleven percent presented intolerance to disaccharides and almost 6% were intolerant to any type of dietary carbohydrate (Lifshitz *et al.*, 1970). The withdrawal of the specific mono-, di-, or polysaccharide from the diet constitutes the only effective treatment.

Infants with diarrhea of less than 1 week's duration usually tolerate disaccharides other than lactose and may therefore be given a lactose-free disaccharide diet (Suharjono *et al.*, 1975a,b). Prolonged, severe diarrhea may require treatment with a disaccharide-free, glucose-containing formula. These diets may be used routinely as a starting formula for all infants hospitalized for severe diarrhea. Glucose is usually given at a concentration of 5%. Carbohydrate-free formulas are necessary only for patients who continue to have diarrhea even on these special diets. To prevent hypoglycemia, parenteral glucose must be provided to all patients not fed carbohydrates (Lifshitz *et al.*, 1970).

The capacity to tolerate carbohydrates is recovered rapidly after diarrhea ceases. An increased number of sugars and more complex carbohydrates may be given to the patient soon afterward. As soon as glucose is tolerated, maltose can be introduced, followed by isomaltose, sucrose, and finally lactose. This can be done by feedings of Dextri-Maltose as a source of carbohydrates in the carbohydrate-free formula. Subsequently, the diet is expanded with cereal and starches, table sugar, and, finally, regular milk formula. Even patients with severe forms of carbohydrate intolerance tolerate lactose within 4 months of the acute stage of the illness. An attempt should be made to diagnose other abnormalities if carbohydrate intolerance persists.

Although several formulas are commercially available in developed countries, they are expensive and difficult to get in developing regions, where diarrhea is more prevalent. However, locally prepared lactose-free or low-lactose formulas, based on local staples, have been developed and shown to be useful (Leake *et al.*, 1974).

In Chile, for example, a formula made up of precooked chickpeas, powdered skim milk, and *dl*-methionine, plus vitamins and minerals, was used to treat marasmic infants with prolonged diarrhea (Barja *et al.*, 1971; Vallejos *et al.*, 1972). A prompt remission of diarrhea was observed, and the average daily weight gain during treatment was 213% of that expected for age (see Table III). Comparison of chickpea formula and a 12% fat powdered milk formula in treatment of acute diarrhea with dehydration demonstrated that normalization of the patients was faster on the chickpea formula (Pak *et al.*, 1974). Other workers have suggested the use of lactose free milk protein, soya bean milk, or vegetable protein mixtures during the period of temporary lactose intolerance often associated with rehabilitation from kwashiorkor (Ghai, 1972; Rao and Ramchandran, 1971; Srikantia and Gopalan, 1960). They have used several high-protein food mixtures which are either milk-free or contain only small amounts of powdered milk (Ghai, 1972). Further research into such diets, and increased documentation of the effectiveness of those already developed, should be encouraged by the international scientific community concerned with the problem.

8. Total Parenteral Nutrition (TPN)

The maintenance of full nutrition for long periods, solely by the intravenous route, is now possible. However, problems arise because of the lack of knowledge of the optimal nutrition requirements and the fundamental differences in the nutritional demands of the growing infant (Dudrick *et al.*, 1972). This form of treatment should be considered only when nutrition cannot be maintained by the oral route (Harries, 1971). Beneficial results have been reported by the use of total parenteral nutrition in chronic intractable diarrhea unresponsive to oral dietary treatment (Ghadimi *et al.*, 1973).

TPN solutions should include calories, water, nitrogen, vitamins, minerals, and "trace metals," in amounts and proportions that will result in their efficient utilization (Forget *et al.*, 1975; Ricour *et al.*, 1975). Nitrogen sources must contain both essential and nonessential amino acids. Protein hydrolysates and mixtures of pure, crystalline amino acids are both available. According to Heird, both "produce growth and development under appropriate conditions," but "neither can be considered ideal for parenteral nutrition in infants" (Heird and Winters, 1975; see also Dudrick *et al.*, 1972).

It is important to provide sufficient nonprotein calories to meet caloric requirements. In the United States, glucose, or a mixture of glucose and fructose, is generally used. In other countries, Intralipid or Lipofundin-S, intravenous fat preparations, have been used in conjunction with carbohydrate sources (Heird and Winters, 1975). Daily inspection of the serum must indicate that the fat is being cleared as the dosage is raised (Harries, 1971).

Table III. Chickpea–Milk Formula in the Treatment of Prolonged Diarrhea in 13 Severely Malnourished Infants[a]

	X	Range
Birth weight	3080 g	2500–3900 g
Age	6 mo 5 days	2 mo 16 days–10 mo 16 days
Weight/age	51.4% of expected weight for age	
Diarrhea's duration	50 days	16–83 days
Hb		
Admission	12.1 g/100 ml	10.5–14.1 g/100 ml
Final[a]	11.5 g/100 ml	9.1–13.3 g/100 ml
Total proteins		
Admission	6.03 g/100 ml	5.44–8.24 g/100 ml
Final	6.15 g/100 ml	5.39–8.13 g/100 ml
Days on formula	25 days	20–41 days
Duration of diarrhea after formula was introduced	2.4 days	0–10 days
Mean weight increment per day as percent of expected weight increment for age	213	90–500

[a] From Vallejos *et al.* (1972).
[b] The decrease may represent an increase in blood volume as dehydration is corrected.

Electrolytes, vitamins, and minerals are provided by various additions to the infusate. However, there is no precise knowledge of parenteral requirements of vitamins and minerals, and in some instances these requirements are likely to be substantially different from the oral ones (Heird and Winters, 1975).

Clearly, total parenteral nutrition is not free of difficulties (Heird and Winters, 1975; Paulsrud *et al.*, 1972). Complications may be due to the presence of the central venous catheter and to the delivery system. Sepsis is the most important of these complications but phlebitis, venous obstruction, and catheter dislodgement with extravasation of fluid also occur (Harries, 1971; Maheshkumar *et al.*, 1971). Everyone with experience in TPN insists that good treatment requires the strictest aseptic techniques (Maheshkumar *et al.*, 1971), intensive nursing care, frequent biochemical monitoring using microtechniques, and a well-trained and dedicated team.

9. Summary

In summary, we have described several alternatives that are now available for the nutritional treatment of diarrhea. Use of the proper diet speeds rehabilitation and can usually help to eliminate digestive impairments more quickly. Even acute episodes associated with primary intolerance can be benefited by the temporary use of such specialized diets. With improved diagnosis, proper allocation of the available facilities, and more individualized treatment, each diarrheal episode will be shorter, recovery will be more complete, and mortality should be greatly reduced.

10. References

Amin-Zaki, L., Taj-Eldin, S., Allos, G., and Al-Rahim, Q., 1971, Infantile diarrhoea in Iraq, *Pakistan Pediatr. J.* **1**:9–16.

Araya, M., Silink, S. J., Nobile, S., and Walker-Smith, J. A., 1975, Blood vitamin levels in children with gastroenteritis, *Aust. N. Z. J. Med.* **5**: 239–250.

Aziz, E. M., 1973, Neonatal penumatosis intestinahis associated with milk intolerance, *Am. J. Dis. Child.* **125**:560–562.

Banwell, J. G., and Sherr, H., 1973, Effect of bacterial enterotoxins on the gastrointestinal tract, *Gastroenterology* **65**(3):467-4977.

Barja, I., Munoz, P., Solimano, G., Vallejos, E., Undurraga, O., and Tangle, M. A., 1971, Fórmula de garbanzo (*Cicer arietinum*) en la alimentación del lactante sano, comunicación preliminar, *Arch. Latinoam. Nutr.* **11**(4):485-492.

Bowie, M. D., Barbezat, G. O., and Hansen, J. D. L., 1967, Carbohydrate absorption in malnourished children, *Am. J. Clin. Nutr.* **20**(2):89–97.

Christie, D. L., and Ament, M. E., 1975, Dilute elemental diet and continuous infusion technique for management of short bowel syndrome, *J. Pediatr.* **87**(5):705–708.

Coello-Ramírez, P., and Lifshitz, F., 1972, Enteric microflora and carbohydrate intolerance in infants with diarrhea, *Pediatrics* **49**(2):233–242.

Coello-Ramirez, P., Gutierres-Topete, G., and Lifshitz, F., 1970, Pneumatosis intestinalis, *Am. J. Dis. Child.* **120**:3–9.

Dahlqvist, A., and Lindquist, B., 1971, Lactose intolerance and protein malnutrition, *Acta Paediatr. Scand.* **60**:488–494.

Danus, O., Urbina, A. M., Valenzuela, I., and Solimano, G., 1970, The effect of refeeding on pancreatic exocrine function in marasmic infants, *J. Pediatr.* **77**(2):334–337.

Danus, O., Chuaqui, B., Vallejos, E., and Solimano, G., 1972, Desnutrición policarencial (kwashiorkor), *Bol. Med. Hosp. Infantil,* **29**(2):145–155.

Dudrick, S. J., Macfadyen, B. V., Jr., VanBuren, C. T., Ruberg, R. L., and Maynard, A. T., 1972, Parenteral hyperalimentation, *Ann. Surg.* **176**(3)259–264.

Dutra de Oliveira, J. E., and Rolando, E., 1964, Fat absorption studies in malnourished children, *Am. J. Clin. Nutr.* **5**:287–292.

Ferguson, A. C., Lawlor, G. J., Newmann, C. G., Oh, W., and Stiehm, E. R., 1974, Decreased rosette-forming lymphocytes in malnutrition and intrauterine growth retardation, *J. Pediatr.* **85**(5):717–723.

Forget, P. P., Fernandes, J., and Begemann, P. H., 1975, Utilization of fat emulsion during total parenteral nutrition in children, *Acta Paediatr. Scand.* **64**:377–384.

Ghadimi, H., Kumar, S., and Abaci, F., 1973, Endogenous amino acid loss and its significance in infantile diarrhea, *Pediatr. Res.* **7**:161-168.

Ghai, O. P., 1972, Prevention and management of kwashiorkor, *Trop. Doc.* **2**:(4):192-196.

Ghai, O. P., and Jaiswal, V. N., 1970, Relationship of undernutrition to diarrhoea in infants and children, *Ind. J. Med. Res.* **58**(6):789-795.

Graham, G. G., Baertl, J. M., Cordano, A., and Morales, E., 1973, Lactose-free, medium-chain triglyceride formulas in severe malnutrition, *Am. J. Dis. Child* **126**:330-335.

Greene, H. L., McCabe, D. E., and Merenstein, G. B., 1975, Protracted diarrhea and malnutrition in infancy: Changes in intestinal morphology and disaccharidase activities during treatment with total intravenous nutrition or oral elemental diets, *J. Pediatr.* **87**(5):695-704.

Harries, J. T., 1971, Intravenous feeding in infants, *Arch. Dis. Child.* **46**:855-863.

Heird, W. C., and Winters, R. W., 1975, Total parenteral nutrition, *J. Pediatr.* **86**:2-16.

Huang, C. T., Woodward, W. E., Hornick, R. B., DuPont, H., and Nichols, B. L., 1976, Effects of shigella diarrhea on fecal bile acid and neutral sterols, *Am. J. Clin. Nutr.* **29**:949-955.

James, W. P. T., 1970, Sugar absorption and intestinal motility in children when malnourished and after treatment, *Clin. Sci.* **39:**305–318.

James, W. P. T., 1971, Jejunal disaccharidase activities in children with marasmus and with kwashiorkor, *Arch. Dis. Child.* **46:**218–220.

Kielmann, A. A., and McCord, C., 1977, Home treatment of childhood diarrhea in Punjab villages, *J. Trop. Pediatr. Environ. Child Health* **23**(4):197–201.

Klish, W. J., Potts, E., Ferry, G. D., and Nichols, B. L., 1976, Modular formula: An approach to management of infants with specific or complex food intolerances, *J. Pediatr.* **88**(6):948–952.

Kumar, V., Ghai, O. P., and Chase, H. P., 1971, Intestinal depeptide hydrolase activities in undernourished children, *Arch. Dis. Child.* **46:**801–804.

Law, D. K., Dudrick, S. J., and Abdou, N. I., 1973, Immunocompetence of patients with protein-calorie malnutrition. The effects of nutritional repletion, *Ann. Int. Med.* **79**(4):545–550.

Leake, R. D., Schroeder, K. C., Benton, D. A., and Oh, W., 1974, Soy based formula in the treatment of infantile diarrhea, *Am. J. Dis. Child.* **127:**374–376.

Levin, B., Abraham, J. M., Burgess, E. A., and Wallis, P. G., 1970, Congenital lactose malabsorption, *Arch. Dis. Child.* **45:**173–177.

Lifshitz, F., 1973, Malabsorption syndrome and intestinal disaccharidase deficiencies, in: *Current Pediatric Therapy,* 6th ed. (S. Gellis and B. M. Kagen, eds.), W. B. Saunders Company, Philadelphia.

Lifshitz, F., Coello-Ramírez, P., and Gutierrez-Topete, G., 1970, Monosaccharide intolerance and hypoglycemia in infants with diarrhea. I. Clinical course on 23 Infants, *J. Pediatr.* **77:**(4):595–603.

Lifshitz, F., Coello-Ramírez, P., Gutierrez-Topete, G., and Cornado-Cornet, M. C., 1971, Carbohydrate intolerance in infants with diarrhea, *J. Pediatr.* **79**(5):760–767.

Lifshitz, F., Coello-Ramírez, P., and Contreras-Gutierrez, M. L., 1971b, The response of infants to carbohydrate oral loads after recovery from diarrhea, *J. Pediatr.* **79**(4):612–617.

Maheshkumar, A. P., Harberg, F. J., Nichols, B. L., Jr., and B. F. Brooks, 1971, The technique and management of parenteral hyperalimentation in children, *St. Joseph's Medical Surg. J.* **6:**97–103.

Mastromarino, A., Rodriguez, J. T., Wilson R., Nichols, B. L., Huang, T. L., and Ordoney, J. V., 1974, Duodenal microflora of Guatemalan children with acute diarrhea, Abstract of paper submitted to the American Society of Microbiology, Chicago, Ill.

Meneghello, J., 1972, *Pediatría,* InterMédica, Buenos Aires.

Musoke, L. K., 1966, A rational therapy of diarrhea; Rec. Exper. in Maternal and Child Health in E. Africa. Monogr. 2; Report of WHO/UNICEF Seminar held at Makerere University College Medical School, Kampala, Uganda, 1966; issued in conjunction with *J. Trop. Pediatr. Afr. Child Health* **12**(3):102.

National Research Council of the National Academy of Sciences, 1975, Immune response of the malnourished child, Position Paper of the Subcommittee on Interactions of Nutrition and Infections (Committee on International Nutrition Programs) of the National Research Council of the National Academy of Sciences, Washington, D. C. (August).

Nelson, W. E., Vaughan, V. C., and McKay, R. J., (eds.), 1975, *Textbook of Pediatrics,* W. B. Saunders Company, Philadelphia.

Pak, N., Bernier, L., Duffau, G., Macaya, J., Soriano, H., Munoz, P., Hidalgo, R., Soto, A., Zerene, N., and Tangle, M. A., 1974, Fórmula de garbanzo (*Cicer arientinum*) en el tratamiento dietético del sindrome diarreico agudo con deshidratación, *Rev. Pediatr. Santiago,* **17:**3–4, 71–74.

Paulsrud, J., R., Pensler, L., Whitten, C. F., Stewart, S., and Holman, R. T., 1972, Essential fatty acid deficiency in infants induced by fat-free intravenous feeding, *Amer. J. Clin. Nutr.* **25**(9):897–904.

Phillips, S. F., 1973, Relationship of diarrhea to maldigestion and malabsorption, *Mayo Clin. Proc.* **48**(9):660–662.

Pinter, K. G., McCracken, B. H., Lamar, C., Jr., and Goldsmith, G. A., 1964, Fat absorption studies in various forms of steatorrhea, *Am. J. Clin. Nutr.* **15:**293–298.

Puffer, R. P., and Serrano, C. V., 1973, Patterns of mortality in childhood, Scientific Publication No. 262, Pan American Health Organization, Washington, D. C.

Rao, G. Purshowtham, and Ramchandran, Padma, 1971, Experience with cotton-seed protein in the therapy of kwashiorkar, *Ind. Pediatr.* **8**(8):380–384.

Ricour, C., Millot, M., and Balsan, S., 1975, Phosphorus depletion in children on long-term total parenteral nutrition, *Acta. Paediatr. Scand.* **64**:385–392.

Rodriguez, J. T., Ordónez, J. V., Alvarado, J., and Nichols, B. L., 1974a, Flora bacteriana intestinal en niños con diarrea crónica, abstract IV, International Congress of Pediatrics, Buenos Aires (October).

Rodriguez, J. T., Huang, T. L., Alvarado, J., Klish, W., and Nichols, B. L., 1974b, Patrón de acidos biliares intestinales en niños guatemaltecos sanos y con diarrea crónica, abstract IV, International Congress of Pediatrics, Buenos Aires (October).

Rodriguez, J. T., Mastromarino, A., Darby, W., Flores, N., Ordoney, J. V., Huang, T. L., Alvarado, J., Wilson, R., Soriano, H., and Nichols, B. L., 1978, Alterations of intestinal flora and function in acute nonspecific diarrhea of infancy, in press.

Solimano, G., Vallejos, E., Danus, O., and Urbina, M., 1972, Niveles de disacaridasa en la mucosa intestinal de lactantes con desnutrición calórico-proteica de tercer grado, *Bol. Med. Hosp. Infantil,* **29**(2):157–163.

Srikantia, S. G., and Gopalan, C., 1960, Clinical trials with vegetable protein foods in kwashiorkor, *Ind. J. Med. Res.* **48**(5):637–644.

Suharjono, S., Boediarso, A., Sutoto, and Dadi, E. M., 1975, Low lactose milk on refeeding of infantile diarrhoea, *Paediatr. Indones.* **15**:247–254.

Suharjono, S., Boediarso, A., Hendardji, H., and Soegiharto, S., 1975b, Refeeding with free lactose milk in children suffering from gastroenteritis and dehydration, *Paediatr. Indones.* **15**:191–197.

Tabaqchali, S., and Booth, C. C., 1967, Relationship of the intestinal bacterial flora to absorption, *Br. Med. Bull.* **23**:285–290.

Tamer, M. A., Santora, T. R., and Sandberg, D. H., 1974, Cholestyramine therapy for intractable diarrhea, *Pediatrics* **53**(2):217–220.

Thomas, M. E. M., and Tillett, H. E., 1975, Diarrhoea in general practice: a sixteen-year report of investigations in a microbiology laboratory, with epidemiological assessment, *J. Hyg.* **74**(2):183–194.

Torres-Pinedo, R., Rivera, C. L., and Fernandez, S., 1966, Studies on infant diarrhea. II. Absorption of glucose and net fluxes of water and sodium chloride in a segment of the jejunum, *J. Clin. Invest.* **45**(12):1916–1922.

Torres-Pinedo, R., Conde, E., Robillard, G., and Maldonado, M., (1968), Studies on infant diarrhea. III. Changes in composition of saline and glucose-saline solutions instilled into the colon, *Pediatrics* **42**(2):303–311.

Torres-Pinedo, R., Rivera, C., and Rodriguez, H., 1971, Intestinal absorptive defects associated with enteric infections in infants, *Ann. N. Y. Acad. Sci.* **176**:284–298.

Vallejos, E. P., Radrigan, M. E., Barja, I., Araya, J., Solimano, G., Tagle, M. A., and Muñoz, P., 1972, Mamadera de garbanzo (*Cicer aretinium*) en el tratamiento de la diarrea prolongada del lactante desnutrido de tercer grado, *Rev. Chile Pediatr.* **43**:17–21.

Vega-Franco, L., 1976, Diarrea y desnutrición, in: *Enfermedades Diarreicas en el Niño,* 2nd ed., Ediciones Médicas del Hospital Infantil de Mexico, Mexico.

Vega-Franco, L., and Suguihara, C. Y., 1976, La dieta elemental en el tratamiento de la diarrea prolongada del lactante, *Boletin Médico del Hospital Infantil* **XXXIII**(2):335–351.

Walker, A. C., and Harry, J. G., 1972, A survey of diarrhoeal disease in malnourished aboriginal children, *Med. J. Aust.* **1**:904–911.

Woodruff, J. F. 1970, The influence of quantitated post-weaning undernutrition on coxsackievirus B_3 infection of adult mice. II. Alteration of host defense mechanisms, *J. Infect. Dis.* **121**(2):164–181.

Index

MIX
Papier aus verantwortungsvollen Quellen
Paper from responsible sources
FSC® C105338

If you have any concerns about our products,
you can contact us on
ProductSafety@springernature.com

In case Publisher is established outside the EU,
the EU authorized representative is:
Springer Nature Customer Service Center GmbH
Europaplatz 3, 69115 Heidelberg, Germany

Printed by Libri Plureos GmbH
in Hamburg, Germany